Agri-Based Bioeconomy

Agri-Based Bioeconomy

Reintegrating Trans-disciplinary Research and Sustainable Development Goals

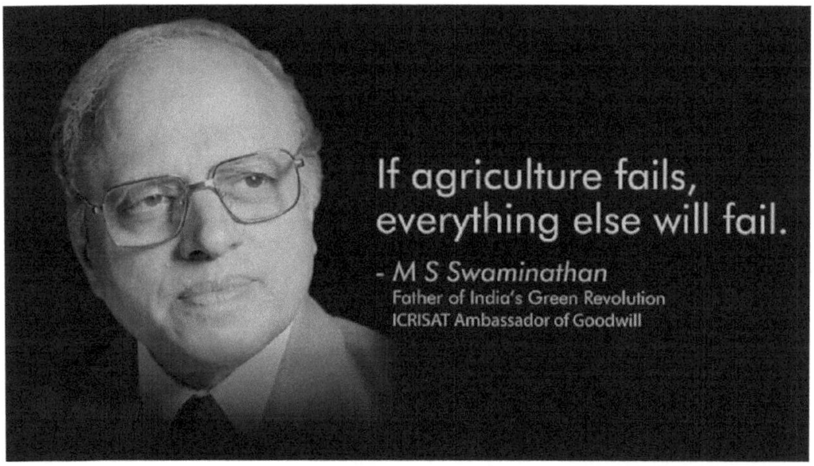

Edited by
Chetan Keswani
Department of Biochemistry
Institute of Science
Banaras Hindu University
India

CRC Press
Taylor & Francis Group
Boca Raton London New York

CRC Press is an imprint of the
Taylor & Francis Group, an **informa** business

First edition published 2021
by CRC Press

6000 Broken Sound Parkway NW, Suite 300, Boca Raton, FL 33487-2742
and by CRC Press
2 Park Square, Milton Park, Abingdon, Oxon, OX14 4RN

Library of Congress Cataloging-in-Publication Data

Names: Keswani, Chetan, editor.
Title: Agri-based bioeconomy : reintegrating trans-disciplinary research and sustainable development goals / edited by Chetan Keswani.
Description: First edition. | Boca Raton : CRC Press, 2021. | Includes bibliographical references and index.
Identifiers: LCCN 2020058428 | ISBN 9780367471002 (hardback) | ISBN 9781003033394 (ebook)
Subjects: LCSH: Agricultural biotechnology--Economic aspects. | Agricultural biotechnology--Environmental aspects. | Sustainable development.
Classification: LCC S494.5.B563 A397 2021 | DDC 338.1--dc23
LC record available at https://lccn.loc.gov/2020058428

ISBN: 978-0-367-47100-2 (hbk)
ISBN: 978-0-367-76890-4 (pbk)
ISBN: 978-1-003-03339-4 (ebk)

Typeset in Times
by Deanta Global Publishing Services, Chennai, India

Contents

Foreword

Over the past few decades, agriculture has undergone a number of technical changes, of which the most evident took place over the course of the late 20th century. Modern agronomy, plant and animal breeding, agrochemicals and mechanization, as well as, more recently, digital technologies, have transformed the sector in a substantial way. Technological developments have led to sharp increases in yields and reduced the need for various inputs (especially labour). But it has, at the same time, caused ecological damage and has, on occasion, led to negative human health effects.

In the context of the bioeconomy, the biotechnology industry is an independent sector producing final products for consumers. More importantly, it is a sector supporting other bio-based sectors at large, which includes technological developments such as resources optimization and bio-based solutions. The links between the bioeconomy and sustainability can also be interpreted through new ecological approaches. A very relevant framework in this direction is that of the socio-ecological systems, largely focusing on suitable representations of the interaction between society and ecosystems to ensure sustainability.

The bioeconomy is being presented as one of the solutions to the various ecological and social challenges, covering all dimensions of sustainability, including climate change mitigation, environmental sustainability through cleaner production processes, economic growth and the creation of new employment opportunities.

Due to the complexity of the interlinkages between the bioeconomy, society and ecosystems, the concept of sustainable development should be regarded as an integrative one. Indeed, it also offers a context within which the concept of the bioeconomy needs to be considered. This line of thinking will become even more important in the future, as the innovation potential of the bio-economy is likely to become more widespread, also helping to minimize adverse socio-economic and environmental impacts. A modern and sustainability oriented bioeconomy may contribute to the implementation of many of the United Nations' Sustainable Development Goals (SDGs), not only in respect of reducing hunger and poverty and addressing climate change, economic inequalities and sustainable consumption, but also in support of goals such as peace and justice.

The current volume emphasizes the potential of the bioeconomy in this direction. It presents examples illustrating its importance and, at the same time, it clarifies some of the potentially negative effects, a trend which largely depends on the ability of socio-economic systems to guide the bioeconomy in suitable directions.

In order to yield the expected benefit, various solutions are assessed on a case-by-case basis in different country and regional contexts, especially in relation to the availability of resources and the need for economic development.

This book focuses on an integrated approach to sustained innovation in various areas of the agri-based bioeconomy. The merits of this book are also based to a great extent on assessments of the industrial and socio-legal implications of bioeconomic advances. I congratulate Dr Keswani for liaising with experts on the subject, gathering a comprehensive body of information that underlines both the potentials and the upcoming challenges related to the development of the bioeconomy and presenting some viable options for translating ideas into easily commercially affordable products and technologies.

Prof. Dr (mult)., Dr h.c. (mult.) Walter Leal
Head of the Research and Transfer Centre "Sustainable Development and Climate Change Management"
Hamburg University of Applied Sciences
Faculty of Life Sciences

Preface

The bioeconomy appears to be one of the most promising yet potentially controversial fields of economic development. In contrast to the high demand for economic solutions and to the widespread use of simplified and sometimes naïve economic concepts by stakeholders and scientists in the bioeconomy, socio-economic research on the transition towards an agri-based bioeconomy remains insufficient and unstructured. In addition, while there are studies in the field of management or for specific sectors, few really examine the agri-based bioeconomy as a whole or look into the specific attributes of this concept. This is the main motivation for this book.

This book is an attempt to bring together frontier scientists in the field on a single platform to provide an overview and a consistent account of the emerging branches that are developing within the agri-based bioeconomy. The book's objective is to reintegrate trans-disciplinary research and sustainable development goals from both an economic and a policy perspective, as well as to identify potential future pathways and issues for this subject.

The book is intended to meet the needs of different user groups, from practitioners to researchers, to get a consistent view of agri-based bioeconomy. Furthermore, the book uses well-established concepts from economics in order to better support the understanding of bioeconomy issues for those not well-versed in economics. Those who are non-experts in economics, life sciences or biotechnology may also be interested in the subject matter of this book. Several parts are purposely kept very simple and are intended as a true introduction to the bioeconomy for the "neophyte".

An effort has been made to avoid considering policy in isolation, but rather to emphasize the interplay between the policy mix, the wider institutional setting, market forces and system organization solutions. Both empirical and conceptual chapters are included, and an attempt has been made to bring them together and yield facts and interpretations for the readers.

Chetan Keswani
Editor

Editor

Chetan Keswani is a post-doctoral fellow in the Department of Biochemistry, Institute of Science, Banaras Hindu University, Varanasi, India. He has a keen interest in regulatory, commercialization and intellectual property issues of agriculturally important microorganisms. He is an elected fellow of the Linnaean Society of London, UK, and received the Best PhD Thesis Award from the Uttar Pradesh Academy of Agricultural Sciences, India, in 2015. He is an editorial board member of several reputed agricultural microbiology journals and has edited nine volumes on the subject.

List of Contributors

Cahyono Agus
Faculty of Forestry
Universitas Gadjah Mada
Yogyakarta, Indonesia

Le Hung Anh
Industrial University of Ho Chi Minh City
Ho Chi Minh City, Vietnam

Adelaide Maria de Souza Antunes
School of Chemistry
Federal University of Rio de Janeiro
Rio de Janeiro, Brazil
and
National Institute of Industrial Property
Rio de Janeiro, Brazil

Cristine Marie B. Brown
National Institute of Molecular Biology and Biotechnology
University of the Philippines
Los Baños, Philippines

Marilyn B. Brown
National Institute of Molecular Biology and Biotechnology
University of the Philippines
Los Baños, Philippines

Dhanay Cadillo Chandler
Faculty of Law
University of Turku
Turku, Finland

Sadhna Chauhan
Department of Plant Pathology
College of Agriculture
G.B. Pant University of Agriculture & Technology
Uttarakhand, India

Manoj Kumar Chitara
Department of Plant Pathology
College of Agriculture
G.B. Pant University of Agriculture & Technology
Uttarakhand, India

Ugo De Corato
Italian National Agency for New Technologies, Energy and Sustainable Economic Development
Department of Bioenergy, Biorefinery and Green Chemistry
Territorial Office of Bari
Bari, Italy

Nasser Delangiz
Department of Plant Biotechnology and Breeding
Faculty of Agriculture
University of Tabriz
Tabriz, Iran

Julio Alves Cardoso Filho
Center of Agricultural Sciences
Sector of Plant Pathology
Federal University of Alagoas
Rio Largo, Brazil

Mansour Ghorbanpour
Department of Medicinal Plants
Faculty of Agriculture and Natural Resources
Arak University
Arak, Iran

Marinus K. Harun
Environment & Forestry Research and Development Institute
Banjarbaru, Indonesia

Kathleen Hefferon
Department of Microbiology
Cornell University
Ithaca, New York

Arash Hemati
Department of Soil Science
Faculty of Agriculture
University of Tabriz
Tabriz, Iran

Akira Homma
Bio-Manguinhos
Oswaldo Cruz Foundation
Rio de Janeiro, Brazil

İrfan Kalaycı
Department of Economics
İnönü University
Malatya, Turkey

Lazaros S. Karaoglanoglou
National Technical University of Athens
School of Chemical Engineering
Athens, Greece

Solmaz Kazemalilou
National Technical University of Athens
School of Chemical Engineering
Athens, Greece

E.P. Kiseleva
Institute of Bioorganic Chemistry
National Academy of Sciences of Belarus
Minsk, Republic of Belarus

Emmanuel G. Koukios
National Technical University of Athens
School of Chemical Engineering
Athens, Greece

Dimitrios P. Koullas
National Technical University of Athens
School of Chemical Engineering
Athens, Greece

Nikolaos Kourakos
National Technical University of Athens
School of Chemical Engineering
Athens, Greece

D. Kołożyn-Krajewska
Warsaw University of Life Sciences – SGGW
Warsaw, Poland

Behnam Asgari Lajayer
Health and Environment Research Center
Tabriz University of Medical Sciences
Tabriz, Iran

Gloria I. Hernández Melchor
Sociedad Iberoamericana de Bioeconomía y Cambio Climático
Asociación Civil
León, Nicaragua

Flavia Maria Lins Mendes
Fiocruz South Mato Grosso
Oswaldo Cruz Foundation
Campo Grande, Brazil

K.I. Mikhailopulo
Institute of Bioorganic Chemistry
National Academy of Sciences of Belarus
Minsk, Republic of Belarus

Robert Nepomuceno
National Institute of Molecular Biology and Biotechnology
University of the Philippines
Los Baños, Philippines

Galina Novik
Belarusian Collection of Microorganisms
Institute of Microbiology
National Academy of Sciences of Belarus
Minsk, Republic of Belarus

Alessandra Moreira de Oliveira
School of Chemistry
Federal University of Rio de Janeiro
Rio de Janeiro, Brazil
and
National Institute of Industrial Property
Rio de Janeiro, Brazil

Sofia Papadaki
National Technical University of Athens
School of Chemical Engineering
Athens, Greece

Cristina Possas
Bio-Manguinhos
Oswaldo Cruz Foundation
Rio de Janeiro, Brazil

Enggal Primananda
Research Center for Plant Conservation and Botanic Gardens
Indonesian Institute of Sciences
Bogor, Indonesia

Pamungkas Buana Putra
Faculty of Forestry
Universitas Gadjah Mada
Yogyakarta, Indonesia
and
Watershed Management Technology Center
Ministry of Environment and Forestry Republic of Indonesia
Surakarta, Indonesia

Diana Ivette Triana Ramírez
División Académica de Ciencias Biológicas
Universidad Juárez Autónoma de Tabasco
Tabasco, Mexico

Claudia Elena Zenteno Ruiz
División Académica de Ciencias Biológicas
Universidad Juárez Autónoma de Tabasco
Tabasco, Mexico

Ángel Sol Sánchez
Colegio de Postgraduados
Area de Agriculture
Tabasco, Mexico

Casiano Alberto Méndez Sánchez
División Académica de Ciencias Biológicas
Universidad Juárez Autónoma de Tabasco
Tabasco, Mexico

Victoria Savich
Belarusian Collection of Microorganisms
Institute of Microbiology
National Academy of Sciences of Belarus
Minsk, Republic of Belarus

Petra Schneider
Magdeburg-Stendal University of Applied Sciences
Magdeburg, Germany

Suzanne de Oliveira Rodrigues Schumacher
School of Chemistry
Federal University of Rio de Janeiro
Rio de Janeiro, Brazil

Rajesh Pratap Singh
Department of Plant Pathology
College of Agriculture
G.B. Pant University of Agriculture & Technology
Uttarakhand, India

Serge Svizzero
Faculté de Droit et d'Economie
Université de La Réunion
Saint Denis, France

Stefano Targetti
Department of Agricultural and Food Sciences
University of Bologna
Bologna, Italy

Davide Viaggi
Department of Agricultural and Food Sciences
University of Bologna
Bologna, Italy

Matteo Zavalloni
Department of Agricultural and Food Sciences
University of Bologna
Bologna, Italy

1

Agricultural Biotechnology and the Sustainable Development Goals

Kathleen Hefferon

CONTENTS

1.1 Introduction

Genetically modified (GM) crops have been commercially available for over 20 years, yet they remain as controversial as ever. According to the International Service for the Acquisition of Agri-biotech Applications (ISAAA), the global area of crops bred using biotechnology has reached approximately 200 million hectares; more than half of this land is used by smallholder farmers in developing countries (Figure 1.1). According to the new International Service for the Acquisition of Agri-biotech Applications report, the global biotech acreage has increased by 3% or 4.7 million hectares, with over half of this increase coming from smallholder farmers in developing countries. First-generation GM crops increased yields and reduced pesticide use by providing resistance traits to biotic stresses, such as insects, weeds and plant pathogens. With the advent of climate change, second-generation biotech crops have targeted abiotic stresses such as heat, drought and flooding. The world is entering a stage of third-generation GM crops, with nutritional traits that are highly valued by consumers and the food and manufacturing industries, such as biofortified, nutritionally enhanced food crops or other crops to reduce food waste.

The Millennium Development Goals (MDGs) were originally set up by the United Nations (UN) between 2000 and 2015 to fight poverty by establishing a series of eight measurable objectives that helped to promote political accountability, improved metrics and global awareness of the world's poor (Sachs, 2012). Great strides were made in the fight against poverty, although much of this advance was a result of a boost in economic growth in China. More recently, the advent of climate change, rising poverty and other pressing global challenges have necessitated a revision of these goals to incorporate overriding sustainability objectives that can be addressed alongside poverty reduction objectives. Rising incomes along with predicted population growth will increase food demand particularly with respect to meat consumption, which, in turn, are expected to exert more pressure on ecosystems. The number of chronically hungry currently exceeds 1 billion, and inequity between the wealthy and the poor continues to grow. Thus, the Sustainable Development Goals (SDGs) were developed (Fullman et al., 2017) and represent a

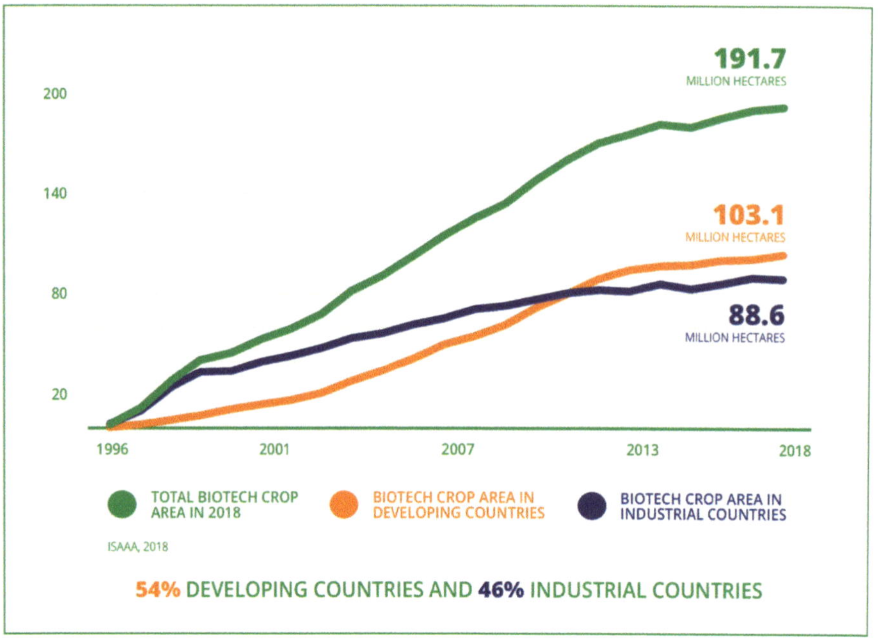

FIGURE 1.1 The International Service for the Acquisition of Agri-biotech Applications (ISAAA) report 2019. (Source: ISAAA.org.)

broader range of sustainable development challenges than the MDGs (Figure 1.2). Technological innovation, including the use of GM crops, is urgently required to reach the SDGs (ISAAA, 2019).

This chapter discusses the development of new technologies associated with agricultural biotechnology, including genetically modified and genome-edited crops. The chapter begins by describing these technologies in more detail, and then describes how these technologies have been used to address food security and poverty, climate change and the environment, food waste and energy, and social welfare.

FIGURE 1.2 United Nations Sustainable Development Goals. (Source: UN.org.)

The chapter ends with a discussion of the challenges remaining regarding consumer acceptance of these technologies.

1.2 Today's Agricultural Biotechnologies

Conventional plant breeding has advanced so that modern molecular techniques such as marker-assisted selection (MAS) are now used as commonplace protocols (Barasachi et al., 2016). Alternatively, agricultural biotechnology that involves the manipulation of recombinant DNA to generate new or improved traits in "transgenic" plants with unique nutritional or agronomic traits is the result of recombinant DNA technologies (Kamthan et al., 2016).

It is important to acknowledge that many of the varieties of crops that are available today have had their genomes altered by a technology that existed long before the advent of recombinant DNA. For example, mutagenesis breeding originates from mutation research that originated in the 1930s and includes the introduction of random mutations to plant cuttings using chemical or irradiation mutagenesis. Explants can then be propagated from these mutation events into new crop varieties. More than 3,000 varieties of crops have been generated through mutagenesis breeding, according to the mutant variety database, including the popular ruby red grapefruit (https://mvd.iaea.org).

Genetic engineering of crops for improved nutritional and agronomic traits is essential and critical for modern plant science (Ricroch and Henard-Damave, 2016). This can include the introduction of a novel trait into a crop either by adding new genes or by making small changes to pre-existing genes that are already part of the crop genome. Novel forms of genetic material are incorporated into the plant genome via Agrobacterium-mediated transformation or particle bombardment (gene gun). A well-known example of a transgenic crop is Golden Rice, which expresses β-carotene and contains genes derived from different species which together reconstitute the β-carotene pathway (Al-Babill and Beyer, 2005).

Cisgenic crops express genes from closely related plants rather than other species. These allow farmers to regain useful traits that have been lost over years of crop breeding. For example, the Wheat Stem Rust Initiative has created cisgenic versions of wheat containing multiple resistance genes to the fungal pathogen Ugg99 using wheat relatives (Singh et al., 2015).

RNA interference (RNAi) technology represents another form of genetic engineering that is gaining in popularity. Plants can be engineered to express the antisense RNA version of a specific plant or pathogen gene, and then expression of the gene targeted is blocked via gene silencing. For example, genetically modified papaya grown in Hawaii confers resistance to papaya ringspot virus by expressing an antisense RNA to the viral genome (Gonsalves, 1998).

The most recent technology used in agricultural biotechnology is genome editing. Genome editing does not require the introduction of new gene sequences; rather, it may direct only one or a few nucleotide changes within a plant genome (Rani et al., 2016; Mao et al., 2018). Genome-editing technologies such as CRISPR-Cas9 make use of various defence strategies derived by bacteria to target specific sequences of DNA and cleave them at these sites with Cas nucleases. The technology is then able to take advantage of DNA repair mechanisms already found in the cells of all organisms. By repairing the sites of cleavage, specialized changes that will be carried through the genome of the "edited" organism to subsequent generations can be produced. Genome editing is an extremely promising new tool for plant breeding, as many complex traits such as crop yield, photosynthesis and drought resistance involve multiple genes, and thus would be difficult to achieve through conventional breeding alone.

Taken together, these technologies can advance agriculture in ways that were never believed possible and at speeds unheard of a few decades ago. The following section describes some recent innovations in genetic engineering to improve crops.

1.3 GM Crops to Address Food Insecurity and Poverty

GM biofortified crops can help meet the nutrition targets of the SDGs to end hunger, improve nutrition and improve livelihoods by increasing crop yields, thereby ending poverty. One avenue to improve yield has been to make use of miRNAs, transcription factors and proteins involved in signalling cascades (Garcia, 2008). Most miRNAs control transcription factor expression (Fu et al., 2012). Pan et al. (2018) showed that overexpression of the precursor of miR408, one of the most conserved plant miRNAs, caused the induction of photosynthetic genes and higher vegetative growth in *Arabidopsis*, tobacco and rice. An increase in grain yield in rice was also observed. Other ways to increase grain width also exist. For example, Shi et al. (2019) identified a ubiquitin-specific protease to increase rice grain size.

Another way to improve plant yield is to increase levels of plant proteins responsible for stimulating cell proliferation or expansion. Some of these result in larger leaf size for biomass (Ciura and Kruk, 2018). Manipulating the levels of phytohormones, especially cytokinins, auxins and gibberellins, can improve plant growth (Mezzetti et al., 2004). For example, transgenic crops that express the *iaaM* gene encoding tryptophan monooxidase, derived from bacteria and involved in auxin synthesis, have been developed to increase fruit size in berries (Mezzetti et al., 2004). Similarly, Kolachevskaya et al. (2015) obtained transgenic potato with increased auxin levels that increased tuber size. Along these lines, Wang et al. (2019) used plant hormones, known as Brassinosteroids (BRs), to increase yield in tomato fruit.

Other approaches include the regulation of plant senescence via changes in cytokinin metabolism. For example, Swartzberg et al. (2006) found that tomato, canola and maize with increased cytokinin synthesis through biotechnology resulted in greater fruit weight (Kant et al., 2015). Alternatively, Bartrina et al. (2011) found that *Arabidopsis thaliana* knockout line *ckx3 ckx5* with reduced cytokine synthesis resulted in increased seed yield. A review on this topic can be found in Zalabák et al. (2013).

Improvements in nitrogen use efficiency in crops can be found in Good et al. (2004), Li et al. (2017b) and McAllister et al. (2012). One achievement would be to generate microbial nitrogenase in higher plants to allow atmospheric nitrogen to be fixed within the plants themselves. To this end, an active Fe subunit of nitrogenase has been expressed in tobacco (Ivleva et al., 2016). Phosphorus is also critical for plant growth. Overexpression of phosphatases has brought about faster growth in plants and increased seed yield (Sun et al., 2012). Similar results were found for *Camelina sativa* plants which overexpressed a phosphatase from *Arabidopsis* (Zhang et al., 2012; Chikara et al., 2018). The authors were able to increase oil yield by up to 13% in seed oil content and up to 52% in seed mass, in comparison to wild-type plants. Finally, Iwamoto and Tagiri (2016) have demonstrated an increase in rice grain number and weight by overexpressing *OsRDD1*, which encodes the transcription factor responsible for inducing genes involved in nutrient transport.

Improvements in yield have also been explored using CRISPR-Cas9 technologies by selecting a variety of target genes (Ricroch, 2019). The first exon of Gn1a and GS3 and the third exon of DEP1 and IPA1 were chosen as targets for the sgRNAs because mutations within these regions have been proven to result in higher yield. The DEP1 mutants increased the number of flowers per panicle (about 50% more than the control) and the IPA1 mutations resulted in improved yield phenotypes. Xu et al. (2016) edited three genes related to grain weight through genome editing, which resulted in significant increases. By deleting 625 bp within the DEP1 gene using CRISPR/Cas9, erect panicles with a higher grain number and a lower plant height were produced (Huang et al., 2009). CRISPR-based sgRNAs designed by Li et al. (2017) have been developed and display early maturation in different rice varieties.

Another way that yield can be increased is by reducing crop waste due to browning. For example, the polyphenol oxidase (PPO) gene is responsible for the oxidative browning of a number of fruits and vegetables including potatoes and apples. Genome editing has been used to block oxidation by polyphenols and reduce browning in both of these crops (Waltz, 2015a,b; Mellidou et al., 2014).

Approximately 2 billion people suffer from one or more micronutrient deficiencies (Hay et al., 2017). Trans-disciplinary strategies are needed to provide micronutrients, such as vitamin A and iron supplementation (Wesseler et al., 2017). The biofortification of crops is an alternative solution to alleviate the burden of malnutrition (Zhao and Shewry, 2011; Masset et al., 2012; Ruel and Alderman, 2013). Progress in conventional biofortification has been made over the past century, with many micronutrient-enriched crops now available in over 50 countries (Harvest Plus, 2015). Evidence is accumulating that indicates such crops are a cost-effective strategy (Meenakshi et al., 2010) to improve micronutrient status among target populations (Low et al., 2007; Haas et al., 2013).

GM biofortified crops such as Golden Rice would be a highly cost-effective investment to reduce target micronutrient deficiencies (De Steur et al., 2017). GM foods with quality traits (such as health traits) are only now being introduced into the food supply chain; for example, low-acrylamide GM potatoes (Rommens et al., 2008; Waltz, 2015) and GM oilseed crops (National Academies of Sciences and Medicine, 2016). Consumers are the reason why there is a growing interest in output traits. For example, a meta-analysis of studies on consumers' willingness to pay revealed general positive reactions towards GM biofortified crops with consumers willing to pay on average 24% more for GM biofortified food as compared to conventional food (De Steur et al., 2017). Consumers' optimism, however, can be significantly influenced by the way that information is provided. While positive information on the nutritional content/benefits or technology increased consumers' intention to purchase, exposure to negative information on GM technology reduced purchasing behaviour. This result explains the significant efforts that have been made to lobby against GM crops, regardless of the scientific basis of given arguments (Wesseler and Zilberman, 2014). Negative publicity on GM crops threatens the lives of malnourished populations in developing regions.

Micronutrient deficiencies in crops can be addressed through the direct fortification of crops in the form of fertilizers and foliar sprays, by the addition of dietary supplements containing vitamins and minerals such as vitamin A, iron, zinc and folate, or via the biofortification of plant varieties using modern breeding techniques (Ricroch and Henard-Damave, 2016). The production of genetically modified, biofortified food crops is potentially advantageous for populations who find it difficult to change their dietary habits or who lack access to nutritional programmes. Staple crops such as rice, banana, cassava and maize are low in micronutrients and sufficient quantities of each must also be bioavailable in order to contribute to a healthy diet. In many cases, the rural poor have little access to the rich source of micronutrients found in fresh fruits and vegetables and cannot afford to diversify their diets. Biofortification using technology could represent a simple answer to their challenges.

Golden Rice, for example, is a biofortified carotenoid-rich variety developed with the aim to improve the dietary value of this major food crop. Carotenoids are plant pigments that also have cancer-fighting and anti-inflammatory properties. The carotenoid β-carotene is converted in the body to vitamin A, which is essential for vision, normal growth and development (Bollineni et al., 2014). Golden Rice is thus a well-known example of using genetic engineering to address a public health problem. Death and blindness of young children due to vitamin A deficiency (VAD) is widespread in economically challenged populations where rice is a primary food source. Natural rice does not express β-carotene in the endosperm, yet is the primary staple of rural poor populations of Southeast Asia.

Initially, Golden Rice was developed through the insertion of the *psy* (phytoene synthase) gene of the daffodil (*Narcissus pseudonarcissus*) and the *crtI* (phytoene desaturase) gene of bacteria (*Erwinia uredovora*) into the nuclei of rice variety BR29 under the control of the glutelin 1 (Gt1) promoter. In 2005, a second construct known as Golden Rice 2 was generated, using maize as the source of the *psy* gene. Golden Rice 2 produces significantly higher amounts of β-carotene compared to the original Golden Rice. The carotene bioavailability of Golden Rice was confirmed in human trials, and Golden Rice has now been approved for human use in the United States, Canada, Australia and New Zealand (Coghlan, 2018).

A randomized trial carried out in the United States demonstrated a high bio-conversion factor of β-carotene to vitamin A in Golden Rice (3.8:1). In this study, it was determined that 100 g of uncooked Golden Rice would provide about 80–100% of the estimated average requirement (EAR) and 55–70% of the recommended dietary allowance (RDA) for adult men and women (Tang et al., 2009). The relatively high micronutrient concentrations in targeted GM biofortified crops such as Golden Rice could close the micronutrient intake gap in human populations. In nearly all cases, a regular portion of the targeted biofortified crop would provide the daily micronutrient requirements. For example, in Asia, Golden Rice could reduce the prevalence of dietary vitamin A inadequacy by up to 30% (children) and 55–60% (women) in Indonesia and the Philippines, and up to 71% (children) and 78% (women) in Bangladesh (De Moura et al., 2016).

Different maize varieties also produce highly variable levels of β-carotene. The β-carotene content can be increased using either conventional breeding or genetic engineering strategies. Transgenic maize biofortified with β-carotene improved health in clinical trials held in North America and Africa. However, biofortified maize is more orange in colour and this may require some effort to change public perception for certain Africans. Orange maize is traditionally associated with animal feed and white maize is considered to be used for human consumption.

Cassava is a staple of 250 million sub-Saharan Africans, but is high in cyanide and low in protein, iron and β-carotene. Transgenic cassava expressing high levels of β-carotene have been demonstrated to successfully increase vitamin A levels in feeding studies, indicating that nutritionally enhanced cassava can be generated.

Banana, the world's most important fruit crop, is a major staple in many African countries, where vitamin A deficiency tends to be greatest (Amah et al., 2018). Bananas are slow to grow, difficult to breed and have low fertility rates, thus improving their nutritional benefits remains a challenge. Conversely, the wide number of varieties and the broad distribution of vitamin A levels make bananas amenable for biofortification through biotechnological strategies. The cooking banana, the East African highland banana (EAHB), that is consumed in Uganda has low vitamin A levels, impacting both infants (children under the age of 5) as well as women of childbearing age. High levels of vitamin A (20 lg/g dry weight) were generated from transgenic banana lines expressing both phytoene synthase (derived from the fruit of the Fe'I banana found in Papua New Guinea, which only grows in small bunches) and under the control of the banana ubiquitone promoter (Ubi). These transgenic lines display a yellow-orange, "golden bunch" phenotype. Biofortified super bananas will improve the vitamin A status of some of the poorest subsistence farmers of Africa (Waltz, 2014; Paul et al., 2017). Unfortunately, to date, no regulatory framework for biotechnology has been set up in Uganda, although the bananas could be ready for release as early as 2021.

In addition to biofortification with vitamins and minerals, crops have also been engineered with enhanced nutritional properties. Tomatoes can be grown worldwide; they are also easy to metabolically engineer for improved properties that can both improve nutrition as well as decrease the risk of chronic diseases, such as hypertension, diabetes and cancer. For example, two transcription factors isolated from snapdragon have been shown to increase anthocyanin, found also in blueberries and cranberries, in tomato plants. Anthocyanin is an antioxidant with clearly demonstrable health benefits, and tomato fruit metabolically engineered to produce high levels of anthocyanins was able to extend the life span of cancer-susceptible mice by up to 30% (Li et al., 2014). Similarly, flavonoid biosynthesis pathways have been metabolically engineered to improve the nutritional value of tomatoes (Lim et al., 2014). A consumer panel reported a higher preference for the flavour and an overall liking of CHI tomatoes over conventional tomato varieties, and several panel members indicated that they would purchase transgenic food if they believed that it would improve their health.

In transgenic tomatoes expressing the gene encoding grape (*Vitis vinifera* L.), stilbene synthase accumulated high concentrations of resveratrol during tomato fruit ripening. Resveratrol, a bioactive compound with beneficial health properties, is found in red grapes and wine, but not in many other common food sources. Since grapes tend to be grown in select geographical regions, the production of resveratrol in commonly produced and well-distributed fruit such as tomato would make it more accessible around the globe.

Omega-3 fatty acids, found in fish and marine life, are beneficial for brain function and cardiovascular health but cannot be made in sufficient quantities for the world's population. Transgenic "designer oilseed" plants such as false flax (*C. sativa*) have been metabolically engineered to synthesize omega-3 fatty acids and are used to feed farmed fishes (Ruiz-Lopez, 2014). Two other examples of traits that have been added to designer oil crops include higher levels of acetyl glyceride and enriched glucoraphanin content, the latter example having preventative effects against cancer.

Plants have also been designed as production platforms for vaccines, antibodies and therapeutic agents. Vaccines derived from plants are easy to scale up, inexpensive, safe and efficacious. Some of the original plant-made vaccines targeted infectious diseases which are the major cause of infant mortality in developing countries, such as enterotoxigenic *Escherichia coli*, rotavirus and cholera. More recently, plant-made vaccines have been developed as a means to stockpile against potential pandemics. For example, a monoclonal antibody generated in tobacco plants to block the Ebola virus has been developed by Mapp Pharmaceuticals. This plant-made antibody cocktail was provided to several US patients as an experimental medicine during the Ebola outbreak of 2015. Another example of a

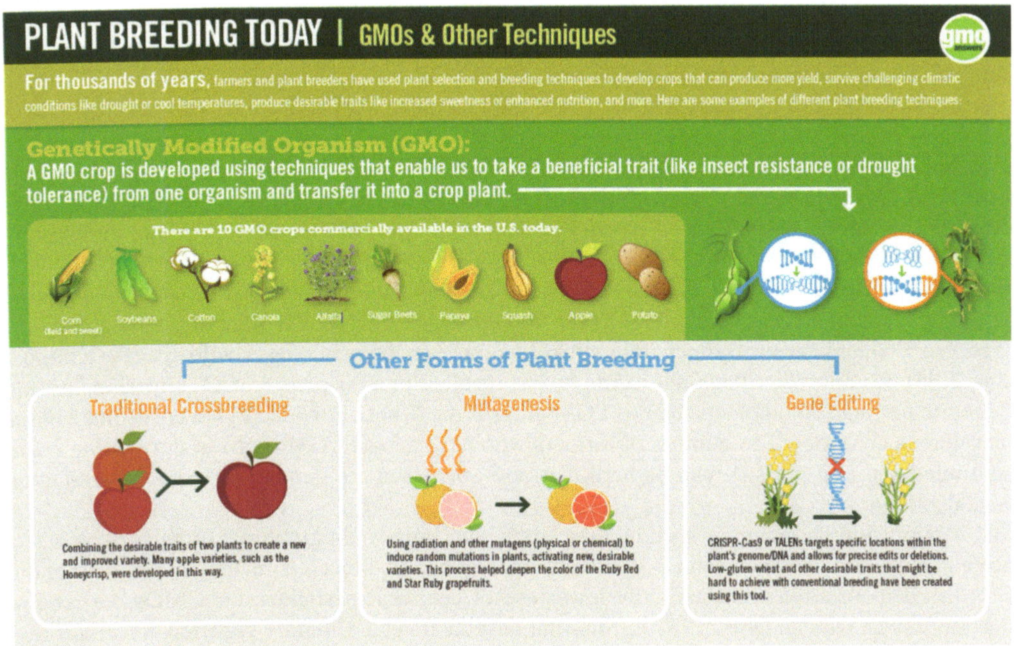

FIGURE 1.3 Plant breeding techniques. (Source: GMO Answers.com.)

plant-made vaccine is the influenza vaccine, that is being produced by the Quebec company Medicago. This vaccine is attractive as it requires no lengthy incubation in eggs as is the case with conventional vaccines. Along these lines, a vaccine against COVID-19 is now under production by the same company. In certain cases, a plant-made vaccine may be superior in efficacy to a conventional vaccine made in mammalian systems. This is the case with the plant-made vaccine for dengue fever virus, which does not cause the same adverse effects of antibody-dependent enhancement that was previously made in an animal system. Another example is the microbicide for HIV, composed of monoclonal antibodies and plant lectins. This microbicide is more efficacious in blocking HIV transmission than its conventional counterpart. Plants can also be used to treat orphan or rarer diseases in a cost-effective manner. An example is the plant-made version of glucocerberosidase, made in carrot suspension cells by the Israeli company Protalix. This plant-made version, known as Taliglucerase alfa, is less than one hundredth the cost of the conventional version of the drug.

While nutritionally enhanced GM crops can be a promising contribution to food security, genetically modified organisms (GMOs) with agronomic traits can also indirectly improve farmers' health by lowering their exposure to insecticides and herbicides (Pray et al., 2001) (Figure 1.3). This along with other environmental concerns will be discussed in the next section.

1.4 GM Crops to Address Climate Change and Environmental Concerns

In addition to their nutritional benefits and indirect economic benefits, GM and genome-edited crops are being developed to address a changing climate and to reduce environmental concerns. This section describes some of the progress that has been made.

The most heavily utilized GM crop technology developed to resist insects is found in transgenic crops which express Bt, a bacterial toxin from the bacteria *Bacillus thuringensis*. This toxin is safe for mammals but creates pores in the alkaline digestive tract of insects, making them susceptible to disease if they feed on a plant expressing Bt. The result has been a tremendous reduction in pesticide use in countries such as India and China, where GM cotton is grown, as well as eggplant in Bangladesh. The health benefits of reduced pesticide use for farming households, especially for women and children, can be significant (Antle and Pingali, 1995; Haggblade et al., 2017). In the case of food crops such as Bt-corn, insect resistance has also led to decreased levels of mycotoxin, which in turn provides health and economic benefits to farmers (Gressel and Polturak, 2018).

Glyphosate is the most well-known herbicide used together with a GM crop. Glyphosate is a broad-spectrum weedkiller that acts by inhibiting the synthesis of three amino acids in plants, yet breaks down easily and is less toxic than most previously used herbicides (Duke et al., 2012). As a result, it was quickly adopted into farming practices the world over. Glyphosate-resistant plants were developed by Monsanto in 1996 and greatly simplified weed management. Today, over 90% of corn, soybean and cotton produced in the United States are genetically modified to be tolerant to glyphosate (Qaim, 2020).

Over 90% of the GM crops available today express traits that confer herbicide and insect resistance. The use of herbicide- and insecticide-tolerant GM crops has enabled farmers to practice conservation tillage, use 37% less chemical input and increase biodiversity by using fewer wild spaces (Klümper and Qaim, 2014). Eliminating tillage, for example, has resulted in a decrease in CO_2 emissions by 1.8 tons per hectare per year. Another advantage of GM crops is the benefit provided to lower-income households in developing countries. The number of farmers who have adopted GM crops in developing countries, where weed and pest pressure can be high, has now surpassed the number in industrialized countries (Lu et al., 2020).

As mentioned in the previous section, GM crops with improved agronomic traits such as pest resistance and herbicide tolerance can also generate indirect health effects. A review of Klümper and Qaim (2014) has demonstrated a substantially lower use of chemical pesticides for GMOs, i.e. an average reduction in pesticide quantity of 37%. Substantial reductions are mainly reported for insect-resistant (*Bt*) crops (Racovita et al., 2015). Declines in insecticide treatments are associated with fewer reported poisonings of farmers. This has been demonstrated with Bt cotton farmers in China (Huang et al., 2002), India (Kouser and Qaim, 2011), Pakistan (Kouser and Qaim, 2013) and South Africa.

Other benefits associated with herbicide-tolerant transgenic plants include a decreased carbon footprint, and a decrease in soil erosion and tractor fuel usage resulting from low-tillage farming practices (Smyth, 2017). In developing countries, this labour-saving feature can shorten the field work time of women who work in agriculture, so that they can focus on other things, such as their children or developing their own businesses. Limitations include the fact that weed management is solely based on herbicides, and as a result, the amount of herbicide used has not lessened (Bonny, 2016), while the toxicity level of the active ingredients emitted into the environment has declined (Wesseler et al., 2011).

Transgenic plants have been developed that exhibit resistance against a variety of other pathogens, including fungi, bacteria and viruses. This has reduced the need for chemical inputs. For example, Lu et al. (2020) found that a powdery mildew resistance gene from a Chinese wheat landrace with two amino acid deletions conferred resistance against *Blumeria graminis* f. sp. *tritici*. Similarly, Hao et al. (2020) expressed a modified thionin (Mthionin) in transgenic plants that exhibited enhanced resistance towards *Fusarium* head blight and several other fungal pathogens.

In the case of bacterial infection, Li et al. (2020) found an approach to confer resistance in citrus from *Xanthomonas citri* by overexpressing wall-associated receptor-like kinases (WAKLs) through a mechanism that uses the reactive oxygen species (ROS) pathway. Niño et al. (2020) addressed *X. oryzae* pv. *oryzae* infection by overexpressing papain-like cysteine proteases in rice, which appears to activate transcription factors and transduction signals that control the upregulation of pathogenesis-related proteins as well as the synthesis of secondary metabolites. Rooney et al. (2020) reported resistance to the plant pathogen, *Pseudomonas syringae* (Ps), by generating transgenic plants which express the bacteriocin, putidacin L1 (PL1). These microbial pathogens could be effective against a broad spectrum of bacteria.

GM crops have been generated which confer resistance to many different types of plant viruses. Much of the earlier research in this field focused on coat protein (CP)-mediated resistance. More recently, however, innovative new ways to create virus-resistant transgenic plants have emerged. Soybean mosaic virus (SMV), for example, is one of the most destructive viral diseases. Luan et al. (2020) generated an inverted repeat or hairpin of the *GmVma12* gene sequence and expressed it in transgenic plants, which were highly resistant to virus inoculation. Pottinger et al. (2020) demonstrated that by exchanging seven amino acids of the PBS1 protein kinase with a cleavage motif of the NIa protease of turnip mosaic virus

(TuMV), the kinase could be activated upon virus infection and resistance induced within the plant. Yang et al. (2020) overexpressed the chloroplast protein ferredoxin 1 (FD1), known to interact with the coat proteins of several plant viruses, in transgenic plants and conferred virus resistance. The authors found that they had interfered with the RNA silencing suppressor function of one of the viral proteins and that this was the cause of virus resistance in the transgenic plants.

Another viable approach to generating pathogen-resistant plants is CRISPR/Cas9 technology. Genome editing has been used to generate glyphosate-resistant plants, for example. Other means to confer resistance include altering target–effector interactions, knocking out genes that make the host plant susceptible and uncoupling defence hormones (Bisht et al., 2019). Genome editing may become the method of choice in the future due to its ease of use, among other things.

1.4.1 GM Crops to Address Climate Change

Climate change has made farming more difficult and yield production less stable. Transgenic crops have been developed to address abiotic stresses such as drought and salt tolerance, as well as heat stress. A few examples are presented here.

In the case of cereals such as wheat and barley, both drought and heat stress have been found to reduce grain size and the expression of genes involved in starch production (Thomas, 2015). Transgenic approaches have also been shown to improve drought tolerance under controlled environment conditions but there are no results to show how well this translates into improved crop performance under field conditions. There are many reports in the scientific literature of transgenic approaches to improve drought tolerance. For example, dehydrins can impact abiotic stress tolerance, as do transcription factors such as the dehydration-responsive element-binding (DREB) protein (Campbell and Close, 1997; Agarwal et al., 2006). The overexpression of DREB genes can improve wheat and barley survival under drought conditions (Morran et al., 2011). Unfortunately, current restrictions on growing transgenic plants in the field generally limit a test of the effectiveness of transgenic approaches to improve drought tolerance (Saint Pierre et al., 2012).

Khan et al. (2019) describe several genes involved in drought resistance, including osmoprotectants, membrane stabilization, detoxification and transport proteins, as well as their signalling proteins and transcription factors. Osmoprotectants can elicit higher tolerance by stabilizing cell membranes. Overexpression of the ferritin gene was demonstrated to induce tolerance to drought stresses (Zang et al., 2017). Transgenic wheat which overexpresses the C_4 enzymes phosphoenolpyruvate carboxylase (PEPC) and pyruvate orthophosphate dikinase (PPDK) demonstrates improved photosynthetic traits and yield as well as drought tolerance (Zhang et al., 2014; Qin et al., 2015). At this point in time (2020), the first drought-tolerant genetically modified wheat is under development by an Argentine company, but has yet to be approved for commercial release (Orrono and Vesprini, 2018).

Transgenic rice has also been developed for drought tolerance using a number of strategies. For example, the transcription factor *OsERF71* was overexpressed in transgenic rice; these transgenic plants suppressed photosynthesis and diverted more energy to survival-critical mechanisms (Ahn et al., 2017). Alternatively, Gao et al. (2018) identified the maize transcription factor ZmPIF3 involved in stomatal closure and water loss. The authors expressed the gene for ZnPIF3 in rice plants and demonstrated increased drought tolerance by reducing stomatal aperture. As in studies with wheat, extensive field trials have not been completed for drought-tolerant rice due to the lack of regulatory structure available (Gaudin et al., 2013).

Groundnut (*Arachis hypogaea* L.) is an oilseed crop and major fodder for the cattle industry in many developing countries. Genetically modified drought-tolerant groundnut varieties are being developed which are resistant to drought and other abiotic and biotic stresses (Gantait and Mondal, 2018). Gong et al. (2020) looked at the NAC transcription factor HaNAC1 from *Haloxylon ammodendron*, a typical woody plant normally grown in desert climes. HaNAC1 was upregulated by osmotic stress and was demonstrated to promote growth and drought tolerance in transgenic *Arabidopsis*, suggesting that this factor could be used as a promising candidate gene for generating crops with tolerance to drought stress. Similarly, Zhao et al. (2019) examined a transcription factor derived from sheep grass, the MYB-related LcMYB2, that upregulates during drought conditions. This transcription factor promotes root growth, the accumulation of osmoprotectants and seed germination in transgenic *Arabidopsis* plants, and could work on food crops as well.

The maize OSCA family is composed of hyperosmolality-gated calcium-permeable channel proteins whose promoter regions contain the elements responsive to abiotic stresses. These gene families were upregulated over 200-fold under drought conditions, and transgenic plants exhibited improved resistance to drought (Cao et al., 2020). The authors speculated that one of these genes, *ZmOSCA2.4*, was regulated by the microRNA miR5054. These results could also be used to generate drought-resistant crops in the future.

Soil salinity and flooding due to climate change can seriously reduce crop productivity. Salt tolerance can also be regulated through the overexpression of specific transcription factors or ion transporters. For example, Sun et al. (2020) discovered that the brassinazole-resistant 1 (BZR1) family of transcription factors, which control plant growth and development, enhanced salt and drought tolerance. Using RNA-seq analysis, the authors found that this transcription factor regulated salt tolerance by inducing the expression of 84 downstream stress-related genes. Wani et al. (2020) also examined the overexpression of endogenous stress mechanisms that can influence salt tolerance. Regulatory and transporter proteins can be used to enhance salt tolerance in crop plants.

By examining plants that thrive under halophilic conditions, scientists can search for additional genes as well as microRNAs that could be utilized for salt tolerance. For example, Zhang et al. (2020) identified the transcriptional activator gene *HbMBF1a* from the halophyte *Hordeum brevisubulatum*, and found that when overexpressed, *HbMBF1a* could enhance the salt tolerance as well as germination in *A. thaliana*. A suite of stress-responsive genes was increased in the transgenic lines under salt conditions. In another study, Nguyen et al. (2019) overexpressed a dehydration-responsive element-binding protein transcription factor in soybean plants and these plants exhibited salt tolerance when compared to non-transgenic control plants. Plant mitochondrial genes can also be used to generate salt tolerance. For example, a gene encoding the alternative oxidase (AOX) in *BnaAOX1b* and overexpressed in rapeseed displayed salt resistance during germination (Yang et al., 2019).

Mangrove trees such as *Avicennia officinalis* that thrive in high salt conditions use specialized ion transporters which can compartmentalize Na^+ ions. Transgenic rice seedlings harbouring these transporters showed enhanced salt tolerance, indicating that this gene can be used to develop salt-tolerant crops (Krishnamurthy et al., 2019). While much emphasis has focused on rice and other cereal crops, other food crops, such as transgenic tomatoes, have also been generated which confer salt tolerance using DREBs and other similar mechanisms (Krishna et al., 2019). In addition, Shah et al. (2018) have developed transgenic *Brassica* which display salt tolerance.

Genetic engineering can be used to make plants more productive, especially under heat stress situations anticipated by climate change and since improvements using traditional breeding methods have reached their limits (Nowicka et al., 2018). Photosynthesis is an important determinant of crop yield. A number of approaches have been taken to increase the rate of photosynthesis in plants. These include using the enzymes from cyanobacteria or green algae to increase carbon sequestration and introducing the C4 pathway into C3 plants. Simkin et al. (2019) have found that photosynthesis may be improved by over 40% using genetic engineering approaches. The overexpression of enzymes involved in photosynthesis, such as sedoheptulose-1,7-biphosphatase (SBPase), can increase both photosynthetic carbon assimilation and biomass yield.

On the other hand, tolerance to cold stress can be another major challenge. Lenke et al. (2019) used the gene *OsRBGD3* derived from a drought-tolerant Indica rice cultivar N22. Constitutive overexpression of the *OsRBGD3* in transgenic plants conferred tolerance to cold stress, as well as significant root development and early flowering in transgenic plants.

1.4.2 Other Environmental Concerns: Conserving Biodiversity

Population growth presents a significant challenge to biodiversity conservation around the world, particularly as population growth increases agricultural production, which contributes to deforestation, the depletion of soil and the loss of biodiversity. Response to this challenge can be addressed using GM technology (Barrows et al., 2014).

A study by Carpenter (2010) analyzed 49 peer-reviewed publications on the adoption of GM crops from 12 countries by comparing the yields between adopters and non-adopters. The author showed that out of 168 results (based on yield outcome), 124 results were positive for GM adopters (most of these farmers came from developing countries) when compared to non-adopters. This result is consistent with the ISAAA report that 18 million poor smallholders have benefited from the adoption of GM crops in the past two decades (James, 2016).

The beneficial role of insects and other arthropods remains crucial in maintaining soil fertility. Several studies have demonstrated the benefit of GM technology (*B. thuringensis-Bt*) on soil biodiversity. Using meta-analysis, Marvier et al. (2007) showed a higher abundance of non-target invertebrates in *Bt* crops compared to non-*Bt* crops. Overall, the adoption of GM crops has created more environmentally friendly farming practices particularly through the enhancement of soil biodiversity.

1.4.3 Phytoremediation Using Agricultural Biotechnology

Heavy metal contamination in the environment can be the result of either anthropogenic activities or natural processes. Heavy metals may occur in the soil in a variety of forms and can be greatly modified with respect to their solubility/bioavailability (Shahid et al., 2015). Bioaccumulation can cause the loss of plants, microbes and animal biodiversity. Phytoremediation therefore is the process of cleaning up environmental pollutants using plants. Genetic engineering technologies can be utilized to develop plants that are capable of bioremediating persistent organic pollutants, heavy metals, petroleum and other pollutants.

Studies have demonstrated that genetic engineering can be effective in producing plants for phytoremediating soils and waters contaminated with pollutants. Many plants harbour genes that enhance the capability of the plants to mobilize, stabilize and/or accumulate metals. The phytoextraction of heavy metals such as cadmium, lead and copper from soil is possible due to their accumulation in plants through metal transporters, the improved production of enzymes of sulphur metabolism, the production of metal-detoxifying chelators and the activation of various antioxidants. This has resulted in the identification of genes involved in increased metal tolerance, metal uptake and hyperaccumulation. In certain instances, transgenic plants with an enhanced phytoremediation capacity are usually produced via the introduction of a gene that is not native to the plant, such as degradative enzymes to achieve the complete mineralization of organic pollutants.

1.5 GM Crops That Address Food Waste and Energy

Plant biotechnology has also played a role in recent innovations in biofuel development. The conversion of agricultural biomass, including residual plant waste, into ethanol could significantly impact our energy needs. The ability to hydrolyze cellulose derived from recalcitrant plant cell walls into fermentable sugars is a significant advance in the generation of affordable renewable energy based on plant material. Currently, this process requires harsh chemical pretreatment of the biomass as well as the use of expensive microbial cellulases and other dissociative enzymes. The cost of these two practices has become a major deterrent to the advent of a strong and sustainable lignocellulosic biofuel industry.

The cost of enzyme production could be lowered substantially by using the plants themselves as production platforms for microbial cellulases. Coordinated efforts have been made to generate transgenic plants which express microbial cellulases. This could reduce the cost burden of enzyme production and thus biomass conversion. Schmidt et al. (2019) conducted a field study of tobacco plants expressing Cel6A, derived from *Thermobifida fusca*, within their chloroplasts. Despite unpredictable growth conditions in the outdoor field, these transgenic crops were able to robustly express cellulases at a price of USD 60 kg, with future projections of USD 80 kg, as compared to conventional microbial bioreactors which produce cellulases at USD 10 kg. On the other hand, recalcitrance can be addressed by modifying the cell walls of plants used for biofuel (Li et al., 2014). Lower recalcitrance of plant cell walls has been achieved by altering cell wall architecture, for example, by manipulating the biosynthesis of hemicellulose without impacting cell wall integrity or plant development. For example, reduced lignin content can be achieved by engineering monolignol pathway genes to decrease lignin content. *Arabidopsis* mutants with lower lignin content displayed a conversion efficiency of 88%, compared to wild-type plants, from which only 18% of cellulose could be converted to glucose. A more complete understanding of cell wall formation and structure will greatly improve lignocellulosic feedstocks and reduce the cost of biofuel production.

The agricultural sector remains the third-largest producer of global greenhouse gas emissions. As mentioned previously, GM crops have been engineered to lower the amount of chemicals using pesticides, thereby lowering GHG emissions. Brookes and Barfoot showed that, on a global scale, the agricultural environmental footprint was reduced by 613 million kilograms of active ingredients among GM-producing countries including the United States, China, India, South Africa, Brazil and Argentina (Brookes and Barfoot, 2017). Also, these authors estimate that 174 million hectares of land were saved through the adoption of GM crops.

Brookes and Barfoot (2017) argue that the reduction in GHG emissions in 2015 alone is equivalent to removing 12 million cars off the road. GM technologies such as drought and salinity tolerance traits can play a crucial role in strengthening resilience and adaptive capacity related to the impact of climate change. Moreover, GM drought-tolerant crops could significantly reduce water usage with huge environmental benefits, particularly in drought-prone regions of Africa, and reduce natural habitat degradation, drought and desertification.

More recently, GM technology has also been used to combat food waste. According to the FAO (2018), one-third of all food is lost before it is consumed. GM technology can halve global food waste at the retail and consumer level. For example, several varieties of late-blight disease-resistant potato have recently been approved for commercialization. The transgenic potato will not only increase yield and reduce the use of fungicides by as much as 45%, but it will also reduce food wastage through its non-browning trait (James, 2017). Similarly, several apple varieties have been genetically modified to be resistant to browning and, oftentimes, bruising, two attributes causing retailer and consumer reluctance (Thompson and Kidwell, 1998; Waltz, 2015). Simplot and Okanagan Specialty Fruits, owned by Intrexon Corporation, have successfully completed field trials and such apples and potatoes are currently on the market. Another example is the non-browning mushroom, which is the first application of the CRISPR-CAS9 technique to be approved in the United States (Waltz, 2016).

1.6 GM Crops, Environmental Justice and Social Welfare

The adoption of GM crops has improved the welfare of many farming households in developing countries. For example, the reduction in pesticide use in cotton production in India and China has generated substantial health benefits (Antle and Pingali, 1995). This not only benefits farm households but also provides positive benefits for farm labourers, particularly women and children (Beckmann and Wesseler, 2004). Herbicide-tolerant crops in Africa can reduce the burden of manual weeding, thus contributing to the empowerment of women and girls. In regions where the use of pesticides is low but pest pressure is high, which is more often observed among poorer households, substantial yield increase has been observed, benefitting in particular poor households (Klümper and Qaim, 2014).

There are additional socio-economic implications for poor households. GM corn resistant to lepidopteran pests such as corn borers has lower levels of mycotoxins, thus generating health benefits for subsistence farm households. Some concerns regarding intellectual property rights relate to the argument that large international seed companies own the rights over transgenic plants and can charge farmers a higher price and restrict their ability to save seeds, thereby undermining farmers' rights (Bonanno et al., 2017). The Nagoya Protocol has been implemented to protect farmer rights and to ensure benefit sharing of plant genetic resources maintained by farmers mainly in the Global South.

1.7 Future Challenges Facing Agricultural Biotechnology

The problems facing modern agricultural biotechnology, especially GM crops, are enormous. Obstacles include limited biotechnology R&D programmes, politics and regulatory delays (Smyth, 2017) and the perception of the potential risks related to GM crops, among others. For example, many developing countries have limited biotechnology R&D programmes, weak laboratory facilities and a paucity of well-trained experts. These problems continue to undermine the application of GM crops in developing countries (Paarlberg, 2008; Adenle et al., 2012). A three-year intensive study determined that a lack of scientific capacity and poorly equipped laboratories are a large constraint on developing new GM products in Africa (Adenle et al., 2013). Other GM products, including GM bananas and GM maize (under R&D programmes and confined field trials), must be sent to industrialized countries (e.g. United States and Australia) for laboratory analysis where there is an ongoing collaboration (Adenle et al., 2017; Wesseler et al., 2017). Consequently, many African countries are unable to regulate and develop GM crops.

International politics controls GM crop decision-making processes around the world (Adenle, 2017). It has been reported that environmental non-governmental organizations found in European countries have spent over tens of billions of US dollars in preventing the use of GM crops (Byrne, 2015), when compared to USD 8.6 billion spent on agricultural biotechnology R&D by US-led multinational companies (Hobbs et al., 2014).

The United States follows a 'substantial equivalence' policy towards regulating innovations in crop biotechnology, whereas the EU follows the 'precautionary principle', initially designed by the United Nations Convention on Biological Diversity under the Cartenega Protocol on Biosafety. These differences in regulatory framework concerning product vs. process have led to serious obstacles in terms of food trade, particularly for those in developing countries. In the case of Golden Rice, for example, release has been hindered in countries such as Bangladesh and the Philippines with field trials destroyed by anti-GMO activists on more than one occasion in both countries (Potrykus, 2017). Golden Rice could make a world of difference by contributing to the reduction in malnutrition in the developing world. After 20 years, Golden Rice has finally been approved in several countries (Wessler and Zilberman, 2014). More recently, a genome-edited version of Golden Rice has also been developed. Whether anti-GM activists will block its release as well remains to be seen.

REFERENCES

Adenle AA (2017). Modern biotechnology for innovation of agricultural development in the developing world: What role can Japan play? *AgBioForum*, 20(1), 54–66.

Adenle AA, Aworh GC, et al. (2012). Developing GM super cassava for improved health and food security: Future challenges in Africa. *Agric Food Secur*, 1, 11.

Adenle AA, Morris EJ, et al. (2013). Status of development, regulation and adoption of GM agriculture in Africa: Views and positions of stakeholder groups. *Food Policy*, 42, 159–166.

Adenle AA, Morris EJ, et al. (2017). *Genetically Modified Organisms in Developing Countries: Risk Analysis and Governance.* Cambridge: Cambridge University Press.

Agarwal PK, Agarwal P, Reddy MK, et al. (2006). Role of DREB transcription factors in abiotic and biotic stress tolerance in plants. *Plant Cell Rep*, 25, 1263–1274.

Ahn H, Jung I, Shin SJ, Park J, Rhee S, Kim JK, Jung W, Kwon HB, Kim S (2017). Transcriptional network analysis reveals drought resistance mechanisms of AP2/ERF transgenic rice. *Front Plant Sci*, 8, 1044. doi:10.3389/fpls.2017.01044. eCollection 2017.

Al-Babili S, Beyer P (2005). Golden rice—five years on the road—five years to go? *Trends Plant Sci*, 10(12), 565–573.

Amah D, van Biljon A, Brown A, Perkins-Veazie P, Swennen R, Labuschagne M (2018). Recent advances in banana (*Musa* spp.) biofortification to alleviate vitamin A deficiency. *Crit Rev Food Sci Nutr*, 12, 1–39.

Antle JM, Pingali PL (1995). Pesticides, productivity, and farmer health: A Philippine case study. In *Impact of Pesticides on Farmer Health and the Rice Environment*, edited by PL Pingali, PA Roger. Dordrecht, Springer Netherlands: 361–387.

Barabaschi D, Tondelli A, Desiderio F, Volante A, Vaccino P, Valè G, Cattivelli L, et al. (2016). Next generation breeding. *Plant Sci*, 242, 3–13.

Barrows G, Sexton S, et al. (2014). The impact of agricultural biotechnology on supply and land-use. *Environment and Development Economics*, 19(6), 676–703.

Bartrina E, Otto M, Strnad T, Werner T (2011). Schmülling cytokinin regulates the activity of reproductive meristems, flower organ size, ovule formation, and thus seed yield in *Arabidopsis thaliana. Plant Cell*, 23, 69–80.

Beckmann V, Wesseler J (2004). How labour organisation may affect technology adoption: An analytical framework analysing the case of integrated pest management. *Environ Dev Econ*, 8(3), 437–450.

Bisht DS, Bhatia V, Bhattacharya R (2019). Improving plant-resistance to insect-pests and pathogens: The new opportunities through targeted genome editing. *Semin Cell Dev Biol*, 96, 65–76. doi:10.1016/j.semcdb.2019.04.008. Epub 2019 May 8.

Bollineni H, Gopala KS, Sundaram RM, Sudhakar D, Prabhu KV, Singh NK, et al. (2014). Marker assisted biofortification of rice with pro-vitamin A using transgenic golden rice lines: Progress and prospects. *Ind J Genet*, 74(4), 624–630.

Bonanno A, Materia VC, et al. (2017). The Plant Protection Products (PPP) sector in the European Union: A special view on herbicides. *Eur J Dev Res*, 29(3), 575–595.

Bonny S (2016). Genetically modified herbicide-tolerant crops, weeds, and herbicides: Overview and impact. *Environ Manage*, 57(1), 31–48.

Brookes G, Barfoot P (2017). Environmental impacts of genetically modified (GM) crop use 1996–2015: Impacts on pesticide use and carbon emissions. *GM Crops & Food*, 8(2), 117–147.

Byrne J (2015). Presentation by President of v-Fluence Interactive at the Agricultural Bioscience International Conference. Melbourne.

Campbell SA, Close TJ (1997). Dehydrins: Genes, proteins, and associations with phenotypic traits. *New Phytol*, 137, 61–74.

Cao L, Zhang P, Lu X, Wang G, Wang Z, Zhang Q, Zhang X, Wei X, Mei F, Wei L, Wang T (2020). Systematic analysis of the maize *OSCA* genes revealing *ZmOSCA* family members involved in osmotic stress and *ZmOSCA2.4* confers enhanced drought tolerance in transgenic *Arabidopsis. Int J Mol Sci*, 21(1). pii: E351. doi:10.3390/ijms21010351.

Carpenter JE (2010). Peer-reviewed surveys indicate positive impact of commercialized GM crops. *Nat Biotechnol*, 28(4), 319–321.

Chhikara S, Abdullah HM, Akbari P, Schnell D, Dhankher OP (2018) Engineering *Camelina sativa* (L.) Crantz for enhanced oil and seed yields by combining diacylglycerol acyltransferase1 and glycerol-3-phosphate dehydrogenase expression. *Plant Biotechnol J*, 16(5), 1034–1045. doi:10.1111/pbi.12847. Epub 2017 Nov 19.

Ciura J, Kruk J (2018). Phytohormones as targets for improving plant productivity and stress tolerance. *J Plant Physiol*, 229, 32–40.

Coghlan A (2018). GM golden rice gets approval from food regulators in the US. *New Scientist*, May 30. Retrieved July 7. https://www.newscientist.com/article/mg23831802-500-gm-golden-rice-gets-approval-from-food-regulators-in-the-us/

De Moura FF, Moursi M, et al. (2016). Biofortified β-carotene rice improves vitamin A intake and reduces the prevalence of inadequacy among women and young children in a simulated analysis in Bangladesh, Indonesia, and the Philippines. *Am J Clin Nutr*. ajcn129270.

De Steur H, Mehta S, et al. (2017). GM biofortified crops: Potential effects on targeting the micronutrient intake gap in human populations. *Curr Opin Biotechnol*, 44, 181–188.

De Steur H, Wesana, J, et al. (2017). Methods matter: A meta-regression on the determinants of willingness-to-pay studies on biofortified foods. *Ann N Y Acad Sci*, 1390(1), 34–46.

Duke SO, et al. (2012). Glyphosate effects on plant mineral nutrition, crop rhizosphere microbiota, and plant disease in glyphosate-resistant crops. *J Agri Food Chem*, 60(42), 10375–10397.

FAO (2018). Food loss and food waste. www.fao.org/food-loss-and-food-waste/en/. Accessed July 17.

Fu LL, Zhao X, Xu HL, Wen X, Wang SY, Liu B, Bao JK, Wei YQ (2012). Identification of microRNA-regulated autophagic pathways in plant lectin-induced cancer cell death. *Cell Prolif*, 45(5), 477–485. doi:10.1111/j.1365-2184.2012.00840.x.Epub 2012 Aug 8.

Fullman N, Barber RM, et al. (2017). Measuring progress and projecting attainment on the basis of past trends of the health-related sustainable development goals in 188 countries: An analysis from the global burden of disease study 2016. *The Lancet*, 390(10100), 1423–1459.

Gantait S, Mondal S (2018) Transgenic approaches for genetic improvement in groundnut (*Arachis hypogaea* L.) against major biotic and abiotic stress factors. *J Genet Eng Biotechnol*, 16(2), 537–544. doi:10.1016/j.jgeb.2018.08.005. Epub 2018 Dec 7.

Gao Y, Wu M, Zhang M, Jiang W, Liang E, Zhang D, Zhang C, Xiao N, Chen J (2018). Roles of a maize phytochrome-interacting factors protein ZmPIF3 in regulation of drought stress responses by controlling stomatal closure in transgenic rice without yield penalty. *Plant Mol Biol*, 97(4–5), 311–323. doi:10.1007/s11103-018-0739-4. Epub 2018 Jun 5.

Garcia D (2008). A miracle in plant development: Role of microRNAs in cell differentiation and patterning. *Semin Cell Dev Biol*, 19(6), 586–595. doi:10.1016/j.semcdb.2008.07.013.

Gaudin AC, Henry A, Sparks AH, Slamet-Loedin IH (2013). Taking transgenic rice drought screening to the field. *J Exp Bot*, 64(1), 109–117. doi:10.1093/jxb/ers313. Epub 2012 Dec 1.

Gong L, Zhang H, Liu X Gan X, Nie F, Yang W, Zhang L, Chen Y, Song Y, Zhang H (2020) Ectopic expression of HaNAC1, an ATAF transcription factor from *Haloxylon ammodendron*, improves growth and drought tolerance in transgenic *Arabidopsis*. *Plant Physiol Biochem*, 151, 535–544. doi:10.1016/j.plaphy.2020.04.008. [Epub ahead of print].

Gonsalves D (1998). Control of papaya ringspot virus in papaya: A case study. *Annu Rev Phytopathol*, 36, 415–437.

Good AG, Shrawat AK, Muench DG (2004). Can less yield more? Is reducing nutrient input into the environment compatible with maintaining crop production? *Trends Plant Sci*, 9(12), 597–605. Review.

Gressel J, Polturak G (2018). Suppressing aflatoxin biosynthesis is not a breakthrough if not useful. *Pest Manage Sci*, 74(1), 17–21.

Haas JD, Finkelstein JL, et al. (2013). Iron biofortified pearl millet improves iron status in Indian school children: Results of a feeding trial. *The FASEB J*, 27(1 Supplement), 355.2–355.2.

Haggblade SM, Smale M, Kergna A, Theriault V, Assima A (2017). Causes and consequences of increasing herbicide use in Mali. *Eur J Dev Res*, 29, 648–674.

Hao G, Bakker MG, Kim HS (2020). Enhanced resistance to *Fusarium graminearum* in transgenic *Arabidopsis* plants expressing a modified plant Thionin. *Phytopathology*, 110(5), 1056–1066. doi:10.1094/PHYTO-12-19-0447-R. Epub 2020 Mar 17.

HarvestPlus (2015). Nutritious staple food crops: Who is growing what? http://bit.ly/CropMap2015.

Hay SI, Abajobir AA, et al. (2017). Global, regional, and national disability-adjusted life-years (DALYs) for 333 diseases and injuries and healthy life expectancy (HALE) for 195 countries and territories, 1990–2016: A systematic analysis for the global burden of disease study 2016. *Lancet*, 390(10100), 1260–1344.

Hobbs JE, Kerr WA, et al. (2014). The perils of zero tolerance: Technology management, supply chains and thwarted globalisation. *Int J Technol Global*, 7, 203–216.

Huang J, Hu, R, et al. (2002). Bt cotton benefits, costs, and impacts in China. *AgBio Forum*, 5(4), 153–166.

Huang X, Qian Q, Liu Z, Sun H, He S, Luo D, Xia G, Chu C, Li J, Fu X (2009). Natural variation at the DEP1 locus enhances grain yield in rice. *Nat Genet*, 41(4), 494–497. doi:10.1038/ng.352. Epub 2009 Mar 22.

ISAAA Brief 46 Executive Summary Global Status of Commercialized Biotech/GM Crops (2019). *Clive James*. International Service for the Acquisition of Agri-Biotech Applications.

Ivleva NB, Groat J, Staub JM, Stephens M (2016). Expression of active subunit of nitrogenase via integration into plant organelle genome. *PLoS One*, 11, Article e0160951.

Iwamoto M, TagiriMicro A (2016). RNA-targeted transcription factor gene *RDD1* promotes nutrient ion uptake and accumulation in rice. *Plant J*, 85, 466–477.

James C (2016). *Global Status of Commercialized Biotech/GM Crops: 2016. The International Service for the Acquisition of Agri-Biotech Applications (ISAAA) Brief No. 52*. ISAAA: Ithaca, NY.

James C (2017). *Global Status of Commercialized Biotech/GM Crops: 2016 ISAAA Briefs*. New York: ISAAA, 49–199.

Kamthan A, Chaudhuri A, Kamthan M, Datta A (2016). Genetically modified (GM) crops: Milestones and new advances in crop improvement. *Theor Appl Genet*, 129(9), 1639–1655.

Kant S, Burch D, Badenhorst P, Palanisamy R, Mason J, Spangenberg G (2015). Regulated expression of a cytokinin biosynthesis gene *IPT* delays leaf senescence and improves yield under rainfed and irrigated conditions in canola (*Brassica napus* L.). *PLoS One*, 10, Article e0116349.

Khan S, Anwar S, Yu S, Sun M, Yang Z, Gao ZQ (2019). Development of drought-tolerant transgenic wheat: Achievements and limitations. *Int J Mol Sci*, 20(13). pii: E3350. doi:10.3390/ijms20133350. Review.

Klümper W, Qaim M (2014). A meta-analysis of the impacts of genetically modified crops. *PloS One*, 9(11), e111629.

Kolachevskaya OO, Alekseeva VV, Sergeeva LI, Rukavtsova EB, Getman I, et al. (2015). Expression of auxin synthesis gene *tms1* under control of the tuber-specific promoter enhances potato tuberization *in vitro*. *J Integr Plant Biol*, 57, 734–744.

Kouser S, Qaim M (2011). Impact of Bt cotton on pesticide poisoning in smallholder agriculture: A panel data analysis. *Ecol Econ*, 70(11), 2105–2113.

Kouser S, Qaim M (2013). Valuing financial, health, and environmental benefits of Bt cotton in Pakistan. *Agri Econ*, 44(3), 323–335.

Krishna R, Karkute SG, Ansari WA, Jaiswal DK, Verma JP, Singh M (2019). Transgenic tomatoes for abiotic stress tolerance: Status and way ahead. *3 Biotech*, 9(4), 143. doi:10.1007/s13205-019-1665-0. Epub 2019 Mar 18.

Krishnamurthy P, Vishal B, Khoo K, Rajappa S, Loh CS, Kumar PP (2019). Expression of AoNHX1 increases salt tolerance of rice and *Arabidopsis*, and bHLH transcription factors regulate AtNHX1 and AtNHX6 in *Arabidopsis*. *Plant Cell Rep*, 38(10), 1299–1315. doi:10.1007/s00299-019-02450-w. Epub 2019 Jul 26.

Low J, Arimond M, et al. (2007). A food-based approach introducing orange-fleshed sweet potatoes increased vitamin A intake and serum retinol concentrations in young children in rural Mozambique. *J Nutr*, 137(5), 1320–1327.

Lu P, Guo L, Wang Z, Li B, Li J, Li Y, Qiu D, Shi W, Yang L, Wang N, Guo G, Xie J, Wu Q, Chen Y, Li M, Zhang H, Dong L, Zhang P, Zhu K, Yu D, Zhang Y, Deal KR, Huo N, Liu C, Luo MC, Dvorak J, Gu YQ, Li H, Liu Z (2020). A rare gain of function mutation in a wheat tandem kinase confers resistance to powdery mildew. *Nat Commun*, 11(1), 680. doi:10.1038/s41467-020-14294-0.

Luan H, Liao W, Song Y, Niu H, Hu T, Zhi H (2020). Transgenic plant generated by RNAi-mediated knocking down of soybean Vma12 and soybean mosaic virus resistance evaluation. *AMB Express*, 10(1), 62. doi:10.1186/s13568-020-00997-6.

Lenka SK, Singh AK, Muthusamy SK, Smita S, Chinnusamy V, Bansal KC (2019). Heterologous expression of rice RNA-binding glycine-rich (RBG) gene *OsRBGD3* in transgenic *Arabidopsis thaliana* confers cold stress tolerance. *Funct Plant Biol*, 46(5), 482–491. doi:10.1071/FP18241.

Li H, Hu B, Chu, C (2017). Nitrogen use efficiency in crops: Lessons from *Arabidopsis* and rice. *J Exp Bot*, 68, 2477–2488.

Li Q, Song J, Peng S, Wang JP, Qu GZ, Sederoff RR, Chiang VL (2014). Plant biotechnology for lignocellulosic biofuel production. *Plant Biotechnol J*, 12(9), 1174–1192. doi:10.1111/pbi.12273. Epub 2014 Oct 20.

Li Q, Hu A, Qi J, Dou W, Qin X, Zou X, Xu L, Chen S, He Y (2020). CsWAKL08, a pathogen-induced wall-associated receptor-like kinase in sweet orange, confers resistance to citrus bacterial canker via ROS control and JA signaling. *Hortic Res*, 7, 42. doi:10.1038/s41438-020-0263-y. eCollection 2020.

Mao Y, Botella JR, Zhu JK (2018). Heritability of targeted gene modifications induced by plant-optimized CRISPR systems. *Cell Mol Life Sci*, 74, 1075–1093.

Marvier M, McCreedy C, et al. (2007). A meta-analysis of effects of Bt cotton and maize on non-target invertebrates. *Science*, 316, 1475–1477.

Masset E, Haddad L, et al. (2012). Effectiveness of agricultural interventions that aim to improve nutritional status of children: Systematic review. *BMJ*, 344, d8222.

McAllister CH, Beatty PH, Good AG (2012). Engineering nitrogen use efficient crop plants: The current status. *Plant Biotechnol J*, 10(9), 1011–1025. doi:10.1111/j.1467-7652.2012.00700.x.Epub 2012 May 18. Review.

Meenakshi JV, Johnson N, et al. (2010). How cost-effective is biofortification in combating micronutrient malnutrition? An ex-ante assessment. *World Dev*, 38(1), 64–75.

Mellidou I, Buts K, Hatoum D, Ho QT, Johnston JW, Watkins CB, Schaffer RJ, Gapper NE, Giovannoni JJ, Rudell DR, Hertog ML, Nicolai BM (2014). Transcriptomic events associated with internal browning of apple during postharvest storage. *BMC Plant Biol*, 14, 328. doi:10.1186/s12870-014-0328-x.

Mezzetti B, Landi L, Pandolfini T, Spena A (2004). The defH9-iaaM auxin-synthesizing gene increases plant fecundity and fruit production in strawberry and raspberry. *BMC Biotechnol*, 4(4), 1–10.

Morran S, Eini O, Pyvovarenko T, et al. (2011). Improvement of stress tolerance of wheat and barley by modulation of expression of DREB/CBF factors. *Plant Biotechnol J*, 9, 230–249.

National Academies of Sciences and Medicine (2016). *Genetically Engineered Crops: Experiences and Prospects*. National Academies Press.

Nguyen QH, Vu LTK, Nguyen LTN, Pham NTT, Nguyen YTH, Le SV, Chu MH (2019). Overexpression of the GmDREB6 gene enhances proline accumulation and salt tolerance in genetically modified soybean plants. *Sci Rep*, 9(1), 19663. doi:10.1038/s41598-019-55895-0.

Niño MC, Kang KK, Cho YG (2020). Genome-wide transcriptional response of papain-like cysteine protease-mediated resistance against *Xanthomonas oryzae* pv. *oryzae* in rice. *Plant Cell Rep*, 39(4), 457–472. doi:10.1007/s00299-019-02502-1. Epub 2020 Jan 28.

Nowicka B, Ciura J, Szymańska R, Kruk J (2018). Improving photosynthesis, plant productivity and abiotic stress tolerance: Current trends and future perspectives. *J Plant Physiol*, 231, 415–433. doi:10.1016/j.jplph.2018.10.022. Epub 2018 Oct 30.

Orroño DD, Vesprini F (2018). Directives and requirements for genetically modified (GM) crop regulation in Argentina. *Crop Breed Appl Biotechnol*, 18, 301–308. doi:10.1590/1984-70332018v18n3r44.

Paarlberg R (2008). Africa's organic farms. Feb. 2008. *The New York Times*. www.nytimes.com/2008/02/29/opinion/29iht-edpaarlberg.1.10576543.html.

Pan J, Huang D, Guo Z, Kuang Z, Zhang H, Xie X, Ma Z, Gao S, Lerdau MT, Chu C, Li L (2018). Overexpression of microRNA408 enhances photosynthesis, growth, and seed yield in diverse plants. *J Integr Plant Biol*, 60(4), 323–340. doi:10.1111/jipb.12634. Epub 2018 Mar 14.

Paul JY, Khanna H, Kleidon J, Hoang P, Geijskes J, Daniells J, Zaplin E, Rosenberg Y, James A, Mlalazi B, Deo P, Arinaitwe G, Namanya P, Becker D, Tindamanyire J, Tushemereirwe W (2017). Golden bananas in the field: Elevated fruit pro-vitamin A from the expression of a single banana transgene. *Plant Biotechnol J*, 15(4), 520–532. doi:10.1111/pbi.12650. Epub 2016 Dec 20.

Potrykus I (2017). The GMO-crop potential for more and more nutritious food is blocked by unjustified regulation. *JIK*, 2(2), 90–96.

Pottinger SE, Bak A, Margets A, Helm M, Tang L, Casteel C, Innes RW (2020). Optimizing the PBS1 decoy system to confer resistance to potyvirus infection in *Arabidopsis* and soybean. *Mol Plant Microbe Interact*, Apr 8. doi:10.1094/MPMI-07-19-0190-R. [Epub ahead of print].

Pray C, Ma D, et al. (2001). Impact of Bt cotton in China. *World Dev*, 29(5), 813–825.

Qaim M (2020). Role of new plant breeding technologies for food security and sustainable agricultural development. *Appl Econ Perspect Policy*, https://doi.org/10.1002/aepp.13044

Qin N, Xu W, Hu L, Li Y, Wang H, Qi X, Fang Y, Hua X (2015). Drought tolerance and proteomics studies of transgenic wheat containing the maize C 4 phosphoenolpyruvate carboxylase (PEPC) gene. *Protoplasma*, 253, 1503–1512. doi:10.1007/s00709-015-0906-2.

Racovita M, Obonyo DN, et al. (2015). What are the non-food impacts of GM crop cultivation on farmers' health? *Environ Evid*, 4(1), 17.

Rani R, Yadav P, Barbadikar KM, Baliyan N, Malhotra EV, Singh BK, Kumar A, Singh D (2016). CRISPR/Cas9: A promising way to exploit genetic variation in plants. *Biotechnol Lett*, 38, 1991–2006.

Ricroch AE (2019). Global developments of genome editing in agriculture. *Transgen Res*, 28(Suppl 2), 45–52. doi:10.1007/s11248-019-00133-6.

Ricroch AE, Henard-Damave MC (2016). Next biotech plants: New traits, crops, developers and technologies for addressing global challenges. *Crit Rev Biotechnol*, 36(4), 675–690.

Rommens CM, Yan H, et al. (2008). Low-acrylamide French fries and potato chips. *Plant Biotechnol J*, 6(8), 843–853.

Rooney WM, Grinter RW, Correia A, Parkhill J, Walker DC, Milner JJ (2020). Engineering bacteriocin-mediated resistance against the plant pathogen *Pseudomonas syringae*. *Plant Biotechnol J*, 18(5), 1296–1306. doi:10.1111/pbi.13294. Epub 2019 Dec 3.

Ruel MT, Alderman H (2013). Nutrition-sensitive interventions and programmes: How can they help to accelerate progress in improving maternal and child nutrition? *Lancet*, 382(9891), 536–551.

Ruiz-Lopez N, Haslam RP, et al. (2014). Successful high-level accumulation of fish oil omega-3 long-chain polyunsaturated fatty acids in a transgenic oilseed crop. *Plant J*, 77(2), 198–208.

Sachs JD (2012). From millennium development goals to sustainable development goals. *Lancet*, 379, 2206–2211.

Saint Pierre C, Crossa JL, Bonnett D, et al. (2012). Phenotyping transgenic wheat for drought resistance. *J Exp Bot*, 63, 1799–1808.

Shah N, Anwar S, Xu J, Hou Z, Salah A, Khan S, Gong J, Shang Z, Qian L, Zhang C (2018). The response of transgenic *Brassica* species to salt stress: A review. *Biotechnol Lett*, 40(8), 1159–1165. doi:10.1007/s10529-018-2570-z. Epub 2018 Jun 1.

Shi C, Ren Y, Liu L, Wang F, Zhang H, Tian P, Pan T, Wang Y, Jing R, Liu T, Wu F, Lin Q, Lei C, Zhang X, Zhu S, Guo X, Wang J, Zhao Z, Wang J, Zhai H, Cheng Z, Wan J (2019). Ubiquitin specific protease 15 has an important role in regulating grain width and size in rice. *Plant Physiol*, 180(1), 381–391. doi:10.1104/pp.19.00065. Epub 2019 Feb 22.

Simkin AJ, López-Calcagno PE, Raines CA (2019). Feeding the world: Improving photosynthetic efficiency for sustainable crop production. *J Exp Bot*, 70(4), 1119–1140. doi:10.1093/jxb/ery445. Review.

Singh RP, Hodson DP, Jin Y, Lagudah ES, Ayliffe MA, Bhavani S, Rouse MN, Pretorius ZA, Szabo LJ, Huerta-Espino J, Basnet BR, Lan C, Hovmøller MS (2015). Emergence and spread of new races of wheat stem rust fungus: Continued threat to food security and prospects of genetic control. *Phytopathology*, 105(7), 872–884. doi:10.1094/PHYTO-01-15-0030-FI. Epub 2015 Jun 29. Review.

Smyth SJ (2017). Genetically modified crops, regulatory delays, and international trade. *Food Energy Secur*, 6(2), 78–86.

Sun F, Suen PK, Zhang Y, Liang C, Carrie C, Whelan J, et al. (2012). A dual-targeted purple acid phosphatase in *Arabidopsis thaliana* moderates carbon metabolism and its overexpression leads to faster plant growth and higher seed yield. *New Phytol*, 194, 206–219.

Sun F, Yu H, Qu J, Cao Y, Ding L, Feng W, Khalid MHB, Li W, Fu F (2020). Maize ZmBES1/BZR1-5 decreases ABA sensitivity and confers tolerance to osmotic stress in transgenic *Arabidopsis*. *Int J Mol Sci*, 21(3). pii: E996. doi:10.3390/ijms21030996.

Swartzberg D, Dai N, Gan S, Amasino RM (2006). Effects of cytokinin production under two SAG promoters on senescence and development of tomato plants. *Plant Biol*, 8, 579–586.

Tang G, Qin J, et al. (2009). Golden rice is an effective source of vitamin A. *Am J Clin Nutr*, 89(6), 1776–1783.

Thomas WT (2015). Drought-resistant cereals: Impact on water sustainability and nutritional quality. *Proc Nutr Soc*, 74(3), 191–197.

Thompson GD, Kidwell J (1998). Explaining the choice of organic produce: Cosmetic defects, prices, and consumer preferences. *Am J Agri Econ*, 80(2), 277–287.

Waltz E (2014). Vitamin A super banana in human trials. *Nat Biotechnol*, 32(9), 857.

Waltz E (2015a). Nonbrowning GM apple cleared for market. *Nat Biotechnol*, 326. http://blogs.nature.com/tradesecrets/2015/03/30/nonbrowning-gm-apple-cleared-for-market. Accessed July 18, 2018.

Waltz E (2015b). *USDA Approves Next-Generation GM Potato*. Nature Publishing Group.

Waltz E (2016). Gene-edited CRISPR mushroom escapes US regulation. *Nature*, 532, 293.

Wang S, Liu J, Zhao T, Du C, Nie S, Zhang Y, Lv S, Huang S, Wang X (2019) Modification of threonine-1050 of SlBRI1 regulates BR signalling and increases fruit yield of tomato. *BMC Plant Biol*, 19(1), 256. doi:10.1186/s12870-019-1869-9.

Wani SH, Kumar V, Khare T, Guddimalli R, Parveda M, Solymosi K, Suprasanna P, Kavi Kishor PB (2020). Engineering salinity tolerance in plants: Progress and prospects. *Planta*, 251(4), 76. doi:10.1007/s00425-020-03366-6.

Wesseler J, Scatasta S, et al. (2011). Environmental benefits and costs of GM crops. In *Genetically Modified Food and Global Welfare* edited by C Carter, GC Moschini, I Sheldon. Bingley: Emerald Group Publishing: 173–199.

Wesseler J, Smart RD, et al. (2017). Foregone benefits of important food crop improvements in sub-Saharan Africa. *PLoS ONE*, 12(7), e0181353. https://doi.org/10.1371/journal.pone.0181353.

Wesseler J, Zilberman D (2014). The econo power of the Golden Rice opposition. *Environ Dev Econ*, 19(6), 724–742.

Xu J, Sheng Z, Wang X, Liu X, Xia J, Xiong P, He B (2016). Enhancement in ionic liquid tolerance of cellulase immobilized on PEGylated graphene oxide nanosheets: Application in saccharification of lignocellulose. *Bioresour Technol*, 200, 1060–1064. doi:10.1016/j.biortech.2015.10.070. Epub 2015 Oct 24.

Yang H, Deng L, Liu H, Fan S, Hua W, Liu J (2019). Overexpression of *BnaAOX1b* confers tolerance to osmotic and salt stress in rapeseed. *G3 (Bethesda)*, 9(10), 3501–3511. doi:10.1534/g3.119.400366.

Yang X, Lu Y, Wang F, Chen Y, Tian Y, Jiang L, Peng J, Zheng H, Lin L, Yan C, Taliansky M, MacFarlane S, Wu Y, Chen J, Yan F (2020). Involvement of the chloroplast gene ferredoxin 1 in multiple responses of *Nicotiana benthamiana* to potato virus X infection. *J Exp Bot*, 71(6), 2142–2156. doi:10.1093/jxb/erz565.

Zalabák D, Pospíšilová H, Šmehilová M, Mrízová K, Frébort I, Galuszka P (2013). Genetic engineering of cytokinin metabolism: Prospective way to improve agricultural traits of crop plants. *Biotechnol Adv*, 31, 97–117.

Zang X, Geng X, Wang F, Liu Z, Zhang L, Zhao Y, Tian X, Ni Z, Yao Y, Xin M, et al. (2017). Overexpression of wheat ferritin gene TaFER-5B enhances tolerance to heat stress and other abiotic stresses associated with the ROS scavenging. *BMC Plant Biol*, 17, 14. doi:10.1186/s12870-016-0958-2.

Zhang Y, Yu L, Yung KF, Leung DY, Sun F, Lim BL (2012). Overexpression of *AtPAP2* in *Camelina sativa* leads to faster plant growth and higher seed yield. *Biotechnol Biofuels*, 5, 19.

Zhang H, Xu W, Wang H, Hu L, Li Y, Qi X, Zhang L, Li C, Hua X (2014). Pyramiding expression of maize genes encoding phosphoenolpyruvate carboxylase (PEPC) and pyruvate orthophosphate dikinase (PPDK) synergistically improve the photosynthetic characteristics of transgenic wheat. *Protoplasma*, 251, 1163–1173. doi:10.1007/s00709-014-0624-1.

Zhao F-J, Shewry, PR (2011). Recent developments in modifying crops and agronomic practice to improve human health. *Food Policy*, 36, S94–S101.

Zhao P, Hou S, Guo X, Jia J, Yang W, Liu Z, Chen S, Li X, Qi D, Liu G, Cheng L (2019). A MYB-related transcription factor from sheepgrass, LcMYB2, promotes seed germination and root growth under drought stress. *BMC Plant Biol*, 19(1), 564. doi:10.1186/s12870-019-2159-2.

2

Exploring New Organizational Concepts for Agriculture in the Bioeconomy

Davide Viaggi, Matteo Zavalloni and Stefano Targetti

CONTENTS

2.1 Introduction and Objectives

The development of the bioeconomy is recognized as a major area of transformation of current economies. The bioeconomy is intended as an economy or a part of an economy based on the use of biological resources, building on a wide range of modern technologies. Current bioeconomy strategies aim explicitly at strengthening the circularity and sustainability features of the bioeconomy. The economic and business literature is working to identify structuring concepts allowing a better understanding of this idea.

While most sectors of the bioeconomy already benefit from a well-established economic literature, the bioeconomy as a whole is struggling to find a clear economic interpretation. One of the main reasons is certainly the complexity of the sector and the need to interface with a number of close disciplines, in particular organization studies and a wide range of social sciences (Viaggi, 2018, 2020). New economic and organizational analyses, on the other hand, are perceived as a major need for the bioeconomy with a view to making the sector efficient and profitable with respect to competing sectors, such as those based on fossil resources.

Agriculture has a special position in the bioeconomy and in the issues illustrated above. Agriculture is the primary source of biomass, used for multiple purposes. Hence, this sector is one of the founding components of the bioeconomy; however, at the same time, it is particularly exposed to the sometimes conflicting development of the different components of the bioeconomy, in particular food and non-food uses. On the other hand, the bioeconomy expands to a much larger view of the management of biological resources, basically expanding the concept of cultivation to the use of ecosystems. Also, resources previously harvested are increasingly cultivated (e.g. fish, algae and forests). Thus, agriculture is expanding its scope and providing its "paradigm" for new areas of human activity.

In the framework of the foregoing observations, agriculture also has a special character with respect to its economic and organizational features. In particular, agriculture is characterized by its diffuse use of ecosystems, potential diffuse effects and pollution, micro-organizational units (farms), relevant role of public policies, high dependence on local resources and climate/weather conditions (hence high risk).

This chapter investigates the potential development of organizational concepts for the bioeconomy, in particular, for agriculture in the bioeconomy. The chapter builds on the literature on socio-ecological systems and explores the concepts of the socio-ecological technological value-enhancing web system

(SETVEWS). In doing so, it aims to explore how the development of these concepts can enable a better analysis and evaluation practice to support private decisions and policymaking in bioeconomy systems.

2.2 Socio-Ecological Systems: Application to Agriculture and the Bioeconomy

According to Ostrom (2009), all humanly used resources can be framed as a social-ecological system (SES). SESs are defined as complex adaptive systems characterized by human–environment feedbacks across multiple interlinked scales (Fischer et al., 2015). As the bioeconomy is the aggregate of sectors using biological resources, SES provides a relevant analytical framework. An important aspect of the framework is its focus on the range of interactions between ecosystem and society that characterizes a specific system and the identification of elements or subsystems composing an SES. Typically, these components include resources (often characterized by common-pool characteristics), its users and the governance system (Janssen & Anderies, 2007).

In an SES, ecological and social components and their related dynamics are interdependent. Together, these interactions concur with the emergence of technologies, institutions, organizations and cultures characterizing an SES (van der Leuuw, 2012). Therefore, an interdisciplinary understanding of these interlinked dynamics is considered necessary to tackle environmental problems such as climate change, biodiversity loss and resource degradation, and eventually to support the development of more sustainable bioeconomies (Binder et al., 2013; Carpenter et al., 2009).

Besides interdisciplinarity, uncertainty is also a central argument of SES studies, which challenges the possibility to fully predict and control (or model) system dynamics (Schlüter et al., 2012) and imposes a reconsideration of the perspective on policy formulation. Several scholars advocate that the goal of SES analyses should not be the identification of "fail-free policies" but rather the design of better policy processes (Anderies & Janssen, 2013). Such processes typically encompass adaptation and learning. For instance, polycentric governance was proposed as a better fit for SES in contrast to traditional command-and-control policy frameworks (Ostrom, 1998). A basic assumption of polycentric governance systems is that the timescale of policy design should adapt to the short- and long-term dynamics between users and ecosystems. To achieve that goal, bottom-up and top-down settings coexist and co-evolve in a polycentric system to ensure the accomplishment of policy objectives.

The application of an SES to agricultural systems is common and has tackled a range of different aspects. Several scholars have considered the SES approach useful as it allows to cope with relevant challenges linked to the sustainability of agricultural systems and assess the transition towards a sustainable bioeconomy (de Schutter et al., 2019). Nevertheless, SESs are complex systems and their assessment is thus challenging. For instance, Schlüter et al. (2012) listed five challenges that can be linked to SES assessments: (i) the evaluation of dynamic social–ecological linkages and their effect on the SES; (ii) the trade-offs derived from different management of multiple resources; (iii) the relevance of timescales; (iv) the impacts of social networks on natural resource problems; and (v) the need to combine different techniques such as modelling with experimental and field research (Janssen & Ostrom, 2006). Some examples of the application of SES approaches concern the disjunction between ecosystem degradation and the perception of the consequences of economic activities (Cumming et al., 2014), the resilience and adaptive management of agro-ecological systems (Walker et al., 2006), the modelling of the drivers that impact on the ecosystem and ecosystem services (Antle & Stoorvogel, 2006), the vulnerability to climate change (Fraser et al., 2011) and the descriptive analyses of different socio-ecological features in different case study areas (Liu et al., 2007). Lescourret et al. (2015) proposed a conceptual SES framework to highlight the delivery of multiple agro-ecosystem services. In that framework, social and ecosystem are composed of structural and functional elements that define the agro-ecosystem management and the range of services provided. A particular aspect that is considered in agricultural SES studies is farmers' perception and conceptualization of the risks emanating from economic, social or ecological changes, and the options that he or she can employ to face them (Darnhofer et al., 2010; Scoones et al., 2007). A common issue tackled by an SES applied to agriculture is the necessity to consider the robustness and the performance of the agricultural systems. This implies combining the economic performance of agricultural activity (usually focusing on short-term dynamics) with the long-term resilience of the system which is affected by uncertainty. Integrating such issues is obviously complicated and information from one case study is often not transferable to different socio-ecological contexts (Liu et al., 2007). To overcome such complexity, building stakeholder organizations for coordinated management planning and the availability of practical instruments for use in participatory approaches are considered essential (Lescourret et al., 2015). That and the development of polycentric governance systems are relevant topics tackled by current research on agricultural SES.

2.3 Beyond SES: Developing the SETVEWS Approach

The relationships that characterize the structure of the bioeconomy are managed not by a single decision maker, but rather by complex societal interactions. With this prospect, the use of the SES framework (Ostrom, 2009) and value webs (Scheiterle et al., 2016; Virchow et al., 2016) appears very promising, with proper modifications, to describe the bioeconomy.

Viaggi (2018, 2020) proposes an overarching concept partially integrating these approaches, called the socio-ecological technological value-enhancing web system (SETVEWS). In addition to SESs and value webs, this concept also integrates ideas from the "technological" views of the bioeconomy and the perspective of ecosystem services and public goods. Bioeconomy systems according with this view are represented as in Figure 2.1.

According to Viaggi (2018), the SETVEWS can be described as

1. "A system of biophysical and social components that interact among each other;
2. The linkages among sub-systems, identified by both material and value flows on multiple product chains and socio-ecological systems;
3. A complex system of adaptation solutions from the interaction between its multiple components, including feedback mechanisms;
4. A set of knowledge and technologies".

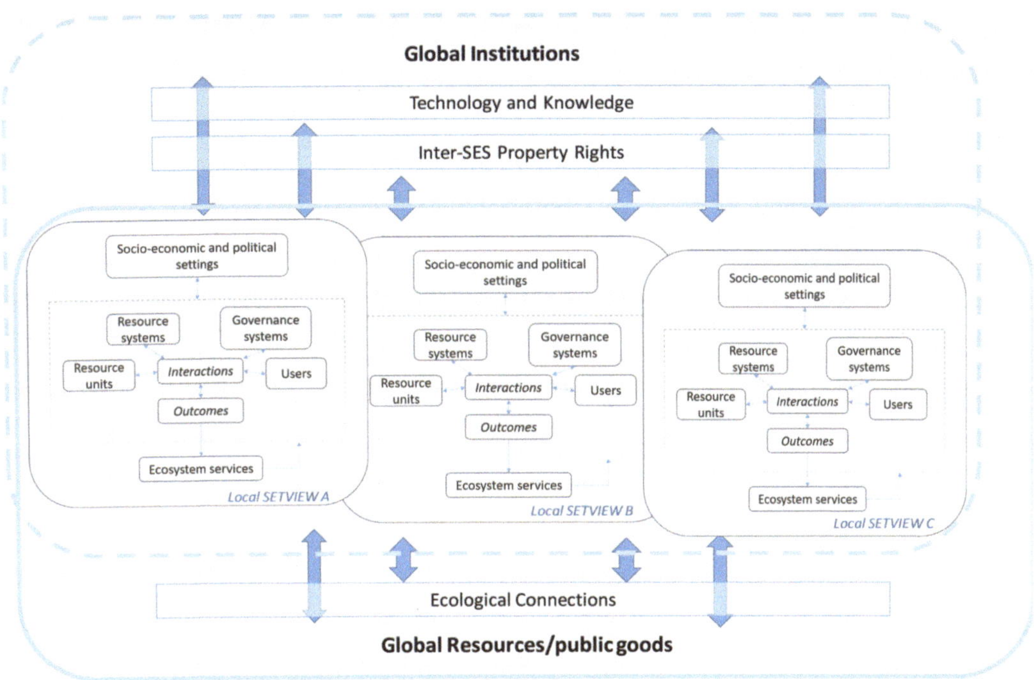

FIGURE 2.1 A simplified structure of the SETVEW.

The two main approaches contributing to the SETVEWS tend to work on different scales, more towards local for SESs and more on a wider scale for biomass webs. Bearing in mind both local and global systems makes it possible to better account for interrelationships that have a global dimension and local effects, such as climate change effects on productivity and the reverse. For this reason, several partial (local) SETVEWSs can be imagined, linked together in a global bioeconomy SETVEWS.

The combination of the bioeconomy with the SES in the SETVEWS framework enables a holistic perspective on the issues at stake. Similarly to the SES approach, the SETVEWS approach emphasizes the interactions between social systems and ecological systems, and highlights the importance of social and economic interactions with the ecological systems in producing value. Compared with an SES, however, the SETVEWS explicitly integrates the chain and inter-chain (web) connections, thereby providing a more technical structure of the flows of value within an economy. Moreover, the bioeconomy perspective emphasizes how public infrastructures shape the relationship and the flow between resources and resource users. Expanding the spatial scope of the SES framework, the bioeconomy filters the flow of resources from the source to the final user through intermediate actors that transform the biological resources. The development of the bioeconomy is based on the technological capacity of infrastructure to transform the biological resources into materials that were once produced through non-renewable resources (or not produced at all). Technology affects the functional links among different components of the system, including the flows of biomass, the relationship between biomass and non-bio-based components and communication. This technological capacity is also an endogenous choice of the system, which purposely invests in the research and in the know-how necessary to ensure the change. In addition, technology is connected with and affects potential organizational solutions. In particular, bioeconomy technologies increase the separability points in the process and allow a broader variety of points where transactions can occur and, in turn, a huge variety of actual value chain/system solutions.

The SETVEWS also expands the concept of an SES to the global bioeconomy. Similarly to the telecoupling framework (Liu et al., 2013; de Schutter et al., 2019), the SETVEWS describes a bundle of relationships between different SESs, explicitly accounting for the global geographical complexity of production processes. In this sense, the SETVEWS reduces the elements that are exogenous to the system with respect to the SES framework.

In common with value webs (and in contrast to global chain analysis), the SETVEWS is not a product-oriented concept. This does not exclude that a final product can be one of the agglomeration criteria and, indeed, this could be an important rationale to link the SETVEWS with the literature on the agricultural value chain, in which the product dimension tends to remain the structuring factor. Indeed, the systemic view brought by the SETVEWS may encourage the understanding of the connection between agriculture, public goods and ecosystems. Viaggi (2018, 2020) proposes to use rationale bundles of attributes (characteristics) as agglomeration, which links with attribute theory and can be linked to either private or public goods. In this way, it also integrates demand-side components (households, etc.) and the institutional context of production and consumption.

By using the term institutions, the approach goes beyond the mere dichotomy between markets and state and builds on the role of intermediate types of institutions identified by contractual relationships with a geographical reference. The firm is also embedded in this category, as the configuration with the higher level of hierarchical relationships. However, the nature of the firm here tends to become rather that of a temporary aggregate of contracts, driven by a value proposition, more similar to hybrid institutions (Williamson, 1991). Hence the importance given by business models to describe economic structures especially when it is necessary to interpret innovation actions. To express this concept, Viaggi (2018) recalls the metaphor of firms as "amoeba objects". This primarily applies not only to firms, but also to other objects (from state to network). Each object (actor) is identified by some stabilizing/identifier mission and topological feature (e.g. location), but is characterized by an observable shape and a set of linkages that are by nature dynamic. The issue of how to define/identify fixes, i.e. characters that allow identifying these entities, becomes a need and a problem at the same time. This is a challenge for agriculture, in which the traditional farm structure is understood through physical features (land, animal heads), while in the new view relationships are more important and access to resources is "just" one of their features.

The SETVEWS is also deemed to be able to understand the governance of complex economic systems in their regional/interregional dimension. This is not only connected to the geographical scale, but also to the need to incorporate concepts such as global common goods and account for flexible boundaries across countries and systems. This can also allow addressing the circular economy concept in a multiscale and multi-regional sense, considering different levels of "circularity" of local ecosystems and addressing the benefits and costs of circularity taking into account the interplay between different resources. Water and nutrients are typical examples of resources that are connected but do not overlap with that of biomass and that, in addition, have their own governance structure.

Over the whole connections among the different actors involved in the SETVEWS, coordination becomes crucial, in both a dynamic and static setting. In a static setting, coordination ensures that the flow of materials between resource providers and resource transformer is smooth. In a dynamic setting, coordination among consumers, government and firms also affects changes over time, intertemporal decisions and resilience.

The SETVEWS can also help explaining value creation by flexibly linking the value generated by consumer preferences and the value generated by the increasing biological capacity of production. This value is expressed through transactions, assets and property rights, taking into account promissory as well as "real" value systems. The link between value and location can also be expressed, for example using indicators of density of value, capital or density of the web itself. At the same time, however, this information remains functionally embedded in a global view and in global value flows.

2.4 Discussion and Final Remarks

The discussion about the bioeconomy in the field of economics, economic organization and policy is largely driven by the search for suitable concepts to interpret current trends and understand the future.

Currently, it is increasingly agreed that approaches to the bioeconomy should become holistic, taking a systemic view and aiming at sustainability and circularity. This may possibly require a new conceptual background, able to inform discussion, instruments and practice. SESs and SETVEWSs can contribute to this process.

Many of these concepts can be of particular importance for agriculture within the bioeconomy due not only to its linkages with the whole sector, but also to its specific features, in light of some major challenges currently on the agenda, such as resource limitations, food needs and climate change.

Clearly, emerging concepts still need to be operationalized into tools and procedures; however, the contribution from a better understanding of current trends and potential evolution through proper concepts should not be downplayed. This reflection corroborates the need for broader approaches, able to endogenize a wider set of variables and allow more comprehensive understanding and support in decision-making, in both the private and public domains.

REFERENCES

Anderies, JM, Janssen, MA, 2013. Robustness of social-ecological systems: Implications for public policy. *Policy Stud. J.* 41, 513–536. https://doi.org/10.1111/psj.12027

Antle, J, Stoorvogel, JJ, 2006. Predicting the supply of ecosystem services from agriculture. *Am. J. Agric. Econ.* 88, 1174–1180.

Binder, CR, Hinkel, J, Bots, PWG, Pahl-Wostl, C, 2013. Research, part of a special feature on a framework for analyzing, comparing, and diagnosing social-ecological systems comparison of frameworks for analyzing social-ecological systems. *Ecol. Soc.* 18, 26. https://doi.org/10.5751/ES-05551-180426

Carpenter, SR, Mooney, HA, Agard, J, Capistrano, D, Defries, RS, Diaz, S, Dietz, T, Duraiappah, AK, Oteng-Yeboah, A, Pereira, HM, Perrings, C, Reid, WV, Sarukhan, J, Scholes, RJ, Whyte, A, 2009. Science for managing ecosystem services: Beyond the millennium ecosystem assessment. *Proc. Natl. Acad. Sci. U. S. A.* 106, 1305–1312. https://doi.org/10.1073/pnas.0808772106

Cumming, GS, Buerkert, A, Hoffmann, EM, Schlecht, E, Von Cramon-Taubadel, S, Tscharntke, T, 2014. Implications of agricultural transitions and urbanization for ecosystem services. *Nature* 515, 50–57. https://doi.org/10.1038/nature13945

Darnhofer, I, Fairweather, J, Moller, H, 2010. Assessing a farm's sustainability: Insights from resilience thinking. *Int. J. Agric. Sustain.* 8, 186–198. https://doi.org/10.3763/ijas.2010.0480

de Schutter, L, Giljum, S, Häyhä, T, Bruckner, M, Naqvi, A, Omann, I, Stagl, S, 2019. Bioeconomy transitions through the lens of coupled social-ecological systems: A framework for place-based responsibility in the global resource system. *Sustainability* 11(20), 5705. https://doi.org/10.3390/su11205705

Fischer, J, Gardner, TA, Bennett, EM, Balvanera, P, Biggs, R, Carpenter, S, Daw, T, Folke, C, Hill, R, Hughes, TP, Luthe, T, Maass, M, Meacham, M, Norström, AV, Peterson, G, Queiroz, C, Seppelt, R, Spierenburg, M, Tenhunen, J, 2015. Advancing sustainability through mainstreaming a social-ecological systems perspective. *Curr. Opin. Environ. Sustain.* 14, 144–149. https://doi.org/10.1016/j.cosust.2015.06.002

Fraser, EDG, Dougill, AJ, Hubacek, K, Quinn, CH, Sendzimir, J, Termansen, M, 2011. Assessing vulnerability to climate change in dryland livelihood systems: Conceptual challenges and interdisciplinary solutions. *Ecol. Soc.* 16, 1. https://doi.org/10.5751/ES-03402-160303

Janssen, MA, Anderies, JM, 2007. Robustness trade-offs in social-ecological systems. *Int. J. Comms.* 1, 43–65.

Janssen, MA, Ostrom, E, 2006. Empirically based, agent-based models. *Ecol. Soc.* 11, 37.

Lescourret, F, Magda, D, Richard, G, Adam-Blondon, AF, Bardy, M, Baudry, J, Doussan, I, Dumont, B, Lefèvre, F, Litrico, I, Martin-Clouaire, R, Montuelle, B, Pellerin, S, Plantegenest, M, Tancoigne, E, Thomas, A, Guyomard, H, Soussana, JF, 2015. A social-ecological approach to managing multiple agro-ecosystem services. *Curr. Opin. Environ. Sustain.* 14, 68–75. https://doi.org/10.1016/j.cosust.2015.04.001

Liu, J, Dietz, T, Carpenter, SR, Alberti, M, Folke, C, Moran, E, Pell, AN, Deadman, P, Kratz, T, Lubchenco, J, Ostrom, E, Ouyang, Z, Provencher, W, Redman, CL, 2007. Complexity of coupled human and natural systems. *Science* 317, 1512–1516. https://doi.org/10.1126/science.1144004

Liu, J, Hull, V, Batistella, M, DeFries, R, Dietz, T, Fu, F, Hertel, T, Izaurralde, RC, Lambin, E, Li, S, Martinelli, L, McConnell, W, Moran, E, Naylor, R, Ouyang, Z, Polenske, K, Reenberg, A, de Miranda Rocha, G, Simmons, C, Verburg, P, Vitousek, P, Zhang, F, Zhu, C, 2013. Framing sustainability in a telecoupled world. *Ecol. Soc.* 18(2), 26. Available at: www.ecologyandsociety.org/vol18/iss2/art26/ [Accessed September 24, 2020].

Ostrom, E, 1998. Scales, polycentricity, and incentives: Designing complexity to govern complexity. In *Protection of Global Biodiversity: Converging Strategies*, eds. LD Guruswamy, J McNeely. Durham, NC: Duke University Press, 149–167.

Ostrom, E, 2009. A general framework for analyzing sustainability of social-ecological systems. *Science* 325, 419–422. https://doi.org/10.1126/science.1172133

Scheiterle, L, Ulmer, A, Birner, R, Pyka, A, 2016. From commodity-based value chains to biomass-based value webs: The case of sugarcane in Brazil's bioeconomy. *J. Cleaner Prod.* 172, 3851–3863. https://doi.org/10.1016/j.jclepro.2017.05.150.

Schlüter, M, McAllister, RRJ, Arlinghaus, R, Bunnefeld, N, Eisenack, K, Hölker, F, Milner-Gulland, EJ, Müller, B, Nicholson, E, Quaas, M, Stöven, M, 2012. New horizons for managing the environment: A review of coupled social-ecological systems modeling. *Nat. Resour. Model.* 25, 219–272. https://doi.org/10.1111/j.1939-7445.2011.00108.x

Scoones, I, Leach, M, Smith, A, Stagl, S, Stirling, A, Thompson, J, 2007. *Dynamic Systems and the Challenge of Sustainability.* STEPS Working Paper 1, STEPS Centre, Brighton.

van der Leeuw, S, 2012. For every solution there are many problems: The role and study of technical systems in socio-environmental coevolution. *Geogr. Tidsskr.* 112, 105–116. https://doi.org/10.1080/00167223.2012.741887

Viaggi, D, 2018. *The Bioeconomy: Delivering Sustainable Green Growth*, CABI, Wallingford (UK), pp. 224, LCCN 2018034273.

Viaggi, D, 2020. Understanding bioeconomy systems: Integrating economic, organisational and policy concepts. In *Bioeconomy for Sustainable Development*, ed. C Keswani. Singapore: Springer, 3–19.

Virchow, D, Beuchelt, TD, Kuhn, A, Denich, M, 2016. *Biomass-based Value Webs: A Novel Perspective for Emerging Bioeconomies in Sub-Saharan Africa, Technological and Institutional Innovations for Marginalized Smallholders in Agricultural Development.* https://doi.org/10.1007/978-3-319-25718-1_14

Walker, B, Gunderson, L, Kinzig, A, Folke, C, Carpenter, S, Schultz, L, 2006. A handful of heuristics and some propositions for understanding resilience in social-ecological systems. *Ecol. Soc.* 11. https://doi.org/10.5751/ES-01530-110113

Williamson, OE, 1991. Comparative economic organization: The analysis of discrete structural alternatives. *Adm. Sci. Q.* 36(2), 269.

3

Assessing the Potential of Agri- and Bio-Based Industries in Transition Economies: Needs and Opportunities in the Case of the Greek Bioeconomy

Emmanuel G. Koukios, Lazaros S. Karaoglanoglou, Sofia Papadaki, Nikolaos Kourakos and Dimitrios P. Koullas

CONTENTS

3.1 Introduction

According to various indications, the new mega-wave of socio-technical change, expected to peak after 2030, will include the emergence of a bioeconomy, i.e. of the whole spectrum of applications of the biological sciences and technologies in all socio-economic areas and sectors, radically transforming our economies, our societies and the world as we know it (Koukios, 2017).

The object of the research carried out by our group in the last few years is to foresee, map and assess the emergence of this great wave we call the bio-tsunami, thus preparing economies and societies to harness its huge potential for health, work, prosperity, environmental balance and quality of life, while avoiding getting crashed by its risks and side effects. Our recent research findings also include critical technological breakthroughs affecting the bio-tsunami's emergence, and the socio-economic clusters describing its main pathways of social and broader change (Koukios & Sacio-Szymańska, 2018).

To promote the applicability of the particular approach, the BIOTOPOS group has put together a series of opinions and other papers, which have initially taken the form of a manifesto, co-authored by scientists and engineers from five southern European Union (EU) countries which, after experiencing a deep and complex systemic crisis, are currently in a slow transition process out of it: Portugal, Spain, Italy, Greece and Cyprus (Koukios et al., 2017).

More recently, these ideas and approaches were tested in a detailed case study of Poland, in collaboration with Polish experts. The case of Poland represents another type of transition economy, i.e. one moving from a centrally planned economy towards a market-dominated one (Koukios et al., 2019).

With the manifesto and the other joint publications, these collaborating authors advocate the adoption of a new development model for such transition economies and societies, focusing on the target of a sustainable bioeconomy. Implementing this model is expected to act as a locomotive to get these countries out of their current difficulties, and into greener growth pastures.

The object of the research presented in this chapter is the application of the targeted approach already developed by the BIOTOPOS group in the case of Greece. The emphasis of the present case study will be on the assessment of the potential of the national agro/bio/eco industries and other relevant systems and sectors, followed by the development of an appropriate national bioeconomy strategy.

3.2 Methodology

A multistep, multifactor and multidiscipline approach is proposed for the assessment of bioeconomy-related developments affecting the pathway of transition economies and societies (Koukios et al., 2019). The particular action steps of this approach include methodological elements from various fields, including future-oriented methods, especially horizon scanning (HSc) (Horizon Scan Report, 2007), in order to identify critical needs and opportunities affecting key drivers and barriers, making possible strategy formulation and scenario-building. These are combined with other relevant tools, such as mapping, techno-economic evaluation, technology assessment and strategy and policy analysis. In this chapter, this approach is applied to the case of Greece, a country currently in the process of preparing its bioeconomy strategy in the frame of regional and national smart specialization efforts.

The methodologies to be applied in the case study of Greece constitute an appropriate selection of the following: an overview and evaluation of the dynamics of bioeconomy-related sectors or bio-sectors; an assessment of the relevant achievements and opportunities, as well as the major risks and challenges of the bio-sectors; a focus on the critical factors resulting from the preceding analysis, and identification of the key factors affecting the emergence of a bioeconomy; and an exploration of the prospects for the development of an optimal bioeconomy strategy.

The Greek bioeconomy consists of 10 sectors listed in Table 3.1. The definition of the particular entities was based on the control of the existing significant interactions and minimization of the resulting overlaps between sectors. A typical example of such sources of conceptual fuzziness is the existence of various bioresource–bioconversion–bioproducts logistic chains.

The criteria employed for the assessment of the above-defined Greek bioeconomy sectors are presented in Table 3.2. The whole approach follows a horizon scanning foresight methodology, appropriately truncated to meet the needs of the particular case study (Koukios & Sacio-Szymańska, 2020). The time horizon of the HSc exercise is that of the year 2030.

- First, the key features of each sector are identified.
- Then, its relative sectoral size is estimated; two factors are adopted at this stage of the analysis, i.e.
  - the sectoral economic income size as defined by the gross domestic product (GDP), expressed in billion Euros per year;
  - the sectoral social significance expressed by the total employment value of the sector;
  - in our estimations of the above factors we have taken into account the primary and secondary income and employment effects of the introduction of a bioeconomy in each sector;
  - as far as the reference year of the estimations, it will be the current year (2020), unless otherwise noted;

TABLE 3.1

Greek Bioeconomy Structure

(A) Agriculture
(B) Agri- and food industries
(C) Forest and wood industries
(D) Traditional bioenergy
(E) Pharmaceuticals—cosmetics
(F) Aquaculture
(G) New biomass—bioenergy
(H) Old and new service sectors
(I) New bio-industries—biorefineries
(J) Inland waters

TABLE 3.2

Bioeconomy Sectors Assessment

- Key features
- Estimated size
  - Income
  - Employment
- Problems
- Opportunities

    - the main sources of the data needed in this analysis were the European and national statistics (Koukios, 2014).
- Finally, the major problems, risks and challenges on the one hand and the key prospects and opportunities on the other hand, are recorded for each sector.

3.3 Assessing Bioeconomy Needs and Opportunities

(A) Agriculture

Key Features

- Agriculture represents the main form (30%) of land use in Greece with 4 Mha out of a national total of 13.2 Mha.
- At the same time, agriculture constitutes the national basis for rural development by offering employment and generating value at the local level.

Estimated Size (2020)

- Income: 8 billion Euro
- Employment: 400,000

Problems

- The performance of Greek agricultural systems depends heavily on the existence of EU and/or national subsidies, which have created a corresponding dependence mentality.
- At the same time, farmers tend to consider research, development and technical change activities as cost items, a trend that has created innovation aversion and blocked growth.

Opportunities

- Greek agriculture could significantly benefit from the ongoing trends towards a new EU common agricultural policy (CAP), as well as reforms in other relevant European and national policy regimes.
- In such a changing context, the emergence of a bioeconomy could provide a new, long overdue, sectoral "identity".

(B) Agri- and Food Industries

Key Features

- The Greek food industry has consistently shown signs of high competitiveness, associated with significant export potential.
- This sector is closely integrated with agriculture; thus, it has the potential and the momentum to mobilize its huge agricultural potential (see above).

Estimated size (2020, includes potential for development)

- Income: 19 billion Euro
- Employment: 150,000

Problems

- The lack of an appropriate bioeconomic strategy might enhance the already present trend of a loss of intellectual capital from past successes.

- The sector partly shares with agriculture a low innovation drive, which coupled with an emphasis on cost-minimization efforts could lead the sector into a low-efficiency loop trap.

Opportunities

- The sector could benefit from the globally emerging new food culture—such as the Mediterranean diet—as well as the growing food/health interfaces.
- The development of novel foods could result in a new spectrum of bio-based and other new products, e.g. from utilizing agro-industrial wastes.

(C) Forest and Wood Industries

Key Features

- Forests in their usual form spread over only 2 Mha of Greek land; however, to have a complete view of the sector, we must also consider the 3 Mha of Mediterranean vegetation areas (maquis, shrub and phrygana areas).
- Due to the type of land use and the transport limitations of the woody biomass and products thereof, this sector is associated with highly decentralized systems and applications.

Estimated size (2020, includes potential for development)

- Income: 6 billion Euro
- Employment: 50,000

Problems

- The phenomena of "human desertification" and regional imbalances lead to a loss of human capital from forest areas that could be irreversible.
- The sector shows significant production phase-out trends as the respective markets are dominated by imported products and bio-product substitutes.

Opportunities

- The sector shows great synergies with the existing as well as the new forms of biomass/bioenergy/biofuel activities.
- The sector could also benefit from the ongoing intensive change in the new materials field, as well as from other innovative products/services.

(D) Traditional Bioenergy

Key Features

- There is a constant and continuous presence of traditional energy-from-biomass in Greece, with the use of 4–5% of the total biomass, equivalent to 1–2 MT/year, mainly from the utilization of fuelwood, mill residues and some wastes.
- Although this sector contributes significant amounts of renewable energy (RES), no respective policy or strategy exists to rationalize it and provide a proper starting frame for RES to build upon.

Estimated size (2020, includes potential for development)

- Income: 2 billion Euro
- Employment: 30,000

Problems

- The sector is characterized by unsustainable and inappropriate use patterns, and makes no use of emerging technological innovations.
- As an example, the use of traditional biomass for fireplaces in urban areas leads to severe air pollution problems due to the emission of small particles, which are dangerous to health.

Opportunities

- The sector could be significantly upgraded by the adoption and use of modern biomass/bioenergy technologies, as well as a rational management system.
- The sector could benefit from its extensive synergies with agriculture, forestry and agro-industry, and become a key step for local and regional development.

(E) Pharmaceuticals—Cosmetics

Key Features

- This is a very competitive sector with great export potential and serious prospects of becoming a flagship sector of the Greek bioeconomy, in close coordination with other bio-sectors, e.g. biorefineries.
- The key for such a bright future is the possibility of the sector generating innovation-based leadership, as some early signs already testify.

Estimated Size (2020)

- Income: 2 billion Euro
- Employment: 15,000

Problems

- Unfortunately, the good prospects of the sector depend on a limited "window of opportunity", as the growth of a bioeconomy favours those with a strategy and a policy—elements still missing from the Greek case.
- The sector has to show a higher research-intensive momentum, especially on future-oriented topics, such as those at the interface of "health–environment–food supplements".

Opportunities

- The sector could especially benefit from the fast-growing quality-of-life and anti-ageing markets (see Problems above).
- For their successful branding, the new Greek industries within this sector should target innovative combinations of ageing with food and "green" products.

(F) Aquaculture

Key Features

- This sector was a success story of the early 1980s, which could become a model for the emergence of new bioeconomy sectors in the current period, i.e. the 2020s.
- The success recipe of the 1980s: Based on new technology, addressed to national markets, the export orientation of some products in combination with regional development efforts.

Estimated size (2020, including potential for development)

- Income: 1 billion Euro
- Employment: 10,000

Problems

- The sector shows loss of momentum and lack of a new identity.
- Lessons from an old story: Land-use conflicts with tourism in coastal areas.

Opportunities

- Future products might include micro-algae and other blue biotech technologies.
- The new sector should be part of a redesign of the maritime bio-economy, and so obtain a new brand that is currently missing.

(G) New Biomass/Bioenergy

Key Features

- The total annual biomass potential of Greece is estimated at 45–50 M dry t/year of many types of biomass, predominantly (>50%) of plant origin.
- To understand the importance of this potential, we estimate (Table 3.3) that if 100% of this biomass was actually converted to bioenergy, it would be enough to substitute 100% of the greenhouse gases (GHG) emitted from the national energy industry of this country.

Estimated size (potential—sustainable—2030)

- Income: 8 billion Euro
- Employment: 80,000

Problems

- Due to the plant origin of most of the Greek biomass, a key constraint for the efficient utilization of this potential will be the optimized conversion of lignocellulosic feedstocks—a topic that is today still under intensive research in many parts of the world.
- To achieve sustainable utilization, soil fertility–related issues and other risks due to highly intensive farming must be addressed.

Opportunities

- A new generation of well selected energy and new industrial crops might emerge after careful identification and cultivation research.

TABLE 3.3

Greek Bioresources Data (OECD, 2009)

• Biomass potential: 45–50 MT/year (4–5% used)
• Energy value of biomass: 15–20 MToe/year (40–50% of national energy use)
• Bioenergy used: 1–2 MToe/year (4–5% of energy use)
• Biomass carbon content: 25 MT/year
• CO_2 equivalent potentially saved: 100 MT/year

- A utilization approach that guarantees both efficiency and sustainability is the one based on locally integrated, innovative bio-communities.

(H) Old and New Service Sectors

Key Features

- The growth trends of bioeconomies should meet the major trends of markets favouring products vs. processes, and services vs. products.
- The range of potential services where bioeconomic inputs would be beneficial includes health, environment, tourism and quality of life, which we hereby aggregate as a composite new sector.

Estimated size (potential, 2030)

- Income: 10 billion Euro
- Employment: 75,000

Problems

This type of market-oriented initiative suffers from two major problems and their associated risks:

- Lack of innovative drive—fear of change.
- Fragmentation of the bioeconomy fields.

Opportunities

This convergence of service businesses could take advantage of two strong "winds of change":

- Possibilities of synergies with other fields and sectors.
- A systematic move towards a new Greek service bioeconomy.

(I) New Bio-Industries—Biorefineries

Key Features

- The growth of a bioeconomy will open the way for new generations of bio-industries, including those with properties unheard of until today.
- Biorefineries constitute an example, offering vital and critical bioresource–sectoral use interfaces.

Estimated size (potential, 2030)

- Income: 5 billion Euro
- Employment: 45,000

Problems

From the long and highly uncertain range of new challenges and risks generated by new bio-industries, we identify:

- The risks of experience transfer from other EU and global regions.
- The need to build new supply chains—role of stakeholders.

Opportunities

From the long and growing list of potential applications, we select two types of approaches:

- The growth of non-energy markets for biomass.
- Targeting new biotech applications.

(J) Inland Waters

Key Features

- On the demand side, inland waters deserve a key position in the emerging bioeconomy landscape as, for example, polluted lakes and wetlands are the result of urbanization.
- On the supply side, water occupies an increasingly critical position due to its merging with other socio-technical factors of change, such as the "water–food–energy nexus".

Estimated size (2030, potential due to pollution control, biomass sources)

- Income: 2 billion Euro
- Employment: 20,000

Problems

- Unsustainable water economy—eutrophication risks.
- Land use conflicts—pressure on resources.

Opportunities (examples)

- New ICT eco-management tools.
- New technologies for soil/water bio-remediation.

3.4 Towards a Greek Bioeconomy Strategy

Table 3.4 summarizes the findings of the previous section. Some key points regarding the prospects for a national bioeconomy strategy include the following:

- The mapping of the bioeconomy terrain permits the identification of two types of bioeconomy: the existing one (BE-2020) and the one in progress (BE-2030), with a time horizon in this study of the year 2030 (Koukios, 2014).
- The relative size of BE-2020 is of the order of 20% of annual GDP, i.e. twice the expected income size, according to the Organization for Economic Co-operation and Development (OECD) and other sources (Koukios, 2014); this difference can be explained on the one hand by the nature of the Greek economy and the role of agriculture, and on the other hand by the enlargement of the bioeconomy concept to cover more sectors and fields.
- The growth potential up to the year 2030 is also very significant: by considering four more sectors, the 2020 size of the Greek bioeconomy can increase by 34–40% depending on the indicator used; thus, a 2030 bioeconomy equivalent to 35% of the Greek GDP is possible, should an appropriate strategy be put in place.

3.4.1 Target Sectors for a Greek Bioeconomy Strategy

The above analysis permits the identification of a few strategic priorities on which a national bioeconomy strategy could be built:

- Agriculture: Leading sector providing 4% of GDP and 0.4 M jobs and covering 30% of the national area, while strongly affecting rural areas and Mediterranean ecosystems.
- Food and agro/bio-industry: A key group of industries (ca. 10% of GDP), aiming at adding value to agricultural and other output; other regionally relevant types of industries such as wood-based industries, bio-industries and biorefineries should be included.
- Biomass and waste: Very large potential (ca. 50 M dry t/a) combined with its limited use offers great possibilities for development, with an emphasis on food and urban wastes and novel energy vectors.
- Strategic approach: The advantages of an agro/bio/eco strategic orientation are based on the high biodiversity of Greek bioresources and the Mediterranean diet value of Greek foods.

3.4.2 Current National Strategic Bioeconomy Actions

The process of putting together the national bioeconomy strategy is ongoing. The following is a list of the relevant actions that have taken place during the last few years:

- Greek bioeconomy forum—platform of actors/stakeholders
- Two MSc programmes running: Piraeus/Athens, Thessaloniki
- Circular economy and forestry strategies launched (May 2018)
- Research and innovation topics funded as part of the national smart specialization strategy—other relevant research results
- Regional strategies (Crete, Thessaly, etc.)—W. Macedonia cluster
- CASA SCAR workshop discussions/conclusions (Athens, May 2018)
- Long and rich experiences from the biomass/bioenergy field
- Studies of various bodies including the Greece 2021 foresight project

TABLE 3.4

Overview of Assessment of Greek Bioeconomy Sectors

Sectors	Income (billion Euro/year)	Employment
Agriculture	8	400,000
Agro- and food industries	19	150,000
Forest and wood industries	6	50,000
Traditional bioenergy	2	30,000
Pharmaceuticals/cosmetics	2	15,000
Aquaculture	1	10,000
BE (bioeconomy)—2020	**38** (20% of GDP 2020)	**655,000**
New biomass—bioenergy	8	80,000
Old and new services	10	75,000
New bio-industries/biorefineries	5	45,000
Inland waters	2	20,000
BE (bioeconomy)—2030	**63** (35% of GDP 2020)	**875,000**
Growth of bioeconomy 2020/2030	**25** (15% of GDP 2020) (40% on BE 2020)	**220,000** (34% on BE 2020)

3.4.3 Responsible Public Bodies and Relevant Stakeholders

The complexity of the issue can be seen from the following list of public bodies and other critical factors including share- and stakeholders:

- Main public bodies: (I) General Secretariat of Research and Innovation; (II) Ministry of Rural Development and Food; (III) Ministry of Environment and Energy.
- Research institutions: (1) Hellenic Agricultural Organization (DIMITER); (2) Centre for Research and Technology—Hellas (CERTH); (3) Hellenic Centre for Marine Research; (4) Centre for Renewable Energy (CRES); (5) Greek universities.
- Stakeholders: Suppliers; downstream processors; logistics agents; marketing actors; business and financial institutions; citizens; worker unions; societal groups and organizations.

Acknowledgments

A part of this chapter was presented by the first author under the title "Overview of State of Play on Bioeconomy in Greece", on behalf of the platform of the Greek bioeconomy forum in collaboration with the BIOTOPOS group of the National Technical University of Athens, at the workshop "Facilitating Development of Bioeconomy Policy—Needs and Gaps", Brussels, May 3, 2019.

REFERENCES

Eurostat, *Agriculture, Forestry and Fishery Statistics*, edition 2019. National Statistical Service of Greece, Athens, 2019.

Horizon Scan Report, *Towards a Future Oriented Policy and Knowledge Agenda*, Eds. R. 't Veld et al., Horizon Scan Publication, The Hague, 2007.

Koukios, EG, "The bio-tsunami project", In: Koukios, E et al. (eds.), *The Emergence of Bioeconomy – Opportunities and Risks – A Forward-Looking Study*, Institute for Sustainable Technologies, Radom, PL, 11–96, 2017.

Koukios, EG, and Sacio-Szymańska, A, "Assessing the emergence of bioeconomy by the radical technology inquirer tool", *European Journal of Futures Research*, 6: 23 (2018) https://doi.org/10.1186/s40309-018-0152-x.

Koukios, E et al., "Targeting sustainable bioeconomy", *Journal of Cleaner Production*, 172: 3931–3941 (2017), https://doi.org/10.1016/j.jclepro.2017.05.020.

Koukios, EG et al., "Assessing the emergence of bioeconomy in transition economies by a future-oriented approach: The case of Poland", In: Keswani, C (ed.), *Bioeconomy for Sustainable Development*, 159–171, Springer, Singapore (2019).

Koukios, EG, and Sacio-Szymańska, A, "The emergence of bioeconomy in the 6th kondratiev wave of change: A horizon scanning-based approach", In: *Bio#Futures – Foreseeing and Exploring the Bioeconomy*, Springer, 2020 (in preparation).

Koukios, EG, "Technology management for the biobased economy: Mapping, dynamics and policies—the case of Greece", In: Trzmielak, DM, and Gibson, DV (eds), *International Cases on Innovation, Knowledge and Technology Transfer*, Centre of Technology Transfer, University of Lodz, Lodz, PL, 129–145, 2014.

OECD, *The Bioeconomy to 2030 – Designing a Policy Agenda*. Report, 2009, http://biotech2030.ru/wp-content/uploads/docs/int/The%20Bioeconomy%20to%202030_OECD.pdf (accessed 17 September 2020).

4

IPRs to Innovation Are as Bioeconomy to Sustainable Development: How Is It All Interconnected?

Dhanay Cadillo Chandler

CONTENTS

Disclaimer: The views expressed in this chapter are those of the author and do not represent any official position by the institution represented by the author.

4.1 Background

4.1.1 What Is the Bioeconomy and How and Why Is It Commonly Related to Biotechnology?

The bioeconomy, as a concept, in its early development referred to the use of biological knowledge for commercial and industrial purposes. One of the components of the concept refers to advancements in the biological sciences and biotechnology that have the potential to transform many industrial production processes.[*] The previous component also fits within the knowledge used to create innovations that are subsequently protected by intellectual property rights (IPRs), i.e. patents. Due to the potential impact as a policy concept, the EU integrated the term knowledge-based economy within the initial bioeconomy conception.[†] Considering that bioeconomy as a "economy" concept relies on renewable natural resources to produce food, energy, products and services, the conservation of biodiversity is paramount. Thus, preventing the loss of biodiversity in a manner consistent with sustainable development goals has been

[*] See Straus, J, Intellectual property rights and bioeconomy, 12 *Journal of Intellectual Property Law & Practice* 7 (2017) 576–590.

[†] Id ut supra Birner.

pointed out to be a fundamental goal. Governing biotechnology and access to genetic resources may be relevant in achieving such objectives.*

Inventions in all fields of technology may be worthy of patent protection, whether they come from biotechnology, biodiversity or green technology, as long as patentability requirements are met. Knowledge-based economy embodies the "vision of achieving economic growth through high-technology industries, which requires investments in innovation and highly skilled labour".† Thereafter, biotech industry fits in well with this definition, thus becoming one of the rationales behind the policy frameworks erected around this industry and, more generally, the life sciences sector. This definition together with the type of industry leads to one of the unavoidable links between a knowledge-based economy, as championed by the EU, and the role of IPRs in achieving innovation.

The concept of a knowledge-based economy seems to represent the evolution from its initial conceptualization, known as the bioeconomy, going beyond biotech and IPRs to a more general aspect that relates to the sustainability and governance of the knowledge and innovations stemming from this, which at times are in conflict with each other. For instance, biotechnology seeks to transform natural processes found in nature with the goal to improve them to satisfy a larger purpose; however, these improvements will require access to traditional knowledge associated with genetic resources.‡ Reviewing both the definition of biotechnology and its international framework reveals why this field falls within the scope of the knowledge-based economy, as the next section analyses. The essential question is not solely about how to innovate, but also about how to ensure continuous access to information without depleting resources or using them in a manner that decreases their value for the alleged proprietor. Admittedly, this idea has more profound ramifications than stated in the example. The relevance was to showcase in a simplistic manner the tension between making use of a resource that represents value quantifiable both in money and in non-pecuniary terms depending on the stakeholder.

4.1.2 Biotechnology, Biodiversity and Access to Genetic Resources in Brief

Biotechnology as a concept has been defined and redefined throughout time. The US Office of Technology, has understood this, as to "include any technique that uses living organisms (or part of organisms) to make or modify products to improve plants or animal, or to develop micro-organisms for specific uses".§ However, the current approach towards biotechnology seems to be broader than in the previous definition. It is not only about techniques applied to living organisms and parts thereof to produce, identify or design substances or to modify organisms for specific applications,¶ but it is also about a multidisciplinary science involving several disciplines, i.e. biochemistry, cell biology and embryology among other.**

The Organization for Economic Co-Operation and Development's (OECD) 2003 working paper on biotechnology points to a common definition among member states, which also serves as a guideline to interpret and furthermore to regulate it:

> the application of Science & Technology to living organisms as well as parts, products and models thereof, to alter living or non-living materials for the production of knowledge, goods

* Genetic resources are commonly associated with traditional knowledge (TK), thereafter falling under the IPR system through a sui generis system of protection. Scholars have pointed to the challenges in looking at both TK and access to genetic resources associated with TK. IPR or not, these are essential for the bioeconomy. See Hamilton, C, Intellectual property rights, the bioeconomy and the challenge of biopiracy, 4 *Genomics, Society and Policy* 3 (2008) 26–45.

† Id ut supra Briner.

‡ See id ut supra Hamilton, and see Leidwein, A, Protection of traditional knowledge associated with biological and genetic resources. General legal issues and measures already taken by the European Union and its member states in the field of agriculture and food production, 9 *The Journal of World Intellectual Property* (2006) 251–275.

§ Wiegele, T, *Biotechnology and International Relations*. University of Florida Press, Gainesville, 1991.

¶ Usha, S and Annadurai, C, The intellectual property rights and pharmaceutical biotechnology in India, 3 *Journal of Pharmacy Research* 10 (2010) 2466–2469.

** Much of this evolution was explored in Cadillo Chandler, DM, Biotechnology and its development in the developing countries: Can IPRs foster innovation in the field? In Singh, HB, Jha, A and Keswani, C (eds.) *Intellectual Property Issues in Biotechnology*, CABI, Wallingford, 2016.

and services. Including, DNA (the coding): genomics, pharmaco-genetics, gene probes, DNA sequencing/synthesis/amplification, genetic engineering; proteins and molecules (the functional blocks): protein/peptide sequencing/synthesis, lipid/protein glyco-engineering, proteomics, hormones, and growth factors, cell receptors/signalling/pheromones; cell and tissue culture and engineering: cell/tissue culture, tissue engineering, hybridization, cellular fusion, vaccine/immune stimulants, embryo manipulation; process biotechnologies: bioreactors, fermentation, bioprocessing, bioleaching, bio-pulping, bio-bleaching, bio desulphurization, bioremediation, and bio filtration; and sub-cellular organisms: gene therapy, viral vectors.*

Despite efforts to consolidate such a definition, for ease of regulation, each agreement or convention seeking to provide a framework for governance leans on redefining biotechnology based on the purpose of the said agreement. Efficiency in governing both risks and opportunities brought upon this field of science developments emerges as one of the most significant challenges. Admittedly, the world as we know it has been revolutionized by biotech innovations, which have not only brought expectations to address fundamental food and health needs globally by fostering R&D in the field of healthcare, agriculture, food production and environmental protection,† but also the need to provide a legal framework.

Acknowledging that biotechnology is "the amalgamation of several technologies ranging from the field of biomedical research and synthetic biology and further transcending to engineering, nanotechnology, and bioinformatics"‡ enhances the possibility to achieve needed innovation for the alleviation of suffering from illnesses and to ensure sustainability and agricultural development, innovations in the fields of biomedical research and biotechnological agriculture are of particular importance to society. Thus, its foreseeable impact relates to access to information, treatment, affordability and sustainability.§

Biotechnology is an important tool in the context of agriculture and food production, thus "(1) access to genetic resources—a key input for agricultural research—and sharing of the benefits from its utilization, (2) intellectual property rights governing the use of proprietary technologies, and (3) biosafety, including environmental and food and feed safety"¶ are three of the main aspects foreseen within the international framework, i.e. the Convention on Biological Diversity (CBD), the International Treaty on Plant Genetic Resources for Food and Agriculture (ITPGRFA), the International Union for the Protection of New Varieties of Plants (UPOV), other conventions related to the environment and the Agreement on Trade-Related Aspects of Intellectual Property Rights (TRIPs) under the World Trade Organization (WTO).

The World Health Organization (WHO), the World Trade Organization, the United Nations Food and Agriculture Organization (FAO), the United Nations Environmental Programme (UNEP), the United Nations Educational, Scientific and Cultural Organization (UNESCO) and the United Nations Industrial Development Organization (UNIDO) are the main organizations working together with countries and lobbyists to establish standards necessary to protect inventions in the field and also to protect the environment from possible negative effects derived from the use of modern biotechnology.

Although biotechnology and the EU framework are relevant for this section, these are explored in depth further in this volume, thereby falling outside the scope of this chapter. As for the international framework, the International Convention for New Plant Varieties drafted in 1961, also known as the

* Devlin, A, *An Overview of Biotechnology Statistics in Selected Countries*. OECD Science, Technology and Industry Working Papers. OECD Publishing, Paris, 2003.

† Horng, D, *International Law on Biotechnology. International Law and Institutions*. Encyclopedia of Life Science Systems, 2009.

‡ Ganguly, N, Croft, S, Singh, L, Sinha, S and Balganesh, T, *Biomedicine and Biotechnology: Public Health Impact*. BioMed Research International, 2014.

§ See id ut supra Cadillo Chandler (2016).

¶ Komen, J, The emerging international regulatory framework for biotechnology, 3 *GM Crops & Food: Biotechnology in Agriculture and the Food Chain* 1 (2012) 78–84.

UPOV Convention, was revisited in 1972, 1987 and 1991.* The UPOV Convention establishes a system for the protection of plant breeders' rights, which is particularly relevant for innovation and sustainability since the convention grants breeders' rights protection through both patents and a special title of protection for one or the same botanical genus or species.†

The Convention on Biological Diversity is said to be an expression of the world community's increasing commitment to sustainable development, the conservation of biological diversity, the sustainable use of its components and the equitable sharing of the benefits emerging from the use of genetic resources (CBD). The CBD was initially envisaged as an environmental law convention* instead of a convention for the promotion and conservation of biological diversity. In its preamble, the convention acknowledges that human activity reduces biological diversity, and at the same time leaves it to each country to decide their own biological resources, thus calling on each state to take measures consistent with both conserving biological diversity and ensuring sustainability when making use of their own biological resources.§ Article 8 (j) of the CBD reconceptualizes the subject matter since it introduces the notion of protection of the knowledge, innovation and practices of indigenous and local communities. It also establishes collaboration practices and benefit-sharing between the holders of the information—indigenous communities—and the third party by requiring the right holder's involvement and approval.¶ Given the scope of the CBD—traditional knowledge and biosafety—two other protocols have been drafted to complement and further guide the implementation of the convention, namely the Nagoya Protocol and the Cartagena Protocol.

The Cartagena Protocol on Biosafety is a complementary agreement to the CBD adopted in Decision EM-I/3 at the Conference of the Parties to the Convention on Biological Diversity from January 2000. This protocol protects biological diversity from potential risks derived from living modified organisms resulting from modern biotechnology that may have adverse effects on the conservation and sustainable use of biological diversity, also taking into account the risks to human health, and specifically focusing on trans-boundary movements.** The main objective of this protocol is to provide an adequate framework of protection in the field for the safe transfer, handling and use of living modified organisms resulting from modern biotechnology that may have adverse effects on biological diversity.

The Nagoya Protocol on Access and Benefits-Sharing (ABS) regulates the use of biotechnological inventions. This protocol is important since it establishes the framework to share the benefits arising from the utilization of genetic resources in a fair and equitable way. Additionally, it allows the ABS clearing house system to achieve its objective. This protocol has binding effects on member states, since it creates the obligation to take appropriate measures at the national level to protect traditional knowledge and genetic resources accessed inside their jurisdictions according to prior-informed consent regulations as established in Articles 15 and 16 of the protocol.††

Prior to the Agreement on Trade-Related Aspects of Intellectual Property Rights, few conventions extended protection to biotechnological developments. Nevertheless, the agreement establishes minimum standards of protection and extends patent rights to all fields of technology, as widely known.

* Each revision incorporated new terms and conditions that reflected the evolution within the field. The revision in the 1978 UPOV Act differed from the previous one by focusing on the subject matter and the system of protection around it, i.e. eligibility requirements, exclusive rights, term of protection and limitation to exclusive rights (Papadopoulou, 2014). An interesting distinction between the revisions from 1978 and 1991 has to do with farmers' right to save and exchange seeds without this constituting an infringement. Within the 1978 revision, saving seeds and exchanging them between farmers did not constitute an infringement, whereas the 1991 revision limited the scope of the exceptions settled in the previous text by limiting the exchange of the seeds that could be saved for use during the following season.

† International Convention for New Plant Varieties, Act of 1961, Diplomatic Conference December 2, 1961.

* Morgera, E, *Faraway, So Close: A Legal Analysis of the Increasing Interactions between the Convention on Biological Diversity and Climate Change Law.* University of Edinburgh School of Law Working Paper No. 2011/05, 2011.

§ Convention on Biological Diversity, 1760 (UNTS 79); 31 ILM 818 (1992), 1992.

¶ Idem.

** Cartagena Protocol on Biosafety to the Convention on Biological Diversity, *Decision EM-I/3*, Conference of the Parties, January 2000.

†† Papadopoulou, F, *Opening Pandora's Box: Exploring Flexibilities and Alternatives for Protecting Traditional Knowledge and Genetic Resources under the Intellectual Property Framework.* Acta Universitatis Stockhomlmiensis, Stockholm, 2014.

Although at times regarded as vague, the language used in the agreement allows member states to interpret its provisions when tailoring the national system to protect plant varieties as mandated either through patents or by a sui generis system or a combination thereof (Art. 27.3.b - TRIPs Agreement). It is crucial that each member state understands its own interests, development objectives and industrial capabilities before effectively tailoring the national legislation under the minimum standards of protection within TRIPs, especially in the fields of biotechnology and biodiversity.*

At the core of the Agreement is the transfer of technology and dissemination of information, which are fundamental elements within the bio-economy concept. In principle, the agreement must be supportive of the CBD. However, its primary objective in the conservation of genetic resources and sustainable use†, somewhat differs from the objectives in TRIPs that seek to ensure the protection and enforcement of intellectual property rights, since these should contribute to member countries' development. Amid the clash and complementariness necessary between the frameworks, a document called Draft Modalities for TRIPs Related Issues, TN/C/W/52, From 19 July 2008 proposed that:

> Members agree to amend the TRIPS Agreement to include a mandatory requirement for the disclosure of the country providing/source of genetic resources, and/or associated traditional knowledge for which a definition will be agreed, in patent applications. Patent applications will not be processed without completion of the disclosure requirement; Members agree to define the nature and extent of a reference to Prior Informed Consent and Access and Benefit Sharing; Text based negotiations shall be undertaken, in Special Sessions of the TRIPS Council, and as an integral part of the Single Undertaking, to implement the above. Additional elements contained in members' proposals, such as PIC and ABS as an integral part of the disclosure requirement and postgrant sanctions, may also be raised and shall be considered in these negotiations.‡

TRIPs do not provide for disclosure on the origin of the genetic resource or the traditional knowledge used in an invention seeking patent application. However, scholars point to this disclosure as providing some degree of certainty in terms of compliance with the prior-informed consent requirement in the CBD,§ thereby complementing the frameworks. Nevertheless, biotechnology and its framework in the context of this chapter are mentioned with the purpose of highlighting their importance and relation to the bioeconomy or the knowledge-based economy.

4.2 Sustainable Development and Intellectual Property Rights

The intersection between IPRs, understood as knowledge, and sustainable development requires from a closer look at the element within "sustainable development" as a concept itself. That is, "development that meets the needs of the present without compromising the ability of future generations to meet their own needs".¶ In this regard, IPRs as a system of exclusive rights that seeks to promote the transfer of technology and the dissemination of knowledge, falls within the grasp of sustainable development as well as the concept of the bioeconomy. That is, in the sense of making available innovation that will lead to future developments. Thereafter, its regulation shall ensure the longevity of the resources leading to new knowledge. When looking into links between sustainable development and IPRs, it is important to

* See id ut supra Cadillo Chandler (2016).

† Id ut supra Papadopoulou.

‡ Communication from Albania, Brazil, China, Colombia, Ecuador, the European Communities, Iceland, India, Indonesia, the Kyrgyz Republic, Liechtenstein, the Former Yugoslav Republic of Macedonia, Pakistan, Peru, Sri Lanka, Switzerland, Thailand, Turkey, the ACP Group and the African Group, TN/C/W/52, July 19, 2008.

§ Id ut supra Papadopoulou.

¶ Young-Gyoo, S, Intellectual property protection of biotechnology and sustainable development in international law, 29 *North Carolina Journal of International Law and Commercial Regulation* (2003) 157, also see United Nations, Our Common Future—Report of the World Commission on Environment and Development, UN Doc. A/42/427—Annex, 4 August 1987, chapter 2, I, also known as the Brundtland Report adopted in the World Commission on Environment and Development (WCED) in 1987.

recognize that international law in this field embraces several aspects, including international cooperation, economic development, the environment and human rights,* all of which are, in one way or another, intertwined with intellectual property rights and its international framework.

Through a closer look into the aforementioned aspects together with the TRIPs Agreement, the link between the two, as part of the bioeconomy is visible. On the one hand, TRIPs Article 8 refers to international cooperation and socio-economic development as the objectives of the treaty itself. On the other hand, the environment is also mentioned within the agreement,† although reference to this is made in the context of biodiversity. In terms of fundamental rights, IPRs in themselves are considered part of fundamental rights by several political constitutions and fundamental rights declarations. In the context of the European Union, Article 17 of the EU Fundamental Rights Charter‡ provides express protection to IPRs without differentiating between them or tangible property. To an extent, the fact that IPRs are also considered to be a fundamental right may be challenging in terms of balancing this as part of economic rights; however, this assertion is not examined within this chapter. Nevertheless, what seems clear is that there is an unavoidable connection or interplay between IPRs and sustainable development as part of the knowledge-based economy. Furthermore, the TRIPs Agreement preamble indicates that IPRs have a social developmental goal when acknowledging that there is an underlying public policy objective in national systems protecting intellectual property rights.§

In addition to the aspects embodied in the concept of sustainable development, other principles are said to be the foundation of this concept, i.e. intergenerational equity, sustainable use, equitable use and integration.¶ Most of these principles resemble those found within the access and benefit-sharing regime established around the CBD and the Nagoya Protocol. Making easy to understand why both biodiversity and biotechnology are areas increasingly governed through the emerging sustainable development "frameworks." The protection of genetic resources in a manner that ensures sustainable use, together with the transfer of this knowledge in a manner that allows for future generations to make use of the resources without depleting them are among the indicators of why biotechnology and biodiversity are at the core of sustainable development.

The principle of sustainable use may be achieved through collaboration models** that allow for the flow of information between interested parties without necessarily putting a cap on the information per se, but instead focusing on the knowledge governance.†† Although knowledge governance is different from the knowledge-based economy, these two concepts relate in the context of sustainability and intellectual property rights for several reasons. Among these, the need to reconceptualize intellectual property (IP) regulation more in line with a collaborative model by "encouraging innovation, engaging in technology transfer, or otherwise ensuring dissemination and diffusion of the results of innovation across borders".‡‡ Interestingly, the aforementioned concepts are found within the TRIPs Articles 7 and 8, establishing the objects and the purposes of the agreement. This has also led to scholars advocating for the policy space found within TRIPs allowing for the concept of sustainability to be included in policies implementing

* Philippe, S, International law in the field of sustainable development: Emerging legal principles, 53 *Sustainable Development and International Law* (1994) 53 (Winfried Lang ed. 1994)

† Art. 27.2—TRIPS "2. Members may exclude from patentability inventions, the prevention within their territory of the commercial exploitation of which is necessary to protect order public or morality, including to protect human, animal or plant life or health or to avoid serious prejudice to the environment, provided that such exclusion is not made merely because the exploitation is prohibited by their law".

‡ European Union: Council of the European Union, Charte des droits fondamentaux de l'Union Européenne (2007/C 303/01), 14 December 2007, C 303/1. www.refworld.org/docid/50ed4f602.html [accessed 11 September 2020].

§ Bannerman, S, The World Intellectual Property Organization and the sustainable development agenda, 122 *Futures* (September 2020), and Chon, M, Intellectual property and the development divide, 27 *Cardozo Law Review* (2006) 2821.

¶ Id ut supra Young-Gyoo at 182.

** Chon, M, Recasting intellectual property in light of the U.N. sustainable development goals: Toward global knowledge governance, 34 *American University International Law Review* (2019) 763.

†† Idem.

‡‡ Chon, M, Charting the triple interface of public–private partnerships, global knowledge governance, and sustainable development goals, 3 *PPPS* 5–6 in (M Chon et al. eds.) *The Cambridge Handbook on Public-Private Partnerships, Intellectual Property Governance, and Sustainable Development.* Cambridge University Press, 2018.

it, or subsequent regional economic partnership agreements.[*] Sustainability as a policy tool to achieve a balance can be foreseen within EU Economic Partnership Agreements, thereafter stressing the imminent link between exclusivity and societal progress.[†] In terms of IP, the wording in TRIPs Article 7 is clear in that IP protection and enforcement needs to be "conducive to social and economic welfare, and to a balance of rights and obligations".[‡] Additionally, Article 8 reiterates the promotion of "public interest in sectors of vital importance to their socio-economic and technological development, provided that such measures are consistent with the provisions of this Agreement".[§] Traditionally, these two articles have given regulatory space for countries to make use of the well-known TRIPs flexibilities.[¶]

4.3 SDG and IPRs at the Core of SDG9: A Roadmap for Development?

4.3.1 Innovation and SDGs

EU policy developments point to the importance of finding alternative mechanisms to promote high-value innovations, while at the same time ensuring that "innovation that matters is actually produced, patients have access to this and that health systems are financially sustainable".[**] Thus, "sustainability" is a concept ever more present in relation to access to healthcare, drug discovery and the financing of R&D. Patents continue to be part of an inventor's reward for contributing an invention to society and, at the same time, society is rewarded with knowledge unknown to them until the patent disclosure takes place.[††] Innovation in the pharmaceutical and biotechnological sector often involves partnerships between the public and private sector, and even when the fruits of the collaboration are protected under intellectual property rights, some aspects may escape the traditional IP regulatory framework. For instance, capacity building and technology sharing for developments across borders are often not discussed; this is where SDGs are highlighted to bridge the gap by demanding that IP regulation be recast into global knowledge governance.[‡‡]

SDGs are recognized as more advanced goals than those initially settled within the UN Millennium Development Goals. Although, to an extent, SDGs seem to challenge the traditional assumption that development occurs as an inevitable and evenly distributed outcome of innovation incentivized by IP, IP could be seen more as an essential element.[§§] This notion is, at the same time, challenged by the OECD

---

[*] Ruse-Khan, HG, The concept of sustainable development in international IP law: New approaches from EU economic partnership agreements? In Kur, A and Vytautas, M (eds.) *The Structure of Intellectual Property Law : Can One Size Fit All?* Cheltenham, Edward Elgar Pub, 2011.

[†] Idem.

[‡] Marrakesh Agreement Establishing the World Trade Organization, Geneva, 15 April 1994, entered into force on 1 January 1995, 1867 United Nations Treaty Series 3, Annex 1C—Agreement on Trade-Related Aspects of Intellectual Property Rights, 1994, "Art. 7: The protection and enforcement of intellectual property rights should contribute to the promotion of technological innovation and to the transfer and dissemination of technology, to the mutual advantage of producers and users of technological knowledge and in a manner conducive to social and economic welfare, and to a balance of rights and obligations".

[§] Idem Art. 8(1): "Members may, in formulating or amending their laws and regulations, adopt measures necessary to protect public health and nutrition, and to promote the public interest in sectors of vital importance to their socio-economic and technological development, provided that such measures are consistent with the provisions of this Agreement.(2). Appropriate measures, provided that they are consistent with the provisions of this Agreement, may be needed to prevent the abuse of intellectual property rights by right holders or the resort to practices which unreasonably restrain trade or adversely affect the international transfer of technology".

[¶] Cadillo Chandler, DM, Achieving sustainable drug development through CSR: Possibility or utopia. In Keswani, C and Singh, SB (eds.) *Bioeconomy for Sustainable Development*, Springer-Nature, Singapore, 2019.

[**] EXPH—Expert Panel on Effective Ways to Investing in Health, *Opinion on Innovative Payment Models for High-Cost Innovative Medicines*. Publications Office of the European Union, Luxembourg, 2018.

[††] Capp, DA, A Propiedade Intelectual Na Constitução. In Fabris, S (ed.) *Límites Jurídicos da Regulação e Defensa da Concorrência* (p. 52), Porto Alegre, 2003.

[‡‡] Id ut supra Chon 2019 at 768.

[§§] Idem at 764.

report acknowledging IP to have a substantial but not determinative role in shaping the bioeconomy,[*] which is seen from the knowledge perspective. The report highlighted other factors likely to have a greater impact on the dissemination of knowledge or the way the knowledge-based economy could be handled. Namely, the general investment climate, public security concerns, regulatory environments and politics, among others.[†] Similarly, scholars have also questioned whether IPRs hinder or further the dissemination of knowledge necessary to contribute to sustainable development.[‡]

As evidence shows, this is not a simple yes or no answer. The level of dissemination furthering innovations varies, and it will be interesting to measure in a few years whether COVID-19 will have either changed the rules for dissemination and openness ensuring sustainable development in the knowledge-based economy, or if instead it will be an industry-based shift from exclusivity to inclusiveness and more openness. Additionally, firms have used reductions in licensing fees to compel developing countries to "choose advanced technologies and to prefer an IPR protection regime",[§] and in doing so the trend leans towards these countries becoming part of the knowledge-based economy. The counterargument for the previous result addresses the need to approach the legal construction of the IPR regime on the basis of its impact on both the utilization of knowledge and the pace of innovation.[¶] In the case of genetically modified organisms (GMOs), Monsanto and Syngenta are two of the most well-known companies fiercely protecting their IPRs, and even when there is nothing illegal about it, scholars have compared this contentious behaviour with that of Cambia's in reporting during 2005 their findings in a similar research to that carried out by Monsanto. The former chose to report and protect their findings through "open source", which would ensure free use of the knowledge provided subsequent improvements would also be made available.[**] This model or approach is highlighted as potentially enhancing sustainability in agriculture, nonetheless, a similar suggestion has been put forward in the context of an inclusive patent system.[††]

Regardless of the model used, innovation is at the heart of the knowledge-based economy and the SDGs themselves. The World Intellectual Property Organization (WIPO) is said to have embraced an agenda for international development by adopting the UN's Millennium Development Goals.[‡‡] As one of the main organizations in the IPR system, the WIPO is a specialized UN agency that works together with governments, businesses and individuals to create a balanced and effective intellectual property system that gives innovation and creativity value. In this regard, the WIPO's director general highlighted that IP as a policy exists not only to prompt investment, but also to promote the creation of an environment leading to innovation that can be traded and shared across borders, while recognizing the fundamental role innovation has in "the process of societal transformation that the Sustainable Development Goals aim to achieve".[§§] To understand further the link between IPRs and SDGs, and the importance of these to the bioeconomy, the next section will briefly assess the objectives and targets in SDG9 as motors paving the way to the current EU policy roadmap.

---

[*] Gold, ER, Herder, M, Trommetter, M, OECD, *The Role of Biotechnology Intellectual Property Rights in the Bioeconomy of 2030*. 2007. www.oecd.org/dataoecd/11/58/40925999.pdf.

[†] Idem.

[‡] Claude, H and Stiglitz, J, Intellectual property, dissemination of innovation and sustainable development, 1 *Global Policy* 3 (2010) 238–239.

[§] Uchida, H, The big push to a knowledge-based economy with intellectual property rights protection, *Review of Development Economics* (2020) 1–9.

[¶] Id ut supra OECD.

[**] Id ut supra Claude and Stiglitz.

[††] See further, Van Overwalle, G, Inventing inclusive patents. From old to new open innovation (December 26, 2014). In Drahos, P, Ghidini, G and Ullrich, H (eds.) *Kritika: Essays on Intellectual Property* (pp. 206–277), vol. 1, Edward Elgar, 2015.

[‡‡] See further WHO, Health in 2015: From MDGs, Millennium Development Goals to SDGs, Sustainable Development Goals (2015). www.who.int/gho/publications/mdgs-sdgs/en/.

[§§] Director General Francis Gurry in WIPO Publication, Innovation driving human progress WIPO and the Sustainable Development Goals (2019).

4.3.2 SDG9 and IPRs

Although several of the SDGs relate to or could be achieved via IPRs, the WIPO seems to be closer to SDG9 through the promotion of innovation. Nevertheless, the focus on this SDG has been criticized to an extent, since the WIPO's current approach towards its own development agenda seems to fall short of the more substantial and wholesome vision it could achieve were this agenda broader by encompassing health, social and cultural well-being, and equity.*

SDG9 addresses the link between IP and development, and is titled "SDG9: Build resilient infrastructure, promote sustainable industrialization and foster innovation"; its goals are† investments in infrastructure (transport, irrigation, energy and information and communication technology); inclusive and sustainable industrial development; and technological progress. These three goals are crucial to achieving sustainable development, because they are conducive to economic and social growth, productivity and, in general, education and health improvements, all of which are largely dependent on investment in infrastructure. Additionally, industrial development has been regarded as the primary source of income generation, allowing at the same time for increasing living standards. Without a doubt, technological progress is at the core of innovation leading to achieving environmental objectives and industrialization necessary for development to take place. The goals in SDG9 are accompanied by a series of targets to be fulfilled:

9.1. "Develop quality, reliable, sustainable and resilient infrastructure, including regional and trans-border infrastructure, to support economic development and human well-being, with a focus on affordable and equitable access for all

9.2. Promote inclusive and sustainable industrialization and, by 2030, significantly raise industry's share of employment and gross domestic product, in line with national circumstances, and double its share in least developed countries

9.3. Increase the access of small-scale industrial and other enterprises, in particular in developing countries, to financial services, including affordable credit, and their integration into value chains and markets

9.4. By 2030, upgrade infrastructure and retrofit industries to make them sustainable, with increased resource-use efficiency and greater adoption of clean and environmentally sound technologies and industrial processes, with all countries taking action in accordance with their respective capabilities

9.5. Enhance scientific research, upgrade the technological capabilities of industrial sectors in all countries, in particular developing countries, including, by 2030, encouraging innovation and substantially increasing the number of research and development workers per 1 million people and public and private research and development spending

9.6. Facilitate sustainable and resilient infrastructure development in developing countries through enhanced financial, technological and technical support to African countries, least developed countries, landlocked developing countries and small island developing states

9.7. Support domestic technology development, research and innovation in developing countries, including by ensuring a conducive policy environment for, inter alia, industrial diversification and value addition to commodities

9.8. Significantly increase access to information and communications technology and strive to provide universal and affordable access to the Internet in least developed countries by 2020"‡

---

* Id ut supra Sara.

† See further UN, Goal 9: Build resilient infrastructure, promote sustainable industrialization and foster innovation. www.un.org/sustainabledevelopment/infrastructure-industrialization/.

‡ Idem.

4.3.3 The EU Roadmap

Knowledge-based economy in the context of sustainable development embodies "the vision of achieving economic growth through high-technology industries, which requires investments in innovation and highly skilled labour",*,† and as such pharma and biotechnology industries are prime motors in the field. The 2000s EU Innovation Policy framework seems to integrate the aforementioned conceptualization that in fact trengthens the links to IPRs.

Sustainable development plays an important role in EU policymaking and bilateral negotiations, which is why SDGs are pursued not only internally, but also through international agreements. For instance, EU Economic Partnership Agreements and, in particular, the EC-CARIFORUM EPA (CEPA)* have been identified as trade partnerships for sustainable development.† In this regard, CEPA includes provisions combining the importance of innovation and sustainable development conducive to what are called the three principles of sustainability: the principle of integration and reconciliation of economic, social and environmental aspects. Even when CEPA has been considered a template for future economic partnership agreements seeking to include sustainable development as a policy goal, there are questions in terms of its implementation and the possibility of it being enforced. Thus, soft law tools may be necessary to achieve such a goal.‡ This is particularly relevant when considering that sustainable innovation requires distinct but interrelated pillars of the environment, the economy and society to ensure that the sustainability of bio-based products exists throughout their entire life cycle.§

Sustainability in the EU has also been present at the regional level. In 2012, the European Commission adopted the Strategy for Innovation for Sustainable Growth: A Bioeconomy for Europe (the Strategy).¶ With this, the EU acknowledges the challenges that a growing population brings and its impact on the environment. The policies or measures in the Strategy emphasize the role that a strong bioeconomy has in the sustainable use of biological resources allowing to produce more for less, limiting the negative impacts on the environment and reducing dependency on fossil resources with the goal of turning Europe into a post-petroleum society.**Additionally, in 2012, the EU's bioeconomy sectors were reported to be worth €2 trillion in annual turnover, roughly representing about 22 million jobs in the single market, hence the importance of developing both coherent policies and allocating sufficient funding to them. Already at that time, the Bioeconomy Strategy Horizon 2020 foresaw an increase in jobs creation of about 130,000 million and about €45 billion in added value by 2025.†† In this regard, the EU budget for Europe 2020 estimates €87.7 billion funding associated with the EU Framework Programme for Research and Innovation "Horizon 2020".‡‡ Research and innovation are key aspects addressed within

* Economic Partnership Agreement between the CARIFORUM states, on the one part, and the European Community and its member states, on the other part (CEPA) signed in Bridgetown (Barbados) on 15 October 2008, OJ 2008 L. CEPA is emphatic in sustainable development being the primary objective of the agreement. See, e.g. Art. 1 CEPA "The objectives of this Agreement are: 1 (a) Contributing to the reduction and eventual eradication of poverty through the establishment of a trade partnership consistent with the objective of sustainable development, the Millennium Development Goals and the Cotonou Agreement;" and Art. 3 (1)" Sustainable development 1. The Parties reaffirm that the objective of sustainable development is to be applied and integrated at every level of their economic partnership, in fulfillment of the overarching commitments set out in Articles 1, 2 and 9 of the Cotonou Agreement, and especially the general commitment to reducing and eventually eradicating poverty in a way that is consistent with the objectives of sustainable development". Also see, Partnership Agreement between Members of the African, Caribbean and Pacific Group of States and the European Community and its Member States (Cotonou Agreement), OJ 2000 L 317/3, amended OJ 2005 L 287/1.

† Id ut supra Russe-Khan at 315.

‡ See further Simon, A, *Corporate Social Responsibility and Biotechnology: Identifying Social Aspects for European Biotechnology Companies*. International Institute for Industrial Environmental Economics, Lund, September 2020.

§ Falcone, PM, González García, S, Imbert, E, et al. Transitioning towards the bio-economy: Assessing the social dimension through a stakeholder lens, 26 *Corporate Social Responsibility and Environmental Management* (2019) 1135–1153. https://doi.org/10.1002/csr.1791

¶ Id ut supra European Commission.

** Idem.

†† Directorate-General for Research and Innovation (European Commission), Commission Staff Working Document. See document accompanying the Communication on Innovation for Sustainable Growth: A Bioeconomy for Europe. https://op.europa.eu/en/publication-detail/-/publication/1f0d8515-8dc0-4435-ba53-9570e47dbd51.

‡‡ European Commission, Council Decision establishing the Specific Programme Implementing Horizon 2020—The Framework Programme for Research and Innovation (2014–2020), Brussels, COM (2011)811/3, 2011.

the aforementioned programme due to their relation to the bioeconomy or the knowledge-based economy. The goal is to create further synergies across sectors to improve or strengthen national and regional research and innovation that translates into increased capacity and skills needed to achieve SDGs. Thus, investments in research, innovation and skills; reinforcement of policy interaction and stakeholders' engagement; and the enhancement of markets and competitiveness in bioeconomy sectors are the three main areas covered by the EU Strategy.

The European roadmap to building a strong bioeconomy has also required a plan of action to achieve its objectives, in which a set of actions and sub-actions include timeframes, the funding models used and the actors involved. For instance, Action N1 seeks to ensure substantial EU and national funding as well as private investment and partnering for bioeconomy research and innovation, and its sub-action A1.1 foresees an increase in EU public funding for research and innovation related to the bioeconomy, with dedicated and enhanced efforts on food security, sustainable agriculture, forestry, fisheries and aquaculture, and bio-based industries.[*] At the EU's national level, the Finnish Bioeconomy Strategy[†] sets similar goals to those of the EU; however, in this context the emphasis is on having sustainable bioeconomy solutions as a base for the well-being and competitiveness of a country. Thereafter the country, among other actions, foresees developing education, training and research to create competence in the field.

4.4 Conclusions

Thus far, IPRs have been a vehicle to achieve innovation since these exclusive rights portray necessary incentives to recoup investments, according with the traditional justification of the rights. Nevertheless, in light of sustainable development and the bioeconomy that require a broader and faster dissemination of knowledge leading to sustainable innovations, scholars began questioning whether far-reaching reforms should be considered in the current IPR regimes while finding alternative ways of incentivizing and financing R&D.[‡] To an extent, the EU strategy assessed above does mention the importance of innovation and IPRs to make use of resources, assigns funding to enhance R&D and also points in the direction of public–private partnerships as models harnessing partnerships. However, even when emphasizing the need to achieve further dissemination, the Strategy does not channel the way or the desired terms and conditions that would ensure such dissemination, which from a cumulative innovation point of view is crucial.

Beyond being a useful concept for policymaking, the bioeconomy or the knowledge-based economy is paramount for sustainable innovation. Thus, recasting the role of IPRs may be needed more than ever in light of the current circumstances, where innovations' exclusive rights are to an extent frowned upon. A case in point is the biotechnological collaboration for the development of a COVID-19 vaccine that has brought into the spotlight the urgency in disseminating knowledge to find a solution.

Although an essential quid pro quo in patents law is exclusivity in exchange for dissemination, there are known limits to this dissemination, for instance, data exclusivity in the EU. On the one hand, knowledge of the patented invention will be disclosed to the public, but this cannot necessarily be used to manufacture a generic product before the period of exclusivity ceases to exist.[§] And, on the other hand, biotechnological innovations from commercialization approval are awarded an additional layer of exclusivity by the health regulatory authorities when the information—knowledge—contained in the marketing authorization application cannot be disclosed or used within a determined time period, known as data exclusivity.[¶] The complexity and interdependency of these IPR and health regulatory systems, at least in the European context, have not been addressed at the policy reform level beyond the acknowledgement

* See Commission Staff Working Document. See document accompanying the Communication on Innovation for Sustainable Growth: A Bioeconomy for Europe, at 42. https://op.europa.eu/en/publication-detail/-/publication/1f0d8515-8dc0-4435-ba53-9570e47dbd51.

† See Finnish Bioeconomy Strategy, 2014. www.bioeconomy.fi/facts-and-contacts/finnish-bioeconomy-strategy/.

‡ Id ut supra Claude and Stiglitz.

§ See TRIPS preamble; and also see id ut supra Claude and Stiglitz.

¶ See Directive 2001/83/EC of the European Parliament and of the Council of 6 November 2001, on the Community Code Relating to Medicinal Products for Human Use (OJ L 311, 28.11.2001, p. 47 [Medicine Code]).

that further dissemination of knowledge is needed. This is why scholars have also pointed towards the need to recast the role of IPRs in a more suitable light to achieve sustainable innovation that is not necessarily based on exclusion but instead inclusion.* Even though this chapter did not intend to review the data exclusivity regime in the EU, its mention at the conclusion is to show that there are regimes in place governing the use of knowledge that while related are not necessarily serving a function in light of sustainability. Admittedly, there are exceptions within the IPR system that enable access to knowledge, e.g. for research purposes. Nonetheless, the data exclusivity regime erected around pharma innovations has strengthened the position of rights holders in a manner that shields them from governmental intervention and even public interest,† which from a sustainability perspective may be counterproductive. Understanding each country's industrial capacity and needs is also a key factor when developing both an industry and a regulatory framework.

BIBLIOGRAPHY

Birner, R, Bioeconomy concepts. In Lewandowski, I (ed.) *Bioeconomy Shaping the Transition to a Sustainable, Biobased Economy* (p. 20), 2018.

Cadillo Chandler, DM, Biotechnology and its development in the developing countries: Can IPRs foster innovation in the field? In Singh, HB, Jha, A and Keswani, C (eds.) *Intellectual Property Issues in Biotechnology*, CABI, Wallingford, 2016.

Cadillo Chandler, DM, Achieving sustainable drug development through CSR: possibility or utopia. In Keswani, C and Singh, SB (eds.) *Bioeconomy for Sustainable Development*, Springer-Nature, Singapore, 2019.

Cadillo Chandler, DM, Uh-Oh We are in trouble! Compulsory licenses vs. data exclusivity in the EU: one more challenge to overcome in the race to find a COVID-19 vaccine? Accepted for publication in *European Intellectual Property Review* (2020).

Capp, DA, A Propiedade Intelectual Na Constitução. In Fabris, S (ed.) *Límites Jurídicos da Regulação e Defensa da Concorrência* (p. 52), Porto Alegre, 2003.

Cartagena Protocol on Biosafety to the Convention on Biological Diversity, *Decision EM-I/3*, Conference of the Parties January, 2000.

Chon, M, Intellectual property and the development divide, 27 *Cardozo Law Review* (2006) 2821.

Chon, M, Recasting intellectual property in light of the U.N. sustainable development goals: Toward global knowledge governance, 34 *American University International Law Review* (2019) 763.

Chon, M, Charting the triple interface of public-private partnerships, global knowledge governance, and sustainable development goals, 3 *PPPS* 5–6 in (Chon, M et al. eds.) *The Cambridge Handbook On Public-Private Partnerships, Intellectual Property Governance, And Sustainable Development* (Cambridge University Press, 2018).

Claude, H and Stiglitz, J, Intellectual property, dissemination of innovation and sustainable development, 1 *Global Policy* 3 (2010) 238–239.

CLIC, *Added Value Materials and Chemicals from Cellulose—A Finnish Approach.* Strategic Research And Innovation Agenda Report, 2018.

Commission Staff Working Document, See document accompanying the Communication on Innovation for Sustainable Growth: A Bioeconomy for Europe, at 42. https://op.europa.eu/en/publication-detail/-/publication/1f0d8515-8dc0-4435-ba53-9570e47dbd51

Communication from Albania, Brazil, China, Colombia, Ecuador, European Communities, Iceland, India, Indonesia, the Kyrgyz Republic, Liechtenstein, the Former Yugoslav Republic of Macedonia, Pakistan, Peru, Sri Lanka, Switzerland, Thailand, Turkey, the ACP Group and the African Group, TN/C/W/52, July 19, 2008.

Convention on Biological Diversity, 1760 (UNTS 79); 31 ILM 818 (1992), 1992.

* See id ut supra Chon (2018).

† See further, Cadillo Chandler, DM "Uh-Oh We are in trouble! Compulsory licenses vs. data exclusivity in the EU: one more challenge to overcome in the race to find a COVID-19 vaccine?", Accepted for publication in *European Intellectual Property Review* 2020.

Devlin, A, *An Overview of Biotechnology Statistics in Selected Countries. OECD Science, Technology and Industry Working Papers.* OECD Publishing, Paris, 2003.

Directive 2001/83/EC of the European Parliament and of the Council of 6 November 2001, on the Community Code Relating to Medicinal Products for Human Use (OJ L 311, 28.11.2001, p. 47 (Medicine Code).

Director General Francis Gurry in WIPO Publication, *Innovation Driving Human Progress WIPO and the Sustainable Development Goals.* 2019.

Directorate-General for Research and Innovation (European Commission), Commission Staff Working Document. See document accompanying the Communication on Innovation for Sustainable Growth: A Bioeconomy for Europe. https://op.europa.eu/en/publication-detail/-/publication/1f0d8515-8dc0-4435-ba53-9570e47dbd5

Economic Partnership Agreement between the CARIFORUM States, of the one part, and the European Community and its Member States, of the other part (CEPA) signed in Bridgetown (Barbados) on 15 October 2008, OJ 2008 L.

European Commission, *Communication from The Commission Europe 2020, A Strategy for Smart, Sustainable and Inclusive Growth.* Brussels, 2010.

European Commission, Council Decision establishing the Specific Programme Implementing Horizon 2020: The Framework Programme for Research and Innovation (2014–2020), Brussels, COM(2011)811/3, 2011.

European Union: Council of the European Union, Charte des droits fondamentaux de l'Union Européenne (2007/C 303/01), 14 December 2007, C 303/1. www.refworld.org/docid/50ed4f602.html

EXPH—Expert Panel on Effective Ways to Investing in Health, *Opinion on Innovative Payment Models for High-Cost Innovative Medicines.* Publications Office of the European Union, Luxemburg, 2018.

Falcone, PM, González García, S, Imbert, E, et al. Transitioning towards the bio-economy: Assessing the social dimension through a stakeholder lens, 26 *Corporate Social Responsibility and Environmental Management* (2019) 1135–1153.

Finnish Bioeconomy Strategy, 2014. www.bioeconomy.fi/facts-and-contacts/finnish-bioeconomy-strategy/

Ganguly, N, Croft, S, Singh, L, Sinha, S and Balganesh, T, Biomedicine and biotechnology: Public health impact, *BioMed Research International* (2014).

Gold, ER, Herder, M, Trommetter, M, OECD, *The Role of Biotechnology Intellectual Property Rights in the Bioeconomy of 2030.* 2007. www.oecd.org/dataoecd/11/58/40925999.pdf

Hamilton, C, Intellectual property rights, the bioeconomy and the challenge of biopiracy, 4 *Genomics, Society and Policy* 3 (2008) 26–45.

Horng, D, *International Law on Biotechnology. International Law and Institutions.* Encyclopedia of Life Science Systems, 2009.

International Convention for New Plant Varieties, Act of 1961, Diplomatic Conference December 2, 1961.

Komen, J, The emerging international regulatory framework for biotechnology. *3GM Crops & Food: Biotechnology in Agriculture and the Food Chain* (2012) 78–84.

Leidwein, A, Protection of traditional knowledge associated with biological and genetic resources. General legal issues and measures already taken by the European Union and its member states in the field of agriculture and food production, 9 *The Journal of World Intellectual Property* (2006) 251–275.

Marrakesh Agreement Establishing the World Trade Organization, Geneva, 15 April 1994, entered into force 1 January 1995, 1867 United Nations Treaty Series 3, annex 1C—Agreement on Trade-Related Aspects of Intellectual Property Rights, 1994.

Morgera, E, *Faraway, So Close: A Legal Analysis of the Increasing Interactions between the Convention on Biological Diversity and Climate Change Law.* University of Edinburgh School of Law Working Paper No. 2011/05, 2011.

Papadopoulou, F, *Opening Pandora's Box: Exploring Flexibilities and Alternatives for Protecting Traditional Knowledge and Genetic Resources under the Intellectual Property Framework.* Acta Universitatis Stockhomlmiensis, Stockholm, 2014.

Partnership Agreement between Members of the African, Caribbean and Pacific Group of States and the European Community and its Member States (Cotonou Agreement), OJ 2000 L 317/3, amended OJ 2005 L 287/1.

Philippe, S, International law in the field of sustainable development: Emerging legal principles, 53 *Sustainable Development and International Law* (1994) 53 (Winfried Lang educ).

Ruse-Khan, HG, The concept of sustainable development in international IP law: New approaches from EU economic partnership agreements? In Kur, A and Vytautas, M (eds.) *The Structure of Intellectual Property Law : Can One Size Fit All?* Cheltenham, Edward Elgar Pub, 2011.

Sara, B, The World Intellectual Property Organization and the sustainable development agenda, 122 *Futures* (September 2020) 102586.

Simon, A, *Corporate Social Responsibility and Biotechnology: Identifying Social Aspects for European Biotechnology Companies.* International Institute for Industrial Environmental Economics, Lund, September 2020.

Straus, J, Intellectual property rights and bioeconomy, 12 *Journal of Intellectual Property Law & Practice* 7 (2017) 576–590.

Uchida, H, The big push to a knowledge-based economy with intellectual property rights protection, *Review of Development Economics* (2020) 1–9.

UN, *Goal 9: Build Resilient Infrastructure, Promote Sustainable Industrialization and Foster Innovation.* www.un.org/sustainabledevelopment/infrastructure-industrialization/

United Nations, Our Common Future—Report of the World Commission on Environment and Development, UN Doc. A/42/427—Annex (4 August 1987), chapter 2, I, also known as the Brundtland Report adopted in the World Commission on Environment and Development (WCED) in 1987.

Usha, S and Annadurai, C, The intellectual property rights and pharmaceutical biotechnology in India, 3 *Journal of Pharmacy Research* 10 (2010) 2466–2469.

Van Overwalle, G, Inventing inclusive patents. From old to new open innovation (December 26, 2014). In Drahos, P, Ghidini, G and Ullrich, H (eds.) *Kritika: Essays on Intellectual Property* (pp. 206–277), vol. 1, Edward Elgar, 2015.

WHO, Health in 2015: From MDGs, millennium development goals to SDGs, sustainable development goals. (2015). www.who.int/gho/publications/mdgs-sdgs/en/

Wiegele, T, *Biotechnology and International Relations.* University of Florida Press, Gainesville, 1991.

Young-Gyoo, S, Intellectual property protection of biotechnology and sustainable development in international law, 29*North Carolina Journal of International Law and Commercial Regulation* (2003) 157–248.

5

Cross-Fertilization between Human, Veterinary and Plant-Derived Vaccines: One Health and Global Patent Landscape in Vaccinology 4.0

Cristina Possas, Adelaide Maria de Souza Antunes, Alessandra Moreira de Oliveira, Flavia Maria Lins Mendes, Suzanne de Oliveira Rodrigues Schumacher and Akira Homma

CONTENTS

5.1 Introduction

Vaccines are widely recognized in the scientific literature and by leading international health organizations as one of the most powerful, low-risk and cost-effective public health interventions, contributing to prevent the spread and to eliminate many infectious diseases worldwide.

In this chapter, we examine from trans-disciplinary "One Health" and global sustainability perspectives, current trends in Vaccinology 4.0, highlighting the main advances and constraints for vaccine development against the main global threats from emerging and resurgent diseases. Our analysis was based on the classification of vaccine categories. In the patent search, we highlighted the main infection diseases with pandemic potential, such as influenza (including avian and swine), severe acute respiratory syndrome (SARS) and Middle East respiratory syndrome (MERS). An analysis is made of new developments in therapeutic vaccines for cancer, identified as a leading disease for vaccine patent deposits and a major challenge. An overview of the patent landscape for these vaccine research, development and innovation (RD&I) fields is also provided.

The growing importance and relevance of the agribusiness has been internationally recognized as a leading economic activity in some emerging economies, such as Brazil. In this scenario, the quality and availability of veterinary vaccines is a crucial component of this economic activity and a key issue for global sustainability.

The increasing international demand for food and meat-derived products and for food safety has contributed to the accelerated expansion of the veterinary vaccine market, which is highly competitive and oligopolistic. In addition, the increasing interaction between humans and domestic animals has led to an exponential increase in the pet market for vaccines and other pet products.

In this highly dynamic economic sector, the capacity for innovation and market differentiation of enterprises is key to ensuring economic competitiveness. The veterinary vaccines market is composed of a few multinational corporations, most of them related to the human vaccine market, which has contributed to reducing the high RD&I costs of vaccine development.

Scientific and technological international collaborations between the fields of human and veterinary vaccines have thus a long and historical tradition, with much common ground in vaccine development and production.

In the market scenario, this cooperation between both fields has significantly accelerated in the last decade. Nevertheless, many gaps still persist in these fields, as discussed here, particularly related to the capacity for innovation and technological development of new vaccines and immunotherapies.

Moreover, the field of plant-derived vaccines is another highly promising and expanding area, with two major products in development: plant-derived antibodies and plant-derived subunit vaccines, as will be described later in the description of the RD&I for these products and related patents.

From this perspective, this chapter presents the current vaccine patent landscape for human and animal vaccines in the last decade, with indicators such as the number of patents including both human and animal vaccines and deposits for top-priority countries according to the type of vaccine and new technological platforms.

5.2 One Health: Need for an Integrated Vaccine Perspective

Nearly 75% of human emerging infectious diseases result from zoonotic transmission (WHO, 2014) and are one of the major global health threats (Bloom and Cadarette, 2019), requiring from policymakers, scientists and technologists an increasing convergence of strategies for innovation, technological development and prevention of these diseases.

Borkenhagen et al. (2019) provide scientific evidence in this direction, in an excellent review based on diverse sources, such as case reports and sero-epidemiological, archeo-epidemiological, environmental and historical studies, considering recent trends in livestock farming.

The aggregated data from this review point to industrialized swine farming as the most likely source of future pandemic viruses, yet the authors stress that the surveillance of influenza A viruses (IAVs) on such farms is remarkably sparse. In the conclusion, this review recommends increased biosafety and biosecurity training for farm administrators and swine workers, supported by a sustainable One Health–oriented virus surveillance throughout industrialized farming and meat production lines.

IAVs are good examples of infective agents that move from swine, avian, feline, equine and canine species to infect humans. Understanding and dealing with the complexity of the scientific and eco-social issues involved in this zoonotic viral circulation and related impacts require trans-disciplinary One Health approaches, with intense collaboration between the fields of human and veterinary influenza vaccines.

There are many other examples of diseases shared between humans and animals, such as rabies, *Salmonella* infection and Q fever (*Coxiella burnetii*). The changing and increasingly complex global eco-social conditions have exponentially increased the detrimental interactions between humans, animals and the environment.

Several conditions have favoured these interactions and closer contacts between humans, wild and domestic animals, such as the disruption of environments and habitats as well as the acceleration of population mobility into new geographic areas, with travel, migration and international trade, and by climate change and deforestation.

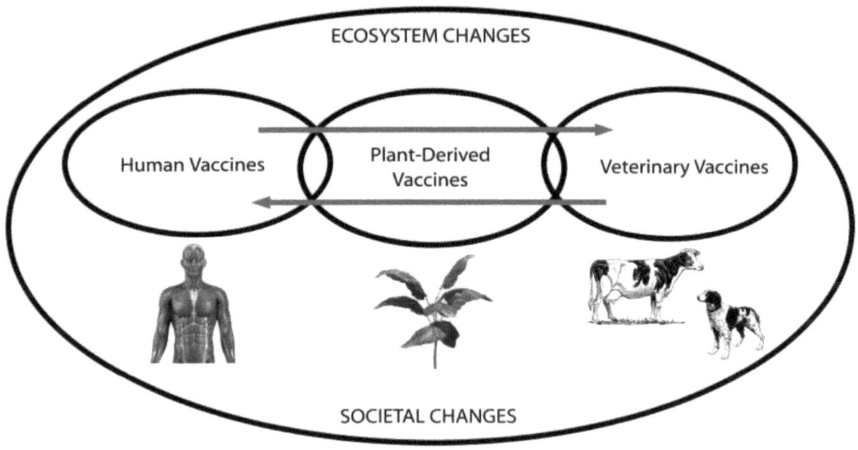

FIGURE 5.1 Vaccines for One Health: Trans-disciplinary collaboration in emerging infectious diseases.

In addition, the global demands for food safety and food security have exponentially increased, stressing the need to avoid harmful bacteria, viruses and parasites in unsafe food that can cause more than 200 diseases. As a consequence, the promising field of plant-derived vaccines has emerged and significantly increased, with several innovative products now in development worldwide.

The complexity of this global scenario requires accelerating vaccine RD&I to deal with disease emergence and resurgence and to achieve the Sustainable Development Goals of the 2030 Agenda (United Nations, 2015).

From this perspective, an integrated approach to the triad human health, animal health and the environment has emerged of particular relevance in the global "One Health" approach to vaccines, as indicated in Figure 5.1.

5.3 A New Innovation Era: Vaccinology 4.0

There are four generations of vaccine development and here we briefly describe the paradigmatic changes from a historical perspective (Plotkin, 2011).

The search for disease prevention with viral inoculation has a long history. Accounts of inoculation against smallpox ("variolation") in China can be found as early as the 10th century and was widely practiced by the 16th century, during the Ming dynasty.

Vaccinology was officially born in 1796, with Edward Jenner (1749–1823), who developed the smallpox vaccine. Vaccinology 1.0 further emerged with the rabies vaccine, which was the first human vaccine developed in 1885 in the laboratory by Louis Pasteur. Other "first-generation" vaccines are the Calmette–Guérin (BCG), plague, inactivated pertussis, tetanus and diphtheria toxoids and anthrax vaccines.

The "second generation" of vaccinology (2.0) emerged from several technological advances, with new technologies for bacterial and viral purification and tissue culture, which resulted in viral vaccines against yellow fever, inactivated and live attenuated poliomyelitis, measles, mumps, rubella, inactivated hepatitis B, polysaccharides meningococcal meningitis, Japanese encephalitis, Lyme disease and rabies in tissue culture.

The third generation of vaccinology (3.0) has emerged from innovations and technological developments in the fields of DNA recombinant, reverse technology, chemical conjugation and genomics and proteomics, such as meningitis meningococcal serogroup B, *Haemophilus influenzae* type b (Hib), pneumococcal and meningitis meningococcal conjugated, hepatitis B, rotavirus and human papillomavirus (HPV) infection.

These breakthroughs had an extraordinary impact on the development of new vaccines and resulted in most of the recent developments in complex combined multi-patented vaccines (Possas et al., 2015).

These advances resulted from the sequencing of microbial genomes and made all potential antigens of each pathogen available for the development of vaccines, thereby increasing potential target vaccines in bacteria, parasites and viruses, revealing their CD4 (+) and CD8 (+) T cell epitopes (Rappuoliet al., 2011).

Recently, significant advances in the field of immunology and immunoinformatics have been made, with the manipulation of large databases on the immune system. The Immunome project allows us to envisage for the next decades, the emergence of a new fourth generation of immunobiologicals (4.0) with vaccines and individualized immunotherapies designed according to individual genetic profiles, thereby significantly reducing their adverse events (Bragazzi et al., 2018; Poland et al., 2011; 2018).

A systems vaccinology approach has led to important breakthroughs in vaccine development, incorporating the contributions of diverse fields such as immunogenetic, immunogenomic, metagenomics, immune profiling and functional studies directed towards understanding and predicting immune responses induced by vaccine exposure (Raeven et al., 2019; Pulendran et al., 2010). These breakthroughs have benefited from advances in molecular biology, using the technologies of mRNA, DNA and new vectors.

This information is directed to engineer and test new vaccine candidates. The main objective of an individualized vaccine is to identify specific immunological profiles to be later customizable and able to provide long-term protection against a pathogen (Chen et al., 2012).

A systems biology approach to decipher novel mechanisms of vaccine response could thus allow the forecast of the immunogenicity and efficiency of candidate vaccines (Pulendran, 2010; Querec et al., 2009). The advent of high-throughput approaches is improving the understanding of the complex interactions permeating the immune system and contributing to accelerating the development of innovative vaccines, enabling the field to advance from the search for correlates of protection to the identification of protection signatures related to an immune protective response (Querec et al., 2009).

Figure 5.2 presents an integrated vaccine diagram for Vaccinology 4.0, indicating the cross-fertilization permeating the fields of human, veterinary and plant-derived vaccines.

5.4 Classification of Vaccines: Categories

Human and veterinary vaccines for farm animals and pets share many common scientific and technological bases. The vaccines currently available can be categorized as: live attenuated, inactivated, recombinant, subunit/conjugate/polysaccharide, toxoid and platform-based vaccines (genomic, vector-based and plant-derived vaccines), as indicated in Figure 5.3.

- Live attenuated: A vaccine that contains an intact but weakened (attenuated) pathogen which stimulates an immune response but does not cause clinical disease. Live vaccines protect against measles, mumps, rubella (MMR vaccine); rotavirus; chickenpox; yellow fever.

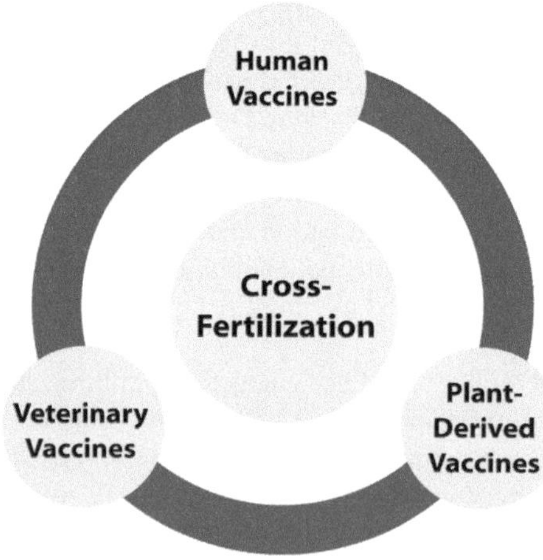

FIGURE 5.2 Vaccinology 4.0: Integrated vaccine diagram.

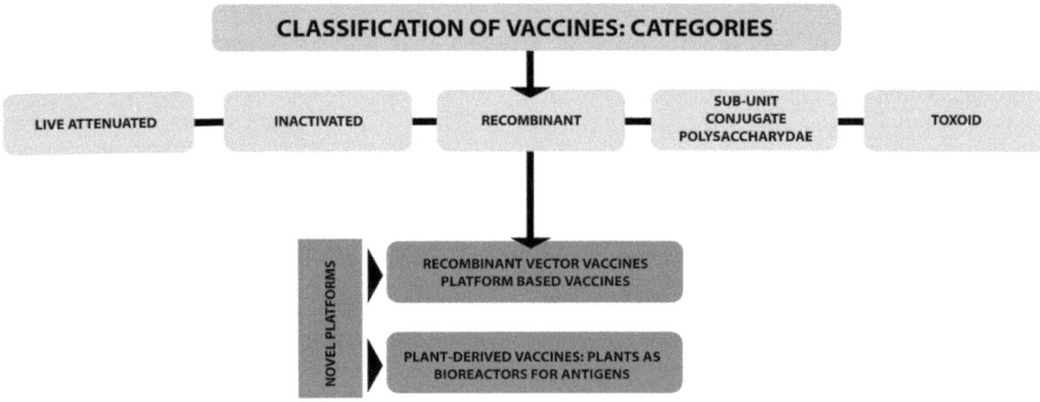

FIGURE 5.3 Classification of vaccines per categories and new vaccine platforms.

- Inactivated: A vaccine that contains a completely inactivated pathogen, which is no longer infectious. These vaccines often contain an adjuvant, which is a compound added to help improve the protective immune response. Inactivated vaccines protect against hepatitis A; influenza; polio; rabies.
- Recombinant: Vaccines that are produced using genetic engineering technology and specific genetic material from a pathogen to produce purified recombinant subunits/proteins which will stimulate an immune response with vaccination. HPV and hepatitis B vaccines are classical examples of subunit vaccines.
- Subunit/conjugate/polysaccharide: Most of the vaccines currently under investigation are based on highly purified recombinant proteins or subunits of pathogens. Conjugate vaccines are somewhat similar to recombinant vaccines as they use a combination of two different components, a weak antigen with a strong antigen as a carrier, so that the immune system has a stronger response to the weak antigen. Most commonly, the weak antigen is a polysaccharide that is attached to a strong protein antigen. These vaccines protect against influenza (injection); *H. influenzae* type b (Hib); pertussis; pneumococcal disease; meningococcal disease.
- Toxoid: A vaccine that is based on inactivated toxins produced by pathogens. These vaccines stimulate immunity and protect the animal against these toxins. Toxoid vaccines protect against diphtheria; tetanus.

Platform-based vaccines—genomic vaccines, recombinant vector vaccines and plant-derived vaccines: Vaccines that are based on innovative technological strategies. The difference between genomic and recombinant vaccines is that the immunogenic protein associated with a recombinant DNA vaccine is made in the laboratory and injected into the vaccine recipient, while the immunogenic protein associated with a DNA vaccine is generated by the cells of the host (McDonnell et al., 2019).

Recombinant DNA vaccines are based on the expression of biological constructs encoding proteins from specific viral pathogens, and are not themselves made of DNA. Instead, they are made of protein or glycoprotein subunits synthesized in the laboratory using recombinant DNA technology.

Antibodies can be induced after one vaccination with DNA, in contrast with recombinant protein vaccinations which generally require a boost. DNA immunization can be used to bias the T-helper (TH) profile of the immune response and thus the antibody isotype, which is not possible with either natural infection or recombinant protein immunization.

Finally, other novel technologies are in development, such as the use of artificial intelligence and machine learning strategies for vaccine development.

5.5 Plant-Derived Vaccines: Concept, Design and Patent Applications

A promising and exciting field of vaccine research is the production of vaccine antigens in genetically modified plants acting as bioreactors, which can then be extracted and purified by conventional methods.

Such human and veterinary plant-based vaccines have several advantages: they can be eaten or applied to mucosal surfaces, avoiding the adverse effects of traditional vaccines resulting from potential contamination with adventitious agents. They can also be produced cheaply in very high amounts through carrier plants such as potatoes, corn, rice and tobacco, and the antigens derived from them are stable and can be stored for long periods of time.

Contamination by a plant virus, which would have an adverse effect on humans, is almost negligible. However, there are several technical and regulatory challenges concerning plant-derived vaccines that must be resolved before they can enter the market (WHO, 2015; Shim et al., 2019; Van der Laan et al., 2006).

Figure 5.4 presents a list of plant-derived vaccines according to a target disease. Infectious diseases account for nearly 96% of these patent-based vaccine applications.

Plant-derived vaccines are developed and produced worldwide in molecular pharming facilities located in several countries, as indicated in Figure 5.5.

5.6 Patent Landscape: Vaccines and Immunotherapies

Our search for global patent applications for vaccines and immunotherapies for the 2010–2019 period resulted in 11,936 patent deposits (4,533 for human vaccines, 5,402 for zoonotic vaccines and 2,001 for veterinary vaccines) from a broad range of applicants in different countries, as indicated in Figure 5.6.

5.6.1 Human Vaccines

In order to initially select exclusively human vaccines, we excluded the terms associated with zoonosis and related etiologic agents from this first step of our search, according to the following sources:

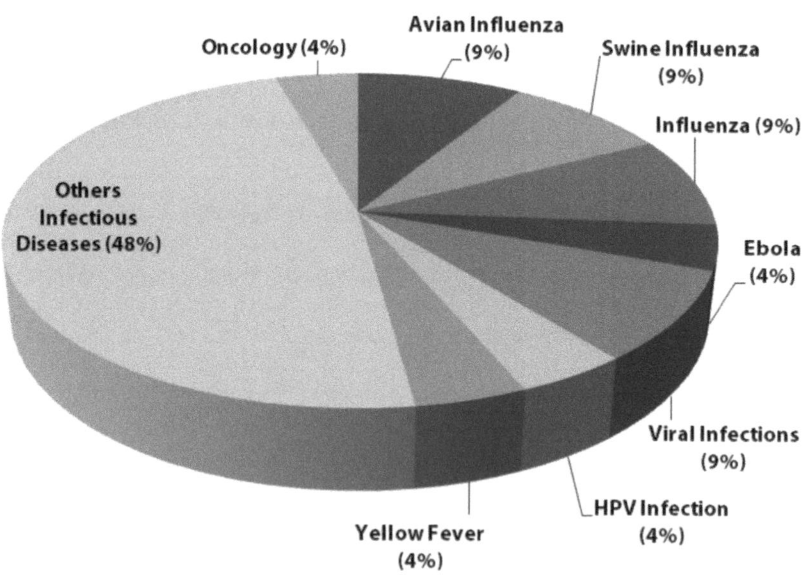

FIGURE 5.4 Plant-derived vaccines in development for target emergent and resurgent diseases. (Source: Elaborated by the authors based on Medtrack® database 2019.)

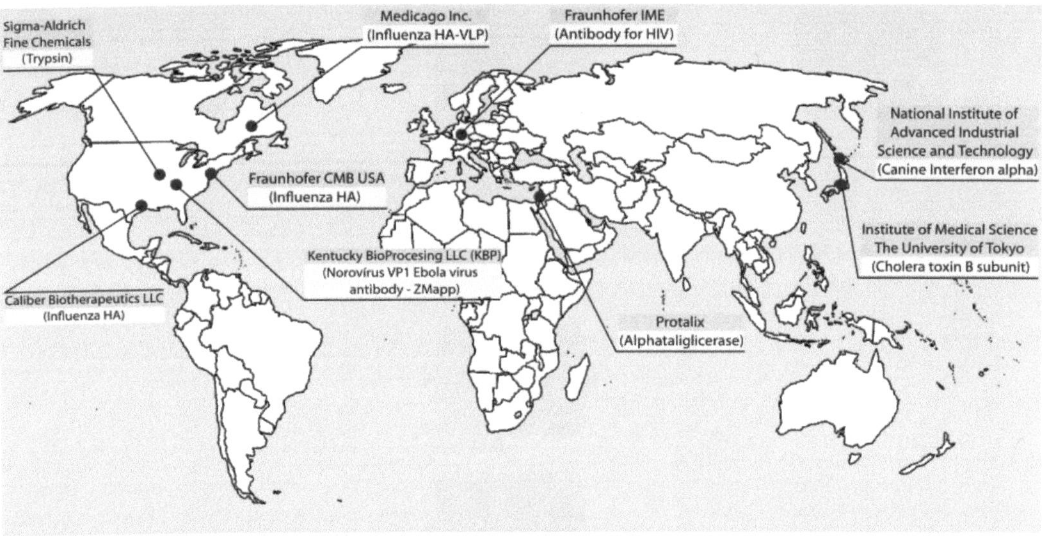

FIGURE 5.5 Molecular farming for vaccines and other biopharmaceuticals: GMP facilities worldwide. (Source: Elaborated by the authors based on Takeyama et al. 2015.)

Oxford Textbook of Zoonoses Biology, Clinical Practice, and Public Health Control—www.nhbs.com/oxford-textbook-of-zoonoses-book (Palmer et al., 2011)

Animal Disease Information: The Center for Food Security & Public Health—www.cfsph.iastate.edu/DiseaseInfo/?transmission[]=006&lang=en

From this approach, names such as "dog, calf, rabbit, duck, bird, piglet, bovine, equine, avian, canine, feline, porcine, hamster, chimpanzee, poultry, animal, non-human, cow, goose, monkey, etc." were excluded.

As a result, we obtained 4,533 patent filings from the Derwent Innovation Index database in the period 2010–2019. This search strategy for human vaccines identified cancer as the main disease in patent documents' abstracts, with 1,478 patent filings. This is probably due to the fact that it is the second leading cause of death in the world (9.6 million deaths in 2018) (WHO, 2018). Table 5.1 illustrates this scenario of great dynamism and promising breakthroughs in cancer therapeutic vaccine research and development.

This result is far from the second and third leading diseases like HIV (227 patent filings) and hepatitis B (148). The described strategy demonstrates that the main priority countries and regions are the United States (with 1,709 patent filings), followed by China (1,589), the European Patent Office (EPO) (336) and Japan (216).

Leading patent applicants: In this patent search strategy, we identified the leading patent applicants with more than 60 patent filings: GlaxoSmithKline Biologicals SA (78), 26 of them from their partnership with Novartis; the US Department of Health & Human Services (63); and Chinese Acad Medica (61).

The development of therapeutic cancer vaccines is a field in which new adjuvants and new immunization schedules will be tested. Therefore, it is important to highlight that this knowledge will be pertinent for other vaccines, providing biotechnology with indications of new molecules that need to be manufactured (Lage, 2006), which reinforces our cross-fertilization approach to vaccine RD&I in this search.

Novel technologies in patent applications: The following examples evidence the notable use of nanotechnology and polymer technology in vaccine development. Surface guanidine-modified nano-adjuvant material is claimed to form a nano-immunizing vaccine and provides good stability and strong immunogenicity (CN106729700); an immunogenic composition which comprises a polymer linked to an adjuvant prodrug and an expression vector to treat a tumour in a subject (EP3347047); and, finally, another immunogenic composition constituted by polymer nanoparticles comprising "first polymer linked to

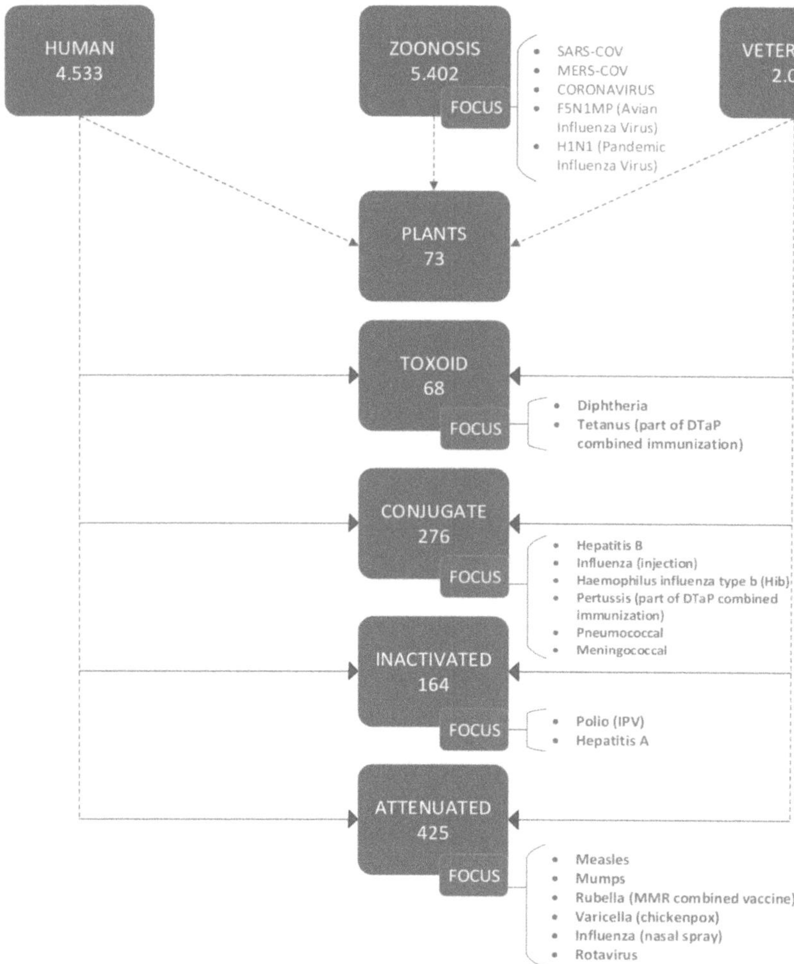

FIGURE 5.6 Number and nature of patent applications: Search for 2010–2019.

conjugate by electrostatic interaction comprising peptide antigen linked to a peptide tag and adjuvant" (EP3419655). These results demonstrate important immunogenic compositions which are useful for immunization as a booster immunization, as observed in the "Immunogenic composition comprises one or more antigens and an adjuvant, which is a toll-like receptor (TLR)-5 agonist" (EP2780038); or the heterologous prime-boost immunization method for the induction of an immune response in humans to a foreign antigen and measles virus and to treat HIV infection (EP2780034).

Genomic and recombinant vaccines (vaccines made from DNA or RNA) are another breakthrough approach (Sette and Rappuoli, 2010; McDonnell et al., 2019). Patent documents for recombinant and genomic vaccines were found for this new product for inducing an immune response, like the "new self-replicating RNA molecule comprising polynucleotide comprising respective nucleotide sequences encoding respective nucleotides that are operably linked to respective sub-genomic promoters" (EP2768530).

Other novel strategies were also identified, such as recombinant vaccines for respiratory syncytial virus (RSV) which was attenuated by modifications to a virus genome for the promotion of an immune response to RSV in a human (EP3387120); a new recombinant polynucleotide encoding the RSV amino acid sequence of a parent RSV for the production of a vaccine (US9957486); and, moreover, a new recombinant one (virus, e.g. RSVcps-1 comprising a large polymerase protein, phosphoprotein, nucleocapsid protein and M2-1 protein) for the production of an immunological response to a viral protein (EP2836231).

TABLE 5.1

Examples of Human Vaccines Patent Documents in Cancer Research

Applicant	Vaccine Formulation	An Indication	Patent Number	Priority Year
US Dept Health & Human Services	Beta-mannosyl ceramide which comprises a sphingosine moiety and a fatty acid moiety having aliphatic hydrocarbon group	Induction of an immune response and treatment of cancer (e.g. melanoma, skin cancer and lung cancer)	US2013039886	2010
US Dept Health & Human Services	Recombinant Listeria strain which comprises a recombinant nucleic acid	Method for eliciting an antitumor T cell response for a subject who has tumour or cancer	US2017080064	2014
GlaxoSmithKline Biologicals SA	Arginine, a counterion, a first immunogenic molecule (Tn group), and a second immunogenic molecule (oligonucleotide)	Treatment of cancer and induction of immunogenic response	BE1022254	2014
Chinese Acad Medica	Composition: e.g. mannose-modified ethylene glycol chitosan as carrier material	Liver cancer immunotherapy	CN108785666	2018
Chinese Acad Medica	Nano graphene oxide-aluminium hydroxide/antigen complex	Prevention of cancer	CN109010810	2018

Source: Elaborated by the authors. Data from the Derwent Index Innovation database.

Recombinant vaccines are also useful to other viruses, such as the new recombinant gL protein of cytomegalovirus (CMV) which is effective in inducing an immune response against CMV (BE1023087) or a new recombinant CMV glycoprotein B with the specific mutation of a hydrophobic surface for inducing an immune response in mammals (BE1024516). They are also useful to bacteria, such as the vaccine that comprises a genetic engineering component: "group B meningococcal factor H-binding protein containing genetic engineering expression" (CN102028941).

5.6.2 Vaccines for Zoonoses

Emerging and re-emerging diseases transmitted from animals to humans (zoonoses) can be controlled and prevented by vaccination and innovative RD&I strategies that are now in development in this field.

A search of the patent documents for these vaccines during 2010–2019 was carried out in the Derwent Innovation Index database using the zoonoses from the following two sources:

Oxford Textbook of Zoonoses Biology, Clinical Practice, and Public Health Control—www.nhbs.com/oxford-textbook-of-zoonoses-book (Palmer et al., 2011)

Animal Disease Information: The Center for Food Security & Public Health—www.cfsph.iastate.edu/DiseaseInfo/?transmission[]=006&lang=en.

The search recovered 5,402 patent filings from China, the United States and the EPO, the leading priority countries and region with 2,419, 1,494 and 327 patent filings, respectively. Our results also demonstrated that the leading applicants were the Chinese Academy of Agricultural Sciences (with 184 patent filings), GlaxoSmithKline Biologicals AS (109) and Pulike Biological Eng Inc (89). The analysis also provided new evidence: the leading zoonoses in the number of patent filings were avian influenza (477), influenza A (363), Newcastle disease (205) and foot-and-mouth disease (189).

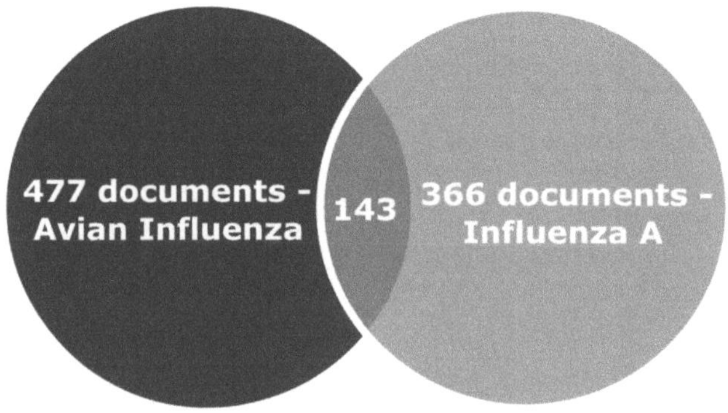

FIGURE 5.7 Patent deposits for avian influenza and/or influenza A.

In this step of our search, we concentrated on the main disease, i.e. influenza, a major global threat (Kile et al., 2017). The search for patent documents on vaccines targeting influenza A and avian influenza was based on keywords related to the two types of influenza associated with the Code Derwent Manual (B14-S11 vaccine [general]), a specific classification of the Derwent Innovation Index database that makes it possible to identify with greater accuracy the technological focus and the International Patent Classification A61K39—medicinal preparations containing antigens or antibodies.

After validating the results and eliminating repeated documents from the number and date of the priority document, 983 patent documents were recorded. Of these, 477 refer only to avian influenza, 363 only to influenza A and 143 to both diseases (Figure 5.7).

According to this strategy, we analyzed, from this general vaccine patent classification, for each type of influenza (influenza A and avian flu), the leading patent applicants, year, country of priority, year and object of patent protection.

5.6.2.1 Influenza A Vaccine

The 506 patent documents related to influenza A vaccine have a diverse focus on protection, such as a new recombinant virus for use in vaccines, preparation of adjuvants, delivery methods, new fusion proteins, in addition to immunogenic compositions and new vaccine compositions.

From our analysis of the applicants, no company or research institution with great relevance in number of deposits was identified. There was also no emphasis on partnerships. The only applicant with more than 10 deposits was Novartis, with 12 deposits, 9 of which were in partnership with Seqirus GmbH, one of the world's leading companies in the production of influenza vaccines.

On average, 50 patent documents were filed per year over the period analyzed. It is noteworthy that 2010 and 2011 had the highest number of deposited documents, 81 and 70, respectively, possibly as a result of the intensification of research that took place in 2009 after the world pandemic that occurred that year.

More than 70% of the technologies related to influenza A vaccine production were developed in the United States, China and South Korea. Most of the 196 American documents focus on new vaccine compositions or immunogenic compositions. As an example, document US2017304429 seeks the protection of a

> method for enhancing an immune response in a subject comprising administering a vaccine, wherein the vaccine comprises an adjuvant and a high-dose antigen. The vaccine is an influenza vaccine, preferably multivalent influenza vaccine comprises preferably 60 mu antigen/strain and an H1N1 strain, an H3N2 strain, a B strain, or their combination.

Most of the 116 Chinese documents are aligned with applied biotechnology and seek to protect new recombinant viruses for use in vaccines. This is the case with Fudan University's document CN101974489, which claims a new recombinant influenza type A H1N1 virus that is useful for preparing a diagnostic reagent and vaccine specific to influenza type A H1N1 virus. A deposit from Shandong University of Traditional Chinese Medicine, CN110305854, claims the recombinant H3N2 subtype influenza virus carrying luciferase comprising two groups of genes.

The same Chinese pattern in patent protection is observed in the 52 South Korean documents. However, these last documents refer to the protection of new fusion solutions such as document KR101525180, which describes a new fusion protein useful to prevent and/or treat viral influenza infection, comprising the antigenic protein from influenza virus matrix protein 2 and hemagglutinin with cholera toxin A1 subunit.

5.6.2.2 ***Avian Influenza Vaccine***

The result of the search for vaccines targeting avian influenza shows that this type of technology is centred on Chinese territory. Of the 620 documents selected, 405 (65%) have China as the country of first deposit, followed by the United States with only 14%. The European territory, considering all countries and the European office itself, does not respond to even 10% of the priority countries in this type of technology.

With the exception of 2019, mainly due to the document's confidentiality period, more than 50 deposits were observed per year from 2010 to 2019, and from 2013 to 2016 this number surpassed the mark of 70 documents per year. The increase in deposits in the mentioned period may be related to the increase in research due to the worsening of avian influenza in people. According to Kile et al. (2017), from 31 March 2013 to 7 August 2017, a total of 1,557 human infections with avian influenza were reported; at least 605 (39%) of these infections resulted in death.

As mentioned, most technologies related to avian influenza vaccines are being developed in China, so the main depositors are universities, companies and research centres based in Chinese territory.

Yangzhou University and South China Agricultural University stand out with 21 deposits. Considering the amount of documents deposited, three companies that focused on the research and development of biological veterinary products, virus vaccines, toxoids, diagnostic kits and other products stand out: Pulike Biological Engineering, Qingdao Yebio Bioengineering and Zhaoqing DaHuaNong Biology Medicine.

It should be noted that Zhaoqing DaHuaNong Biology Medicine is the appointed enterprise by the Ministry of Agriculture of China (MOA) for the manufacturing and exporting of the avian influenza vaccines.

Considering the other applicants, Sanofi Pasteur, the vaccine division of the pharmaceutical multinational Sanofi, and the University of Georgia Research Foundation are the largest patent holders outside of Chinese territory with six documents each.

In the titles of the patent applications in this search, it is possible to notice a diverse range of protection products aimed at the production of vaccines for avian influenza, such as new adjuvants, the preparation of immunogenic compositions, new fusion proteins and polypeptides, in addition to the preparation of vaccines.

However, the most important result to highlight is related to gene vaccines, with protection for new genes, recombinant viruses, recombinant DNA and recombinant strains. Table 5.2 presents some examples of these deposits.

Considering the importance of human outbreaks caused by specific zoonotic viruses, new strategy searches were performed for specific zoonoses: MERS and SARS. For MERS research, the search identified 21 patent applications. The priority countries were China (9 patent filings), the United States (5), South Korea and the UK (2 each), Holland, Japan and the EPO (1 each). The leading patent applicant with six patent filings (all filed in China) was the Institute for Viral Disease Control and Prevention, as indicated in Figure 5.8. The strategy also provided evidence of the recovery documents which were filed in countries other than their priority country (Table 5.3). In a SARS strategy search, few patent filings were obtained (10), seven from China. The leading applicant was Tefei Tianjin Biomedical Technology Co with three patent documents. All of them were vaccines comprising, respectively: (1) N-antigen of SARS virus (CN104292338); (2) SARS virus including the receptor-binding domain (RBD)

TABLE 5.2

Patent Deposits for Recombinant and Genomic Avian Influenza Vaccines

Applicant	Title	Patent Number	Priority Year
Harbin Veterinary Res Inst Caas	New recombinant avian influenza virus type H1N1 inactivated vaccine strain useful for preventing animal or human influenza prepared by gene recombination of A/swine/Jiangsu/40/2011(H1N1) with A/PR 8/34 (H1N1)	CN104232594	2014
YanglingLvfang Biological Eng Co Ltd	New H9 subtype avian influenza virus SY strain useful for preparing vaccine for preventing H9 subtype avian influenza virus and for preparing diagnostic reagent and therapeutic agent	CN104928259	2015
Hubei Acad Agric Sci Animal Husbandry	Recombinant Newcastle disease virus strain used in the preparation of recombinant virus, and H5 avian influenza and Newcastle disease heat-resisting live vaccine, named as rTS-hemagglutinin-1/H5	CN105950572	2016
Harbin Veterinary Res Inst Caas	New gene encoding hemagglutinin protein of avian influenza virus useful for preparing avian influenza virus DNA vaccine, subunit vaccine or genetically engineering vaccine, and monoclonal antibody, has specific base pair sequence	CN107630024	2017
China Animal Health & Epidemiology Centre	New H7N9 subtype avian influenza virus HA gene mutant useful for constructing recombinant plasmid of H7N9 subtype avian influenza virus HA gene mutant	CN108913701	2018

Source: Elaborated by the authors. Data from the Derwent Index Innovation database.

antigen recombinant protein (CN104292339); and (3) SARS bivalent antigen recombinant baculovirus (CN104293740).

The Public Health Emergency of International Concern determined by the World Health Organization (WHO) is now again a reality because of the documented cases of COVID-19 in China followed by other countries. For this reason, we believe that it is of most relevance to understand the scenario of vaccines' patent documents for the treatment/prevention of COVID-19. Terms such as Coronaviridae or Coronavirinae or Coronavirus or Coronaviruses or HCOV or nCoV were used for the strategy search in the Derwent Innovation Index database. Ten documents were recovered during 2010–2019. China and the United States were the leading priority countries in number of documents (3 each).

5.6.3 Veterinary Vaccines

In order to select the documents on veterinary vaccines in the Derwent Innovation Index database during the period 2010–2019, selected words were used in the strategy search to find the main applicants and their documents. A combined strategy with words such as pig, fish, chicken, sheep, cattle, horse, dog, calves, rabbit, duck, bird, piglets, bovine, equine, avian, canine, feline, porcine, hamster, chimpanzee, poultry, animal, non-human, cows, goose, monkey, etc. was used, and 2,001 documents were recovered, with China the main priority country (1,292 patent filings) followed by the United States (253) and South Korea (only 76). The main applicants were the Chinese Academy of Agricultural Sciences (141 patent filings) followed by Qingdao Yebio Bioengineering Co Ltd (56) and Pulike Biological Eng Inc (50). The main diseases related to the strategy were porcine circovirus (113 documents), swine fever (110) and porcine reproductive and respiratory syndrome (102). Like human vaccines, different types of vaccines are available to control infectious veterinary diseases. Inactivated vaccines were observed for different pathogens such as Seneca Valley virus (CN109679927), porcine parvovirus (CN102719407), rabbit viral haemorrhagic disease virus (CN108588034) and *Riemerella anatipestifer* (CN102908616). Live attenuated vaccines were observed for preventing canine distemper in racoon dogs (CN107115525), African swine fever (CN110093324) and foot-and-mouth disease (WO2018090994). Furthermore, a vaccine

Main Applicant

Institute for Viral Disease Control and Prevention- Chinese Center For Disease Control and Prevention (CDC) (6 patent filings)

1. CN103724406 (2013) - Synthetic peptide vaccine useful for preparing drug for preventing or treating Middle East respiratory syndrome caused by corona virus, where drug is inducing drug against hepatitis C virus antibodies in humoral immune response.

2. CN105169384 (2015) - New recombinant protein useful for expressing Baculovirus-insect cell and purifying N-terminal domain of Middle East respiratory syndrome-Coronavirus (MERS-CoV) spike protein, and preventing and treating MERS-CoV infection.

3. CN105273067 (2015) - New recombinant 41-type Adenovirus vector vaccine coding middle east respiratory syndrome coranavirus (MERS-CoV) spike protein, useful for preventing and treating MERS-CoV, and enhancing immunity.

4. CN105543248 (2015) New optimized coding Middle East respiratory syndrome-coronavirus S protein used in preparing recombinant adenovirus type 5 vector vaccine, i.e. common vaccines in preventing and treating Middle East respiratory syndrome-coronavirus.

5. CN107298712 (2016) New fully humanized anti-middle east respiratory syndrome neutralizing antibody for preparing agent, vaccine, reagent for diagnosing, treating or preventing middle east respiratory syndrome coronavirus.

6. CN106119213 (2016) MERS-CoV chimeric virus-like particles used alone or in combination with other vaccines for preventing or treating MERS-CoV infection.

FIGURE 5.8 Patent filings of the Institute for Viral Disease Control and Prevention. (Source: Elaborated by the authors. Data from the Derwent Index Innovation database.)

from inactivated toxins was recovered: "*Clostridium perfringens* epsilon toxin attenuated mutant useful for preventing and/or treating *Clostridium perfringens* disease" (CN104560780). Another key point to remember is the use of recombinant antigens in immunologic adjuvants as described in the following patent document: "recombinant plasmids pVAX-interleukin (IL)21-IL15 formed by fusing interleukins IL21 and IL15 by using pVAX1 plasmids as carriers" (CN103083660) or the use of a nucleotide sequence useful for the preparation of an immunologic adjuvant to prepare a foot-and-mouth disease virus vaccine (CN103382480). Not to mention the genetically engineered *Pichia pastoris* used to prepare a subunit vaccine (avian metapneumovirus vaccine) with the advantage that *P. pastoris* is able to express soluble protein product and can be used for the vaccine's industrial production (CN105861349). Pet vaccine is part of this important global animal vaccine market and helps to fuel its growth. For this reason, a new strategy was developed to summarize the contributions and findings made by pet vaccines (especially canine and feline ones). In total, 80 documents were found, 50 from China, and from the document analysis it was clear that the main applicant was the Chinese Academy of Agricultural Sciences with 12 patent filings, followed by Univ Qingdao Agric (5 documents) (Table 5.4). This analysis found evidence for the importance of animals' healthcare as a target for investment in RD&I.

Finally, in our patent search, we identified the main patent applicants in each field (human, veterinary and zoonotic vaccines) and also applicants operating in the intersections between these fields, as indicated in Figure 5.9. For human vaccines, the main patent applicants were the Chinese Academy of Medicine, Novartis and the University of Pennsylvania; for veterinary vaccines, the main applicants were the Chinese Academy of Fishery Sciences, Qingdao Yebio Bioengineering Co Ltd and the University of Shandong Agriculture. For zoonotic vaccines, Sanofi. In the intersection between human and zoonotic vaccines, GlaxoSmithKline Biologicals SA and the US Department of Health and Human Services. In the intersection between veterinary and zoonotic vaccines, the Chinese Academy of Agricultural Sciences and Pulike Biological Eng Inc.

TABLE 5.3

Examples of Patent Documents Filed in Other Countries Out of Their Priority Countries from Search Strategy for MERS

Priority Country	Patent Number	Patent Title	Priority Year	Applicant	Location of Second Filings
EUA	US9889194	New protein comprising a Middle East respiratory syndrome coronavirus (MERS-CoV) spike protein and an immunopotentiator, useful for inducing a protective immune response against MERS-CoV and for preventing infection with MERS-CoV	2013	New York Blood Cent Inc	European Patent Office, Israel, United States
EUA	US2016206729	Immunogenic composition useful to produce high affinity antibody for inducing immune responses against Middle East respiratory syndrome coronavirus (MERS-CoV) infection, comprises MERS-CoV nanoparticle comprising trimer of spike polypeptide	2013	Novavax Inc	Brazil, Canada, China, European Patent Office, Hong Kong, Indonesia, India, Japan, South Korea, Mexico, Russia, Singapore, United States
EUA	US10016497	Immunogenic composition useful for inducing immune response against Middle East respiratory syndrome coronavirus (MERS-CoV) for treating MERS-CoV infection, comprises nucleic acid molecule comprising specific sequences	2013	Univ Pennsylvania; Inovio Pharm Inc	Australia, Canada, China, European Patent Office, Hong Kong, India, Japan, South Korea, United States
EUA	US10301377	Inducing immune response to Middle East respiratory syndrome coronavirus (MERS-CoV) S protein in subject, comprises administering prime-boost vaccination or by administering nucleic acid molecule encoding MERS-CoV S protein to subject	2015	US Dept Health & Human Services	European Patent Office, South Korea, United States

Source: Elaborated by the authors. Data from the Derwent Index Innovation database.

5.6.4 Plant-Derived Vaccines

In order to obtain the patent results for the most promising plant-derived vaccines with regard to the concept of inventions, summaries and claims were analyzed in three patent databases, the Derwent Innovation Index database, Espacenet and SciFinder, with the same search strategy. Seventy-three priority patent applications were obtained between 2010 and 2019 (Figure 5.10).

From the results, the great versatility in the use of plant systems in the production of vaccine became evident. In light of the evidence from the analysis of patents, a better understanding was gained of how

TABLE 5.4
Examples of Pet Vaccines Patent Documents

Applicant	Immunogenic Composition	An Indication	Patent Number	Priority Year
Harbin Veterinary Res Inst Caas	Canine adenovirus I velogen strain (attenuated)	Prevention or treatment of diseases caused by type I canine adenovirus	CN101914502	2010
Chinese Academy of Agricultural Sciences	Sputum parvovirus attenuated vaccine strain HB-F69 and canine distemper virus attenuated vaccine strain CDV3-CL	Prevention of fever and parvovirus enteritis	CN108066758	2017
Chinese Academy of Agricultural Sciences	Canine distemper virus attenuated vaccine strain CDV-OR12	Prevention of canine distemper in racoon dogs	CN107115525	2017
Chinese Academy of Agricultural Sciences	Raccoon dog fever virus attenuated vaccine strain	Preparation of a medicine for prevention of canine distemper	CN107158370	2017
Chinese Academy of Agricultural Sciences	Raccoon dog parvovirus (RDPV)-HeB17 strain	Prevention and/or treatment of parvoviral enteritis	CN108690834	2018

Source: Elaborated by the authors. Data from the Derwent Index Innovation database.

the characteristics and the variety of the plants make them eligible for the production of immunobiologicals and other products, as will be shown in the following examples.

Plants such as lettuce (*Lactuca sativa*), apple, tomato and carrot were edible plant cells to use its plastid comprising a transformed plastid genome to express tuberculosis antigens (patent number US2015196627). Tobacco, soybean or pisum plants were used to produce an oligomeric protein in eukaryotic cells to use in vaccines (WO2018115305). The strawberry fruit was the tissue chosen to provide a propagation-blocking vaccine for oral administration against malaria infection (WO2018043741).

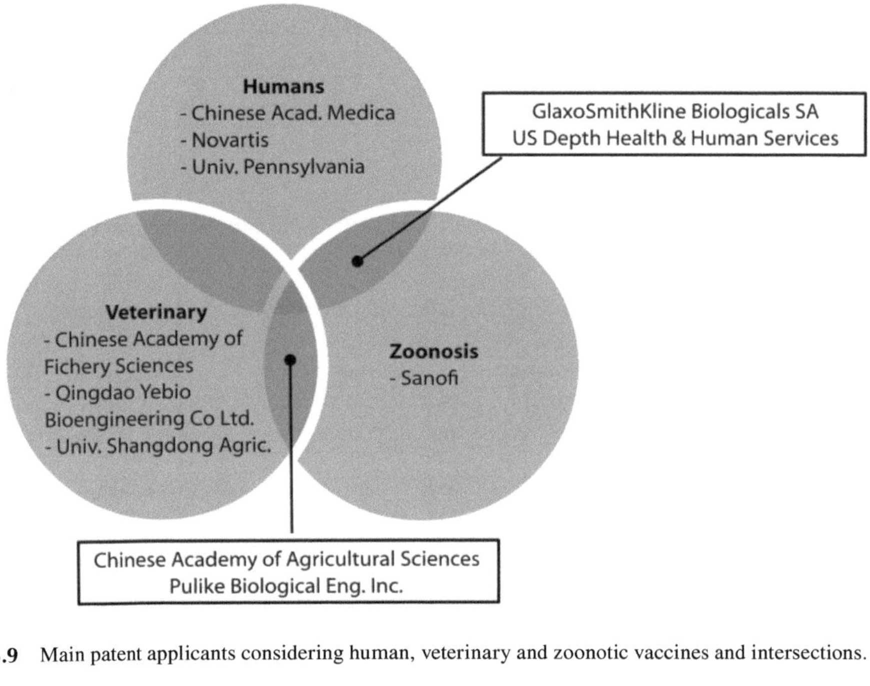

FIGURE 5.9 Main patent applicants considering human, veterinary and zoonotic vaccines and intersections.

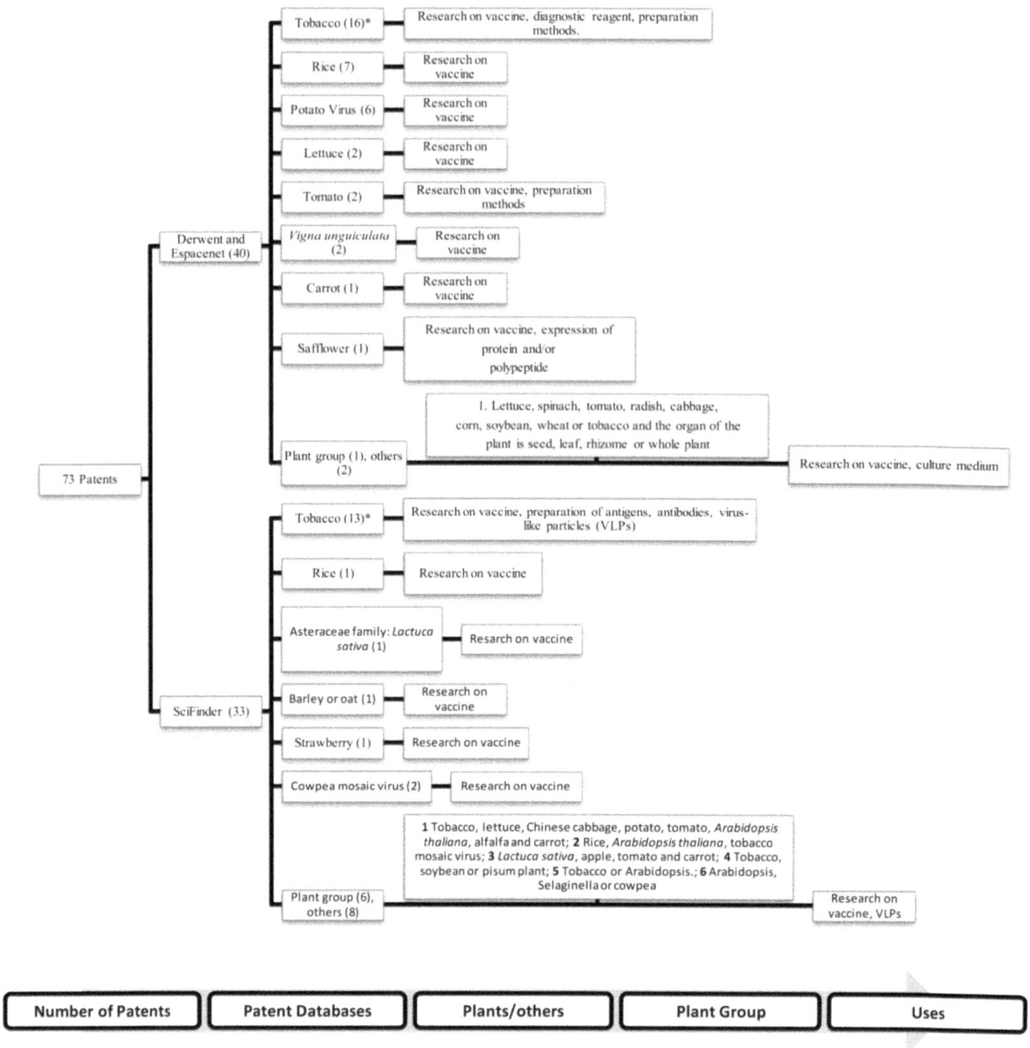

FIGURE 5.10 Patent database search of plant-derived vaccines.

Tobacco, lettuce, Chinese cabbage, potato, tomato, *Arabidopsis thaliana*, alfalfa and carrot were useful in the production of a recombinant rotavirus used to treat or prevent rotavirus (KR2013066128; KR1428843) (Scifinder).

Another key point to note is the greater number of patents filed for some species. This is very evident for tobacco (*Nicotiana benthamiana*), whose uses range from plant cultivation to vaccine production, in agriculture for the prevention and control of viruses in leaf production, for expressing a foreign gene in a plant, in the production of vaccines to protect plants, in addition to its use as a reactor for the production of vaccines (Derwent Innovation Index database & Espacenet database) (Table 5.5).

The search also highlighted the relevance of the use of rice in the production of vaccines for a range of diseases with an important impact on health systems. For the prevention and/or treatment of Alzheimer's disease, a vaccine composition was developed wherein the prolamim of rice could be used (JP2013112668). Additionally, a method was developed for the mass production of a recombinant dengue virus vaccine using a plant virus expression system (KR2017053223) (Scifinder Scholar) (Table 5.6).

TABLE 5.5
Tobacco Plant Cells–Derived Vaccines and Other Inventions

Applicant	Invention	An Indication	Patent Number	Priority Year
Tobacco Res Inst Caas	Damage-induced type promoter, and the nucleotide sequence of the promoter is SEQ ID NO: 1	Expression of a foreign gene in a plant	CN103205429	2013
Univ Jiangxi Agricultur Al	Production of a multi-epitope vaccine of hepatitis C virus	Production of the anti-hepatitis C virus vaccine	CN104353071	2014
Univ Shandong Agriculture	Method for inoculating plant viruses using agrobacterium	Vaccines to protect plants	CN104152487	2014
Korea Inst Sci & Tech. Res Inst Ind Science & Tech	Hydroponic cultivation method of tobacco plants	Cultivation of tobacco plant for the production of a pig fever antigen vaccine	KR101926384	2015
Mudanjiang Tobacco Science Res Inst; Heilongjian G Co Ltd; China Nat Tobacco Corp; Univ Shandong Agricultur Al	Preparation method of an anti-tobacco potato Y virus vaccine	Prevention and control of tobacco potato Y virus	CN105420264	2015

Source: Elaborated by the authors. Data from the Derwent Index Innovation database.

TABLE 5.6
Rice Plant Cells–Derived Vaccines and Other Inventions

Applicant	Invention	An Indication	Patent Number	Priority Year
Hokko Chem Ind Co, Univ Hirosaki	The vaccine composition with prolamin of rice	Prevention and/or treatment of Alzheimer's disease	JP2013112668	2011
Gyeonggi Do	A marker gene for identifying a transgenic rice plant with diarrhoea vaccine activity	Detection of an insertion of a gene with vaccine activity in chromosome of a rice plant	KR20130070032	2011
Ventria Bioscience Inc	Compositions and methods of making a monocot seed product expressing high levels of recombinant Osp protein	Manufacture of a Lyme disease vaccine formulation	US2016213766	2013
Industrial Cooperation Foundation Chonbuk National University, South Korea	Method for mass-producing a recombinant dengue virus vaccine using a plant virus expression system	Vaccine against dengue virus at a low production cost	KR2017053223; KR101783677	2015
Univ Arizona State	A poultry vaccine comprising an antigenic protein	Prevention of necrotic enteritis in poultry	WO2019010135 (Priority Country: United States)	2017

Source: Elaborated by the authors. Data from the Derwent Index Innovation database, Espacenet database and SciFinder Scholar.

5.6.5 Conventional Vaccines

The investigation of the main patents of conventional vaccines was based on a search strategy in the Derwent Innovation Index database using selected diseases/etiologic agents for each kind of vaccine between 2010 and 2019. For a definition of conventional vaccines, we based our search on the link: www.historyofvaccines.org/index.php/content/articles/different-types-vaccines

For toxoid vaccines, the search selected the terms diphtheria, tetanus and pertussis, with 68 priority patent applications recovered. The main priority countries were China (43%) and the United States (26%). These results also highlight the main patent applicants in the world in number of patent filings. The leading applicant was Novartis AG (CH) with 11 patent filings, 6 of them from their partnership with GlaxoSmithKline Biologicals SA (BE). The second was Chengdu Olymvax Biotechnology Co Ltd (CN) with six patent filings. Together, the main applicants primarily deposited their patent applications outside their countries of origin. GlaxoSmithKline Biologicals SA and Novartis AG's first filings were in the United States (5 patent filings) and the EPO (5). Chengdu Olymvax Biotechnology Co Ltd has deposited in China (6) and Novartis AG in Germany (5), the United States (2), the EPO (1) and the UK (1).

For conjugated vaccines analysis, 276 patent filings were identified, including combined vaccine formulations, using the terms of some diseases and etiologic agents such as hepatitis B, influenza, *H. influenza* type b, pertussis, *Streptococcus pneumoniae*, *Neisseria meningitidis* and *Human papillomavirus*. This analysis evidenced key findings: China and the Unites States were the leading priority countries of conjugate vaccines with 43 and 22%, respectively, of all patent filings. They were followed by India and South Korea with 8% each. It also indicated the leading patent applicants of conjugate vaccines: Bravovax Co Ltd (China) (with 19 patent filings), MSD Wellcome Trust Hilleman Lab Pvt Ltd (India) (10) and GlaxoSmithKline Biologicals SA (Belgium) (8). However, an additional four patent filings were found from GlaxoSmithKline Biologicals SA/Novartis AG (Switzerland). They were followed by Beijing Zhifei Green Bamboo Bio-Pharm Co (China), Chengdu Olymvax Biotechnology Co Ltd (China), Serum Inst India Ltd (India) and the US Department of Health & Human Services (United States), all with eight patent filings each.

For whole-killed vaccines, poliomyelitis, hepatitis A and rabies were the terms used in the search strategy. The search resulted in 164 patent filings between 2010 and 2019. The results indicate that China and the United States together accounted for 76% of all priorities. The top applicants with Chinese and American priorities were INST Medical Biology Chinese Acad Medica (from China, with 7 patent filings); Novartis AG (Switzerland, 6); Pulike Biological Eng Inc (China, 6); GlaxoSmithKline Biologicals SA (Belgium, 6); Chengdu Kanghua Biological Prod Co Ltd (China, 5); and Merck Sharp & Dohme Corp (United States, 5).

Finally, for live attenuated vaccines, the search used the terms measles, mumps, rubella, varicella, influenza, rotavirus, zoster (shingles) and yellow fever. The results led to the identification of 425 patents filed between 2010 and 2019. The main applicant countries were the United States (with 149 patent filings), followed by China (122), the EPO and Russia (35 each). The top patent applicants were Sanofi Pasteur (France, 16 patent filings), Pulike Biological Eng Inc (China, 15), Intervet Int Bv (Netherlands, 11) and the US Department of Health & Human Services (United States, 11). Influenza was the leading disease in claimed documents (306 patent filings), followed by measles (74), yellow fever (57), rotavirus (54), varicella (46), mumps (41), rubella (40) and zoster (38).

5.7 Technological, Market and Regulatory Constraints

The flow from vaccine initial concept to production is complex, with several steps, requiring an average of 10–15 years and US $2.6 billion to develop a new vaccine or immunobiological product (Plotkin et al., 2017). For this reason, the instruments and strategies used in the vaccine development process need to start adding value from the moment they are implemented.

In a recent study, Hong et al. (2018) discuss these constraints from a technological perspective and provide a new approach to improving control and operations procedures in biopharmaceutical manufacturing. Challenges and opportunities are described for (1) microscale technologies for continuous high-speed processing; (2) operations of modular plug-and-play units with integrated monitoring and control systems; (3) dynamic modelling of unit operations and entire biopharmaceutical plants to support the development of processes and control of the entire plant; and (4) model-based control technologies to optimize start-up, change and shutdown.

The authors point out that the main challenge is the ability to simultaneously address uncertainties, nonlinearities, time delays, non-minimum phase behaviour, restrictions, spatial distributions and continuous and discrete mixed operations that arise in biopharmaceutical operations. The design of adaptive and hybrid control strategies is also discussed.

Moreover, from the market perspective, one of the main gaps is related to the constraints that still limit cooperation between the human and veterinary vaccine fields, particularly for market reasons. There has been a long-standing difficulty in convincing human vaccine producers to enter the veterinary vaccine market. This constraint has been attributed to several reasons, such as low prices. Veterinary vaccines must be cheap as they are often paid directly by farmers (Possas et al., 2016).

Nevertheless, evidence provided here indicates that this scenario has significantly changed in the last decade, with leading pharmaceutical companies accelerating the development of innovative vaccines against several viral diseases affecting domestic and farm animals, in rapidly growing farm and pet markets.

The global acceleration of the emergence and resurgence of zoonoses affecting farm animals has also contributed to change this market scenario for vaccine development, with many farmers now convinced of the need to vaccinate and protect their animals from these new threats and pay the higher prices required for these vaccines.

Another gap is related to the need to overcome the scientific, technological and legal patent barriers constraining access to vaccines, particularly for the development of new adjuvants by developing countries' manufacturers. Vaccine adjuvants play a crucial role in boosting the immune response for vaccine protection and is a major constraint limiting the development of innovative vaccines. There is an urgent need for the development of new adjuvants and for novel strategies to minimize the potential adverse effects of adjuvanted vaccines.

Finally, another major gap is related to the need to overcome the current obstacles in the development of plant-derived vaccines. In the last three decades, some important plant-derived vaccines have been developed, but innovative technological trends in this area have not yet been consolidated. Most of them are still in the first stages of development (Phase I/Phase II) and no vaccine product has yet been commercialized. There are many reasons, related to public acceptance, access and economic feasibility and to regulatory issues related to potential adverse effects. Regulatory issues are very complex and, despite their great social relevance, are often important constraints on vaccine development (Pickering et al., 2020).

In addition, there is still relatively low investment from multinational pharmaceutical enterprises in this area, in spite of its extraordinary advances, probably due to the need for long-term development, high risk and uncertainty, when compared to alternative platforms. This low investment contrasts with an urgent need to accelerate vaccine development and access (Stevens et al., 2017) in a scenario permeated by increasing complexity from emerging infectious diseases.

These gaps contrast with the enormous potential benefits of plant-derived vaccines: animal components–free products; large-scale, lower investments; diversified technological platforms; frontier technologies; long-term innovation prospects (WHO, 2015).

5.8 Preparedness: Accelerating Vaccine RD&I for Emerging Diseases and Disease X

Extraordinary breakthroughs in vaccine development were evidenced in our patent search,.Novel vaccine technologies have been incorporated into SARS-CoV-2 vaccines at unprecedented speed, such as mRNA, a revolution in Vaccionology. These technologies have allowed thousands of people to start being vaccinated worldwide with innovative SARS-CoV-2 vaccines with very high efficacy and minimal adverse events, such as the mRNA vaccines produced by Pfizer/BioNTEch and Moderna. Despite these advances, many financial and regulatory barriers persist, preventing access to vaccines and timely vaccine RD&I responses to emergent and resurgent diseases, such as the COVID-19 pandemic, a major global governance issue.

Despite important advances in vaccine development evidenced in our patent search, the above-mentioned constraints prevent timely vaccine RD&I responses to emergent and resurgent diseases, a major global governance issue. From this perspective, the Coalition for Epidemic Preparedness Innovations (CEPI) has taken on the challenge to develop vaccines against neglected diseases that primarily strike the poor, such as MERS, Lassa, Nipah, Chikungunya and Rift Valley fever. The Coalition for Epidemic Preparedness Innovations is a public–private coalition that aims to derail epidemics by speeding up the development of vaccines up to Phase II clinical trials (Brende et al., 2017). It has been created as a global model by the Norwegian government, funding agencies and stakeholders. Conceptually, the CEPI supports platform technologies, seeking expertise worldwide to put them together, in order to expedite the required vaccine development.

A CEPI study estimated the costs of developing vaccines for diseases that could escalate into global humanitarian crises. The study focused on 11 diseases that cause relatively few deaths at present and primarily strike the poor, including MERS, Lassa, Nipah, Chikungunya, Rift Valley fever, and preparedness for Disease X. They estimated that it would cost between US $2.8 billion and US $3.7 billion to develop at least one vaccine for each of them. This should be set against the potential cost of an outbreak. The 2003 SARS outbreak in East Asia cost US $54 billion.

The CEPI is being funded by the Welcome Trust, the Bill and Melinda Gates Foundation, the European Commission and several governments (Norway, Japan, Germany, Canada, Australia, the UK, Belgium) and other donors.

In spite of the CEPI's extraordinary contributions to global vaccine RD&I, these initiatives are still limited in their scope. Governments from developed and developing countries should provide substantial and sustainable contributions to the CEPI and create coordinated local CEPI-like initiatives integrated to the CEPI.

5.9 Conclusion

The results presented here indicate that despite the extraordinary advances in vaccine RD&I in the last decade, many gaps still persist. In an epidemiological scenario with an increasing number of new diseases such as COVID-19 and the threat of a future new "Disease X" anticipated by the increasing complexity of global eco-social conditions, there is an urgent need to accelerate innovation and access to vaccines.

In order to deal with these new challenges, the future of vaccine innovation relies on the ability of policymakers to translate discoveries into public value, from a new global sustainability perspective. This new perspective should contribute to anticipating challenges and opportunities and to integrating diverse knowledge and disciplines to deal with an increasingly complex global epidemiological scenario. From this perspective, the development of 21st-century human and veterinary vaccines requires strong scientific, technological and academic integration and collaboration.

Furthermore, an important aspect of this collaboration in the vaccine field is that both animals and humans are exposed and susceptible to common disease-causing pathogens, thus approaches to immunization can use interrelated knowledge and technologies.

In the patent scenario for the immunobiologicals in our study, the cross-fertilization of knowledge and technologies is clearly observed. Our patent search provides many examples of the growing importance of this interdisciplinary cross-fertilizing approach.

The cross-fertilization processes permeating the fields of human, veterinary and plant-derived vaccines are thus increasingly blurring the disciplinary boundaries between these three areas of vaccine RD&I. The challenges posed by the complexity of this new scientific and technological scenario require the ability of research policymakers to conceive and implement new trans-disciplinary approaches to new developments in Vaccinology 4.0.

One example is the use of the most varied types of plants as bioreactors, such as tobacco, or the use of transgenic plants (from leaves, seeds or fruits) for the treatment/prophylaxis of human and veterinary diseases. A second example is the development of plant-derived vaccines for chronic degenerative diseases, such as Alzheimer's disease, with soybean, kidney bean and rice in their composition. Finally, another example is a veterinary foot-and-mouth disease vaccine which has the antigen expression in a plant or microalgae. Our results indicate that plant technology is a very important technological platform for vaccine development, with enormous potential for vaccine innovation due to its speed and low cost, but it remains little explored.

The exponential increase in the global number of patent applications identified in our search for conventional, recombinant, genomic, vectored and personalized vaccines for infectious and chronic degenerative diseases is impressive, with nearly 12,000 patent claims in the last decade. This scenario evidences a positive change in the priorities of public health RD&I investments aiming at the control of infectious and non-infectious diseases. These scientific and technological advances, favoured by incentives to collaborate in different knowledge interfaces, can result, if supported in a sustainable way and translated into effective public policies, in important global economic and social impacts.

Nevertheless, in spite of all the scientific and technological advances described here, some crucial issues constraining vaccine development for emerging and resurgent diseases need to be overcome in order to meet the Sustainable Development Goals of the 2030 Agenda (United Nations, 2015).

The first issue refers to the urgent need to accelerate the flow of the global vaccine pipeline for emerging neglected diseases, overcoming the so called "valley of death" in this pipeline. As mentioned previously, from 240 candidate vaccines in the global pipeline for neglected and emerging infectious diseases that mainly affect the poorest countries, only two have reached the product stage and are widely used in these countries: one conjugate vaccine for serogroup A meningitis and a vaccine against the Japanese encephalitis virus (Kaslow et al., 2018). Globalization and the intensification of international travel, as illustrated by the outbreak of COVID-19, influenza, dengue, measles and other emerging and resurgent neglected infectious diseases in many developed and developing countries are increasingly challenging territorial boundaries.

The recent aggravation of COVID-19 pandemic with emergence of severe and lethal variants in UK (B.1.1.7), South Africa (B.1.351) and Brazil (P1), the potential threats of influenza pandemic and the possibility of a new emerging pandemic, a future "Disease X", are evidence of the urgent need for a new paradigm for RD&I and preparedness. This new paradigm should be incorporated into RD&I-based surveillance structures, with local CEPI-like initiatives, coordinated by governments and integrated into CEPI.

The third issue is related to the need for greater investments in recombinant and genomic vaccine and in the Immunome project, promoting increased awareness of the role that the information sciences can play in vaccine development. Besides the use of patent information for technological foresight of vaccines, the incorporation of new advances in artificial intelligence, machine learning and big data will be of utmost relevance for accelerating vaccine development and supporting more effective preparedness initiatives.

Finally, in the long term, sustainable investments and funding mechanisms supporting novel technological vaccine platforms and international collaborations will be crucial in accelerating vaccine RD&I from One Health and sustainability perspectives, in order to meet the urgent global epidemiological challenges of the 21st century.

REFERENCES

Bloom, DE and Cadarette, D (2019). Infectious Disease Threats in the 21st Century: Strengthening the Global Response. *Frontiers in Immunology*, 10, 549. doi:10.3389/fimmu.2019.00549

Borkenhagen, LKS, Salman, MD, Ma, MJ and Gray, GC (2019). Animal Influenza Virus Infections in Humans: A Commentary. *International Journal of Infectious Diseases*, 88, 113–119. doi:10.1016/j.ijid.2019.08.002

Bragazzi, NL et al. (2018). Vaccines Meet Big Data: State-of-the-Art and Future Prospects. From the Classical 3Is ("Isolate–Inactivate–Inject") Vaccinology 1.0 to Vaccinology 3.0, Vaccinomics, and Beyond: A Historical Overview. *Frontiers in Public Health*, 6, 62. doi:10.3389/fpubh.2018.00062.

Brende, B et al. (2017). CEPI—A New Global R&D Organisation for Epidemic Preparedness and Response. *The Lancet*, 389(10066), 233–235. doi:10.1016/S0140-6736(17)30131-9.

Chen, H, Wang, P, Chen, S et al. (2012). Shikonin Induces Immunogenic Cell Death in Tumor Cells and Enhances Dendritic Cell-Based Cancer Vaccine. *Cancer Immunology Immunotherapy*, 61(11) 1989–2002. doi:10.1007/s00262-012-1258-9

Hong, MS, Severson, KA, Jiang, M, Lu, AE, Love, JC and Braatz, RD (2018). Challenges and Opportunities in Biopharmaceutical Manufacturing Control. *Computers & Chemical Engineering*, 110, 106–114. doi:10.1016/j.compchemeng.2017.12.007

Kaslow, DC, Black, S, Bloom, DE, Datla, M, Salisbury, D and Rappuoli, R (2018). Vaccine Candidates for Poor Nations Are Going to Waste. *Nature*, 564, 337–339. doi:10.1038/d41586-018-07758-3

Kile, JC et al. (2017). Update: Increase in Human Infections with Novel Asian Lineage Avian Influenza A (H7N9) Viruses during The Fifth Epidemic-China, October 1, 2016–August 7, 2017. *MMWR*, 66(35), 928–932. doi:10.15585/mmwr.mm6635a2

Lage, A (2006). On the Cross-Fertilization Between Biotechnology and Immunology: Current Situation in Cuba. *Vaccine*, 24(Supplement 3), S3–S6. doi:10.1016/j.vaccine.2005.01.101

McDonnell, WM and Askari, FK (2019). The Emerging Role of DNA Vaccines Medscape, Monday March 2 2020. Medtrack Plant-Derived Vaccines According to Application. *Medtrack Database*. http://medscape.com/viewarticle/715527_8

Palmer, SR Soulsby, L, Torgerson, PR, Brown, DW, Soulsby, L and Torgerson, PR (2011). *Oxford Textbook of Zoonoses. Biology, Clinical Practice, and Public Health Control*, 2nd ed. Oxford University Press, Oxford.

Pickering, LK, Meissner, HC, Orenstein, WA and Cohn, AC (2020). Principles of Vaccine Licensure, Approval, and Recommendations for Use. *Mayo Clinic Proceedings*, 95(3), 600–608. doi:10.1016/j.mayocp.2019.11.002

Plotkin, SA (Ed.). (2011). *History of Vaccine Development*. Springer, New York.

Plotkin, SA, Robinson, JM, Cunningham, G, Iqbal, R and Larsen, S (2017). The Complexity and Cost of Vaccine Manufacturing–An Overview. *Vaccine*, 35(33), 4064–4071. doi:10.1016/j.vaccine.2017.06.003

Poland, GA, Kennedy, RB and Ovsyannikova, IG (2011). Vaccinomics and Personalized Vaccinology: Is Science Leading Us Toward a New Path of Directed Vaccine Development and Discovery? *PLoSpathogens*, 7(12), e1002344. doi:10.1371/journal.ppat.1002344

Poland, GA, Ovsyannikova, IG and Kennedy, RB (2018). Personalized Vaccinology: A Review. *Vaccine*, 36(36), 5350–5357. doi:10.1016/j.vaccine.2017.07.062

Possas, C, Antunes, AMS, Mendes, FML, Martins, R and Homma, A (2016). Innovation and Intellectual Property Issues in the Decade of Vaccines: A Brazilian Perspective. In: Singh, HB, Jha, A and Keswani, C (Ed.), *Intellectual Property Issues in Biotechnology*. CABI, London, 181–192.

Possas, C, Antunes, AMS, Mendes, FML, Schumacher, SOR, Martins, RM and Homma, A (2015). Access to New Technologies in Multipatented Vaccines: Challenges for Brazil. *Naturebiotechnology*, 33(6), 599–603. doi:10.1038/nbt.3244

Pulendran, B, Li, S and Nakaya, HI (2010). Systems Vaccinology. *Immunity*, 33(4), 516–529. doi:10.1016/j.immuni.2010.10.006

Querec, TD, Akondy, RS, Lee, EK, et al. (2009). Systems Biology Approach Predicts Immunogenicity of the Yellow Fever Vaccine in Humans. *Natureimmunology*, 10(1), 116. doi:10.1038/ni.1688

Raeven, RH, Van Riet, E, Meiring, HD, Metz, B and Kersten, GF (2019). Systems Vaccinology and Big Data in the Vaccine Development Chain. *Immunology*, 156(1), 33–46. doi:10.1111/imm.13012

Rappuoli, R, Black, S and Lambert, PH (2011). Vaccine Discovery and Translation of New Vaccine Technology. *The Lancet*, 378(9788), 360–368. doi:10.1016/S0140-6736(11)60440-6

Sette, A and Rappuoli, R (2010). Reverse Vaccinology: Developing Vaccines in the Era of Genomics. *Immunity*, 33(4), 530–541. doi:10.1016/j.immuni.2010.09.017

Shim, BS, Hong, KJ, Maharjan, PM and Choe S (2019). Plant Factory: New Resource for the Productivity and Diversity of Human and Veterinary Vaccines. *Clinical and Experimental Vaccine Research*, 8(2), 136–139. doi:10.7774/cevr.2019.8.2.136

Stevens, H, Debackere, K, Goldman, M, Mahoney, RT, Stevens, P and Huys, I (2017). *Vaccines: Accelerating Innovation and Access*. World Intellectual Property Organization, Geneva.

Takeyama, N, Kiyono, H and Yuki, Y (2015). Plant-Based Vaccines for Animals and Humans: Recent Advances in Technology and Clinical Trials. *Therapeutic Advances in Vaccines*, 3, 139–154.

United Nations (2015). *Transforming Our World: The 2030 Agenda for Sustainable Development. General Assembly 70 Session*. World Intellectual Property Organization, Geneva.

Van der Laan, JW, Minor, P, Mahoney, R, Arntzen, C, Shin, J and Wood, D (2006). WHO Informal Consultation on Scientific Basis for Regulatory Evaluation of Candidate Human Vaccines from Plants, Geneva, Switzerland, 24–25 January 2005. *Vaccine*, 24(20), 4271–4278. doi:10.1016/j.vaccine.2005.12.063

World Health Organization, Regional Office for South East Asia (2014). A brief guide to emerging infectious diseases and zoonosis. WHO, Geneva.

World Health Organization/WHO (2015). *Plant Derived Vaccines*. WHO, Geneva, February 11.

World Health Organization/WHO. (2018). *Cancer*. www.who.int/news-room/fact-sheets/detail/cancer

Patent References

Ai, D, Lu, Q, Luo, L, Luo, Q, Pan, Z, Qiao, L, Shao, H, Wang, H, Wang, H, Wen, G, Zhang, R and Zhang, T. Recombined Newcastle Disease Heat-Resisting Vaccine Strain for Expressing H5 Subtype Avian Influenza Virus Truncated HA Protein and Preparation Method. CN105950572, filed June 15, 2016.

Bai, X, Chai, X, Cheng, Y, Deng, X, Hu, B, Liu, H, Lu, R, Luo, G, Shi, N, Xu, S, Yan, X, Zhang, H, Zhang, L, Zhao, C and Zhao, J. Bivalent Live Vaccine of Canine Distemper-Parvovirus Enteritis of Racoon Dog, and Preparation Method and Application of Bivalent Live Vaccine. CN108066758, filed December 26, 2017.

Bai, X, Deng, X, Hu, B, Lian, S, Liu, H, Lu, R, Lyu, S, Shi, N, Wang, Y, Xu, S, Xue, X, Yan, X, Zhang, H, Zhang, L, Zhao, J and Zhu, Y. Racoon Dog Canine Distemper Living Vaccine, Preparation Method and Application Thereof. CN107115525, filed May 16, 2017.

Bai, X, Deng, X, Hu, B, Lian, S, Liu, H, Lu, R, Lyu, S, Shi, N, Wang, Y, Xu, S, Xue, X, Yan, X, Zhang, H, Zhang, L, Zhao, J and Zhu, Y. Raccoon Dog Distemper Freeze-Dried Live Vaccine and Preparation Method and Application Thereof. CN107158370, filed May 16, 2017.

Berzofsky, JA, Besra, GS, Illarionov, P, O'konek, JJ and Terabe, M. Beta-Mannosylceramide and Stimulation of NKT Cell Anti-Tumor Immunity. US2013039886, filed March 11, 2011.

Berzofsky, JA, Chen, Z, Khleif, S, Petit, R and Wallecha, A. Methods and Compositions for Increasing A T-Effector Cell to Regulatory T Cell Ratio. US2017080064, filed March 5, 2015.

Bi, Y, Gao, D, Ji, Q, Li, J, Li, Z, Peng, S and Xu, W. *Neisseria Meningitidis* Group B Recombinant Protein Chimeric Vaccine and Preparation Method Thereof. CN102028941, filed February 11, 2011.

Black, S, Del Giudice, G, Nicolay, U and Rappuoli, R. Vaccination of Immunocompromised Subjects. US2017304429, filed September 25, 2015.

Bolton, D, Custers, JHHV, Radosevic, K and Roederer, M. Heterologous Prime-Boost Immunization Using Measles Virus-Based Vaccines. EP2780034, filed November 13, 2012.

Brock, LG, Buchholz, UJ, Collins, PL, Dinapoli, JM, Le Nouën, C, Mueller, S and Wimmer, E. Attenuation of Human Respiratory Syncytial Virus by Genome Scale Codon-Pair Deoptimization. US9957486, filed February 7, 2014.

Bu, Z, Chen, W, He, X, Liu, J, Liu, R, Zhao, D. Gene Deletion Attenuated African Swine Fever Virus and Application Thereof as Vaccine. CN110093324, filed April 26, 2019.

Buchholz, UJ, Collins, PL, Luongo, CL and Murphy, BR. Genetically Stable Live Attenuated Respiratory Syncytial Virus Vaccine and its Production. EP2836231, filed March 13, 2013.

Cai, S, Hong, R, Jiao, T, Li, H and Li, J. H9 Subtype of Avian Influenza Virus Inactivating Vaccine and Preparation Method Thereof. CN104928259, filed May 22, 2015.

Cao, F, Ma, G, Wang, X and Yan, M. Tumor Vaccine Based on Nano-Graphene Oxide-Al(Oh)3/Antigen Composite and Preparation Method Thereof. CN109010810, filed August 31, 2018.

Carfi, A, Chandramouli, S and Settembre, E. Antigenes Du Cytomegalovirus. BE1024516, filed February 14, 2017.

Carfi, A, Ciferri, C and Xing, Y. Antigenes Du Cytomegalo Virus Et Leurs Utilisations. BE1023087, filed January 22, 2016.

Chang, J, Jiang, Z and Yu, L. Toxin Attenuation Mutant for Epsilon Toxin of *Clostridium Perfringens* and Application of Toxin Attenuation Mutant. CN104560780, filed November 28, 2014.

Chen, H, Chen, P, Deng, G, Jiang, Y, Li, Y, Liu, J, Shi, J and Zeng, X. Gene for Coding H5 Subtype Avian Influenza Virus Hemagglutinin, and Application Thereof. CN107630024, filed September 13, 2017.

Chen, H, Deng, Y, Guo, X, Hong, T, Lan, J, Lu, Z and Tan, W. MERS-CoV Spike Protein Coding Gene Optimization-Based Recombinant 5-Type Adenovirus Vector Vaccine. CN105543248, filed March 13, 2015.

Chen, H, Deng, Y, Guo, X, Hong, T, Lan, J, Lu, Z and Tan, W. Recombinant 41-Type Adenovirus Vector Vaccine Coding MERS-CoV Spike Protein. CN105273067, filed March 13, 2015.

Chen, H, Deng, Y, Huang, B, Lan, J, Qin, K, Tan, W and Wang, W. MERS-CoV Chimeric VLPs Vaccine Taking Influenza Virus M1 as Matrix. CN106119213, filed June 20, 2016.

Chen, H, Chen, Y, Qiao, C and Yang, H. Recombinant Homologous Avian H1N1 Influenza Virus Inactivated Vaccine Strain (JS40/PR8) as Well as Preparation Method and Application of Inactivated Vaccine Strain. CN104232594, filed September 11, 2014.

Chen, J, Gai, Q, Shu, T, Yan, J and Zhang, Y. Recombinant Protein Containing SARS Virus N Antigen and Baculovirus Displaying N Protein. CN104292338, filed July 18, 2013.

Chen, J, Gai, Q, Shu, T, Yan, J and Zhang, Y. Recombinant Protein Containing SARS Virus RBD Antigen and Baculovirus Displaying RBD Protein. CN104292339, filed July 18, 2013.

Chen, J, Gai, Q, Shu, T, Yan, J, Zhang, Y and Zhao, H. Recombinant Baculovirus with Surface Displaying SARS Bivalent Antigen, and Preparation Method and Application Thereof. CN104293740, filed July 18, 2013.

Chen, J, Hou, G, Jiang, W and Li, J. Method for Increasing Proliferation Titer of Vaccine Strain of H7N9 Subtype Avian Influenza. CN108913701, filed August 13, 2018.

Chen, J, Huang, S, Li, Z, Song, H, Xu, M, Zhou, D and Zhu, X. Immunologic Adjuvant for Preventing Toxoplasma Infection and Application of Immunologic Adjuvant. CN103083660, filed January 25, 2013.

Chen, R, Guo, S, Guo, Z, He, G, Huang, L, Jiao, Y, Li, H, Li, L, Li, R, Li, X, Li, Z, Liu, D, Qiao, C, Qiu, E, Sun, H, Sun, J, Wan, X, Wang, C, Wei, J, Yan, P, Yuan, Y and Zhang, H. Preparation Method of Anti-Tobacco Potato Y Virus Vaccine. CN105420264, filed July 7, 2015.

Cheng, S, Cheng, Y, Feng, E, Guo, L, Luo, G, Wang, Z and Yi, L. Raccoon Dog Parvovirus Strain, Application of Raccoon Dog Parvovirus Strain, Raccoon Dog Parvovirus Inactivated Vaccine and Preparation Method of Raccoon Dog Parvovirus Inactivated Vaccine. CN108690834, filed May 31, 2018.

Chung, IS, Kim, HU, Kim, JB, Kim, SH, Kim, WY, Lee, EY, Lee, HS, Lee, KR, Lee, YH, Rho, KH and Shon, DH. The Transformant for Using Vaccination Against Rotavirus and Vaccine Composition Comprising the Same. KR20130066128, filed December 12, 2011.

Chung, IS, Kim, HU, Kim, JB, Kim, SH, Kim, WY, Lee, EY, Lee, HS, Lee, KR, Lee, YH, Rho, KH and Shon, DH. The Transformant for Using Vaccination Against Rotavirus and Vaccine Composition Comprising the Same. KR101428843, filed December 12, 2011.

Collins, PL and Mccarty, TC. Recombinant Respiratory Syncytial Virus Strains Comprising Ns1 And Ns2 Gene Shifts. EP3387120, filed December 12, 2016.

Cong, Q, Gao, R, Li, X and Liu, J. Optimized Plant Virus Inoculation Method. CN104152487, filed June 13, 2014.

Conrad, U, Ho, TT and Phan, H. Oligomeric Vaccines from Plants by S-TAG-S-Protein Fusions. WO2018115305, filed December 21, 2017.

Cui, Q, Du, R, Rong, L and Wang, L. Recombinant H3N2 Subtype Influenza Virus Carrying Luciferase, Construction Method and Application. CN110305854, filed July 18, 2019.

Cui, S. Porcine Parvovirus BQ-C Strain and Application of Porcine Parvovirus BQ-C Strain to Preparation of Inactivated Porcine Parvovirus Vaccine. CN102719407, filed May 25, 2012.

Daniell, H. Plastid-Expressed Mycobacterium Tuberculosis Vaccine Antigens ESAT-6 and MTB72F Fused to Cholera Toxin B Subunit. US2015196627, filed October 24, 2012.

Deng, Y, Gao, F, Hu, Y, Lan, J, Tan, W and Yan, J. Recombinant NTD Protein Subunit Vaccine for Resisting MERS-CoV Infection. CN105169384, filed July 22, 2015.

Deng, Y, Geng, H, Meng, X, Tan, W, Wang, H, Wen, B and Yang, Y. MERS-CoV (Middle East Respiratory Syndrome Coronavirus) Synthetic Peptide Vaccine with Neutralization Activity and Application of MERS-CoV Synthetic Peptide Vaccine. CN103724406, filed November 20, 2013.

Deng, Y, Liu, X, Lu, S, Qin, K, Shu, Y, Tan, W, Wang, W, Yu, S and Zhou, J. Fully Humanized Neutralizing Antibody Against Middle East Respiratory Sydrome Coronavirus. CN107298712, filed April 14, 2016.

Diamos, A, Mason, H and Roland, K. Vaccine for Prevention of Necrotic Enteritis in Poultry. WO2019010135, filed July 2, 2018.

Ding, C, Han, X, Qiu, X, Song, C, Tan, L, Wang, X and Yu, S. *Riemerella Anatipestifer* Tervalent Inactivated Vaccine and Preparation Method Thereof. CN102908616, filed September 13, 2012.

Du, L, Jiang, S, Zhao, G, Zhou, Y. Immunogenic Composition for MERS Coronavirus Infection. US9889194, filed February 28, 2014.

Du, Y, Li, L, Liu, X, Ma, S, Song, X and Wang, X. Recombinant Strain for Producing B-Type Acian Metapneumovirus F Protein. CN105861349, filed April 14, 2016.

Fang, P, Li, J, Li, Y, Liu, X, Lyu, J, Ma, Z, Pan, L, Wang, Y, Wu, X, Yan, H, Zhang, Y, Zhang, Z and Zhou, P. Seneca Valley Virus, Preparation Method of Seneca Valley Virus Inactivated Vaccine, Seneca Valley Virus Inactivated Vaccine and Application Thereof. CN109679927, filed February 25, 2019.

Fang, Y, Jiang, S, Liu, X, Lv, J, Pan, L, Wang, Y, Zhang, Y, Zhang, Z and Zhou, P. Asia 1 Type Foot and Mouth Disease Virus Antigen and Preparation and Application Thereof. CN103382480, filed July 31, 2013.

Garcon-Johnson, NMJ, Plante, M, Van, M and Marcelle, P. Vaccine Comprising A Tlr-5 Agonist as Adjuvant for Use in Cutaneous Immunisation. EP2780038, filed November 19, 2012.

Gong, D, Sun, Y, Wang, G, Xia, Q and Xie, M. Promoter and Applications Thereof. CN103205429, filed April 24, 2013.

Gotanda, T, Ikadai, H, Itchoda, N, Ito, A, Kagaya, U, Matsumura, T, Sasaki, H and Tabayashi, N. Malaria Propagation-Blocking Vaccine. WO2018043741, filed September 4, 2017.

Graham, B, Joyce, MG, Kanekiyo, M, Kong, WP, Mascola, J, Modjarrad, K, Shi, W and Wang, L. Middle East Respiratory Syndrome Coronavirus Immunogens, Antibodies, and Their Use. US10301377, filed February 24, 2016.

Guo, Y, Huang, M and Kong, L. Expression Vector PVX-6His-CTBt-Bt for Producing Multi-Epitope Vaccine of Hepatitis C Virus. CN104353071, filed October 23, 2014.

Han, YH, Kim, HD, Kim, SK, Lee, GY, Lee, HJ, Lim, JW and Soh, HS. Marker Gene for Expressing Genes Having Vaccine Activity of Diarrhea and Identification Method Using the Same. KR20130070032, filed December 19, 2011.

Hasegawa, H, Kawarabayashi, T, Shimada, Y, Shoji, M, Takahashi, A and Terakawa, T. Vaccine Composition for Prevention and/or Treatment of Alzheimer's Disease. JP2013112668, filed November 30, 2011.

Hong, SP, Kim, CJ, Lee, JS and Rho, JB. New Fusion Protein Useful for Preventing and/or Treating Influenza Viral Infection, Comprising Antigenic Protein of Influenza Virus Matrix Protein 2, and Haemagglutinin Having Cholera Toxin A1 Subunit. KR101525180, filed October 6, 2014.

Huang, P, Jin, Z, Kong, D and Wang, W. Nano Adjuvant Material with Guanidyl Modified Surface as Well as Preparation Method and Application of Nano Adjuvant Material. CN106729700, filed December 8, 2016.

Huang, N, Johnson, BJB, Macmanus, C and Zhang, D. OspA Fusion Protein for Vaccination Against Lyme Disease. US2016213766, filed September 23, 2014.

Jang, YS, Kim, DH, Kim, TG and Yang, MS. Method for Mass Producing Dengue Virus Vaccine Using Plant Viral Expression System. KR20170053223, filed November 6, 2015.

Jeong, SW and Lim, JT. Method for Hydroponics of Nicotina Benthamiana. KR101926384, filed December 17, 2015.

Jiang, L, Jiang, Q, Ju, L and Lu, XA. H1N1 Influenza Virus and Application Thereof. CN101974489, filed June 30, 2010.

Kong, D, Sun, H and Zhang, C. Dc Targeted Nano Vaccine Applied to Immunotherapy of Liver Cancer and Preparation Method Thereof. CN108785666, filed June 5, 2018.

Lilja, A and Mason, P. Recombinant Self-Replicating Polycistronic Rna Molecules. EP2768530, filed October 11, 2012.

Liu, D, Liu, L, Qu, L, Wang, M and Zhang, H. I-Type Canine Adenovirus Attenuated Vaccine Strain and Application Thereof. CN101914502, filed July 30, 2010.

Liu, G, Miao, Q, Yin, D and Zhu, J. Variant Rabbit Haemorrhagic Disease Virus and Applications Thereof in Preparing Inactivated Vaccine. CN108588034, filed March 22, 2018.

Liu, Y, Massare, M and Smith, G. Immunogenic Middle East Respiratory Syndrome Coronavirus (MERS-CoV) Compositions and Methods. US2016206729, filed September 19, 2014.

Lynn, G and Seder, R. Peptide Vaccines Comprising Self-Assembling Polymer Nanoparticles. EP3419655, filed February 27, 2017.

Lynn, G, Seder, R and Seymour, L. Expression Vector Delivery System and Use Thereof for Inducing an Immune Response. EP3347047, filed September 9, 2016.

Muthumani, K, Sardesai, NY and Weiner, D. MERS-CoV Vaccine. US10016497, filed November 26, 2014.

Wang, H, Yang, D, Yu, L and Zhou, G. Temperature-Sensitive Attenuating Strain of Foot-and-Mouth Disease Virus, Construction Method Therefor and Uses Thereof. WO2018090994, filed November 20, 2017.

Weston, E and Torossian, K. Stable Vaccine Composition Used to Treat Cancer and Induce Immunogenic Response, Comprises Arginine, Counterion, First Immunogenic Molecule, and Second Immunogenic Molecule Comprising Oligonucleotide. BE1022254, filed August 1, 2014.

6

Potential of Ecosystem Services in an Agri-Based Bioeconomy: The Traditional Vietnamese Integrated Farming System VAC

Le Hung Anh and Petra Schneider

CONTENTS

6.1 Introduction to the Principles of Agri-Based Bioeconomy

The development of mankind is closely connected with the consumption of natural resources and the availability of affordable energy. The evolution of human energy systems is closely related to land use. In Western Europe, a qualitatively different energy system was dominant 150–200 years ago, which was described as a "controlled solar energy system" by Sieferle (1995). It was named the solar energy system because it was essentially based on biomass, which ultimately represented stored solar energy (Haberl, 2006). The total amount of biomass produced on a given area in one year is called the net primary production. The use of stored solar energy was mainly through agriculture and forestry, which is why such societies were called agrarian society (Haberl, 2006), and most of the socially used energy was utilized to feed humans and livestock (Fischer-Kowalski & Haberl, 2007).

From an ecological point of view, the energy systems of agrarian societies are understood as systems in which human societies engage in ecological energy flows, in which they make use of part of the net primary production (Haberl, 2006). The limits of the controlled solar energy system in that period were reached through population growth and the need to ensure food security. Maintaining soil fertility was a key problem that led to limits on crop growth and a decrease in crop yields (Krausmann et al., 2009). Furthermore, transport was limited, since land transport was associated with high energy costs. These factors led to the formation of largely autonomous agricultural systems in the form of village communities, which produced only a small surplus available to landlords and towns (Haberl, 2006). For this reason, urbanization and spatial differentiation were also limited (Krausmann et al., 2009). These kinds of self-sustaining agrarian societies declined from around the 18th century, when the use of coal led to the first energy transition: the share of biomass in energy sales dropped and coal lifted the growth barriers. According to Geißler (2002), there are two meaningful economic indicators for the development of an agrarian society into an industrial society: the shares of different production sectors in the value added and in the employees. According to Clark (1940) and Fourastié (1954), the conceptual basis is the classification of the economy into primary (agriculture, forestry and fisheries), secondary (industry, craft and mining) and tertiary (services) sectors (Geißler, 2002). Towards the end of the 19th century, the transition from agrarian society to industrial society took place widely in the economic sphere. An important driving force for industrialization was the population explosion from the mid-18th century until the late 19th century. An agrarian revolution that shifted to the much more productive crop rotation and higher-yielding crops provided more food. The mechanization of agriculture associated with the Industrial Revolution, as well as biological innovations such as new crops, not only added to the food supply, but also accelerated population growth. The industrialization of agriculture completed the process of decoupling energy and land (Krausmann et al., 2009). Mechanization resulted in the substitution of livestock as a means of production. Agrochemicals temporarily led to the overcoming of nutrient barriers (Krausmann et al., 2009), resulting over the past 50 years in the so-called Green Revolution (Pingali, 2012). Despite nutritional gains, the Green Revolution also caused environmental impacts due to the degradation and exhaustion of the soil and land (Gomiero, 2016) as well as the substitution of small farms for large commercial farms (Sebby, 2010). Several experts have recommended remembering traditional agrarian principles and soil ecological cycles in order to re-establish a self-sustaining productive agriculture (Gomiero, 2016; Padmavathy & Poyyamoli, 2011). Some countries, including Germany, have even promoted an energy transition from fossil to renewable energies (Hedberg, 2017; Valdes et al., 2019). In Vietnam, a traditional agrarian system is still in use, and was even upgraded with modern techniques especially biogas technology, based on the system of closed material cycles according to Braungart and McDonough (2002). The combination of improving resource productivity with ensuring sustainable resource and materials management based on the 3R principle (reduce, reuse, recycle) is a central element of green growth policies (OECD, 2017), which has been widely applied in South East Asia as a resource management strategy. This contribution investigates the potential of VAC as a circular agrarian approach and the balancing of environmental impacts through the application of ecological engineering principles. VAC is considered an integrated agriculture–horticulture–aquaculture system.

Agro-ecosystems play a major role in maintaining biodiversity and various functions and ecosystem services. In addition to crop products, such services also include carbon fixation and water retention. Further, agricultural landscapes which are a scene of human–environment relationships not only guarantee an income for farmers, but also represent a place of cultural identity. However, this multifunctionality of agricultural landscapes is affected by an increasingly one-sided use, which in turn causes a reduction in biodiversity.

On the other hand, land use is being intensified in many regions of the world, which is often associated with the increased use of agrochemicals and a reduction in the structural diversity of the landscape. By optimizing agricultural production and the associated simplification of landscape diversity, the agricultural landscapes only focus on one specific ecosystem service, namely the agricultural yield, which can, however, result in a reduction in the potential of the landscape to provide a wide variety of ecosystem services. According to the Millennium Ecosystem Assessment (2005), ecosystem services are a collective term for the diverse contributions that ecosystems make to human well-being. Biodiversity is a natural capital with diverse benefits for agriculture, the economy and society. This capital can be used through ecosystem services, which make a significant contribution to securing agricultural yields. Ecosystems are complex units which consist of diverse abiotic and biotic components that interact with each other and which humans change, use and depend on. Ecosystem services can be divided into supply services, regulatory services and cultural services. In terms of agriculture, for instance, the following ecosystem services are relevant: pollination, natural pest regulation, soil conservation, the maintenance of nutrient cycles, the regulation of the water supply as well as the storage and conversion of carbon dioxide. The present contribution discusses the traditional Vietnamese VAC approach and illustrates the potential of its ecosystem services (Figure 6.1).

FIGURE 6.1 Traditional small gardens in Vietnam.

6.2 VAC—The Cage Garden Approach: VAC Models (Garden–Pond–Livestock)

6.2.1 General Overview

The VAC system refers to a form of domestic agriculture in which food gardening, fish farming and animal husbandry are fully integrated, developed in the Red River Delta of Vietnam. As such, VAC is a traditional Vietnamese integrated farming ecosystem. Vườn–Ao–Chuồng is the Vietnamese name for the integration system of garden–pond–livestock, in which the three main elements of the farm model not only appear in the scale of a farm but also have a very close relationship with each other. "V" stands for Vườn, being horticulture crop vegetable gardens; "A" stands for Ao, representing aquaculture fish shrimp ponds in lakes, rivers, streams or the sea with aquatic animals such as fish, shrimp, crabs, frogs, turtles, baba, etc.; and "C" stands for Chuồng, illustrating animal husbandry livestock as well as animal pens, sheds and stables. The cage is a symbol of the breeding activities of cattle and poultry such as pigs, cows, chickens, ducks, goats and sheep on farms. VAC represents a more or less closed system through the integration of fish farming with livestock production and the farming of agricultural crops and vegetables. In the English version, VAC is assumed as a cage garden model. VAC is a highly bio-intensive approach, creating a comprehensive farming system that efficiently utilizes land, water and solar resources to help farmers achieve high economic efficiency. VAC is characterized by low investment levels, at the same time contributing to minimize environmental pollution problems in livestock and protecting the health of those around them.

6.2.2 General System Configuration

The general VAC system can be illustrated as shown in Figure 6.2.

The VAC system is a circular system where all loops are closed (Figure 6.3). Ponds (A) are usually constructed as a central part of the VAC system. They deliver water to irrigate the garden. In the ideal case, the water volume of the pond and the precipitation are sufficient to irrigate the garden. Products from the garden (V) are used to feed fish. Sediment that is enriched with nutrients from the pond is used to fertilize the garden. Water from the pond is used for domestic purposes and to produce aquatic weeds for feeding pigs in the livestock pen (C). Fishes can be further used as nutritious animal feed. Pig and other manures are used as fertilizers for crops and rice in the fields. Both types of manures and crops are also used as food for fish in the pond.

Expanded integrated VAC farming systems are V-A-C-R, where R has the following meanings:

- in the highlands (mountainous areas): "R" is the abbreviation of Rừng, which is forestry, and
- in rice production areas: "R" is the abbreviation of Ruộng, which is rice paddy field (Mekong area, Red River Delta).

6.2.3 Types of VAC Approaches

Developed from age-old agricultural production practices, VAC farming now takes place in many regions of Vietnam, with various models according to the terrain and the climate. Since the 1980s, many

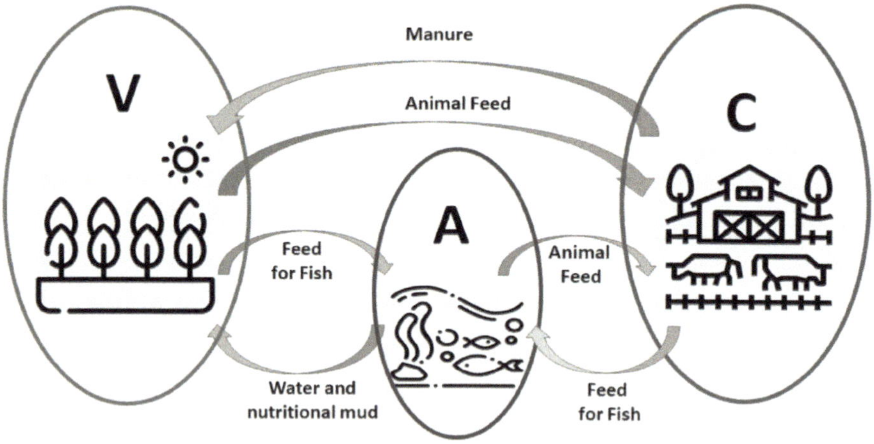

FIGURE 6.2 General configuration of the VAC system. (Source icons: www.flaticon.com/.)

FIGURE 6.3 Illustration of the VAC system in practice.

households have developed VAC systems. Van Huong et al. (2018) identified four VAC types in aquaculture households: the traditional VAC, new VAC, animal fish (AF) and commercial fish (FS):

1. Traditional system: Intensive orchard, very low input aquaculture within villages.
2. New system: Intensive orchard, low input aquaculture outside residential areas.
3. Semi-intensive system (AF system): Orchard medium input aquaculture outside residential areas (animal + fish production).
4. FS system: Extensive orchard high input aquaculture with commercially intensive fish production.

According to Van Huong et al. (2018), the systems are classified and characterized as follows:

***Traditional VAC system*:**

- Most of the traditional system is like a self-subsistence model because fish is harvested regularly for home consumption. It effectively utilizes resources available in the agricultural practices on the household level.
- Most households perform the practice based on their own traditional experience in aquaculture production in connection with rice production, medium-scale animal husbandry and small-scale horticulture.

New VAC system:

- Larger-scale fish ponds are developed and the systems focus on fruit production. These systems are usually found outside villages with less residential density.
- The systems not only apply more technology and advanced techniques for fish farming but also diversify more husbandry products than the traditional VAC systems.
- Pig and poultry production are the most common, but cattle may also be added. However, the number of cattle is not high, just one to three cattle per farm.

Animal fish system:

- This system mainly focuses on a combination of fish and larger pig or poultry production.
- Fruit orchards and rice production fields are less likely due to inadequate labour. The area of paddy land (2,220 m^2) is less than in the traditional VAC.
- Major production inputs (pellet feed, veterinary chemicals) are from materials and feeding shops/suppliers in local markets.
- Households rely more on markets for both the input and output of their production.

Fish system:

- This system is the most commercially intensive production of fish compared to other systems.
- With the largest scale of farm size, it sets a high priority on fish production.
- Fish is a high market value species for the best economic return.
- The area of paddy land is smallest because most of it has been converted for aquaculture purposes.

Many collective VAC gardens have been established by cooperatives, businesses, kindergartens, schools, churches and pagodas. Gardens are often established around centres for orphans, the aged and the disabled, where people provide free labour and, in turn, are provided with food. They can then share their VAC production with family members, or sell products on the local market. Integrated farming is a traditional approach to family food production in poor rural regions of Vietnam. The farming system is managed by family members and practically all labour comes from the household. This can be a partial or a full VAC system depending on the financial and labour conditions, for example, 85–90% of rural families maintain a garden and livestock pen, with 30–35% of these having fish ponds. In many villages, 50–80% of families have the full VAC system, which provides the main income for the whole family.

Furthermore, agroforestry is a well-known method of production focusing on combining agricultural crops with forestry trees based on economic-ecological principles in order to improve land use efficiency, diversify products and increase income to improve the living standards of people living in mountainous areas (Nguyễn & Lê Đồng, 2012). This farming method has a long history and was originally borne from the people's production experience, making it quite local. In Vietnam, the original model had two separate areas, namely "garden" and "forest"; where there are favourable conditions for water resources, there is an additional third area——Ao|| for fish stocking; to take advantage of the space, in addition to crops, people build "cages" to raise a number of animals including poultry (chickens, ducks, backyard ducks, geese, etc.) and cattle (pigs, buffalo, cows, goats, etc.). Therefore, the garden–forest land contains a fourth system called "Chuồng" (Nguyễn & Lê Đồng, 2012). With these items, production models in mountainous areas are called agroforestry as follows: garden–forest (abbreviated as VR), garden–barn–forest (VCR), forest–garden–pond–shed (RVAC) and garden–pond–shed (VAC). The following section presents the current status of the models based on Nguyễn and Lê Đồng (2012).

VAC farms can be found in various agro-ecological conditions: (1) irrigated lowlands, (2) rainfed uplands and (3) peri-urban areas. Lowland systems are significantly different from upland systems in terms of resource availability and use: (a) suburban, (b) intensive (i.e. rice production) and (c) lowland (i.e. flood-prone environments).

6.2.4 Regional Variations of VAC Approaches

6.2.4.1 VAC Systems in Northern Vietnam

Northern Vietnam comprises lowlands and uplands, and the VAC systems are adapted according to the local conditions. An upland VAC is usually found in mountainous regions. The pond is constructed close to the house garden and livestock pens so that domestic and kitchen waste is drained into the fish pond. A lowland application is usually found in regions such as Ha Noi, Hai Duong, Hung yen, Ha Nam, Nam Dinh and Hai Phong (Le Thanh Luu, 2001). Typically, houses are constructed close to the pond. In sandy regions, the houses are often built at some distance from the pond for hygienic reasons. Figures 6.4 and 6.5 provide an illustration of the VAC upland system.

6.2.4.1.1 Upland VAC and Its Characteristics

According to Le Thanh Luu (2001), the characteristics are as follows:

- **Garden:** An area of 1,000–5,000 m^2 includes fruits (bananas, oranges, peaches and apricots) and various vegetables (green onions, sweet potatoes and other crops such as sugarcane, tea and cassava).

FIGURE 6.4 Illustration of the VAC upland system in practice.

FIGURE 6.5 Illustration of the typical VAC system in northern Vietnam.

- **Livestock:** Most families keep various animals on the farm such as one or more water buffalo and cattle, pigs, ducks and chickens. A portion of the livestock manure is used for manuring trees and vegetables. Trees are manured once or twice a year and vegetables are manured according to their needs. Animal manure is applied twice a month at 0.05–0.15 kg/m into the pond. The large ruminant animals are allowed to graze or are fed farm byproducts. Pigs and poultry are usually fed kitchen waste, or cassava, rice bran, sweet potatoes, banana trunks and water hyacinth on other farms.
- **Fish pond:** Fish ponds with areas of 100–1,500 m^2 and 1 m depth are usually allocated a more central part of the farm for better management. Pond silt is removed every three to four years and used as fertilizer. Domestic washing and kitchen residue are channelled into the pond daily. Ponds are often drained after the final harvest, usually in February. The pond bottom is kept dry for one to three weeks after being cleaned, limed and manured, and is then filled with water for restocking. Three months after stocking, farmers begin to harvest on a weekly basis using small nets and continuously restock and harvest the pond.

6.2.4.1.2 Lowland VAC and Its Characteristics

According to Le Thanh Luu (2001), the characteristics are as follows:

- **Garden:** Usually small, 400–500 m^2. Cultivated fruit trees are bananas, oranges, papayas, peaches, litchi, longan and apples. Ornamental trees and flowers are planted as a main income source in many suburban family gardens. Cultivated vegetables are green onions, sweet potatoes, cress, tomatoes and water spinach. The leaves of legumes, peanuts and green beans are used for manuring ponds. Both perennial and annual crops are planted to provide year-round food to the house and products for the market.
- **Livestock:** Pens for pigs, water buffalo and cows are constructed at the corner of the garden close to the pond. Other common poultry running freely might be ducks and chickens. Large ruminant animals are allowed to graze. Pigs and poultry are usually fed kitchen waste, or cassava, rice bran, sweet potato, banana trunks and water hyacinth on other farms. Livestock manure is used to fertilize vegetables.
- **Ponds:** Usually 50–400 m^2 of different shapes with an average depth of 1.0–1.2 m. Pond mud is annually removed and used to manure fruit trees. Pond water is used for irrigating the garden, particularly vegetables. Ponds are drained after the final harvest (usually in January or February). The pond is then kept dry for a few days, limed, manured and refilled with rainwater or irrigation water (early rains may start at the end of March). Domestic washing and kitchen residue may be channelled into the pond, with a small part of the manure from livestock used to manure the pond (according to the farmer's experience).

6.2.4.2 VAC Systems in Central Vietnam

Central Vietnam where more than 10 million people live is dominated by sandy soil, which is 63% of the Vietnamese sandy soil. The main products are rice, potatoes, peanuts, oranges, peppers, cassava, pineapples, soybeans, sugar, sesame, tea, coffee, rubber, livestock and forest products (Le Thanh Luu, 2001). Typically, the products are of quite low quality, with low added value. The main crops promoted through governmental agricultural extension programmes are tea, peanuts, rubber and oranges.

In recent years, the model of agro-tourism has been developed in association with the development of high-tech agriculture in the Central Highland, for instance in Dalat and Lam Dong. Furthermore, in many provinces of central Vietnam, the garden economy has been intensively developed, especially fruit gardens, resulting in further diverse systems beyond the conventional VAC, particularly VAC-forest and VAC-biogas. Moreover, the large potential of eco-tourism gardens and organic gardens has been developed in the last years. This development has also improved the landscape and created a beautiful view together with the speciality of fruit gardens for local people and tourists. It provides more economic benefits; meanwhile, the average family income comprises 50–60% of eco-tourism gardens and organic

FIGURE 6.6 Illustration of the typical VAC system in central Vietnam.

gardens. This is also an environmentally friendly development as it ensures the ecosystem balance, restores the soil fertility and supports clean water resources. The core of the implemented system here is also the reuse of human and animal waste in order to increase the soil nutrition.

Figure 6.6 illustrates typical VAC systems in central Vietnam. In several provinces such as Thanh Hoa, Nghe An and Ha Tinh there are large fruit orchard gardens with an average area of over 10,000 ha (Le Thanh Luu, 2001).

On mound hills, farmers raise livestock for manure to improve the soil and cultivate pomelos and lemons, and the resulting income in the Thanh Hoa province is 400–500 million VND/ha per year. In Hue, there is an area of over 8,680 ha of garden soil, accounting for 15% of the area of agricultural soil throughout the whole province. The income from the garden economy accounts for 70% of the total income of a family, from 35 to 100 million VND/year, mainly from key fruit trees such as grapefruits, oranges, bananas and mangoes, as well as tea and peppers.

On hollow lands, the government has encouraged local farmers to build ponds for raising fish and softshell turtles, combined with growing fruit trees and raising pigs, chickens and ducks. As a result, the living conditions have greatly improved, and the villages renovated. Local farmers have converted sandy areas in Ha Tinh, Quang Binh, Quang Nam, Quang Ngai, Binh Đinh and Ninh Thuan into agricultural lands. Furthermore, fish ponds were dug and the water drained from the pond is used for irrigating the fruit gardens. The main products are pitahaya (dragon fruits), lemons and mangoes. The cultivation of fruits and vegetables has caused an increase in income of up to 40–50 million VND/family per year. Thus, the VAC system is an illustrative example of sustainable subsidence agriculture.

6.2.4.3 VAC Systems in Southern Vietnam

In southern Vietnam, agriculture is practiced mainly in the Mekong Delta. VAC in the Mekong Delta is very diverse, dynamic and commercially oriented (Dang et al., 2005). The VAC system in the Mekong Delta is characterized by five common farm resource systems: homestead, orchard dikes, ponds, orchard trenches and rice fields. The farming is characterized by particular agro-ecological conditions:

- **At macro scale:** Large-scale farming is commonly practiced in central parts of the delta where the flood depth is medium, freshwater is available throughout the year and the soil is alluvial.
- **At micro scale:** Small-scale farming is usually practiced in areas located along rivers, canals and roads where the topography is higher, the soil is more fertile and access to freshwater is easier.

Before 1999, rice farming was the first priority to ensure food security and export, followed by aquaculture, animal husbandry and fruit production, which became increasingly important for agricultural development, including the agro-processing industry and exports. According to Dang et al. (2005), the general configuration of the VAC system in the Mekong Delta and its resource flows are illustrated in Figure 6.7.

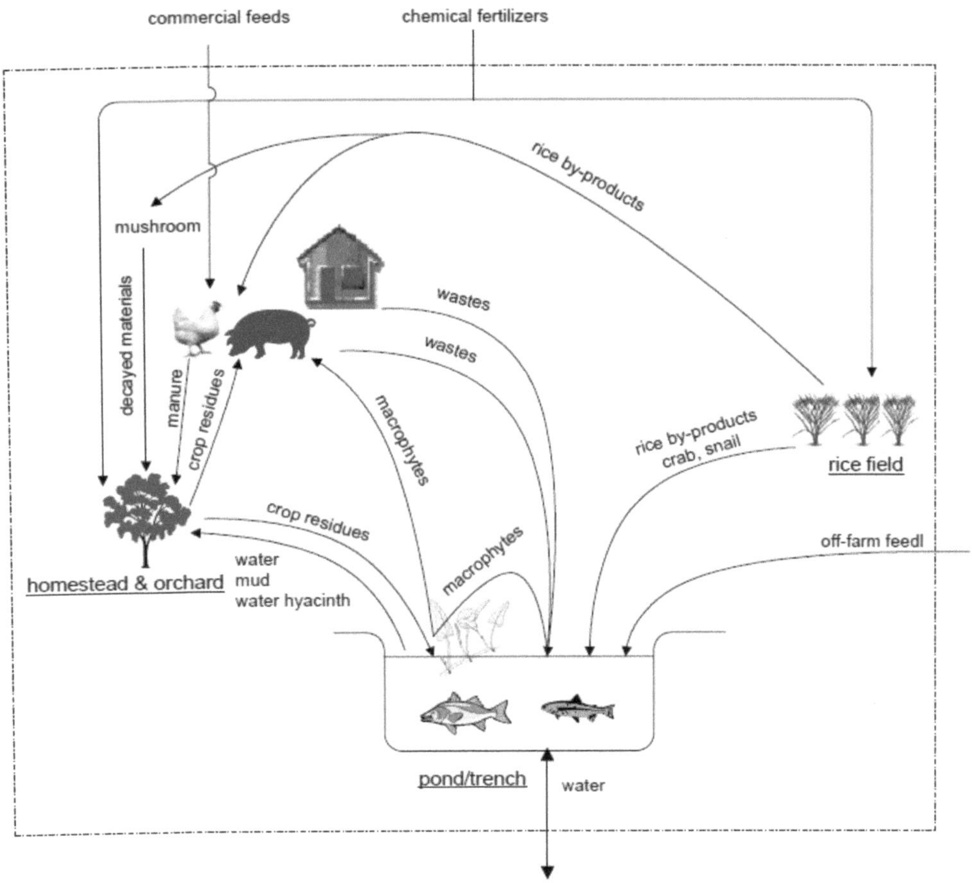

FIGURE 6.7 General configuration of the VAC system in the Mekong Delta. (Source: Dang et al. 2005.)

Agricultural restructuring and diversification in the Mekong Delta from 1990 to 2001 led to changes in agricultural land uses and crops, aquaculture and animal production methods. For example, the decline in rice productivity was mainly caused by the coastal shrimp culture and partly by intensive *Pangasius* catfish production. The same situation occurred in fruits and aquaculture production.

The annual growth rates in pig and poultry husbandry declined, then increased together with crop productivity, aquaculture and animal farming thanks to the comprehensive policy innovations of the government in 1986 (Doi Moi Policy). Figures 6.8 and 6.9 illustrate typical VAC systems in southern Vietnam, characterized by rice fields.

According to Dang et al. (2005), seven common VAC farming systems can be classified based on the uses and intensity levels of orchards and ponds:

1. ***Intensive fruit culture low input aquaculture:***

   The main fruit crops are longan, lemon, mango, apple and sapodilla.

   - Poultry is free-range farmed, and pigs are raised on a very small scale (one or two pigs per household).
   - Hatchery-produced fingerlings in fruit orchard trenches.
   - Fish polyculture (<2 tons/year) with low production, mainly relies on natural food in orchard trenches: silver barb (*Barbodes gonionotus*), Indian carps (*Labeo rohia*, *Cirrhina mrigalla*), silver carp (*Hypophthalmichthys molitrix*), grass carp (*Ctenopharyngodon idella*), giant gouramy (*Ospronemus exodon*), Colossoma (*Colossoma brachypomum*) and *Pangasius* catfish (*Pangasianodon hypophthalmus*).

FIGURE 6.8 Illustration of the typical VAC system in southern Vietnam.

FIGURE 6.9 Typical VAC system in the Mekong Delta.

2. ***Intensive fruit culture medium input aquaculture:***

   Fruit cultivation is the particular focus of this system.

   - The animal component is quite commercial such as pigs and poultry at small–medium scales.
   - Omnivorous fish species (2–10 tons/ha per year) are giant gouramy (*Pangasius* catfish), hybrid walking catfish (*Clarias macrocephalus*, *C. gariepinus*), climbing perch (*Anabas testudineus*) and detritivorous fish (snakeskin gouramy *Trichogaster pectoralis*).

6.2.5 VAC Off-Site Integrations

From a broader view, VAC-integrated farming also includes off-site integrations: the use of off-farm resources, such as agro-industrial products and byproducts (Dang et al., 2005). A more recent definition is based on a human livelihood perspective and an expected increase in expansion towards peri-urban areas. Linkages could take place between different farms and specialized agro industries and integrated farming involving aquaculture. There are concurrent or sequential linkages between two or more human activity systems (one or more of which is aquaculture), directly on site, or indirectly through off-site needs and opportunities, or both. More integration (diversification) provides chances for more nutrient flows and the possibility to meet increased nutrient requirements to intensify production.

- **At farm level**: Open avenues for on-farm or concurrent integration, both on small- and large-scale commercial agro businesses, with manure and fish production taking place on the same farm.
- **At community or regional level**: Open opportunities have been created for off-farm integration (between farms) such as the sale of chicken manure to specialized fish farms.

In integrated resources management, this is an evolving approach to the sustainable use of natural resources.

6.3 Advanced VAC System

In the last decade, huge progress has been made in integrating modern innovative biotechnological systems into the VAC approach. The new components are composting and biogas production facilities. For instance, in the Biorist project (February 2016–January 2019), biogas from manure and rice straw was produced. Biogas is used for cooking. For larger production, electricity is produced for household use. Digest from the biogas facility might be used as liquid organic fertilizers on rice fields, thereby reducing the amount of mineral fertilizer needed. Figure 6.10 provides a schematic illustration of an advanced VAC system.

In Vietnam, the traditional use of rice straw as a construction material or cattle feed has largely decreased and thus most of the rice straw is either combusted close to the harvested areas, emitting high amounts of particulate matter, or it remains in the paddies, where it is mainly converted to carbon dioxide and methane. Since 1991, the production of paddy rice has doubled and, depending on the rice species, the amount of rice straw has increased to 58 million tons/year, which is equivalent to 15 million tons of methane if it were fully processed to biogas. In accordance with the Vietnamese master plan for power development, the production of biogas will increase over the next two decades by 500 MW until 2020 and a further 200 MW until 2030 (Government of Vietnam, Decision 1208/QD-TTg). Particularly in southern Vietnam, advanced VAC systems are needed, as the volume of residual biomass is very high due to rice production. The economy of Vietnam is dominated by agriculture with paddy rice as the major crop cultivated on 4.5 million ha. Vietnam is the third-largest rice producer in the world and the main rice-producing region is the Mekong River Delta (1.9 million ha, 50% of national rice production of which 90% is exported).

Rice straw can also be used for bio-ethanol production to replace fossil fuel. Before fermentation to produce bio-ethanol, conversion of lignocelluloses into fermentable sugars including glucose, xylose, arabinose, galactose and mannose takes place using pretreatment methods, including physical, chemical and biological methods. Binod et al. (2010) reported that bio-ethanol may yield as much as 0.42 L/kg. Following bio-ethanol production, the residue can be used for compost production, which can be added to marginal soils to improve paddy rice productivity. Rice straw and husk can also be used to produce biochar, another organic amendment to increase crop productivity. Because biochar is stable in soil, as indicated by a long soil mean residence time of up to thousands of years (Kuzyakov et al., 2014), the addition of biochar to flooded soils may reduce biochar-C decomposition and thus greenhouse gas emissions. As biochar was reported to increase available water-holding capacity, the addition of the amendment is expected to improve rice yield in the dry season and increase water-usage efficiency.

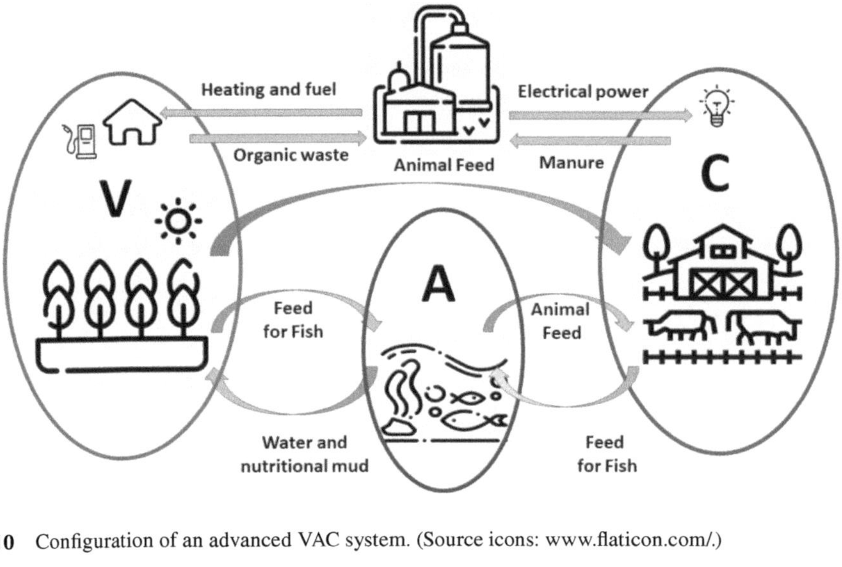

FIGURE 6.10 Configuration of an advanced VAC system. (Source icons: www.flaticon.com/.)

6.4 Ecosystem Services of the VAC System

Due to its extensive land use, agriculture influences numerous ecosystem services such as water supply, water quality, soil protection and climate protection. Agriculture is both a provider and a user of ecosystem services. The relationship between ecosystem services and agriculture is complex and the demarcation is blurred because natural processes play an important role in agriculture and the services of agriculture (out of the sphere of influence of humans) often connected to the services of the ecosphere. The Millennium Ecosystem Assessment (2005) grouped ecosystem services into four broad categories, which are illustrated for the particularities of the Vietnamese VAC system as follows:

Provisioning: Production of food, energy, water, wood, etc.
Regulating: Climate extremes buffer, disease and pest control
Supporting: Nutrient cycles, oxygen production, water purification
Cultural: Educational (traditions) and recreational benefits

Agriculture benefits from ecosystem services such as fertile soil, water availability and pollination and also provides services such as securing the biodiversity of the cultural landscape. Agriculture and related land management therefore not only contribute to the preservation of ecosystem services, but can also jeopardize their continued existence by impairing ecosystems. The benefits that these services represent for the population are divided into the group's health, safety, natural diversity and economic performance. This takes into account the well-being of people and the economic aspect of the final ecosystem services as well. The inventory serves to raise awareness and make visible the achievements of nature in connection with agricultural production. The VAC output has been assessed and recognized to be of economic, social and environmental benefit. VAC enhances nutrition and health benefits: the VAC system improves nutrition standards in families' diets by providing vegetables and fruits of high nutritional value (spinach, legumes, carrots, bananas, papayas, lemons, etc.), and eggs, fish and meat for consumer needs in families.

VAC is based on a scientific approach for sustainable development. It links "traditional" elements such as indigenous plants and animals with "modern" elements such as advanced and modern biotechnology, and uses the solar system as well. In the garden, there are many different types of high and low plants to make the most of photosynthesis from sunlight. In the pond, many kinds of fish and shrimps could take advantage of food sources in many layers by depth of water. Furthermore, VAC technology is based on the regeneration strategy: sunlight is regenerated through the photosynthesis of leaves to create food sources for human needs, providing raw materials for the processing industry. Waste and residues from animal husbandry and farming activities are regenerated as the source of input materials (biogas, compost) for subsequent production cycles and to keep the environment clean. The role of people in promoting reciprocal relationships in VAC: the system is operated by an important factor, which are people. Indeed, people have played an important role in promoting reciprocal relationships in VAC. People at the centre consume VAC products to survive and develop and vice versa, they influence and adjust the reciprocal relationships between components of the system with complementary factors from outside such as biological fertilizers, animal feed, etc., and control the waste treatment process of VAC, in order to achieve the highest efficiency for the system and environmental protection. VAC has expanded its concept to develop hundreds of hectares of hill gardens, farms, forests, swamps, lakes and other wetlands. The breeding area has also developed on the farm with thousands of animals and livestock. The definition of VAC has therefore expanded:

- **V:** A symbol of farming activities, because "gardening" is not only in the home garden but also in the forest and on the fields.
- **A:** A symbol of diversified aquaculture to exploit surface water resources in ponds, lakes, ditches, rivers, streams and seas with fish, shrimp, crab, frogs, seaweed products, algae, tortoises, etc.

- **C:** A symbol of livestock activities on large-scale sites such as chickens, ducks, pigs, goats, buffalo, cows, etc. In some places, C also includes raising horses, elephants and some specialties such as deer and pythons.

The development of the VAC system can contribute to limiting pollution and improving the environment. Another important factor is that VAC helps to optimize the use of land, water and solar resources; all wastes go through a closed production cycle (biogas, fertilizer, biochar). The sustainability of VAC is particularly characterized by the fact that the core system is still based on the principle of agrarian solar societies.

6.5 Conclusions

The VAC economic model is an integrated farming system that combines tradition and modernity, and ensuring a stable economic income for farmer households is a major constituent element. However, historically during the development of the Collective Agricultural Cooperative (Loos et al., 2019), the VAC farming system was almost forgotten. VAC was restored and developed in the period when the policy of "Doi Moi" was enacted, which encouraged household economic development. VAC is a highly bio-intensive model in which farming, aquaculture and cattle raising are the main activities. These activities have close ecological relationships and loops, creating a comprehensive farming system, helping to better use the land, water and solar energy more rationally to achieve effective high economic results with low investment.

Basically, agriculture is determined to be sustainable when its development not only meets the needs of the current generation, but also satisfies the needs of future generations. Sustainable agriculture is mainly based on the diversity and abundance of sustainable agricultural systems with high economic potentials, meeting the human need of food, in which the development does not negatively affect social resources, natural resources or the environment. In the different ecological regions of Vietnam, VAC systems can meet the requirements of sustainable agriculture with basic economic, social and environmental aspects, particularly:

- VAC ensures a biological balance and improves the soil to ensure high economic efficiency and stability over the long term.
- VAC contributes to hunger eradication and poverty alleviation by increasing incomes, creating local jobs and improving living standards for farmers.
- VAC exploits natural resources to create diverse food sources without exhausting these resources and contributes to creating a cleaner and nicer environment.
- VAC development is to establish an important part of bio-agriculture, a clean and sustainable agriculture.

By nature, the VAC system is a dynamic and flexible system. Depending on local and regional conditions, VAC biological systems may include all three components V, A, C or just two components such as VA, CV or AC. In such cases, human interaction with the components of the system still plays a decisive role. The growth of plants and animals depends on the important effects of human adjustments such as fertilizer or animal feed provided according to the biological characteristics of each crop or animal. However, the condition of the surrounding environment is an important factor determining their growth process. Therefore, soil and climate conditions are the basic factors forming VAC models for different ecological pools.

REFERENCES

Binod, P, Sindhu, R, Singhania, RR, Vikram, S, Devi, L, Nagalakshmi, S, Kurien, N, Sukumaran, RK, Pandey, A (2010). Bioethanol production from rice straw: An overview. *Bioresour Technol* 101(13), 4767–4774. doi:10.1016/j.biortech.2009.10.079. Epub 2009 Nov 26.

Clark, C (1940). *Conditions of Economic Progress*. London: Macmillan. x + 504. 25s.

Dang, KN, Le Thanh, D, Nguyen, VS, Verdegem, M (2005). Development of "VAC" integrated farming systems in the Mekong Delta, Vietnam: A view of a system and a participatory approach. In R Yamada (ed.), *Development of Integrated Agriculture Farming Systems in the Mekong Delta. Tuoitre*, 101–125.

Fischer-Kowalski, AM, Haberl, H (eds.) (2007). *Socioecological Transitions and Land-Use Change, Socioecological Transitions and Global Change: Trajectories of Social Metabolism and Land Use.* Cheltenham: Edward Elgar Publishing, ISBN 978-1-84720-340-3

Fourastié, J (1954). Die große Hoffnung des 20. *Jahrhunderts.* ("The Great Hope of the 20th Century"). Köln-Deutz.

Geißler, R (2002). Die Entstehung der Industriegesellschaft: sozioökonomischer Wandel im 19. Jahrhundert. In *Die Sozialstruktur Deutschlands.* VS Verlag für Sozialwissenschaften, 23–48. https://doi.org/10.1007/978-3-322-95667-5_2, ISBN 978-3-531-32923-9.

Gomiero, T (2016). Soil degradation, land scarcity and food security: Reviewing a complex. *Chal Sustain* 8, 281. doi:10.3390/su8030281

Haberl, H (2006). Wandel von Kulturlandschaften: Von der Biomasse zur Fossilenergie – und wieder zurück? *Informationen zur Raumentwicklung, Heft* 1/2, 111–123.

Hedberg, A (2017). Germany's energy transition: Making it deliver. *Discussion Paper, The King Baudouin Foundation Is the Strategic Partner of the European Policy Centre*, http://aei.pitt.edu/91830/1/pub_7997_germanyenergytransition.pdf, accessed 29.03.2020.

Krausmann, F, Gingrich, S, Eisenmenger, N, Erb, K-H, Haberl, H, Fischer-Kowalski, M (2009). Growth in global materials use, GDP and population during the 20th century. *Ecol Econ* 68(10), 2696–2705. doi:10.1016/j.ecolecon.2009.05.007.

Kuzyakov, Y, Bogomolova, I, Glaser, B (2014). Biochar stability in soil: Decomposition during eight years and transformation as assessed by compound-specific C-14 analysis. *Soil Biol Biochem* 70, 229–236. doi:10.1016/j.soilbio.2013.12.021

Le Thanh, L (2001). The VAC system in Northern Viet Nam. In *Integrated Agriculture-Aquaculture: A Primer, FAO Fisheries.* Technical Paper 407, www.fao.org/3/Y1187E/y1187e10.htm#l, ISBN 92-5-104599-2, accessed 30.04.2020.

Loos, J, Batáry, P, Grass, I, Westphal, C, Bänsch, S, Bosem Baillod, A, Hass, AL, Rosa, J, Tscharntke, T (2019). Vulnerability of ecosystem services in farmland depends on landscape management. In M Schröter, A Bonn, S Klotz, R Seppelt, C Baessler (eds.), *Atlas of Ecosystem Services.* Cham: Springer. https://doi.org/10.1007/978-3-319-96229-0_15

McDonough, W, Braungart, M (2002). *Cradle to Cradle: Remaking the Way We Make Things.* New York: North Point Press. Print.

Millenium Ecosystem Assessment (2005). *Millennium Ecosystem Assessment (Program). Ecosystems and Human Well-Being.* Washington, DC: Island Press.

Nguyễn Bình, L, Lê Đồng, T (2012). *HIỆN TRẠNG MÔ HÌNH NÔNG LÂM KẾT HỢP Ở HUYỆN PHÙ NINH TỈNH PHÖ THỌ.* http://iebr.ac.vn/database/HNTQ7/1669.pdf, accessed 30.04.2020.

OECD, (2017). Agricultural Policy Monitoring and Evaluation 2017, OECD Publishing, Paris. http://dx.doi.org/10.1787/agr_pol-2017-en. ISBN 978-92-64-27563-8.

Padmavathy, K, Poyyamoli, G (2011). Alternative farming techniques for sustainable food production. In E Lichtfouse (ed.), *Genetics, Biofuels and Local Farming Systems, Sustainable Agricultural Reviews 7*, ©Springer Science+Business Media B.V, 13, doi:10.1007/978-94-007-1521-9.

Pingali, PL (2012). Green revolution: Impacts, limits, and the path ahead. *PNAS* 109(31), 12302–12308. https://doi.org/10.1073/pnas.0912953109

Sebby, K (2010). *The Green Revolution of the 1960s and its Impact on Small Farmers in India, Partial Fulfillment of Requirements for the Bachelor of Arts and Sciences* Thesis, Faculty of The Environmental Studies Program at the University of Nebraska-Lincoln.

Sieferle, RP (1995). Naturlandschaft. *Kulturlandschaft. Industrielandschaft. Comparativ* 4, 40–65.

Valdes, J, Poque González, AB, Ramirez Camargo, L, Valin Fenández, M, Masip Macia, Y, Dorner, W (2019). Industry, flexibility, and demand response: Applying German energy transition lessons in Chile. *Energy Res Soc Sci* 54, 12–25, https://doi.org/10.1016/j.erss.2019.03.003, ISSN 2214-6296.

Van Huong, N., Huu Cuong, T., Thi Nang Thu, T., Lebailly, P. Efficiency of Different Integrated Agriculture Aquaculture Systems in the Red River Delta of Vietnam. *Sustainability* 2018, 10, 493.

Weixiang, W, Min, Y, Qibo, F, Kim, M, Hailong, W, Haohao, L, Yingxu, C (2012). Chemical characterization of rice straw-derived biochar for soil amendment. *Biomass Bioenergy* 47, 268–276.

7

On the Sustainability of Direct-Seeded Rice

Serge Svizzero

CONTENTS

7.1 Introduction

As highlighted by Sustainable Development Goal #2, a deep transformation of our current agricultural system will be a major issue at the global level in the coming decades. The purpose of the transformation is to shift from conventional agriculture to agroecology, the latter being not only more productive but also more resilient, sustainable and inclusive. Indeed, food security is a major issue because the world population is still increasing (9 billion people in 2050) with developing countries becoming wealthier, leading to a modified diet. It is expected that to achieve food security by 2050, food production needs to increase by 70–100% (Godfray, 2010). Since about 60% of the world's food energy supply is provided by a handful of crops (maize, rice, wheat, potato), attention focuses on the sustainability of their cultivation.

Several initiatives aim to improve the sustainability of rice cultivation (e.g. SRP, 2015; Winrock International, 2018; FAO, 2020a, 2020b) given its importance for food security, and its dependence and influence on the environment. On the one hand, rice is the daily staple for more than 3.5 billion people, accounting for 19% of dietary energy globally. It is produced by 144 million smallholders and provides livelihoods for over 1 billion people. However, the goal of feeding 9 billion people is facing a 44% shortfall in global rice production by 2050. On the other hand, rice is produced on 160 million hectares, uses 34–43% of the world's irrigation water for production and is responsible for up to 10% of global methane emissions. Rice farmers, especially in South East Asia, are among the most vulnerable to climate change impacts.

In order to improve the sustainability of rice cultivation, many scholars and organizations support a technological switch from the traditional—and dominant in Asia—hand transplanting (transplanted rice [TPR]) method to direct seeding (DSR) (Pandey et al., 2002; Pandey and Velasco, 2002; Kumar and Ladha, 2011; Joshi et al., 2013, Liu et al., 2015; Marasini et al., 2016; Chakraborty et al., 2017; Devkota et al., 2019, 2020). They consider that DSR is a new technology for climate change and food security that improves water and environmental sustainability, and profitability. Indeed, DSR reduces labour and tillage costs, improves water management and reduces methane emissions.

However, in addition to its economic and ecological advantages, DSR has a major disadvantage, namely it fosters the proliferation of weeds, especially weedy rice. As stated by the American Society of Agronomy:

> In agronomic and non-crop systems, weeds are the number one pest problem. Dollar losses associated with weedy and invasive plant species are in the billions. Weeds have invaded and threaten many ecosystems and are increasing in their importance for many state and federal programs.
>
> **(ASA, 2020)**

There is an abundance of academic literature on weeds in rice fields, given that they are present on all continents and that their adverse effects can be up to 100% of yield (Balasubramanian and Hill, 2002; Pandey and Velasco, 2002; Delouche et al., 2007; Marambe, 2008; Kumar and Ladha, 2011; Rathore et al., 2013; Singh et al., 2013; Matloob et al., 2015; Nadir et al., 2017). Even in the late 1990s—as indicated by the title of their article: "Weeds: A Looming Problem in Modern Rice Production"—Olofsdotter et al. (1998) were already aware of the weed problem, and especially of weedy rice, related to the shift from TPR to DSR.

The exacerbated weed problem under DSR has severe economic and environmental consequences. Farmers may end up using most of the labour saved by direct seeding to control weeds. In addition, the chemical cost of weed control tends to be higher than that of transplanted rice and can be as high as 20% of the total pre-harvest cost. More of the chemical weed control methods used in direct-seeded rice can also be potentially damaging to human health and the environment. In this chapter, our aim is thus to provide an overview of the advantages and disadvantages caused by the switch to direct-seeded rice in order to assess the sustainability of DSR. It appears that economic and ecological short-term advantages could be balanced by longer-term disadvantages implied by the proliferation of weeds under DSR.

7.2 Assessment of Rice Cultivation Sustainability

The Sustainable Rice Platform (SRP, 2015) is a multi-stakeholder partnership (established in 2011 and co-convened by the UN Environment and the International Rice Research Institute [IRRI]) to promote resource-use efficiency and sustainability both on-farm and throughout the rice value chain. The SRP is developing a range of tools to promote sustainable rice cultivation, including standards for sustainable rice cultivation that are based on priorities defined in a set of 12 performance indicators (PIs) (Table 7.1).

In 2017, Winrock International Institute for Agricultural Development, a US non-profit organization, launched an initiative titled "Adapting a Global Sustainability Standard for U.S. Rice Production" (Winrock International, 2018). Like the SRP, this US-sustainable rice certification (US-SRC) applies to

TABLE 7.1
Performance Indicators (PIs) for Sustainable Rice Cultivation as Defined by the Sustainable Rice Platform Standards

Indicator No.	Indicator Name	Category	Guiding Principles
1	Profitability: net income from rice	Economic	Improved livelihood
2	Labour productivity	Social	Improved livelihood
3	Productivity: grain yield	Economic	Improved livelihood
4	Food safety	Social	Consumer needs
5	Water-use efficiency	Environmental	Resource use efficiency and sustainable intensification
6	Nitrogen-use efficiency	Environmental	Resource use efficiency and sustainable intensification
7	Phosphorus-use efficiency	Environmental	Resource use efficiency and sustainable intensification
8	Pesticide-use efficiency	Environmental	Resource use efficiency and sustainable intensification
9	Greenhouse gas emission	Environmental	Climate change mitigation
10	Health and safety	Social	Labour conditions
11	Child labour	Social	Labour conditions
12	Women empowerment	Social	Social development

Source: SRP (2015).

rice production, but unlike the SRP—which focuses on smallholder rice growers in South East Asia—the US-SRC aims to recognize the differences between US practices and those common in other parts of the world. The goals of the US-SRC are to create a platform from which US rice growers could objectively claim the sustainability of their production practices and provide metrics to verify that these practices conform to internationally recognized standards.

In addition to the tools (SRP and US-SRC) used to assess the sustainability of rice cultivation, some initiatives (including funding) support the sustainable production of rice to assist farmers and supply chains adversely impacted by climate change in South East Asia. As an example, the Sustainable Rice Landscapes Initiative (SRLI), launched in 2018 (FAO, 2020b), aims to deliver smart and sustainable agri-solutions for rice farmers and supply chains supported by a USD 50 million funding proposal. The main objective of the partners in this initiative is to harness multiple opportunities to meet the growing global demand for sustainable rice, using a public–private partnership approach towards achieving the UN Sustainable Development Goals (SDGs). The initiative builds on the efforts of the Sustainable Rice Platform to drive transformative change in the rice sector, especially to contribute to meeting national greenhouse gas (GHG) reduction targets under the Paris Agreement, as well as to restore degraded landscapes and conserve biodiversity.

7.3 Rice Ecosystems and Establishment Methods

Figures for rice area and productivity in different ecosystems in different countries show that more than 90% of rice is cultivated under the irrigated ecosystem in China, Japan, Korea and Pakistan. On the contrary, in countries such as Bangladesh, Bhutan, Myanmar, Nepal and Thailand, about 50% of rice is produced in rainfed lowland and flood-prone ecosystems. Rice land ecosystems are categorized into four types: irrigated, rainfed lowland, upland and deep water rice (also called floating or flood-prone areas) (Greenland, 1997; FAO, 2020a). Rice area and production at the world level between the four rice ecotypes are distributed as follows (Matloob et al., 2015).

Irrigated rice covers 55% of the world's rice-growing area and provides 75% of total rice production. Rainfed lowland rice has a central position in the continuum of rice ecosystems. It is usually transplanted, and is grown in levelled, bunded fields that are shallowly flooded with rainwater. It is grown on about 25% of the world's rice area (about 48 million ha) and contributes about 18% of the global rice supply. One-third of South and South East Asian rice lands lie within this ecosystem. Upland rice cultivation is reported to be grown on 9.2% of the rice area and accounts for about 3.8% of total rice production. Flood-prone rice is defined as either that growing on inland areas where the sustained water depth exceeds 50cm or tidal (coastal) wetland rice. It accounts for 4% of world rice production.

7.3.1 Rice Establishment Methods: Traditional Puddled Transplanting Rice (TPR) and Direct-Seeded Rice (DSR)

There are two principal methods of rice establishment: traditional puddled transplanting (or, equivalently, traditional transplanted-flooded rice) and direct seeding (Balasubramanian and Hill, 2002). Transplanting involves replanting (usually by hand, but it can also be done mechanically) rice seedlings grown in nurseries to puddled soils. Puddling benefits rice by facilitating easy seedling establishment, creating anaerobic conditions to enhance nutrient availability, reducing water percolation losses and controlling weeds.

Direct seeding can be either dry or wet, based on the physical condition of the seedbed and seeds (pre-germinated or dry). In any case, it is "direct" because seeds (dry or pre-germinated) are directly sown (broadcasted or drilled). Under dry direct seeding, dry seeds are sown on dry (unsaturated) soils. This establishment method is traditionally practiced in most Asian countries, in rainfed upland and lowland ecosystems, as well as in flood-prone areas. Wet direct seeding involves sowing (by broadcasting or sowing in line) pre-germinated seeds in wet (saturated) puddled soils. It is a common practice in irrigated areas, and it is further subdivided into three sub-methods—aerobic wet seeding (when the pre-germinated seeds are sown on the puddled soil), anaerobic wet seeding (when the pre-germinated seeds are sown into the puddled soil) and water seeding—based on the level of oxygen in the vicinity of the germinating seed or the depth of the floodwater at seeding. Water seeding is widely practiced in the United States, Australia and Europe, and is mainly carried out by aerial water seeding.

7.4 Switch from Transplanting to Direct Seeding

Although DSR is the oldest crop establishment method, by the 1950s transplanting had become the dominant method of crop establishment in most of Asia while DSR remained dominant—from the 1920s or so—in Europe and North America. From the mid-1960s and the beginning of the Green Revolution, the importance of DSR increased in Asia. In 2000, it was estimated that the direct-seeded area in Asia was about 29 million ha, which was approximately 21% (28% for India, 5–9% for China) of the total rice area in the region (Pandey and Velasco, 2002). In some Asian countries, such as Sri Lanka and Malaysia, more than 90% of the rice-cultivated area was under DSR in the early 2000s (Kumar and Ladha, 2011).

Several authors and studies foster the switch from TPR to DSR (Pandey et al., 2002; Pandey and Velasco, 2002; Kumar and Ladha, 2011; Joshi et al., 2013, Liu et al., 2015; Marasini et al., 2016; Chakraborty et al., 2017; Devkota et al., 2019, 2020) because they consider DSR as low-input demanding; DSR is considered a new technology for climate change and food security that improves water and environmental sustainability, and profitability. In the conclusion of their global meta- and mixed-model analysis (based on data from 323 on-station [1980–2016] and 9 farm studies [2009–2016]), Chakraborty et al. (2017) stated that "traditional rice transplanting (...) can effectively be substituted by direct-seeding with no or little negative impact on rice grain yield but, more importantly, with significant reductions in water input use and methane emissions, and higher net income". For the FAO (2020a), in addition to the genetic improvement—namely the adoption of high-yielding and early maturing rice varieties—the application of combinations of existing technologies would save time, land and water for the intensification of rice production in the future. Among the six agronomic practices recommended by the FAO (2020a), because they are supposed to sustainably increase rice production, the following four have a direct or indirect link with DSR:

- Minimum and/or zero tillage.
- Direct seeding in lowland rice production.

- Aerobic rice or irrigated upland rice.
- Land levelling using laser beam.

Indeed, zero tillage cannot be performed in TPR, as well as aerobic rice. Land levelling, which is essential for good water management and also for weed control and the efficiency of nitrogen use, is relevant for both rainfed and irrigated rice systems, as well as for DSR and TPR.

By using the sustainable rice platform (SRP, 2015), Devkota et al. (2019, 2020) provide recommendations for agronomic practices in order to improve the sustainability of rice cultivation in different countries. They favour the switch to DSR: indeed, in their studies of six Asian countries, these authors recommend "not flooding the fields for >30 days before planting" (Devkota et al., 2019). Their study of rice cultivation in eastern India is even more explicit; they recommend the adoption of DSR (rather than TPR) since it has the lowest production cost (provides a reduction of USD 130/ha) and the highest GHG reduction, even though it provides a lower yield (Devkota et al., 2020).

7.5 DSR Economic Advantages: Higher Profitability through Costs Reduction and/or Higher Revenue

It is often claimed in the academic literature that due to manpower limitations and the problematic water management of traditional puddled rice transplantation, rice cultivations have increasingly switched to DSR. This is so because, generally, water availability and the opportunity cost of labour are the major determinants of crop establishment methods. In fact the switch from TPR to DSR is the result of a combination of rice growers' different motives related to economic profitability and risk management.

The economic motive for farmers is to achieve higher profitability or net economic returns, defined as the difference between the gross economic returns and the total cost. With conventional tillage, farmers get a higher net economic return under either wet or dry DSR compared to TPR, with 13% higher for wet DSR (Chakraborty et al., 2017). Under dry DSR with zero tillage, the economic returns are even higher by 25.9% compared to TPR, due to high savings in water input and cost of cultivation. Two reasons explain such an increase in farmers' profitability: a reduction in production costs and/or an increase in total revenue (gross economic returns).

The production costs include two parts: the cost of inputs (seeds, agro-chemicals fertilizers and pesticides) and the cost incurred by the use of production factors (human labour and machines) in field operations (such as land preparation, sowing, irrigation, weeding, agro-chemicals applications, harvesting, threshing). An overall analysis of 77 published studies in Asian countries shows that various methods of direct seeding reduced the cost of production by USD 9–125 ha compared with conventional practice (TPR) (Kumar and Ladha, 2011). According to Chakraborty et al. (2017), wet DSR incurred a reduction in cost of 2.4–8.8% compared to TPR. All these cost reductions were largely due to lower labour cost or tillage cost or both under DSR.

7.6 Reduced Tillage and Cropping Intensification

7.6.1 Reduced Tillage

According to the chosen rice establishment method, the necessity of tillage can be relatively relaxed. Under TPR, rice seedlings are transplanted into puddled soil after land preparation with wet intensive tillage, a method which is very energy intensive. Tillage operations account for 15% of the total production cost of irrigated rice and farmers in the Indo-Gangetic Plains are assumed to spend USD 50–60 ha on land preparation (Matloob et al., 2015). On the contrary, under dry DSR, seeds are sown on unpuddled soil, either after dry tillage or conservation tillage. The latter involves either no tillage or minimal tillage, and is called "conservation agriculture" when crop residues are used as mulch. Of course, reduced tillage and a fortiori zero tillage has a significant positive impact on economic profitability because it reduces the production cost (USD 37–92/ha; Kumar and Ladha, 2011). In fact, several options exist for tillage (conventional, reduced and zero tillage) as well as for crop establishment (TPR, dry DSR, wet DSR). According to Chakraborty et al. (2017), under conventional tillage, wet DSR and dry DSR have a higher grain yield than TPR, with wet DSR having the largest yield advantage (1.3–4.7%). Even though wet DSR seems the best practice, this conclusion can be slightly modified given the influence of soil texture (clayey, loamy, sandy) and season (dry, wet). Beyond the strictly economic benefits, the main benefits of minimum and/or zero tillage practices are the conservation of organic matter and soil moisture, the reduction in water and wind erosion, the reduction in fuels and animal and human energy and the time and water required for land preparation, and the possible provision of a favourable environment for biological activity (FAO, 2020a).

7.6.2 Cropping Intensification

DSR is also associated with higher profitability brought about by an increase in total revenue (or gross economic returns), the latter defined by multiplying the grain yield by the standard price of rice. Indeed, DSR allows an increase in cropping intensity, from unique (under TPR) to double or even triple cropping per year. To be sure, the DSR establishment method requires less time than TPR, partly due to earlier crop maturity by 7–10 days for wet DSR (Singh et al., 2013), and the availability of high-yielding, short-duration varieties (95–105 days), fostered by cropping intensification under DSR.

7.7 Reduced Labour Cost and Drudgery

DSR requires less labour than TPR (1–2 person/day per hectare for wet DSR versus more than 30 person/day per hectare for TPR; Balasubramanian and Hill, 2002); therefore, the labour costs incurred by farmers are substantially reduced. A reduction in labour input of 11–75%, depending on the type of direct seeding, location, season and weed infestation, has been documented in DSR systems (Pandey and Velasco, 2002). DSR also avoids the drudgery of hand transplantation, the latter being considered "a cruel and inhuman procedure, mainly inflicted to women" (Gressel, 2005). Compared to hand transplanting, direct-seeding methods save on labour and can reduce the pre-harvest labour requirement by as much as 50%, depending on the nature of the production system (Pandey and Velasco, 2002), and by as much as 60%, with an average saving of 27%, according to Kumar and Ladha (2011, 342). Such difference in labour use is explained by the fact that DSR (dry and wet) avoids several stages of the production process under TPR, such as nursery raising, seedling uprooting and transplanting. However, the total labour saving mainly depends on whether weeds are present and how they are controlled. Indeed, labour use for controlling weeds is higher (12–200%) in DSR than in TPR (Kumar and Ladha, 2011). One may also notice that even though the shift to DSR is mainly explained by the increase in labour cost, the latter is directly dependent on the rate of growth of the entire economy, i.e. including all sectors beyond agriculture.

7.7.1 Labour Bottleneck

Another disadvantage of the hand transplanting method is that because transplanting must be completed within a short period of time, it requires a huge amount labour. This can lead to seasonal labour shortages, especially in developing countries in which farming activities are mainly family based. Indeed, for poor rural families it is almost economically impossible to hire people other than from within the family. The resulting labour bottlenecks during the hand transplanting period hinder the growth cycle of rice and ultimately the yield. Indeed, since seedling transplanting is often delayed because of manual transplanting and, as all fields are not transplanted simultaneously, this also results in above-optimal age of rice seedlings, producing fewer tillers per plant in addition to enhancing transplanting shock (Matloob et al., 2015). Under the DSR system, the demand for labour is spread out over a longer time period because agricultural pre-harvest operations (such as land preparation, laddering and weeding) in this system are spread over several months. Therefore, farmers can make full use of family labour, thereby avoiding labour bottlenecks.

7.8 More Efficient Water Use

Globally, the share of water for agriculture is declining fast, especially in Asia due to rising population levels. Several reasons explain this decline: the competition for other water uses (e.g. domestic, industrial) and the lower water quality (due to chemical pollution, salinization). As a result, physical and economic scarcity of water is often viewed as a driver for DSR.

Transplanted rice is a gigantic user of water with consumption that is usually three times higher than other upland crops. In conventional TPR, the production of 1 kg of rice requires on average 2,500 L of water (ranging from 800 to more than 5,000 L) (Kumar and Ladha, 2011). Of the total available fresh water, 90% is consumed by the agricultural sector and more than 50% of this fraction is used to irrigate paddy fields in Asia. These large quantities of water required by TPR are explained by the fact that the water input to rice fields has several uses. Indeed, it not only serves in land preparation, but it also compensates for evapotranspiration, seepage and percolation losses during crop growth.

Under DSR, the risk related to water management is reduced, i.e. a crop failure is less likely to occur. This is because DSR is less dependent on water resources than TPR, i.e. it has greater efficient water use and a higher tolerance to water deficit, either in rainfed areas or for irrigated rice cultivation. According to a review of 44 studies performed by Kumar and Ladha (2011), irrigation water use under DSR is 12–33% lower than under flooded TPR (12% for wet DSR, 21–25% for dry DSR after either tillage or zero tillage, 33% for dry DSR on raised beds). These water savings are confirmed by Chakraborty et al. (2017); whatever the tillage option, DSR has 15.6–38.3% lower water use than TPR. The largest water saving (38.3%) is associated with dry DSR and conventional tillage.

Under DSR, rice cultivation is possible even though water resources are highly variable or very low. When rice cultivation is within rainfed areas and rainfall is highly variable, direct dry seeding reduces the production risk. Thus, rainfall uncertainty can explain (i.e. even if there is no labour scarcity) the shift to DSR in some regions. Similarly when irrigation water supply is low, successful rice cultivation (especially direct dry seeded) can be established using pre-monsoon showers. Even water use in direct wet-seeded rice is substantially lower than in transplanted rice.

7.9 Reduction of Methane Emissions (CH_4) and Other GHGs

The switch to DSR is also explained by the reduction in methane emissions (CH_4) that are induced by flooded paddy fields. However, the extent to which such a reduction in methane emissions is a real motive for farmers to switch from TPR to DSR is unclear. It seems more likely that the reduction in methane emissions is an unintended consequence of the switch to DSR.

In 2005, it was estimated that of global anthropogenic emissions, agriculture accounted for 60% of N_2O and 50% of CH_4 (Smith et al., 2007). More recently, it is thought that food production could account for as much as one-third of man-made methane emissions. A total of 13% comes directly from farming activities, including flooded paddy fields (in addition to manure decomposition, digestion of forage by ruminants and application of nitrogen applications) (Bradshaw, 2016). Flooded rice culture with puddling and transplanting is considered one of the major sources of methane emissions (Sass, 2002)—and to a lesser extent of carbon dioxide (CO_2) and nitrous oxide (N_2O)—because prolonged flooding results in an anaerobic soil condition. Flooded rice fields emit CH_4 due to a methanogenesis process that occurs in anaerobic conditions during which organic matter undergoes decomposition. In fact, flooding rice fields blocks oxygen penetration into the soil, which allows bacteria capable of producing methane to thrive. Field methane emissions from rice cultivation are estimated according to the 2006 IPCC guidelines for national greenhouse gas inventories (IPCC, 2006). These estimations take into account the rice ecosystem types and water regimes during a cultivation period. Most N_2O emissions derive from nitrogen fertilizer application, while CO_2 emissions are mainly from farmers' activities on their land, particularly when crop residues are burned and machines use energy (i.e. emissions from fossil fuel utilization).

Several mitigation practices in rice management exist that reduce net GHG emissions. For instance, draining wetland rice once or several times during the growing seasons reduces CH_4 emissions (Smith et al., 2007). Rice–wheat rotation is another practice. In fact, factors affecting CH_4 emissions are numerous and include weather conditions, the water regime, soil properties, land practices about irrigation, organic amendments, fertilization and rice varieties. According to Kumar and Ladha (2011), under continuous flooding, the reduction in CH_4 emissions ranged from 24% to 79% in dry DSR and from 8% to 22% in wet DSR, whereas under intermittent irrigation, the reduction ranged from 43% to 75% in dry DSR compared with TPR. These conclusions are confirmed by Chakraborty et al. (2017); whatever the tillage option, methane emissions are reduced by 43.9–63.4% under DSR (compared to TPR), with the largest reduction (63.4%) under dry DSR with zero tillage.

7.10 Weeds of Rice

Agricultural weeds are weeds that invade and persist in agricultural fields. They have evolved to survive in the artificial setting of crop fields, but they are not intentionally selected by humans and evolve through unintentional human-mediated selection. Thus, agricultural weeds have occurred under the same forces that guide evolution in natural settings, including genetic drift, gene flow and natural selection. There are two broad categories of agricultural weeds. On the one hand are weeds—such as weedy rice, a conspecific species of rice with an AA genome—closely related to domesticated species, which are known as conspecific and congeneric weeds. Since gene flow and introgression from and to crops are possible for weeds with crop relatives, these forces influence the mechanisms of weed evolution. On the other hand are weeds that lack crop relatives; therefore, the evolution of these weeds is restricted to new mutations or standing variation.

7.10.1 The Agricultural Weed Syndrome

Agricultural weeds are neither wild nor domesticated; they compose a unique evolutionary state that has developed in parallel with crop domestication. Indeed, their adaptive traits are, for some, shared with crops (such as an erect growth habit or disease resistance), and for others, similar to those favoured in wild species (e.g. seed dormancy, seed shattering). Adaptive traits enabling agricultural weeds to survive and thrive vary from species to species. However, many are related to the competitive ability and persistence of weeds in agricultural settings. By analogy to the domestication syndrome (Hammer, 1984), this suite of traits can be defined as the "agricultural weed syndrome" (Vigueira et al., 2013). It mainly includes traits such as crop mimicry, seed dispersal, seed dormancy, rapid growth, high nutrient use efficiency and herbicide resistance.

TABLE 7.2

Important Weeds of Rice

SI. No.	Scientific Name	English Name
1.	*Echinochloa colonum*	Wild rice
2.	*Echinochloa crusgalli*	Barnyard grass
3.	*Cyperus rotundus*	Purple nut sedge
4.	*Cyperus irri*	Yellow sedge
5.	*Cyperus difformis*	Common sedge
6.	*Eclipta alba*	False daisy
7.	*Celosia argentia*	Cock's comb
8.	*Dactylotenium aegypticum*	Crowfoot grass
9.	*Setaria glauca*	Cattail millet
10.	*Scripus* spp.	Club rushes
11.	*Panicum* spp.	—
12.	*Paspalum* spp.	—

7.10.2 Crop Weeds and Weedy Crops

The damage caused by weeds can be seen in various ways and seriously affects various agricultural processes. Weeds cause problems due to (FAO, 2006):

- Competing with crops for space, water, light and nutrients.
- Releasing root exudates and foliar leachates toxic to crops.
- Creating a favourable habitat for the proliferation of other pests (e.g. arthropods, nematodes, mites and pathogens), serving as hosts for them.
- Interfering with the normal harvesting process and contaminating produce.

Weedy crops are not completely wild, i.e. they are an intermediate form between wild and domesticated forms. Weedy rice illustrates this since itreverses to wild traits such as seed retention, seed dormancy or red pericarpcolor. However, some domesticated traits are kept, such as an erect growth habit and a tall height. In fact, the traits of weedy rice are the result of natural selection, i.e. the adaptation of plants that have to survive and reproduce in agricultural fields. In the latter, weedy crops have some advantages such as easier access to arable soil, fertilizers, irrigation and lesser predation (due to pesticides and other devices used by farmers to eradicate/monitor predators and pests other than weedy crops). However, they also have to compete with the crops cultivated in the field, as well as with the farmer. The competition with the former is about access to soil nutrients, water as well as light; this may explain why weedy crops have kept an erect growth habit and a tall height. Since most cultivated crops have nowadays a shorter stature (due to the introduction of dwarf genes since the Green Revolution), weedy crops that have a taller height and more tillers have a serious competitive advantage for light capture. Weedy crops also compete with the farmer since the latter may, unintentionally or consciously, eliminate them. Indeed, the farmer might unconsciously eliminate weedy crops if he was able to harvest them. To avoid such issue, weedy crops have adopted an asynchronous and shorter growth cycle compared to domesticated crops. They ripen more rapidly and shatter before harvesting. Seed dormancy is also an advantage for weedy crops. Indeed, when the farmer intends to suppress them, for instance by spraying herbicides or by rototilling/rotavaring the field before sowing, he is able to suppress only the seedlings of weedy crops that have already sprouted. All the other weedy seeds present in the soil seed bank remain safe and thus are able to germinate in subsequent months or years, due to their relative dormancy rates. In addition, crop mimicry—especially during the seedling stage—is also an adaptive strategy of weeds, including weedy species, in order to avoid human removal (i.e. by hand weeding) by physically resembling the crop in which weeds grow. All these adaptations, resulting from natural selection, explain why weedy crops thrive in cultivated fields, and not in the wild, and also why they represent a major problem for farmers.

Weedy rice (*O. sativa* f. *spontanea*), also known as "red rice", is an important weed of cultivated rice (*O. sativa*). The presence of weedy rice is well attested from an historical point of view. It has been an important weed of cultivated rice for hundreds of years (Ellstrand et al., 2010). It was identified in 1846 in the United States (Delouche et al., 2007) and in 1916 in South Korea (Nadir et al., 2017). In the literature, the reference to weedy rice in Italian fields occurred in the early 19th century (Grimm, 2013). As weedy rice is a conspecific species of rice with an AA genome, it has traits of both cultivated and wild rice. Vegetatively, weedy rice possesses several traits of the agricultural weed syndrome. It is a crop mimic and has a taller height, its infructescence shatters and its seeds exhibit dormancy. Beyond these common traits, weedy rice exhibits high genetic and phenotypic diversity (Tseng, 2013; Olajumoke et al., 2016; Rathore et al., 2016).

7.11 Rice Establishment Methods and Weeds

The traditional rice cultivation method is to flood the fields and then transplant the rice seedlings. The transplant culture was designed and adopted millennia ago in order to access important benefits, including weed control, resulting from the impact of flooding, hand transplantation and the crop–weed competition. Indeed, puddling of the soil with the retention of a film of water on the surface destroys weed seedlings that have emerged and results in anaerobic conditions that are unfavourable for the germination of weed seeds. In addition, the use of relatively well-developed seedlings for transplanting provides a competitive advantage to the crop. It also facilitates the identification of weed seedlings that do emerge for removal during either seedling transplanting—indeed, when rice is planted by hand, each seedling is scrutinized and weedy ones discarded—or post-transplant weeding operations.

7.11.1 Weed Proliferation under DSR

Even though it has several advantages, direct seeding also has several potential disadvantages. In Asian countries, the yield of direct-seeded rice under farmers' field conditions tends to be slightly lower than that of transplanted rice (Pandey and Velasco, 2002; Kumar and Ladha, 2011) (Table 7.3). The two main reasons for the lower performance of DSR are poor and uneven establishment and the weed problem that is exacerbated under DSR.

There are several reasons for poor and uneven establishment. First, surface-sown seeds (in wet DSR) can be damaged, i.e. either by predation or desiccated by dry weather or damaged by floods. Second is the problem of lodging at maturity because of poor root anchorage on the soil surface. Moreover, higher spikelet sterility is also observed in DSR compared to TPR.

The latter, i.e. the weed problem, is even worse. Weeds are a major constraint on the success of DSR in general and on dry DSR in particular. Changes in rice establishment methods as well as water, tillage and weed management practices in DSR lead to changes in weed composition and diversity (Juraimi et al., 2013). Thus, the adoption of DSR has exacerbated the weed problem, including grasses—such as weedy rice (*Oryza sativa* f. *spontanea*) and *Echinochloa* spp. and sedges—due to the absence of the suppressive effect of flooding on the weeds that emerge before or with the crop (Matloob et al., 2015). Indeed, in DSR the cultivated rice does not have a head start due to transplanting from nurseries. The simultaneous emergence of weeds with rice seedlings makes weed control in DSR a complex phenomenon due to the overlapping of planting and the need for weed control. Furthermore, the hot and humid conditions of tropical Asia are conducive to year-round luxuriant growth of almost all weed species, with numerous flushes of weeds during the entire rice-growing season. While cultivated rice, as well as weedy rice, is a C_3 plant, most of the weeds in rice are C_4 plants that have high adaptability and faster growth—due to high drought tolerance, water use and photosynthetic efficiency—and thus dominate the crop. Moreover, the crop–weed competition is now more favourable to weeds; indeed, as cultivated rice was dwarfed to increase the harvest index, the taller weeds now have a competitive advantage. Under DSR, the non-chemical control of weeds is less efficient than under TPR and so the weed problem is potentially exacerbated. However, the increasing availability of chemical weed control methods (except for weedy rice) has favoured the shift to DSR.

7.12 Economic Consequences of Weeds

The economic consequences of weeds present in rice depend on several parameters, including the rate of infestation, the type of weeds as well as the cultivated varieties and the measures taken to control them. Farmer's income is reduced due to a combination of the following factors:

- A "quantity effect", i.e. substantial reductions in yield.
- A "price effect", i.e. lower grain prices owing to discounts due to the presence of weeds seeds (including red rice) in the grain.
- A "cost effect", i.e. increased costs of production incurred by control measures such as labour, equipment, chemicals and energy. Since weedy rice is a conspecific and a congeneric of cultivated rice, it has a similar morphology and physiology; thus, it is very difficult to control by either hand weeding or chemical means.

When the reduction in a farmer's income becomes overwhelming, this can lead to the eventual abandonment of the area for rice production.

TABLE 7.3

Indicators of Weedy Rice Density and Extent of Infestation

Extent of Infestation	Density (number of mature plants per square metre [Durand-Morat et al., 2018])	Density (number of panicles per square metre [Chauhan, 2013])
Light	1	—
Moderate	4	15–20
Heavy	>10	21–30

7.12.1 Yield Reduction

Significant reductions in world rice production are estimated at 35, 24 and 16% due to weeds, pests and pathogens, respectively (Schreiber et al., 2018). When weeds are present in agricultural fields, the crop suffers depressed yields because the weeds compete with crops for space and access for both above and below ground resources such as light, soil nutrients and water.

In order to demonstrate that the weedy rice problem is very serious, different indicators (infestation rate, weedy rice density, yield loss, profitability threshold) are provided for various regions and countries. For instance, as many authors do, Nadir et al. (2017) quoted various studies about the infestation rate in different countries: Cuba (80%), Europe and the United States (30–70%), Malaysia (74%) and Senegal (50%). However, this information is not sufficient to assess the extent of the weedy rice problem. It is necessary to add another indicator, the density of weedy rice, either the number of plants per square metre or the number of panicles per square metre (Table 7.3).

7.12.2 The Yield Loss Function

The next step is to define the yield loss function, i.e. to provide the relationship between the weedy rice density and the induced reduction in yield. For the most part, the yield loss function relates the percentage of rice yield loss to the density of mature weedy rice plants per square metre (Zhang et al., 2014; Durand-Morat et al., 2018). Although it seems obvious that the yield loss curve is an increasing function of the weedy rice density, several scenarios are possible, i.e. this function may have various shapes and properties as can be deduced from Table 7.4.

All the different scenarios depicted in Table 7.4 are likely to exist because the yield loss function depends on several factors, including:

- Abiotic factors, such as temperature, rainfall.
- Biotic factors, e.g. the presence of weeds other than weedy rice.

TABLE 7.4

Variations in the Yield of Cultivated Rice Under Different Weedy Rice Densities

Weedy Rice Densities (plants/m²)	Induced Yield Reduction (%)	References
4	13	Kumar and Ladha (2011)
16	37	
25	48	
40	58	
215	82	
3–10	30–60	Rathore et al. (2013)
35–40	60/90 (for tall/short cultivars)	
2	24.94	Xu et al. (2018)
4	31.40	
6	33.69	
8	60.14	
5	19–22	Delouche et al. (2007)
22	35	
60	30–50	
215	83	

- The varieties of cultivated rice.
- The biotypes of weedy rice.
- The farming practices: seed density, method of establishment, amount of fertilizer applications.
- The eventual methods of weed management (application of herbicides, hand weeding, stale seed bed, previous crop rotation).

7.12.3 Thresholds Associated with the Yield Loss Function

Two worthwhile thresholds are associated with the yield loss function. The first is the threshold density of weedy rice that prevents yield losses; Smith (1988) reports that it is equal to one to three plants per square metre in the southern United States. This useful indicator is quoted in many research articles (Delouche et al., 2007; Rathore, 2013; Chauhan, 2013); however, one may wonder the extent to which it can be applied to regions or countries where the conditions for rice cultivation (as shown by the various and numerous factors influencing the yield loss function) are very different from those prevailing in the southern United States.

The second threshold is economic profitability, i.e. the yield reduction (in percentage) necessary to reduce a producer's profitability below the level that covers variable costs. When such a situation occurs, it means that it is more profitable for the rice producer that his rice field either rotates to another crop or goes into fallow. Durand-Morat et al. (2018) reported that on average this threshold equals a yield loss of 38%, and ranges from 24% to 51%. In addition, Durand-Morat and Nalley (2019) have estimated the number of years necessary to reach the profitability threshold. It is equal to 1.5, 2.5 and 4 years when the infestation rate is heavy, moderate or light, respectively (as defined in Table 7.3).

7.12.4 Beyond the Field: Lower Quality and Grain Price

Several defects appear in rough rice, which according to the US standards is defined as rice (*O. sativa* L.) that consists of 50% or more of paddy kernels. When they are co-harvested with cultivated rice, the seeds of weedy rice—which are edible but unpalatable (Vigueira et al., 2013), and red in most cases—are an additional defect that degrades the quality of the harvested grain. Indeed, it is not desirable for marketing and packaging purposes to have red kernels comingled with white rice, and their presence leads to a reduction in the value of the rice. This is because the weedy rice seed mimics rice seed, and thus it is difficult to mechanically separate it from rice seed. Removing it requires additional milling and separation through sorting machines; the dockage selects, segregates and weighs the foreign matter. The red layer of the weed grains harvested with the crop can be removed with extra milling, but this operation results in broken grains and grade reduction (Ferrero, 2003). When the red rice contamination of milled rice exceeds 2.5%, a dockage penalty is applied and its level increases with the rate of contamination (as shown in Table 7.5, based on Durand-Morat et al., 2018; Durand-Morat and Nalley, 2019).

TABLE 7.5

Grades and Grade Requirements of Milled Rice for US Department of Agriculture in 2019[a] and Dockage Penalty Applied by Riceland Foods Inc. in 2014

Grade	Weedy "Red" Rice (maximum limit, %)	Discount ($/metric ton)
US #1	0.5	—
US #2	1.5	—
US #3	2.5	6.6
US #4	4	13.22
US #5	6	26.43
US #6	15	48.95

[a] www.gipsa.usda.gov/fgis/standards/ricestandards.pdf.

7.13 An Integrated Weed Management (IWM) Strategy for Weedy Rice Control

Weeds of rice are difficult to control because of their genetic (for weedy rice), morphological and phenological similarities with rice. Thus, what seems an effective strategy against weeds in DSR is integrated weed management, i.e. using several methods—preventive, cultural, mechanical, biological and chemical—that target different phases of the weedy rice cycle (Labrada, 2002; Ferrero, 2003; FAO, 2006; Chauhan, 2013; Juraimi et al., 2013; Singh et al., 2013; Mahajan, 2014; Matloob et al., 2015; Mispan et al., 2019). Using a diversity of weed management methods is more important than striving to exclude any single method (Harker and O'Donovan, 2013). Indeed, any weed management method that is continuously repeated provides heavy selection pressure for weed adaptation and resistance to that practice. In theory, IWM aims to provide sustainable weed control, i.e. involves the selection, integration and implementation of effective weed control means with due consideration for economic, sociological and environmental consequences.

However, weed control methods that can be applied under DSR are far from being as sustainable as is usually thought. They are expensive and time-consuming, and are not environment friendly since they require more fuel energy, water and herbicide. Among the 12 performance indicators of the SRP, #8 is about "pesticide use efficiency", and concerns various pests (weeds, insects, rodents, birds, molluscs and pathogens). The assumption is that improved pesticide management leads to increased farm profitability (through increasing yield and/or decreasing input costs), decreased environmental contamination by pesticides (leading to improved biodiversity and human health) and greater food security. The scorecard associated with the weed management indicator is presented in Table 7.6.

In countries where DSR is widely adopted, herbicide use has steadily increased, resulting in the appearance of weed resistance to certain herbicides. The selective control of weedy rice has never been achieved to a satisfactory level with herbicides, such as the application of non-selective herbicides (paraquat, glyphosate, glufosinate) before and after the emergence of weedy rice, but before planting rice. Indeed, the application of pre-emergence (anti-germinative) herbicides provides weedy rice control but it is phytotoxic if applied less than at least 15 days before crop sowing. Moreover, the physiological similarity between weedy rice and cultivated rice makes it difficult to use post-emergence herbicides in rice fields. Since some weeds, such as weedy rice, cannot be effectively controlled either through non-chemicals methods or with herbicides, the scorecard (see Table 7.6) implicitly recognizes that these weeds imply the lowest level of performance (associated with a score equal to zero), i.e. their presence

TABLE 7.6

Scorecard of Weed Management Indicator

Requirement	Level(s) of Performance	Score
Nonchemical methods of weed control include: • Good land preparation • Flooding • Mechanical weeding • Manual weeding • Biological control agents Appropriate herbicide application follows the principles of IPM and meets all of the following criteria: • Herbicide is applied during the early crop growth stage before the rice canopy closes and when weeds are small. • An appropriate herbicide is used for the type of weed problem (choice of mode of action). • Local information about herbicide-resistant weeds is used when choosing an appropriate herbicide.	a) Weeds are controlled without herbicides.	/10
	b) Weeds are controlled with a combination of physical and chemical techniques, with a maximum of one herbicide application per season, which must meet the criteria for appropriate use.	/5
	c) Weeds are controlled with up to four herbicide applications (must have distinct modes of action) per crop cycle, meeting all criteria for appropriate use.	/2
	d) Weeds are not effectively controlled or are managed with inappropriate herbicide use.	0

Source: From SRP Performance Indicator #8 "Pesticide Use Efficiency" (SRP, 2015).

is strongly unfavourable to rice sustainability. Moreover, weed control methods do not usually lead to the total eradication of a weed infestation. Incomplete control of the weed for a given year could lead to eliminating the results of several years of good control. For instance, weedy rice escapes of 5% or less can produce enough seeds to restore the original soil seed bank population levels (Ferrero, 2003).

7.14 Conclusion

The increasing scarcity of labour and water resources has been the main driver in the ongoing shift from TPR to DSR. This shift has induced several short-term advantages such as higher profitability of rice production, better management of water resources and a reduction in methane emissions. However, weed proliferation is the main long-term disadvantage associated with DSR. Since the scarcity of labour and water resources is increasing with the passage of time, it is unlikely that the shift to DSR will be reversed to TPR in the coming years. Thus, weed control in DSR, and particularly the control of weedy rice, should be at the centre of any strategy aiming to improve the sustainability of rice cultivation. Weed control methods must be sought that are friendlier to the environment and substantially reduce the cost of weed management to farmers. Given the drawbacks associated with either IWM or the use of herbicide-resistant rice varieties, one should target the development of competitive and allelopathic rice varieties for increased weed competitiveness. Their use can minimize yield losses and herbicide dependence because these cultivars can suppress weed seed production, limit future weed infestations and fit easily into current agronomic practices.

REFERENCES

American Society of Agronomy (2020), Weedy and invasive plant species community, available online at www.agronomy.org/membership/communities/weedy-and-invasive-plant-species (accessed on February 15, 2020).

Balasubramanian, V, JE Hill (2002), Direct seeding of rice in Asia: Emerging issues and strategic research needs for the 21st century. In Pandey, S, Mortimer, M, Wade, L, Tuong, TP, Lopez, K, Hardy, B editors, *Direct Seeding: Research Issues and Opportunities*, pp. 15–39. Proceedings of the International Workshop on Direct Seeding in Asian Rice Systems: Strategic Research Issues and Opportunities, 25–28 January 2000, Bangkok, Thailand. Los Baños (Philippines): International Rice Research Institute, 383 p.

Bradshaw, JE (2016), *Plant Breeding: Past, Present and Future*. Switzerland: Springer.

Chakraborty, D et al. (2017), A global analysis of alternative tillage and crop establishment practices for economically and environmentally efficient rice production, *Scientific Reports* 7: 9342.

Chauhan, BS (2013), *Management Strategies for Weedy Rice in Asia*. Los Baños: International Rice Research Institute, 16 p.

Delouche, JC, NR Burgos, DR Gealy, G Zorrilla de San Martin, R Labrada (2007), *Weedy Rices – Origin, Biology, Ecology and Control*. FAO Plant Production and Protection Paper 188. Rome: Food and Agriculture Organization of the United Nations (FAO), p. 144.

Devkota, KP et al. (2019), Economic and environmental indicators of sustainable rice cultivation: A comparison across intensive irrigated rice cropping systems in six Asian countries, *Ecological Indicators* 105: 199–214.

Devkota, KP et al. (2020), Assessing alternative crop establishment methods with a sustainability lens in rice production systems of Eastern India, *Journal of Cleaner Production* 244.

Durand-Morat, A et al. (2018), The implications of red rice on food security, *Global Food Security* 18: 62–75.

Durand-Morat, A, LL Nalley (2019), Economic benefits of controlling red rice: A case study of the United States, *Agronomy* 9: 422.

Ellstrand, NC et al. (2010), Crops gone wild: Evolution of weeds and invasives from domesticated ancestors, *Evolutionary Applications* 3: 494–504.

FAO of the United Nations (2006), Recommendations for improved weed management, available online at www.fao.org/3/a0884e/a0884e00.htm (accessed on January 8, 2020).

FAO of the United Nations (2020a), Sustainable rice systems, available online at www.fao.org/agriculture/crops/thematic-sitemap/theme/spi/scpi-home/managing-ecosystems/sustainable-rice-systems/en/ (accessed on January 5, 2020).

FAO of the United Nations (2020b), The Sustainable Rice Landscapes Initiative (SRLI), available online at www.fao.org/asiapacific/partners/networks/rice-initiative/fr/ (accessed on January 9, 2020).

Ferrero, A (2003), Weedy rice, biological features and control. In Labrada, R, editor, *Weed Management for Developing Countries. Addendum 1*, FAO, available online at www.fao.org/3/Y5031E/y5031e09.htm#bm9 (accessed on January 8, 2020).

Godfray, HC (2010), Food security: The challenge of feeding 9 billion people, *Science* 327(5967): 812–818.

Greenland, DJ (1997), *The Sustainability of Rice Farming.* CAB International, in association with IRRI, Wallingford, UK and New York, NY, USA, 273 p.

Gressel, J, editor (2005), Introduction – the challenges of ferality. In *Crop Ferality and Volunteerism*, pp. 1–8. Boca Raton, FL: CRC Press.

Grimm, A et al. (2013), Microsatellite markers reveal multiple origins for Italian weedy rice, *Ecology and Evolution* 3(14): 4786–4798.

Hammer, K (1984), Das Domestikationssyndrom. *Die Kulturpflanze* 32: 11–34.

Harker, KN, JT O'Donovan (2013), Recent weed control, weed management, and integrated weed management, *Weed Technology* 27: 1–11.

IPCC (Intergovernmental Panel on Climate Change) (2006), Guidelines for national greenhouse gas inventories 2006, available online at www.ipcc-nggip.iges.or.jp/public/2006gl/pdf/4_Volume4/V4_05_Ch5_Cropland.pdf (accessed on January 1, 2020).

Joshi, E et al. (2013), Management of direct seeded rice for enhanced resource-use efficiency, *Plant Knowledge Journal* 2(3): 119–134.

Juraimi, AS et al. (2013), Sustainable weed management in direct seeded rice culture: A review, *Australian Journal of Crop Science* 7(7): 989–1002.

Kumar, V, JK Ladha (2011), Direct seeding of rice: Recent developments and future research needs, *Advances in Agronomy* 111: 297–413.

Labrada, R (2002), Major weed problems in rice—red/weedy rice and the *Echinochloa* complex. In *FAO Rice Information*, Vol 3. chapter 2.

Liu, H et al. (2015), Dry direct-seeded rice as an alternative to transplanted-flooded rice in Central China, *Agronomy for Sustainable Development* 35: 285–294.

Mahajan, G et al. (2014), Integrated weed management in rice. In Chauhan, S, Mahajan, G, editors, *Recent Advances in Weed Management*, pp. 125–154. New York: Springer.

Marambe, B (2008), Weedy rice: Evolution, threats, and management, *Tropical Agriculturist* 157: 43–64.

Marasini, S et al. (2016), Direct seeded rice cultivation method: A new technology for climate change and food security, *Journal of Agriculture and Environment* 17: 30–38.

Matloob, A et al. (2015), Weeds of direct-seeded rice in Asia: Problems and opportunities, *Advances in Agronomy* 130: 291–336.

Mispan, MS et al (2019), Managing weedy rice (*Oryza sativa* L.) in Malaysia: Challenges and ways forward, *Journal of Research in Weed Science* 2: 149–167.

Nadir, S et al. (2017), Weedy rice in sustainable rice production. A review, *Agronomy for Sustainable Development* 37(46).

Olajumoke, B et al. (2016), Competitive ability of cultivated rice against weedy rice biotypes—A review, *Chilean Journal of Agricultural Research* 76(2): 243–252.

Olofsdotter, M, A Watson, C Piggin (1998), Weeds: A looming problem in modern rice production. In Dowling, NG, Greenfield, SM Fischer, KS, editors, *Sustainability of Rice in the Global Food System*, pp. 165–173. Davis, CA (USA): Pacific Basin Study Center, and Manila (Philippines): International Rice Research Institute, 404 p.

Pandey, S, Mortimer, M, Wade, L, Tuong, TP, Lopez, K, Hardy, B, editors (2002), *Direct Seeding: Research Issues and Opportunities.* Proceedings of the International Workshop on Direct Seeding in Asian Rice Systems: Strategic Research Issues and Opportunities, 25–28 January 2000, Bangkok, Thailand. Los Baños (Philippines): International Rice Research Institute, 383 p.

Pandey, S, L Velasco (2002), Economics of direct seeding in Asia: Patterns of adoption and research priorities. In Pandey, S, Mortimer, M, Wade, L, Tuong, TP, Lopez, K, Hardy, B editors, *Direct Seeding: Research Issues and Opportunities*, pp. 3–14. Proceedings of the International Workshop on Direct Seeding in

Asian Rice Systems: Strategic Research Issues and Opportunities, 25–28 January 2000, Bangkok, Thailand. Los Baños (Philippines): International Rice Research Institute, 383 p.

Rathore, M et al. (2013), Weedy rice: An emerging threat to rice cultivation and options for its management, *Current Science* 105(8): 1067–1072.

Rathore, M et al. (2016), Characterization of functional trait diversity among Indian cultivated and weedy rice populations, *Scientific Reports* 6: 24176.

Sass, RL (2002), CH_4 emissions from rice agriculture. In Background Papers. *IPCC Expert Meetings on Good Practice Guidance and Uncertainty Management in National Greenhouse Gas Inventories*, pp. 399–417. Published by Institute for Global Environmental Strategies (IGES) for the IPCC, available online at www.ipcc-nggip.iges.or.jp/public/gp/bgp/4_7_CH4_Rice_Agriculture.pdf (Accessed on January 8, 2020).

Schreiber, F et al. (2018), Competitive ability of rice cultivars in the era of weed resistance. In Dunea, D, editor, *Plant Competition in Cropping Systems*. Intechopen.

Singh, K et al. (2013), Weedy rice: An emerging threat for direct-seeded rice production systems in India, *Rice Research* 1(1): 106.

Smith, RJ (1988), Weed thresholds in southern US rice (*Oryza sativa*), *Weed Technology* 2: 232–241.

Smith, PD et al. (2007), Agriculture. In Metz, B et al., editors, *Climate Change 2007: Mitigation*. Contribution of the Working Group III to the Fourth Assessment Report of the Intergovernmental Panel on Climate Change. Cambridge: Cambridge University Press.

SRP (2015), The Sustainable Rice Platform (SRP). Bangkok, available online at www.sustainablerice.org (accessed on January 6, 2020).

Tseng, TM (2013), *Genetic Diversity of Seed Dormancy and Molecular Evolution of Weedy Red Rice*. Theses and Dissertations, p. 595, available online at http://scholarworks.uark.edu/etd/595

Vigueira, CC et al. (2013), The red queen in the corn: Agricultural weeds as models of rapid adaptive evolution, *Heredity* 110: 303–311.

Vigueira, CC, W Li, KM Olsen (2013), The role of *Bh4* in parallel evolution of hull colour in domesticated and weedy rice, *Journal of Evolutionary Biology* 26(8): 1738–1749.

Winrock International (2018), Adapting a global sustainability standard for U.S. rice production, available online at www.winrock.org/ms/winrock-sustainable-rice-initiative/?pid=18789 (accessed on January 9, 2020).

Xu, XM et al. (2018), Effect of weedy rice at different densities on photosynthetic characteristics and yield of cultivated rice, *Photosynthetica* 56(2): 520–526.

Zhang, Z et al. (2014), A model of the relationship between weedy rice seed-bank dynamics and rice-crop infestation and damage in Jiangsu province, China, *Pest Management Science* 70: 716–724.

8

Integrated Biocycles Management for Harmonization of Agri-Based Bio-Economy in Tropical Natural Resources

Cahyono Agus, Pamungkas Buana Putra, Marinus K. Harun and Enggal Primananda

CONTENTS

8.1 Introduction

The tropical ecosystem has high temperature, rainfall, moisture and light intensity and rapid organic cycling throughout the year, thus it has the highest biodiversity and net primary production (Agus, 2018). As part of the Ring of Fire, the land is constantly fertile because it is continuously supplied with new volcanic material which contains a large number of essential nutrients, and the earth's plate has resulted in the accumulation of valuable mine minerals (Agus, 2019). Tropical natural resources are a major potential resource but have less economic value because they are mismanaged and are not based on nature norms. We must shift our paradigm from natural resource–based development to knowledge-based development (Agus, 2019). The paradigm shift from mere exploitation to efforts to empower natural resources will provide benefits and a shift from the red and green economy concept to the concept of a blue economy that will be more smart, global, focused and futuristic for sustainable development (Agus, 2018).

Indonesia, an archipelago country situated on the equator, has extensive geographic areas consisting of tropical land and seas. Indonesia occupies a land area of 1.9 million km^2 (thirteenth widest in the world) composed of 17,504 islands, and a total maritime area of 6.4 million km^2 (BPS-Statistics Indonesia, 2020; Biro Perencanaan dan Informasi, 2018). It contains abundant natural resources and varied commodities that meet human needs such as water, oxygen, food, medicine, fibre, wood, energy and industrial raw materials.

Indonesia's biomass productivity is classified as the highest in the world (750 g C.ha^{-1}.th^{-1}) due to the high amount and distribution of rainfall, air temperature, soil temperature and air humidity and the soil moisture regime (Agus, 2018). Although tropical soils are classified as old and nutrient-poor because they are supported by highly active microorganisms and fast closed cycles, the growth of vegetation is relatively fast, 10 times that of temperate regions. However, economic productivity in the tropics is much lower than in temperate regions because biological management is not yet efficient or effective, so food must be imported from countries with much lower land productivity.

Currently, around 60% of Indonesia's poverty is concentrated in rural areas, and more than 70% of rural poverty is related to agriculture. Improving agricultural conditions in the form of strategies, regulations, implementations, technology, management and institutions will improve all aspects of the lives involved. The food and energy independence programme is unable to be implemented due to the mismanagement of the 6M's (man, money, material, machine, method, management). The program is constrained systematically, with program ambiguity, project mafia, economic rent-seeking, prioritization of personal gain and group interests, neglecting tasks, project orientation, administrative constraints, weak oversight and climate change all providing problems.

The excessive exploitation of natural resources has resulted in tragedy due to repeated environmental and humanitarian disasters, which have become a painful reality. Natural resources are still considered as ancestral heritage in the concept of resource-based development, so they must change and adopt a new paradigm of knowledge-based natural resource management. Programs must apply science and technology, innovations and sustainable development concepts that are smart, global, focused and futuristic (Agus, 2018). Thus, a new paradigm is needed to create natural and human resource management that can support a sustainable environment and a dignified life.

The new paradigm of agriculture must empower all aspects of agriculture as the leading supplier of clothing, food and shelter, as part of a sustainable environment and as provider of environmental tourism (agro-tourism), bio-farm production and bioenergy production. The management of the agricultural sector must be integrated by eliminating sectoral egos. It must be developed in harmony with other land uses such as forestry, plantation, horticulture, settlement, mining, infrastructure, industry and tourism in one ecological management landscape unit, both horizontally and vertically.

Natural energy and mineral resources tend not to be renewable because of their protracted formation processes, while the speed of their exploitation is very high. Indonesia has experienced a golden era through the excessive exploitation of oil, coal, gold, tin, etc., generating abundant foreign exchange and supporting national development. However, these natural resources have run out, leaving behind environmental, social, cultural, humanitarian and, more significantly, economic burdens.

The biological resources in the tropics are renewable and can be sustainably managed because they are supported by climate, rainfall, temperature, sunlight and rapid cycles throughout the year. The management of biological resources with a focus on biodiversity in the form of plants, animals and humans in the tropics plays a strategic role in the realization of a sustainable environment and a dignified life. A resource management system based on nature will support a successful, harmonious and sustainable blue earth. This chapter discusses the opportunities for land (water, mineral and microclimate) and biological (plant, animal and human) resources with the superior characteristics of natural-based tropical ecosystems in providing added value to sustainable economic, environmental, social, cultural and welfare aspects. Environmental site engineering, genetic manipulation and technical and socio-cultural engineering contribute significantly to the initial sustainable development of land rehabilitation in a deserted tropical land (Agus et al., 2019c, 2019d, 2019e, 2020a, 2020b).

8.2 Tropical Forest Management

In the 1970s, the Indonesian government reasoned that forests as development capital should be utilized for economic development (Pemerintah Republik Indonesia, 1967). Timber exploitation is carried out in natural forests based on the system of the Forest Concession Rights and the Right to Collect Forest Products based on Government Regulation Number 21 of 1970 (Kartodihardjo and Supriono, 2000). However, forest exploitation practices are not carried out sustainably so the capacity and the quality of forests continue to be degraded (Pirard et al., 2016; Kartodihardjo and Supriono, 2000).

Once the supply of wood from natural forests was less than the needs of the forestry industry and because the potential and quality of natural forests had been degraded and damaged, the government created and implemented an Industrial Plantation Forest (IPF) programme in addition to developing silvicultural system practices in natural production forests (Suhendang, 2020; Budi, 2008). The concept of

IPF is to rehabilitate degraded and unproductive natural forests and maintain them based on the principle of benefits, the principle of sustainability and the principle of the company (Indrawan, 2008; Pemerintah Republik Indonesia, 1990).

Conceptually, IPF is carried out in a degraded area of a production forest. It plays an essential role in increasing the area of reforestation and reducing the deforestation of natural forests so that the environmental quality and supply of materials for the forestry industry can be maintained. However, Pirard et al. (2016) mention the controversy over IPF, which turned out to have negative impacts, such as social and economic conflicts (the tenure, the limited local community access, dirty practices of corporations and businesses, as well as the foreign exchange value of industrial products). Additionally, the extent of the rehabilitation of damaged forest production land cover is insignificant and environmental improvements have not been achieved. Several ecological disasters have occurred due to the use of exotic species, monoculture systems and clear-cutting systems (Ali et al., 2018; Pirard et al., 2016; Syahadat and Irawanti, 2014; Kartodihardjo and Supriono, 2000).

Silvicultural systems for the management of natural production forests in Indonesia continue to be developed and refined to support forest sustainability (Suhendang, 2020; Budi, 2008). In the 1990s, various modifications to the practice of the Indonesian selective logging and planting (TPTI) silvicultural system for Dipterocarpaceae natural production forests based on a combination of the regeneration practices of human-made forests and selective cutting began to be trialled. The various silvicultural systems are the selective cutting and line planting system (TPTJ), the Indonesian selective cutting line system (TJTI), shelterwood cutting, intensive silviculture (Silin) and the silvicultural multisystem (Suhendang, 2020; Suryatmojo et al., 2014; Panjaitan et al., 2013; Indrawan, 2002).

The era of ecosystem-based forest management has become a necessity. The extent of forest degradation and the trend for environmentally friendly forest management are a strong impetus (Suhendang, 2020). Currently, the Indonesian government is changing its forest management paradigm to development capital with an emphasis on economic aspects through timber exploitation and ecological and socio-cultural aspects (Ministry of Environment and Forestry Republic of Indonesia, 2018). The forest is a whole ecosystem and plays an integral part in sustaining life, both micro and global. The balance of the forest ecosystem is maintained through the existence of biodiversity and its natural resources that are used as part of a life support system and sustainable development.

Ecosystem-based forest management with a close-to-nature concept approach can be adapted to the management of natural production forests in Indonesia and to rehabilitating and restoring them. This system is not new because the principle of close-to-nature silviculture (CNS), as described by Jacobsen (2001) and Schütz (1999), cit Colak et al. (2003), is also practiced in several silvicultural systems in TPTI, except concerning the regeneration approach. Some silvicultural systems in TPTI use human-made regeneration in intensive planting, even in intensive silvicultural systems that use seedling from tree breeding strategies (Pamoengkas et al., 2019; Indrawan, 2002). The intensive silvicultural systems has also conducted environmental engineering and plant protection from pests and diseases (Karmilasanti and Wahyuni, 2018; Widiyatno et al., 2011). This system is an effort to increase forest productivity through comprehensive management by working to produce commercially targeted species while still maintaining the forest cover.

On the other hand, IPF in Indonesia should be developed from an environmental perspective (Suhendang, 2020). Exotic species should be mixed with and gradually replaced by the native species, and the use of the clear-cutting and human-made regeneration silviculture system should be paired with the selective cutting silviculture system (Raharjo et al., 2009). Multisystem silviculture should be tried first before it is broadly transformed, taking into account the current IPF management situation. The performance of native species is less than that of exotic species, thus it has not been able to replace the exotic species as plant material in IPF. The use of short-cycle intolerant species is more appropriate when applying a clear-cutting silviculture system than a selective cutting system.

The opinion of Möller (1921, 1922), and several other silviculturists who support CNS (Remeš, 2018; Schütz et al., 2016), on the regeneration and maintenance of production forests using natural automation is that difficulties will be experienced if applied in Indonesia without modifications. The condition of Indonesian forests that have edaphic diversity, the high biodiversity of tropical ecosystems, the variations in the actual conditions of the forest cover (primary forests, secondary forests, logged-over area, open land) and the variations in the social conditions have consequences in the formulation of prescriptions for adaptive local silvicultural systems (Suhendang, 2020). Similarly, the opinion of Schütz et al. (2016) and Jacobsen (2001) is that the processes chosen to manage multi-strata mixed forests should be based on forest dynamics information, not necessarily allowing nature to work alone. The implementation of a silvicultural multisystem in a forest area is more profitable than a single silvicultural system.

8.3 Nature-Based Forest Management

Indonesia has tropical forests with an incredible wealth of resources, much biodiversity and much functional ability (Kementerian Lingkungan Hidup dan Kehutanan, 2017). Various types of forest formations with unique characteristics due to physiographic, topographic, climatic and edaphic variations can be found in Indonesia, such as coastal forests, mangrove forests, peat forests, swamp forests, heat forests, lowland forests, monsoon forests, highland forests and mountain forests (Wahyudi, 2013; Steenis, 2010). These forests generally have the same functions and benefits, namely as a life support system (Pemerintah Republik Indonesia, 1999) for habitat and landscape ecosystems (Yan et al., 2011), but their resilience is based on each type of forest ecosystem (Oliver et al., 2015).

Forest resilience is crucial because of the dynamics of a forest's interaction with the environment and anthropogenic activities related to how the forest is able to maintain its function and structure while experiencing disturbance and change (Scheffer, 2009; cit Reyer et al., 2015). In Indonesia, the pressure on forests has passed a critical threshold where forests are experiencing complex degradation in quantity and quality either due to local disturbances or the impact of global climate change.

Reyer et al. (2015) stated that forests react to these complex conditions, as can be seen in Indonesia's forest ecosystems. Production forests in Indonesia have a forested cover of only 60% of the total area (Ministry of Environment and Forestry Republic of Indonesia, 2018). This condition has an impact on the productivity of the production forest sector, as seen from a decline in state revenue from year to year (Yasman et al., 2016). Based on environmental and biodiversity parameters, this impact on productivity is demonstrated by a decrease in the quality of the growing areas, such as a decrease in the area of peatlands (Lisnawati, 2012; Agus et al., 2004; Hedin et al., 2003); an increase in pests and diseases in plantations (Barber, 2004); changes to and loss of the ability of hydrological and nutrient cycles (Lisnawati, 2012; Agus et al., 2004; Hedin et al., 2003); decreased deposits and increased carbon release by forests because of the process of change and degradation (Rochmayanto et al., 2013); and a decrease and change in the local community life around the forest in farming and the search for non-timber forest products (NTFPs) such as forest honey and agarwood (Siburian, 2016; Gunawan et al., 1998). In addition to these reactions, there is a broad and mutually triggering impact in the form of an increase in hydrometeorological disasters that occur both on a local and a global scale due to forest damage and loss at a regional scale (Pramono and Savitri, 2018; Pramono and Putra, 2017; Susanti et al., 2017).

Indonesia's population, which continues to grow with a projected 311 million by 2045 from 255 million in 2015 (Subdirektorat Statistik Demografi, 2018), will undoubtedly result in increased pressure on forest resources. Demands on the forest will grow, related not only to the function of humans but also the benefits and interests of humans. How will forests survive? Will forest land be used for other purposes? Will forests still function and provide an ecosystem with their unique structure and biodiversity? This concept is based on a series of questions for forests and forestry in Indonesia in the future.

The challenges to forest management in Indonesia are increasingly complex, having to deal simultaneously with environmental, productivity and anthropogenic issues. Therefore, forest management in Indonesia must be directed towards a new forest management paradigm based on forests' multifunctional and multi-beneficial ecosystems. Management must be comprehensive and consolidate theory and practice from various scientific disciplines by looking at the real issues and problems. Mitigation, adaptation and resilience form a sound basis for ecological production and anthropogenic sustainability.

The restoration of the ecosystem is one approach to managing and sustaining forests in the future. Restoration goals include scenarios that take into account new environmental conditions and future ecosystem trajectories (Hobbs et al., 2006); building adaptive forest management through restoration that emphasizes functional perspectives (ecosystem, biodiversity, economic and social culture) and the landscape (Remm et al., 2019; Stanturf et al., 2014). This links small-scale forest communities with the variable functions of large-scale ecosystems through rebuilding the population of plant species, increasing forest habitat extension, increasing landscape connectivity and enabling biodiversity to thrive (Rother et al., 2019).

Silviculture is oriented towards the broad management of natural resources, one of which is CNS. Silviculture is not only about forest exploitation through logging but has also developed dynamically due to research findings on social change and ecological disturbance (O'Hara, 2016). A CNS approach can be integral to forest management efforts.

CNS is a concept of forest management formulated by Gayer in the 19th century (O'Hara, 2016). This concept was developed after Bernhardt opposed the management of same-age monocultures using the clear-cutting method. Gayer subsequently supported it with concepts and ideas published in his seminal book *The Mixed Forest* (Remeš, 2018; Schütz, 1998). This system is a criticism of the traditional management concept of healthy forests in the form of a monoculture forests, such as agricultural systems (Jacobsen, 2001).

Gayer argues that the idea of forest management that "returns to nature" is principled and practiced biologically, with regeneration such as natural forests, small-scale logging (shelterwood), planting stands using native species and uneven-aged trees that are suitably located (Remeš, 2018; Schütz, 1998). The close-to-nature concept is based on restoring forest management to the basic premise of forestry, namely providing a range of benefits and forest ecosystem services that are broader and that can be achieved without management (O'Hara, 2016).

The paradigm was developed to shift the paradigm of management of plantations with clear-cut systems to management efforts that mimic nature through the creation of multi-strata mixed forest stands and high biodiversity with various forms of regeneration on a narrow or a broad scale. The processes chosen for forest management do not necessarily allow nature to work alone (Schütz et al., 2016; Jacobsen, 2001).

The term close-to-nature silviculture has developed based on the improvisation and orientation of its approaches, such as ecologically oriented silviculture (Frivold, 1992 cit. Remeš, 2018), ecologically sound silviculture (Poleno, 1993, 1994; cit Remeš, 2018), diversity-oriented silviculture (Lähde et al., 1999; cit Remeš, 2018) and nature-oriented silviculture (Koch and Skovsgaard, 1999; cit Remeš, 2018). There is also ecologically justified silviculture and an alternative approach using the terms holistic forestry, eco-forestry, common-sense forestry, continuous cover forestry and ecosystem management (O'Hara, 2014; cit Remeš, 2018).

Whatever the improvisation and orientation of foresters in applying the CNS paradigm to manage forests, they wish to carry out management based on ecological principles to create a stable and sustainable forest function (Wang et al., 2018; Jacobsen, 2001). Jacobsen (2001) stated that the focus of management is not only on standing structures but also on forest structures, the principles of which are (i) to maintain soil productivity through efforts to create permanent forest cover while maintaining litter and logged-over waste on the forest floor; (ii) to create stable stands to support the optimization of wood productivity; (iii) to ensure age rotation and the management of every tree; (iv) to continuously use a natural regeneration approach and its phases; and (v) to eliminate the clear-cutting system to maintain a permanent forest cover.

8.4 Agroforestry

The International Council for Research in Agroforestry (ICRAF) defines agroforestry as follows:

> a land management system based on sustainability, which increases overall land yield, combines the production of crops (including tree crops), forest plants and/or animals simultaneously or in sequence on the same land unit and applying management methods that are appropriate to the culture of the local population
>
> **(King and Chandler, 1978)**

Agroforestry is based on two fundamental premises: biological and socio-economic (King, 1979). Biologically, tree components in agroforestry can absorb nutrients far from the soil (outside the reach of agricultural crop roots), process them and utilize them for biomass growth (Harun, 2014). Furthermore, tree components can be efficiently recycled through falling litter, which is decomposed into humus and is available again in the soil. The contribution of agroforestry in the socio-economic field is more varied than pure agriculture or pure forestry because its business components are more diverse, and the combination of yields is more stable compared to traditional agriculture. Additionally, agroforestry can fulfil three economic segments of society: short-, medium- and long-term needs (Sabarnurdin, 2000).

Agroforestry consists of three main components: forestry, agriculture and animal husbandry (Harun, 2014). However, these systems are generally aimed at the production of a typical commodity or a similar product group. The combination of these three components can produce several combinations of multiproduct. Agroforestry is a dynamic planting pattern in which each combination of the different constituent elements forms a different system. These different systems add to the wealth of the agroforestry system. The scope of agroforestry is more comprehensive and the community will have an alternative system to apply according to the characteristics of the region and the interests and objectives of land use.

The nature and function of agroforestry systems are diverse and sophisticated (Harun, 2014). Therefore, classification of various categories in the agroforestry systems is required to evaluate, understand and improve the existing systems. Agroforestry systems are mostly implemented in areas of constant land use. Agroforestry technology is used to implement improvements and innovations derived from research results and to develop good management in agroforestry systems. Thus, agroforestry systems include original forms of agroforestry practices, while agroforestry technology produces improved agroforestry forms, such as improved fallow, alley cropping and multipurpose trees on farmlands. The forms of agroforestry can be grouped as follows (Nair, 1985; King and Chandler, 1978):

1. Agro-silviculture: the deliberate use of land to produce wood, as well as agricultural and forestry products
2. Sylvopastural: a system of forest land management to produce wood and to raise livestock
3. Agro-sylvopastural: a land management system for producing agricultural and forestry products and simultaneously raising livestock
4. Multipurpose forest trees production system: the management and planting system of various types of wood, their leaves and fruits, which can be used as human food or animal feed
5. Sylvo-fishery: a land management system designed to produce wood and simultaneously function as a fishpond/pond
6. Apiculture: a land management system that plants trees as a source of honeybee food. In addition to producing wood, the trees also produce honey with high-selling value and medicinal properties
7. Seri-culture: a land management system aimed at planting trees as a silkworm feed source

The intensive silvicultural systems has also

1. arranging its components according to the time and place of the structure
2. the importance and role of function components
3. the purpose of production or output system output
4. the socio-economic character of the socio-economic basis
5. the ecological basis is the basis of ecology

Patterns of agroforestry systems combine their constituent components into one space and time (Harun, 2014), avoiding negative interactions between the constituent components. Negative interactions that do occur can be in the form of competition for nutrients, sunlight, water and growing space. Tree canopies that are too dense prevent sunlight from reaching the strata below, where crops grow. Tree roots that extend and occupy the soil horizon to a depth of less than 50 cm can disturb the roots of crops resulting in a struggle for nutrients that harms crops. Vergara (1982) classifies agroforestry planting patterns into several forms:

a. Trees as fences (trees along border). This pattern involves planting trees along the border and crops in the middle of the land. Trees planted along the border function as fences or boundaries.
b. Pattern of intermittent alternation of trees and annual crops (alternate rows). This pattern is an agroforestry planting model that places trees and crops alternately in rows. This agroforestry pattern is possible on relatively flat land.
c. Alley cropping: This pattern places trees on the right and left edges of crops. An array of trees is grown east to west, which allows the crops to get full sunlight in the morning and the evening. The alley cropping pattern is often referred to as tunnel shaped because when viewed from the end of the land, it resembles a cave tunnel. This pattern is similar to the trees along a border pattern when both sides are planted with trees.
d. Block system (blocking system): This pattern is the planting between forestry plants and agricultural crops that are planted side-by-side to form two different plots. This pattern is intended to make it easier for farmers to treat each block because of its uniform nature.
e. Random mix patterns: This pattern is the random planting between forestry plants and crops that are planted irregularly. This pattern is formed because there has been no initial planning of the location of the plants. Farmers arbitrarily plant by choosing a space on their land.

8.5 Patterns for Agroforestry

Agroforestry is a combination of woody plants/forestry components, such as trees, shrubs, palms, bamboo or other woody plants, with agricultural plants (annual crops) and/or animals (livestock), both in terms of time (temporal arrangement) and spatial arrangement (Harun, 2014). According to von Maydell (1988), the ideal combination occurs when all agroforestry components are continuously on the same land. Naturally—or often for economic reasons—the combination between components is closely related to the dynamics of seasonal changes' equilibrium by following the annual rhythm and the inevitable succession due to periodic or sporadic human disturbance or treatment.

The combination of various components in an agroforestry system produces a variety of reactions that individually or concurrently can be found in one management unit (von Maydell 1987).

a. Competition: Trees, shrubs, agricultural crops and animals compete for light, water, nutrients, living space, work inputs, land, capital and so on. There is no direct evidence for this competition but it is suspected. For instance, certain plants are parasitic intermediaries for other plants, for example, trees as nests for birds, which can result in reduced crop gains. Often, competition is evident, for example, a reduction in weeds due to the protective canopy of trees.
b. Complementary: Such parallel reactions can be time, space or quantity. Utilization of all biotopes spatially and quantitatively so that one land can produce multi-production that can be obtained simultaneously, including vegetable and animal proteins. Availability of broadleaf or fruit such as *Acacia albida* in Africa as animal feed for which grass is not available during the dry season.
c. Dependency: Some mushrooms can only grow on individual trees. Certain types of animals can only live on grazing land. For example, in Africa, it is well known that the system would break down if there was no balance between long and short weeds. Short grass–feeding animals only approach their food if the grass is not too high.

Due to a combination of various components in agroforestry, reactions can occur in monoculture and mixed systems. A cropping system with a robust monoculture farming system, both with annual crops and woody trees, will reduce the production per unit area. This occurs due to competition for light, water and nutrients. If the spacing is widened, the magnitude of the competition decreases. In practice in the field, farmers manage their crops by adjusting the cropping patterns and spacing. In a mixed system of various types of plants or mixed cropping (trees with annual crops or only other trees), each type of plant can change its environment. Many branched plants will overshadow other plants, but some plants that are not too close will benefit; this process is often called "facilitation".

Suprayogo et al. (2003) explained that three main factors need to be considered in compiling agroforestry components: (1) the process of interaction; (2) the factors causing interactions; and (3) the types of interactions. These are explained further as follows:

1. In mixed farming systems, competition between plants that are planted side by side on the same land often occurs when there are limited plant life sources. This competition usually manifests in the form of growth barriers to other plants, both directly and indirectly. Direct barriers, for example, can be through the effect of allelopathy (rarely found in the field). Indirect barriers can be reduced light intensity due to tree shade, or depleted nutrient and water availability due to the proximity of two adjoining plant species. Plants sometimes affect other plants through a "third party", that is, if they become hosts for pests or other plant diseases. In addition to their negative effects, indirect interactions can have positive effects, for example, the influence of *Dadap* trees on coffee plants; the *Dadap* tree functions as an N enhancer as well as a shade.
2. Factors causing interactions. In general, negative interactions can occur due to (1) the limited carrying capacity of the land which determines the maximum population size that can grow on the land and (2) limited growth factors. The concept of natural carrying capacity is important. This concept illustrates the maximum number of species in an area, both as a monoculture and a mixed system. A species may be able to grow in abundance in a field. If two species grow together on the land, one species can be more competitive than the other. This condition is likely to result in the second species becoming extinct. The concept is also applied in the agricultural business, in the hope that staple crops will grow better. The limited competition for growth factors are water, nutrients and light. Plant growth deteriorates if there is a decrease in the availability of one or more factors. Nutrient deficiencies in the land may occur due to natural fertility or due to the process of nutrient loss from the land, for example, due to evaporation and washing. Water shortages can occur due to low water-holding capacity, uneven distribution of rainfall or a large amount of water loss (surface flow). Knowledge of growth factors, such as the amounts needed by plants and their availability in a field, is required for optimal plant growth in the agroforestry system.
3. Types of tree–soil–plant interactions. In an agroforestry system, planting various plants on the same land will cause various forms of interaction between the plants. In principle, there are three types of interactions in the agroforestry system: (1) Interaction (complementarity = mutual benefit): if an increase in the production of one crop is followed by an increase in the production of other plants. (2) Neutral interaction: if the two plants do not influence each other (an increase in annual crop production does not affect tree production or vice versa). (3) Negative interactions (competition = mutual disadvantage): when a decrease in the production of one crop follows an increase in the production of other plants or a decrease in both.

The key to the success of an agroforestry system is highly dependent on tree management, which can reduce any adverse effects and maximize the beneficial effects (with the number of workers as needed). Tree–soil–plant interactions depend on the growth and the specific shape of a tree, both in the canopy and the plant roots. Ways to reduce the detrimental effects of trees include the following:

a. Set the canopy: the height of a seasonal plant is usually lower than that of a tree because the tree creates shade, reducing the amount of light that annual plants need to grow. Farmers usually adjust the spacing between the plants, as well as pruning several branches of the tree to reduce the detrimental effect of the tree on the annual crop. The shade is reduced by pruning tree branches during the growing season but they are allowed to grow in the dry season to suppress weed growth (e.g. *Imperata*).

b. Regulate root growth: root growth can be regulated by pruning. Pruning the tree canopy causes reduced root activity. The growth of new roots follows the regrowth of the tree canopy. The first pruning is carried out if the tree is at least two years old. Two things need to be considered when pruning tree branches: the height of pruning from the ground surface and the frequency of pruning. Pruning stems that are too close to the soil surface will trigger the formation of fine roots in the topsoil providing greater opportunity for competition for water and nutrient requirements with annual crops—trim the height to at least 75 cm from the ground. Pruning lower than 75 cm will cause the growth of the tree roots to be concentrated in the topsoil, leading to competition with annual crops. The same will occur if trimming is too frequent. Pruning should be undertaken not more than three times a year. Frequent trimming of the crown will trigger the formation of fine roots in the upper layer. The shallow root system of trees due to improper tree management will also harm the growth of the tree. Shallow rooting results in trees that are less resistant to drought during the dry season. Root growth can be regulated by choosing tree planting techniques. Tree planting techniques are carried out according to the availability of planting material and energy by direct seedling in the field, cuttings or rooted seeds. The planting of cuttings is faster and reduces the weed population. How trees are planted in the field also determines the depth of the roots. Tree seedlings planted directly from seeds usually have a deeper root system than those planted in the form of stem cuttings or through a seedbed in a polybag.

c. Increase the positive influence of trees through the selection of shade plants: The magnitude of the influence of shade trees in agroforestry means that not all types of plants can be planted with trees. Therefore, the selection of shade-tolerant plant species in agroforestry is needed.

Incorporating constituent components and agroforestry activities must also be followed by a timely integration (Harun, 2014). It is necessary to consider the harmonization between the time and duration of interaction between components of forestry, agriculture, fisheries and/or livestock. This combination does not always appear in the field, leading to a misconception that a form of land use cannot be categorized as agroforestry. The timeframe and the process of sustainable land use are essential to consider in agroforestry, which is often not as simple as a monoculture. This understanding is often not as simple as a monoculture. Huxley (1999) and Nair (1993) categorize combinations of time as follows:

1. Co-incident: combinations during the cultivation period of agroforestry species/components
2. Concomitant: a combination at the beginning or end of the cultivation time of a type/component of agroforestry
3. Overlapping: a combination overlapping between the end and the beginning of two (or more) types/components of agroforestry
4. Interpolated: a combination inserted in the period of cultivation of species/components of agroforestry

The combination of components can be periodic (intermittent) or continuous (continuous). If the combination of agroforestry components is simplified, then the combination can be broadly divided into two, namely, a permanent combination and a temporary combination. The two combinations are explained by von Maydell (1987) as follows:

1. Permanent combination: This combination of agroforestry components can consist of a forestry component with at least one agricultural and one livestock component. This permanent combination can be found as follows:
   i. A combination of forestry, agricultural and livestock components that are sustainable depending on the land used (co-incident), for example, various forms of home gardens that can be found in many areas of the archipelago.
   ii. Permanent maintenance of stands/trees on agricultural lands as a means of improving the land; protective or water-retaining plants, for example, planting *Turi* trees (*Sesbania*

grandifora) on rice paddies in Java, and protective trees on commercial plantations (coffee, cocoa).

iii. Permanent maintenance/grazing of cattle (annual time) on forested woodlands, regardless of age of stands.

2. Temporary combination: Livestock grazing or the presence of animals in wooded areas only occurs during certain seasons (continuously interpolated). Another example is the presence of a variety of forest animals (especially bird species) in forest and home gardens during the fruit season (especially December to March).

8.6 Landscape Agroforestry

Landscape agroforestry is a science that studies agroforestry systems on a landscape scale. Spatial and temporal aspects are very influential in this study (Arifin et al., 2009). Landscape agroforestry studies differ from agroforestry studies that are commonly carried out on a site or plot scale (Pastur et al., 2012). Agroforestry landscapes are landscape objects utilized in agroforestry-patterned activities (Arifin et al., 2009). Land-use systems at the landscape scale use ecological boundaries. The ecological boundary is in the form of a watershed boundary, peat layer thickness, water depth and length of submergence (inundation), whether or not tidal waters flow, etc. Each unit of land has a structure with functions that follow its biophysical conditions. The assessment of allotment and land use, ranging from peat ecosystem protection to peat ecosystem cultivation, is carried out using an integrated or segregated approach (Baran-Zgłobicka and Zgłobicki, 2012). Analysis of agroforestry landscapes can be accessed using a system of interactions between objects and elements in them, which provide integrated interaction results. In agroforestry landscape analysis, the patterns of a system that can be repeated both spatially and temporally can be studied. Patterns that can be repeated are called knowledge. This knowledge can be seen in the study of links between activities (Pastur et al., 2012).

Landscape ecology aims to study and enhance relationships between spatial patterns and ecological processes at various landscape scales and at the level of organization or ecosystem unity. A landscape is an area that contains two or more ecosystems. Agroforestry landscapes consist of four main principles, namely (Arifin et al., 2009):

1. Heterogeneous development and spatial dynamics
2. Interaction and exchange of heterogeneity in the existing landscape
3. Influence of spatial heterogeneity on the biotic and abiotic process
4. Occurrence of management of spatial heterogeneity

These principles indicate that the emphasis of the agroforestry landscape is on the relationship between patterns, processes and scale. Some factors to consider in discussing and conducting research on agroforestry landscapes are ecological flows in mosaic agroforestry landscapes, land use, land cover changes, scale and other ecological processes. Agroforestry landscapes include the landscape of conservation and the conservation of the ecosystem.

Factors that characterize and simultaneously compose agroforestry landscapes are taken into consideration in classifying the characteristics of the agroforestry landscape, namely (Arifin et al., 2009; Pastur et al., 2012):

1. Scale and heterogeneity (including composition, structure and function)
2. Patches and mosaics
3. Borders and edges
4. Interference and fragmentation

Heterogeneity is the diversity in the size and type of various landscapes. Agroforestry landscapes look at how spatial scales affect the abundance of organisms at each level of a landscape's area, as well as the behaviour and function of a landscape's organisms on the overall ecosystem. The scale components that must be considered in determining the characteristics of an agroforestry landscape are composition, structure and function (Pastur et al., 2012). Composition refers to the number of patch types in a landscape and their relative abundance. Structure is also determined by the composition, configuration and proportion of different patterns along a landscape. Function refers to how each element interacts in a landscape based on the life cycle events in an ecosystem. Agroforestry landscape theory emphasizes the role of humans on the structure and function of the landscape. The integrity of the constituent factors of a landscape's characteristics helps maintain its resilience to external threats, including the development and transformation of land by human activities (Moser et al., 2002). Landscape analysis evaluates the final results of natural and cultural systems. This analysis standardizes the results of these two systems in the form of maps and sketches. Landscape functional units can be evaluated both qualitatively and quantitatively in agroforestry (Arifin et al., 2009; Moser et al., 2002).

8.7 Agro-Biotechnology

Technology has brought much progress, including to the agricultural sector, as represented by the development of biotechnology, which is implemented from the very start of cultivation to post-harvest handling. Biotechnology is defined as "any technological application that uses biological systems, living organisms, or derivatives to make or modify products or processes for specific use" (Ghasemi et al., 2015; FAO, 2004). Moreover, it has the most significant potential impact on agriculture, livestock, fisheries, forestry, industry, medicine, pharmacy and environment science (FAO, 2011). Its vast application outside genetic engineering includes genomics and bioinformatics, markers-assisted selection, micropropagation, tissue culture, cloning, artificial insemination and embryo transfer. However, genetic engineering, mainly in the crop sector, is where biotechnology most directly affects agriculture in developing countries (Khan et al., 2017). Its application is continually improving and has been a breakthrough for industrial sectors in line with technological advancements, the presence of new species and varieties of microorganisms, plant genetic diversity and dynamic climate change.

The various benefits of the application of agricultural biotechnology include its use as a tool for increasing food production, saving time and cost, reducing the use of chemicals and making agriculture more sustainable from an environmental perspective (Björnberg et al., 2015; Ghasemi et al., 2015; Hansson and Joelsson, 2013). In general, several aspects of biotechnology contribute to sustainable agriculture programmes through a reduction in the use of chemicals, i.e. pesticides and insecticides; genetic deployment conferring tolerance or resistance to biotic and abiotic stresses; enhancement of nitrogen fixation; enhancement of nutrient uptake and use; improvement in crop productivity; technological advancement to produce biomass energy; and increase in plant nutrition, especially in seeds and fruits (Persley, 2000).

Agricultural biotechnology is often linked to the role of microorganisms. Countries situated in the tropical rainforest regions are so-called paradises of flora and fauna with a higher level of biodiversity than other parts of the world. The role of microorganisms, including bacteria and fungi, is also beneficial to humans and is needed in biotechnology. Their role in the nutrient cycle in tropical rain forests is vital due to the very favourable climatic conditions, namely high rainfall, moisture, temperature and light intensity throughout the year (Agus et al., 2019c). The closed nutrient cycling system in typical tropical rainforests is rapid and continuous. It begins with the litter fall of leaves, branches and stems to the forest floor that is immediately decomposed by microorganisms, turning the litter into fragmented plant residue and humus. Humus is decomposed organic material with the characteristics of soil; hence, its rich nutrients can be reabsorbed by plants.

Many soil microorganisms play an essential role in nutrient cycling and environmental interactions. Nearly 90% of plant species, including flowering plants, bryophytes and ferns, have connections with arbuscular mycorrhizal fungi (AMF) (Zhu et al., 2010; Ahanger et al., 2014). Diverse microorganisms in the soil are essential in the decomposition and recycling process of plant organic matter into humus. It has been recognized that, aside from being a decomposer, soil microorganisms also help plants absorb more nutrients for their growth.

Microbes have a role in helping plants to harness essential energy sources through symbiosis with plant roots, thus expanding the root absorption area (Bowles et al., 2016). In return, plants donate their waste byproducts as food for microbes. Furthermore, by employing modern technology, scientists can integrate two components, namely biological agents and technology, into biotechnology to optimize these friendly microorganisms to develop biofertilizers.

Among the benefits of biotechnology is the development of biofertilizers in supporting plant growth and soil health. Biofertilizers are a mixture of natural ingredients, including organic matter and microorganisms, formulated into fertilizers containing essential organic matter for plants and enhancing soil fertility (Sadhana, 2014). There are two forms of biofertilizers, i.e. liquid and solid. For their application, they can be mixed into the seeds or applied directly into the soil. Principally, such fertilizers help plants use all the food available in the soil and air, allowing farmers to minimize the number of chemical fertilizers. It is an attempt to preserve the environment (Yang et al., 2009). The continuous use of fertilizers, herbicides, fungicides and insecticides will adversely affect the quality of soil fertility, plants, animals, insects, microorganisms, humans and the environment (Yang et al., 2004). According to El-Habbasha et al. (2007), the application of organic biofertilizers to replace inorganic fertilizers is vital to protect the environment from the adverse effects of inorganic fertilizers. In developing countries, biofertilizers (mycorrhizae/bacteria) have been applied in various sectors of the agricultural industry to support various agricultural commodities, i.e. paddy and secondary crops, horticulture and fruits, estate crops and timber plant species (Table 8.1).

Another benefit of biotechnology is related to arbuscular mycorrhizal fungi, which are very useful in the reclamation of degraded land resulting from open mining activities. The application of biotechnology techniques with AMF has been carried out by Wulandari et al. (2014) usin *Glomus clarum* inoculant collected from coal mining areas.. The AMF inoculant was propagated in sterilized zeolite media with sorghum as the surrogate plant for three months in a greenhouse. AMF inoculation in the nursery was performed during the weaning period of the seedlings. The AMF inoculum was inserted through a hole directly into the plant's roots as proposed by Schenck (Wulandari et al., 2014; Kurniaty and Damayanti, 2011). The use of AMF is crucial in the reclamation of former mining areas because the soil has been severely deprived of nitrogen and phosphorus (Bücking and Shachar-Hill, 2005). The revegetation of post–coal mining sites with AMF can improve the absorption of soil nutrients, particularly phosphorus, support the growth of plants and enhance the sites (Taylor et al., 2010; Sushardi et al., 2006). Revegetation by the fast-growing pioneer legume species of *Pongamia pinnata* and the application of AMF could improve some chemical soil properties making them suitable for rehabilitation programmes in post-mining areas. Additionally, AMF has been proven to raise the soil pH toward the ideal or close to neutral, as well as increase soil organic C, total soil N and plant-available P contents. The AMF inoculation of *P. pinnata* successfully helps bioremediation by increasing the uptake of heavy metals such as Fe by 11.7% and by its accumulation in the plant root zone by 90.4% (Agus et al., 2019; Sun et al., 2018). Plants inoculated with AMF can be tolerant of abiotic stresses, such as salinity, drought, nutrients, alkali, cold stress and extreme temperatures, thereby increasing per hectare the yield of a large number of crops (Begum et al., 2019).

Biotechnology also successfully combines organic matter and technology through the compaction and modification of organic matter into organic pots, which can be used to rehabilitate degraded land. These organic pots are composed of a mixture of leaf litter, compost, manure, zeolite adhesives and clay with a specific ratio. Their application is simple, merely placing them with plant seedlings on critical lands (Agus et al., 2019a). Old, acidic, nutrient-poor oxisols characterize the critical soils of tropical forests. Post–coal mining sites are characterized by the accumulation of heavy metals, soil compaction and high erosion; sandy lands are characterized by high temperature, low organic matter and high salinity; and peatlands are characterized by poor aeration and drainage and an acidic soil. Therefore, the rehabilitation of degraded tropical lands requires the combination of AMF and the application of organic pots. The ideal amount of soil organic matter is 3–5% (Hardjowigeno, 1987). Despite this small amount, organic matter has many functions on soil properties and plant growth, for example to improve the soil structure; increase the sources of N, P, S and other nutrients; increase the ability of the soil to retain water; increase the ability of the soil to retain nutrients (CEC); and as a source of energy and nutrition for both micro- and macroorganisms (Hardjowigeno, 1987). The use of an organic pot ameliorant on

TABLE 8.1

Biofertilizer Application (Mycorrhizae and Bacteria) on Five Top Agricultural Commodities of Tropical Developing Countries

No.	Agricultural Commodities	Biofertilizer/FAM-AM/ Bacteria	Dose Fertilizer Application	Observed Responses	References
A	*Paddy and Secondary Crops*				
1	Wetland Paddy var. Ciherang on (latosol soil, pH 4.9)	Biofertilizer (Biost) containing *Azotobacter* sp., phosphate solvent microbial, *Bacillus* sp, *Trichoderma* sp.	Inorganic fertilizer 50% (125 kg urea ha^{-1}, 37.5 kg SP-36 ha^{-1}, 25 kg KCl ha^{-1}) + 200 kg biofertilizer Biost	Increased total of fresh and dry weight, fresh and dry root weight, number of seedlings, 1,000 grains weight and absorption of N, P, K	Bakrie et al. (2010)
2	Paddy var. Inpago LIPI Go2 (dryland, podsolic soil, pH 3.74)	Organic fertilizer (BioVam *Glomus agregratum*; Bioplus bacteria *Azospirillum* sp. and *Azotobacter* sp.) inorganic fertilizer NPK	Organic + inorganic fertilizer (pH + NPK 50% of recommended dosage)	Reduced use of inorganic fertilizers by up to 50%, moreover without fertilizing, and increasing productivity	Mulyaningsih et al. (2015)
3	Maize (inceptisol soil, pH 7.52)	FMA *Glomus* sp.	30 spore of FMA + microbial consortium 0.8 g/polybag	Increased plant height, fresh weight, total root length, number of spores, mycorrhizal colonies, bacterial populations, soil pH, P-available and C-organic	Nafiah and Prasetya (2019)
4	Soybean var. Grobogan (ultisol soil)	A mixture of FMA *Glomus* sp., *Gigasspora* sp., *Entrophospora* sp.	±1,000 spore of FAM and manure dose 20 ton/ha	Increased production, number of pods per plant, pod weight per plant, number of seeds per plant and weight of 20 seeds, height of plants, number of branches, dry root weight, dry crown weight, P uptake of plants	Malik et al. (2017)
5	Peanut: *Arachis hypogaea* L. var. Domba	FMA *Glomus fasciculatum*	Application of mycorrhizal 50 g/plant (±five spores/g) and inorganic fertilizer NPK 1 g/polybag	Increased plant height, dry root weight and dry crown weight	Prasasti et al. (2013)
B	*Horticulture and Fruits*				
1	Shallot: *Allium ascalonicum*	FAM *Glomus* sp. black colour, *Glomus* sp. and *Gigaspora* sp.	FAM 2.5–5.0 g/plant + NPK 2.5–5.0 g/plant	Increased tuber weight, nutrient content N, P, K, plant growth, percentage root infection	Sumiati and Gunawan (2007)

(Continued)

TABLE 8.1 (CONTINUED)

Biofertilizer Application (Mycorrhizae and Bacteria) on Five Top Agricultural Commodities of Tropical Developing Countries

No.	Agricultural Commodities	Biofertilizer/FAM-AM/ Bacteria	Dose Fertilizer Application	Observed Responses	References
2	Chili: *Capsicum annum* L. var. Lado F1 (andisol soil)	FAM *Glomus* sp. and *Gigaspora* sp.	*Glomus* sp. and *Gigaspora* sp. (10 g)/plant	Fresh and dry weighted weights, the fresh and dry root weight and the vegetative root phase length	Hadianur et al. (2018)
3	Tomatoes: *Solanum lycopersicum*	FAM *Glomus fasciculatus*	FAM *Glomus fasciculatus* dose 10 g/plant	*G. fasciculatus* was able to suppress *F. oxysporum* infection and improved tomato plant growth significantly and increased dry weight and production	Nurhayati (2010)
4	Banana: *Musa paradisiaca* cultivar Barangan	FAM *Glomus* type-1 dan *Acaulospora* type-4 and *Glomus fasciculatum*	50 g inoculant FMA (±100 spores)/plant	AMF increased Barangan banana resistance to blood disease bacteria (BDB)	Suswati et al. (2013)
5	Citrus: var. *Japanche Citroen* (JC)	FAM *Glomus manihotis*, *Scutellospora heterogama*, *Gigaspora margarita*, *Gigaspora roseae*, *Glomus etunicatum*, *Acaulospora tuberculata*	Inoculant FMA 500 spores/ plant	Increased seedling growth, height and diameter plant, leaf number	Anwarudin et al. (2005)
C	*Estate Crops*				
1	Oil palm: *Elaeis guineensis* Jacq.	Mycorrhizal *Glomus* sp. mv 15 isolate	37.5 g/polybag in seedling	Increased seedling height, stem diameter, leaf area total, fresh and dry weight of seedling	Lubis et al.. (2019); Palasta and Rini (2017)
2	Rubber: *Hevea brasiliensis* (Muell. Arg.) clone: PB 260, 1-year-old after planting	Mycorrhizal biofertilizer	Combination of (inorganic fertilizer 25% + 100 g mycorrhizae)/plant, estimated 150 spores/100 g	Increased plant height and diameter of the rubber	Ferry et al. (2015)
3	Sugar cane: *Saccharum officinarum* L.	Mycorrhizal biofertilizer	FMA fertilizer 10 g + 36 g organic fertilizer/ polybag	Increased diameter, available P of plant tissues, leaf number, percentage of AMF infection	Raziqin et al. (2015)
4	Coffee: *Coffea conephora*	Mycorrhizal biofertilizer	20–60 g/plant of FMA + 20–6% NPK inorganic fertilizer	Increased plant height, stem diameter, number of primary branches, a crown's diameter, number of segments per branch and fresh weight of 100 grains	Ferry and Rusli (2014)

(Continued)

TABLE 8.1 (CONTINUED)

Biofertilizer Application (Mycorrhizae and Bacteria) on Five Top Agricultural Commodities of Tropical Developing Countries

No.	Agricultural Commodities	Biofertilizer/FAM-AM/ Bacteria	Dose Fertilizer Application	Observed Responses	References
5	Cocoa: *Theobroma cacao* L .Clone UAH (Upper Amazone Hybrid) in entisol soil	FMA *Glomus* sp.	FAM *Glomus* sp. 10 g/plant + NPK inorganic fertilizer 5 g	Increased leaf area, percentage of root infection, height, diameter, fresh root weight, dry biomass weight, root length	Nasrullah et al. (2018)
D	*Timber-Producing Plants*				
1	Teak *Tectona grandis* Linn. (limestone post-mining soil)	FMA *Acaulospora* sp. (A), *Gigaspora* sp.	FMA *Gigaspora* sp. 5 g/ seedling + compost 5%; FMA *Acaulospora* sp. 5 g/seedling + compost 15%	Increased top-root ratio, diameter, high, index of seedling quality	Prayudyaningsih and Sari (2016)
2	Sengon: *Paraserianthes falcataria* (L) Nielsen (ultisol soil pH 5.15)	Mycorrhizal *Glomus* sp. and *Rhizobium* bacteria	Double inoculation of *Glomus* sp. 5 g + *Rhizobium* of density 5×10^{-9} cell/seedling	Increased height, diameter, leaflets number, dry weight, top-root ratio, root nodules, seed vigour index, percentage infection, N and P plant tissues	Nusantara (2002)
3	Mahoni: *Swietania macruphylla* King	FAM *Glomus mosseae*	FMA (*Glomus mosseae*) 20 g/ polybag + wood and coconut charcoal	Increased plant height and stem diameter, leaf number	Lisda et al. (2016)
4	Eucalyptus: *Eucalyptus pellita* F. Muell	Ectomycorrhizal fungi *Pisolithus arrhizus* and FMA *Glomus* sp.	Ectomycorrhizal fungi *Pisolithus arrhizus* 5 g/plant + FMA *Glomus* sp. 5 g/plant	Increased plant height, diameter, dry weight, absorption of N and P by plant	Irianto (2009)
5	Akasia: *Acacia mangium* (ex-limestone mining quarry)	Ectomycorrhizal fungi *Scleroderma* sp.	Mix 3 g of spores to 1 L of water, and the application 10 mL of spores/plant	Increased plant height, diameter, total dry weight, shoot dry weight, root dry weight, root length, leaf number and leaf area	Aprillia et al. (2019)

dry and nutrient-poor sand can serve as a compact growing medium, thereby saving water, reducing heat and evaporation, providing sustainable nutrients, improving rooting systems and enhancing plant growth (Agus et al., 2019a; Agus, 2018). Moreover, the application of organic pots from chicken manure shows better results in supporting the growth of plant seedlings in post–coal mining sites compared to the performance of those made from cow manure and leaf litter (Veronika, 2016).

8.8 Integrated Biocycle Farming System

The Integrated Biocycle Farming System (IBFS) refers to the concept of sustainable development through the optimal management of land resources (land, water, air, temperature, etc.), biological resources (animals, plants, humans and other living things) and environmental resources (relationships between creatures, etc.). This programme is an embodiment of the Education for Sustainable Development programme, which pays attention to increases in economic value, environmental sustainability, social and cultural justice, both synergistically and optimally. A balance between production and consumption must be achieved so that in one unit, the land can produce food, feed, shelter, fertilizer, water, oxygen, herbal medicine, tourism and others (Agus, 2018).

Land is used in a harmonious, holistic and integrated way for various purposes, namely: (i) biomass production (agricultural sector), (ii) environment, (iii) biological habitat and gene conservation, (iv) infrastructure space, (v) natural resources and (vi) aesthetics and culture, which are the main characteristics of the IBFS. Each landscape element should not emphasize each other's sectoral interests but must also be interrelated and support each other in harmony. Whole system output and outcomes take precedence over the output of each of its constituent elements. The IBFS is the integration of the agricultural and non-agricultural sectors, through the recycling of organic materials between these sectors in an integrated manner. Food produced by villages that is sent to cities results in a deficit of organic matter in villages, so the organic material collected in the city needs to be returned to the agricultural system in the village (Agus, 2018).

The IBFS promotes environmental, aesthetic, social, cultural and economic values in harmony, without any one value dominating. IBFS is not concerned with the economic value alone, which can be detrimental and is forced to eliminate other factors, as in large agricultural enterprises, but must harmonize all existing aspects. The IBFS is also carried out using a system of rotation and a diversity of plants so that biodiversity and crop cycles are maintained to support the growth and development of life and the environment. Significant innovations in the fields of artificial and functional biotechnology, nanotechnology and probiotics must be made to bring about a better life and an increasingly sustainable environment. A higher quality of life and a better environment can be achieved by empowering functional bio-nano-technology probiotics (Agus, 2019; Agus, 2018).

The management of organic closed cycles and the integration of integrated crop management (ICM), integrated nutrient management (INM), integrated moisturemanagement (IMM), integrated pest management (IPM) are the main characteristics of the IBFS so that an independent and quality closed-cycle system occurs. Management of biological resources and management of healthy and integrated ecosystems is an absolute requirement so that the universe can be reorganized to increase productivity, quality of life and guarantee the sustainability of a sustainable environment. Integrated landscape ecological management and the concept of agri-politicalism is one strategy to eliminate the egocentric box in the IBFS so that it no longer emphasizes the ego of the sector but must work together. Management of plants and organic matter in specific must also be managed in one land unit, because each plant species has its own character of life and productivity. Furthermore, IBFS must be managed using a holistic and integrated system (Agus, 2018, 2019).

8.9 Conclusions

The development of integrated biocycle management (IBM) with a nature-based ecosystem approach in the management of all natural resources, including land resources (land, minerals, water, air, microclimate) and biological resources (flora, fauna, humans), is a multifunctional and multiproduct system. This integrated farming system can produce food, feed, fibre, fertilizer, wood, energy, water, oxygen, medicine, mysticism and tourism, and has higher environmental, economic, socio-cultural and health values. This system is essential for the sustainable and productive management of tropical natural resources.

REFERENCES

Agus C (2018) Development of blue revolution through integrated bio-cycles system on tropical natural resources management. In: *World Sustainability Series*. pp 155–172.

Agus C (2019) Integrated bio-cycle system for sustainable and productive tropical natural resource management in Indonesia. In: Singh HB (ed) *Bioeconomy for Sustainable Development*. Springer-Nature, Singapore, pp 201–216.

Agus C, Karyanto O, Kita S, et al. (2004) Sustainable site productivity and nutrient management in a short rotation plantation of *Gmelina arborea* in East Kalimantan, Indonesia. *New For* 28:277–285. https://doi.org/10.1023/B:NEFO.0000040954.27630.2f

Agus C, Anggari PE, Faridah E, et al. (2019a) Role of watering interval and media composition on the growth of exotic fast growing species on coal mining soil. In: *IS BIOREV 2018*. IOP Publishing.

Agus C, Primananda E, Faridah E, et al. (2019b) Role of arbuscular mycorrhizal fungi and *Pongamia pinnata* for revegetation of tropical open-pit coal mining soils. *Int J Environ Sci Technol* 16:3365–3374. https://doi.org/10.1007/s13762-018-1983-5

Agus C, Wulandari D, Cahyanti PAB, et al. (2019c) Environmental site engineering and integrated bio-cycles management for rehabilitation of degraded tin mining land in tropical ecosystem. *IOP Conf Ser Earth Environ Sci* 398:012013. https://doi.org/10.1088/1755-1315/398/1/012013

Agus C, Hendryan A, Harianja V, Faridah E, Atmanto WD, Cahyanti PAB, Wulandari D, Pertiwiningrum A, Suhartanto B, Bantara I, Hutahaean BP, Suparto B, Lestari T (2019d) The role of soil organic amendment of humus paramagnetic and compost for remediation of post tin mining tailing media and their growth of *Reutealis trisperma* seedling. *Int J Smart Grid Clean Energy* 8(5):556–561. doi:10.12720/sgce.8.5.556-561

Agus C, Anggari PE, Faridah E, Wulandari D, Suginingsih, Ratnaningrum YWN, Winarni WW, Lestari T, Sunaryo Y, Pertiwingrum A, Napitupulu RP, Primananda E (2019e) Role of watering interval and organic pot on the growth of exotic fast-growing species on coal mining media. *IOP Conf Ser Earth Environ Sci* 308:012055. https://iopscience.iop.org/article/10.1088/1755-1315/308/1/012055

Agus C, Ilfana FF, Azmi AI, Fahreza WD, Wulandari P, Harun MK, Yuwati TW (2020a) The effect of tropical peat land-use changes on plant diversity and soil properties. *Int J Environ Sci Technol* 16:3365–3374. https://doi.org/10.1007/s13762-018-1983-5

Agus C, Kusuma MGCA, Faridah E, Dina A, Wulandari D, Bantara I, Hutahaean BP, Lestari T (2020b) Paramagnetic humus and callophyllum inophyllum for rehabilitation of tropical anthropogenic deserted tin-mined soil. *Int J Adv Sci Technol* 29(7):2931–2941.

Ahanger MA, Tyagi SR, Wani MR, Ahmad P (2014) Drought tolerance: role of organic osmolytes, growth regulators, and mineral nutrients. In: *Physiological Mechanisms and Adaptation Strategies in Plants Under Changing Environment*. Springer, New York, pp 25–55.

Ali M, Setyo OY, Fitria N, et al. (2018) *Perubahan Peraturan Perundang-Undangan Terkait Hutan Tanaman Industri (PP dan Produk Hukum KLHK) Sejak 1996–2017.* Pekanbaru

Anwarudin MJ, Jumjunidang F, Riska D (2005) Pengaruh Inokulasi Cendawan Mikoriza Arbuskula terhadap Pertumbuhan Bibit Jeruk Varietas Japanche Citroen. *J Hortik* 15:171–176. https://dx.doi.org/10.21082/jhort.v15n3.2005.p%25p

Aprillia D, Riniarti M, Bintoro A (2019) The application of ectomycoriza in ex-limestone mining growth media to assist the growth of mangium (*Acacia mangium*). *J Sylva Lestari* 7:332–341.

Arifin HS, Wulandari C, Pramukanto Q, Kaswanto RL (2009) *Analisis Lanskap Agroforestri: Konsep, Metode, dan Pengelolaan Agroforestri Skala Lanskap Dengan Study Kasus Indonesia, Filipina, Laos, Thailand, dan Vietnam.* IPB Press, Bogor (ID)

Bakrie MM, Anas I, Sugiyanta IK (2010) Application of inorganic and bio-organic fertilizer on system of rice intensification. *J Tanah Lingkung* 12:25–32. https://doi.org/10.29244/jitl.12.2.25-32

Baran-Zgłobicka B, Zgłobicki W (2012) Mosaic landscapes of SE Poland: should we preserve them? *Agrofor Syst* 85:351–365. https://doi.org/10.1007/s10457-011-9436-x

Barber PA (2004) Forest pathology: the threat of disease to plantation forests in Indonesia. *Plant Pathol J* 3:97–104. https://doi.org/10.3923/ppj.2004.97.104
Begum N, Qin C, Ahanger MA, et al. (2019) Role of arbuscular mycorrhizal fungi in plant growth regulation: implications in abiotic stress tolerance. *Front Plant Sci* 10:1068. https://doi.org/10.3389/fpls.2019.01068
Biro Perencanaan dan Informasi (2018) Menko Maritim Luncurkan Data Rujukan Wilayah Kelautan Indonesia. In: *Kementeri. Kemaritiman dan Investasi Republik Indones.* https://maritim.go.id/menko-maritim-luncurkan-data-rujukan-wilayah-kelautan-indonesia/. Accessed 29 Feb 2020
Björnberg KE, Jonas E, Marstorp H, Tidåker P (2015) The role of biotechnology in sustainable agriculture: views and perceptions among key actors in the Swedish food supply chain. *Sustainability* 7:7512–7529. https://doi.org/10.3390/su7067512
Bowles TM, Barrios-Masias FH, Carlisle EA, et al. (2016) Effects of arbuscular mycorrhizae on tomato yield, nutrient uptake, water relations, and soil carbon dynamics under deficit irrigation in field conditions. *Sci Total Environ* 566:1223–1234. https://doi.org/10.1016/j.scitotenv.2016.05.178
BPS-Statistics Indonesia (2020) *Statistical Yearbook of Indonesia.* Jakarta.
Bücking H, Shachar-Hill Y (2005) Phosphate uptake, transport and transfer by the arbuscular mycorrhizal fungus *Glomus intraradices* is stimulated by increased carbohydrate availability. *New Phytol* 165:899–912.
Budi SW (2008) Penerapan Multisistem Silvikultur Pada Areal Hutan Produksi (IUPHHK) Di indonesia: Tinjauan Aspek Teknik Silvikultur. In: Indrawan A, Istomo WC, et al. (eds) *Lokakarya Nasional Penerapan Multisistem Silvikultur Pada Pengusahaan Hutan Produksi Dalam Rangka Peningkatan Produktivitas Dan Pemantapan Kawasan Hutan.* Fakultas Kehutanan IPB, Bogor (ID).
Colak AH, Rotherham ID, Calikoglu M (2003) Combining 'Naturalness Concepts' with close-to-nature silviculture. *Forstwissenschaftliches Cent* 122:421–431. https://doi.org/10.1007/s10342-003-0007-1
El-Habbasha SF, Abd El Salam MS, Kabesh MO (2007) Response of two sesame varieties (*Sesamum indicum* L.) to partial replacement of chemical fertilizers by bio-organic fertilizers. *Res J Agric Biol Sci* 3:563–571.
FAO (2004) *The State of Food and Agriculture: Meeting the Needs of the Poor?* Food and Agriculture Organization of The United Nations (FAO), Rome (IT).
FAO (2011) Biotechnologies for agricultural development: proceedings of the FAO international technical conference on agricultural biotechnologies in developing countries: options and opportunities in crops, forestry, livestock, fisheries and agro-industry to face. In: *Proceedings of the FAO International Technical Conference on Agricultural Biotechnologies in Developing Countries: Options and Opportunities in Crops, Forestry, Livestock, Fisheries and Agro-industry to Face the Challenges of Food Insecurity and Climate.* Food and Agriculture Organization of the United Nations, Rome (IT).
Ferry Y, Rusli R (2014) The effects of Mycorrhizal and NPK fertilization dosage for growth and production of robusta coffee among productive coconut plantation. *J Littri* 20:27–34.
Ferry Y, Rusli R, Towaha J (2015) Effect of fertilizer and mycorrhiza on growth of young rubber plant in gradually rejuvenation models. *J Tanam Ind dan Penyegar* 2:85–90.
Ghasemi HD, Navab FK, Ghavidel RA (2015) Biotechnology in agriculture and its relationship to the principles of sustainable agriculture. In: *First National Conference on Modern Achievements of Biosciences and Agriculture.* Zabol University, p 2019.
Gunawan R, Thamrin J, Suhendar E (1998) *Industrialisasi Kehutanan dan Dampaknya Terhadap Masyarakat Adat, Kasus Kalimantan Timur.* AKATIGA, Bandung.
Hadianur H, Syarifuddin S, Kesumawati E (2018) Pengaruh Jenis Fungi Mikoriza Arbuskular Terhadap Pertumbuhan dan Hasil Tanaman Cabai Merah Besar (Capsicum annum L.). *J Agrotek Lestari* 3.
Hansson SO, Joelsson K (2013) Crop biotechnology for the environment? *J Agric Environ eth* 26:759–770.
Hardjowigeno S (1987) *Ilmu Tanah.* Mediyatama Sarana Perkasa, Jakarta.
Harun MK (2014) *Agroforestry Berbasis Jelutung Rawa : Solusi Sosial, Ekonomi, dan Lingkungan Pengelolaan Lahan Gambut / Marinus Kristiadi Harun. -- Cet. 1.* FORDA Press, Bogor.
Hedin LO, Vitousek PM, Matson PA (2003) Nutrient losses over four million years of tropical forest development. *Ecology* 84:2231–2255. https://doi.org/10.1890/02-4066
Hobbs RJ, Jentsch A, Temperton VM (2006) Restoration as a process of assembly and succession mediated by disturbance. In: *Linking Restoration and Ecological Succession.* pp 150–167.
Huxley PA (1999) *Tropical Agroforestry.* Blackwell Science Ltd, London.
Indrawan A (2002) Penerapan Sistem Silvikultur Tebang Pilih Tanam Indonesia (TPTI) Pada HUtan Dipterocarpaceae, Hutan HUjan Dataran Rendah di HPH PT. *Hugurya, Aceh. J Manaj Hutan Trop* VIII:75–88.

Indrawan A (2008) Sejarah Perkembangan Sistem Silvikultur Di Indonesia. In: Indrawan A, Istomo WC, et al. (eds) *Prosiding Lokakarya Nasional Penerapan Multisistem Silvikultur Pada Pengusahaan Hutan Produksi Dalam Rangka Peningkatan Produktivitas Dan Pemantapan Kawasan Hutan*. Fakultas Kehutanan IPB dengan Direktorat Jenderal Bina Produksi Kehutanan Departemen Kehutanan Republik Indonesia, Bogor, pp 1–12.

Irianto RSB (2009) Inokulasi Ganda *Glomus* sp. dan *Pisolithus arrhizus* Meningkatkan Pertumbuhan Bibit *Eucalyptus pellita* F. Muell. *J Penelit Hutan dan Konserv Ala Med* 6:159–167.

Jacobsen MK (2001) History and principles of close to nature forest management : a central European perspective. In: *Integrated Forest Management: Tools for Preserving Woodland Biodiversity.*

Karmilasanti WT (2018) Evaluasi Kegiatan Penerapan Sistem Silvikultur Tebang Pilih Tanam Jalur (TPTJ) Dengan Teknik Silvikultur Intensif (Silin) Di Kalimantan. *J Penelit Ekosist Dipterokarpa* 4:83–94.

Kartodihardjo H, Supriono A (2000) *Dampak Pembangunan Sektoral Terhadap Konversi dan Degradasi Hutan Alam: Kasus Pembangunan HTI dan Perkebunan di Indonesia*. Center For International Forestry Research (CIFOR), Bogor.

Kementerian Lingkungan Hidup dan Kehutanan (2017) *Statistik Lingkungan Hidup dan Kehutanan Tahun 2016*. Kementerian Lingkungan Hidup dan Kehutanan, Jakarta.

Khan FF, Ahmed K, Ahmed A, Haider S (2017) Application of biotechnology in agriculture: review article. *World J Biol Biotechnol* 2:139–142. https://doi.org/10.33865/wjb.002.01.0013

King KFS (1979) Concept of agroforestry. In: Chandler MT, Spurgeon D (eds) *Proceeding of an International Conference in Agroforestry*. ICRAF, Nairobi (KE), pp 1–14.

King KFS, Chandler MT (1978) *The Watershed Lands*. ICRAF, Rome (IT).

Kurniaty R, Damayanti RU (2011) Penggunaan mikoriza dan pupuk P dalam pertumbuhan bibit mimba dan suren umur 5 bulan. *J Penelit Hutan Tanam* 8:207–214.

Lisda L, Umar H, Yusran Y (2016) PENGARUH MIKORIZA DAN ARANG PADA MEDIA TUMBUH TERHADAP PERTUMBUHAN SEMAI MAHONI (*Swietenia macrophylla* King.). *War Rimba* 4:119–124.

Lisnawati Y (2012) Perubahan Hutan Alam Menjadi Hutan Tanaman Dan Pengaruhnya Terhadap Siklus Hara Dan Air. *Tekno HUtan Tanam* 5:61–71.

Lubis YH, Panggabean EL, Azhari A (2019) Pengaruh Pemberian Pupuk Kandang dan Mikoriza terhadap Pertumbuhan Tanaman Kelapa Sawit (*Elaeis guineensis* Jacq.) di Pembibitan Pre-Nursery. *Agrotekma J Agroteknologi dan Ilmu Pertan* 3:85–98. https://dx.doi.org/10.31289/agr.v3i2.1123

Malik M, Hidayat KF, Yusnaini S, Rini MV (2017) Pengaruh Aplikasi Fungi Mikoriza Arbuskula dan Pupuk Kandang dengan Berbagai Dosis Terhadap Pertumbuhan dan Produksi Kedelai (*Glycine max* [L.] Merrill) Pada Ultisol. *J Agrotek Trop* 5:63–67.

Ministry of Environment and Forestry Republic of Indonesia (2018) *The State of Indonesia's Forests 2018, 2018th Edn*. Kementerian Lingkungan Hidup dan Kehutanan, Jakarta (ID).

Moser D, Zechmeister HG, Plutzar C, et al. (2002) Landscape patch shape complexity as an effective measure for plant species richness in rural landscapes. *Landsc Ecol* 17:657–669.

Mulyaningsih ES, Sukiman H, Ermayanti TM, et al. (2015) Respon padi gogo terhadap pupuk hayati di lahan kering Kabupaten Konawe Selatan, Sulawesi Tenggara. *J Pengkaj dan Pengemb Teknol Pertan* 18:251–261.

Nafiah BI, Prasetya B (2019) The effect of biofertilizer microbial consortia and arbuscular mycorrhizae on growth of maize in an inceptisol. *J Tanah dan Sumberd Lahan* 6:1325–1332. https://doi.org/10.21776/ub.jtsl.2019.006.2.13

Nair PKR (1985) Classification of agroforestry systems. *Agrofor Syst* 3:97–128. https://doi.org/10.1007/BF00122638

Nair PKR (1993) *An Introduction to Agroforestry*. Kluwer Academic Publishers in Cooperation with International Centre for Research in Agroforestry (ICRAF), Dordrecht.

Nasrullah N, Nurhayati N, Marliah A (2018) Pengaruh dosis pupuk NPK (16: 16: 16) dan mikoriza terhadap pertumbuhan bibit kakao (*Theobroma cacao* L.) pada media tumbuh subsoil. *J Agrium Unimal* 12.

Nurhayati (2010) Effects of inoculation time of mycorrhiza vesicular arbuscular on fusarium oxysporum infection and tomatoes growth. *J Agrivigor* 9:280–284.

Nusantara A (2002) Response of Sengon (*Paraserianthes falcataria* (L) Nielsen) seedlings on double inoculation of arbuscular Mycorrhizae Fungi and *Rhizobium* sp. *J Ilmu-Ilmu Pertan Indones* 4:62–70.

O'Hara KL (2016) What is close-to-nature silviculture in a changing world? *Forestry* 89:1–6. https://doi.org/10.1093/forestry/cpv043

Oliver TH, Heard MS, Isaac NJB, et al. (2015) Biodiversity and resilience of ecosystem functions. *Trends Ecol Evol* 30:673–684. https://doi.org/10.1016/j.tree.2015.08.009

Palasta R, Rini MV (2017) Pertumbuhan bibit kelapa sawit dengan aplikasi fungi mikoriza arbuskular dan beberapa dosis pupuk fosfat. *J Agro Ind Perkeb* 5:97–106.

Pamoengkas P, Zamzam A, Dwisutono A (2019) Vegetation recovery of logged-over Dipterocarp forests in central Kalimantan, Indonesia. *Floresta e Ambient* 26:1–11. https://doi.org/10.1590/2179-8087.123917

Panjaitan S, Lazuardi D, Rahmanto B (2013) Kajian Efektivitas Sistem Silvikultur Tebang Rumpang Terhadap Peningkatan Produktifitas Dan Kelestarian Hutan Alam Produksi. *J Penelit Dipterokarpa* 7:109–122. https://doi.org/10.20886/jped.2013.7.2.109-122

Pastur GM, Andrieu E, Iverson LR, Peri PL (2012) Agroforestry landscapes and global change: landscape ecology tools for management and conservation. *Agrofor Syst* 85:315–318. https://doi.org/10.1007/s10457-012-9496-6

Pemerintah Republik Indonesia (1967) *Ketentuan-Ketentuan Pokok Kehutanan.*

Pemerintah Republik Indonesia (1990) *Hak Pengusahaan Hutan Tanaman Industri.* Republik Indonesia.

Pemerintah Republik Indonesia (1999) *Kehutanan.* Republik Indonesia.

Persley G (2000) Agricultural biotechnology and the poor: promethean science. In: Persley GJ, Lantin MM (eds) *Proceedings of an International Conference Washington, DC, 21–22 October 1999.* Consultative Group on International Agricultural Research. Washington DC (US), p 5.

Pirard R, Petiti H, Baral H, Achdiawan R (2016) *Dampak Hutan Tanaman Industri di Indonesia: Analisis Persepsi Masyarakat Desa di Sumatera, Jawa dan Kalimantan, Occacional.* CIFOR, Bogor.

Pramono IB, Putra PB (2017) Tipologi Daerah Aliran Sungai Untuk Mitigasi Bencana Banjir Di Daerah Aliran Sungai Musi. *J Penelit Pengelolaan Drh Aliran Sungai* 1:143–165. https://doi.org/10.20886/jppdas.2017.1.2.143-165

Pramono IB, Savitri E (2018) Flash flood in Arau watershed, West Sumatera: a mitigation study. In: *International Conference on Disaster Management (ICDM) 2018. MATEC Web of Conferences,* EDP Sciences, pp 1–6.

Prasasti OH, Purwani KI, Nurhatika S (2013) Pengaruh mikoriza *Glomus fasciculatum* terhadap pertumbuhan vegetatif tanaman Kacang Tanah yang terinfeksi patogen Sclerotium rolfsii. *J Sains dan Seni ITS* 2:E74–E78.

Prayudyaningsih R, Sari R (2016) Aplikasi Fungi Mikoriza Arbuskula (FMA) Dan Kompos Untuk Meningkatkan Pertumbuhan Semai Jati (*Tectona grandis* Linn. f.) Pada Media Tanah Bekas Tambang Kapur. *J Penelit Kehutan Wallacea* 5:37–46.

Raharjo SAS, Butar T, Kosasih AS (2009) Prospek Pengembangan Hutan Tanaman Campuran Dalam Pembangunan HTI (Prospect of mixed-species plantation in industrial plantation forest). *Tekno Hutan Tanam* 2:1–7.

Raziqin WA, Soedradjad R, Hartatik S (2015) The effect of Mycorrizhal fungi and organic fertilizer on the growth of sugarcane (*Saccharum officinarum* L.) seedlings using Single Bud Planting (SBP) method. *Berk Ilm Pertan* 1.

Remeš J (2018) Development and present state of close-to-nature silviculture. *J Landsc Ecol Republic* 11:17–32. https://doi.org/10.2478/jlecol-2018-0010

Remm L, Lõhmus A, Leibak E, et al. (2019) Restoration dilemmas between future ecosystem and current species values: the concept and a practical approach in Estonian mires. *J Environ Manage* 250:1–8. https://doi.org/10.1016/j.jenvman.2019.109439

Reyer CPO, Rammig A, Brouwers N, Langerwisch F (2015) Forest resilience, tipping points and global change processes. *J Ecol* 103:1–4. https://doi.org/10.1111/1365-2745.12342

Rochmayanto Y, Darusman D, Rusolono T (2013) *Hutan Rawa Gambut dan HTI Pulp Dalam Bingkai REDD +, Pertama.* FORDA Press, Bogor.

Rother DC, Liboni AP, Magnago LFS, et al. (2019) Ecological restoration increases conservation of taxonomic and functional beta diversity of woody plants in a tropical fragmented landscape. *For Ecol Manage* 451:117538. https://doi.org/10.1016/j.foreco.2019.117538

Sabarnurdin MS (2000) Agroforestry untuk agribisnis. *Bul Kehutan* 42

Sadhana B (2014) Arbuscular mycorrhizal fungi (AMF) as a biofertilizer-a review. *Int J Curr Microbiol Appl Sci* 3:384–400.

Schütz JP (1998) *Close-to-nature Silviculture : Is This Concept Compatible with Favouring Species Diversity in Forests ? Dans for.*

Schütz JP, Saniga M, Diaci J, Vrška T (2016) Comparing close-to-nature silviculture with processes in pristine forests: lessons from Central Europe. *Ann For Sci* 73:911–921. https://doi.org/10.1007/s13595-016-0579-9

Siburian R (2016) Kehidupan Masyarakat Sekitar Hutan dan Ketahanan Sosial pada Ekologi Hutan yang Berubah. *J Masy dan Budaya* 18:467–486.

Stanturf JA, Palik BJ, Dumroese RK (2014) Contemporary forest restoration: a review emphasizing function. *For Ecol Manage* 331:292–323. https://doi.org/10.1016/j.foreco.2014.07.029

van Steenis CGGJ (2010) *Flora Pegunungan Jawa, Bahasa Ind.* LIPI Press, Jakarta.

Subdirektorat Statistik Demografi (2018) *Proyeksi Penduduk Indonesia 2015–2045 - Hasil SUPAS 2015.* Badan Pusat Statistik, Jakarta.

Suhendang E (2020) *Meretas Sketsa Jalan Menuju Pengelolaan Hutan Berbasis Ekosistem Dalam Pengurusan Hutan Berperspektif Lanskap Di Indonesia.* IPB Press, Bogor (ID).

Sumiati E, Gunawan OS (2007) Aplikasi pupuk hayati mikoriza untuk meningkatkan efisiensi serapan unsur hara NPK serta pengaruhnya terhadap hasil dan kualitas umbi bawang merah. *J Hortik* 17.

Sun Z, Song J, Xin X, et al. (2018) Arbuscular mycorrhizal fungal 14-3-3 proteins are involved in arbuscule formation and responses to abiotic stresses during AM symbiosis. *Front Microbiol* 9:9–19. https://doi .org/10.3389/ fmicb.2018.00091

Suprayogo D, Hairiah K, Wijayanto N, et al. (2003) Peran Agroforestri pada Skala Plot: analisis Komponen Agroforestri sebagai Kunci Keberhasilan atau Kegagalan Pemanfaatan Lahan. *World Agrofor Cent.*

Suryanto NDR, Priyono H, et al. Multisistem Silvikultur Menjadikan Pemanfaatan Kawasan Hutan Produksi Menjadi Lebih Baik. *Policy Br.*

Suryatmojo H, Supriyo H, Purnomo S (2014) Dampak Penerapan Sistem SilvikulturTebang Pilih Tanam Jalur Terhadap Kelestarian Kesuburan Tanah Dalam Menunjang Kelsetarian Pengelolaan Hutan Alam (The impact of selective cutting and strip planting system implementation toward sustainability of soil). *J Mns dan Lingkung* 21:50–59. https://doi.org/10.22146/jml.18511

Susanti PD, Miardini A, Harjadi B (2017) Analisis Kerentanan Tanah Longsor sebagai Dasar Mitigasi di Kabupaten Banjarnegara. *J Penelit Pengelolaan Drh Aliran Sungai* 1:45–59.

Sushardi FE, Iskandar E, Rahayu S (2006) Mycorrhizal formation and growth of Shorea leprosula in Bukit Suharto after using charcoal and rockphosphate. In: *Plantation Technology in Tropical Forest Science.* Springer, Tokyo, pp 175–183.

Suswati N, Azwana N (2013) Peningkatan Ketahanan Tanaman Pisang Barangan Terhadap Blood Disease Bacterium (BDB) Dengan Aplikasi Fungi Mikoriza Arbuskular Indigenus. *J Hama dan Penyakit Tumbuh Trop* 13:96–104.

Syahadat E, Irawanti S (2014) *Kebijakan Pembangunan HTI di Indonesia Salahkah?* 8:1–4.

Taylor MD, Kim ND, Hill RB, Chapman R (2010) A review of soil quality indicators and five key issues after 12 yr soil quality monitoring in the Waikato region. *Soil Use Manag* 26:212–224.

Vergara NT (1982) *New Directions in Agroforestry: The Potential of Tropical Legume Trees. Improving Agroforestry in the Asia-Pacific Tropics.* EAPI, Honolulu, HI.

Veronika S (2016) *Pertumbuhan Semai Kranji (Pongamia Pinnata) Dengan Aplikasi Pupuk Pot Organik Pada Berbagai Jenis Tanah Marginal.* Universitas Gadjah Mada Yogyakarta.

von Maydell HJ (1987) Editorial: international research in agroforestry. *Agrofor Syst* 5:193–195. https://doi .org/10.1007/BF00119122

von Maydell HJ (1988) *Agroforestry (Lecture Notes).*

Wahyudi (2013) *Sistem Silvikultur di Indonesia teori dan Implementasi.*

Wang X, Lu Y, Xing H, et al. (2018) Effects of close-to-nature conversion on *Pinus massoniana* plantations at different stand developmental stages. *Trop Conserv Sci* 11. https://doi.org/10.1177/1940082918767953

Widiyatno W, Soekotjo S, Naiem M, et al. (2011) Pertumbuhan Meranti (*Shorea* spp) Pada Sistem Tebang Pilih Tanam Jalur Dengan Teknik Silvikultur Intensif (TPTJ_SILIN). *J Penelit Hutan dan Konserv Alam* 8:373–383. https://doi.org/10.20886/jphka.2011.8.4.373-383

Wulandari D, Cheng W, Tawaraya K (2014) Arbuscular mycorrhizal colonization enhanced early growth of *Mallotus paniculatus* and *Albizia saman* under nursery conditions in East Kalimantan, Indonesia. *Int J For Res* 2014. https://doi.org/10.1155/2014/89849

Yan H, Zhan J, Zhang T (2011) Resilience of forest ecosystems and its influencing factors. In: *Procedia Environmental Sciences.* Elsevier B.V., pp 2201–2206.

Yang J, Kloepper JW, Ryu C-M (2009) Rhizosphere bacteria help plants tolerate abiotic stress. *Trends Plant Sci* 14:1–4.

Yang S, Li F, Malhi SS, et al. (2004) Long-term fertilization effects on crop yield and nitrate nitrogen accumulation in soil in northwestern China. *Agron J* 96:1039–1049. https://doi.org/10.2134/ agronj2004.1039

Yasman I, Benyamin R, Siswoyo H, et al. (2016) *Road Map Pembangunan Hutan Produksi Tahun 2016–2045*. Asosiasi Pengusaha Hutan Indonesia, Jakarta.

Zhu X-C, Song F-B, Xu H-W (2010) Arbuscular mycorrhizae improves low temperature stress in maize via alterations in host water status and photosynthesis. *Plant Soil* 331:129–137. https://doi.org/10.1007/s 11104-009-0239-z

9

A Paradigm Shift towards Sustainability: Utilizing Agricultural Waste as a Valuable Resource in Various Agricultural Endeavours

Robert Nepomuceno, Cristine Marie B. Brown and Marilyn B. Brown

CONTENTS

9.1 Introduction

An estimated 1.6 billion hectares of the Earth's global land area is allotted to agriculture. This represents around 12% of the world's total land area of 13.2 billion hectares. Sugarcane, corn and rice represent the top three most produced commodities in the world with a combined production of 3.6 billion tons (FAOSTAT, 2013). However, every system is inherently inefficient—any endeavour, system or engine is accompanied by the production of waste byproducts. This is especially relevant in agriculture since waste from farming endeavours almost always exceeds the yield. For instance, for every 660 million tons of rice produced, approximately 800 million tons of dry agricultural waste, primarily in the form of rice straw, is generated (Dominguez-Escriba and Porcar, 2010). This is further exacerbated by the decreasing trend in productivity and intensive farming practices. Rice, for instance, is regarded to have reached its yield barrier, such that traditional breeding efforts are only expected to have minimal gains. The decrease in productivity, brought about by the combined effects of long-term unsustainable farming practices and the plant structural limitation, then results in intensive farming practices to compensate for the low harvests. Thus, instead of two cropping seasons per year, an additional cropping season is squeezed in. This practice results in the rapid depletion of nutrients and the generation of massive amounts of agricultural wastes after harvesting, in the form of biomass. In 2012, worldwide maize production was estimated to have reached 875,226,630 tons with the United States, China and Brazil contributing the largest share (Ranum et al., 2014). Today's corn production is primarily utilized for the generation of biofuels, industrial products, animal feed and human consumption, and since corn harvest normally only comes in grain form, it results in a high quantity of residue. Corn stover, a collective term for the residual stalk, leaf, cob and husk tissues, usually remains in the field as it is of little use following harvesting (Dhugga, 2007). In the Philippines, corn stover amounts to ~4.4 million tons of dry weight each year and is considered the leading producer of agricultural residue in the country. In 2018, the Philippines ranked as the eighth largest rice producer in the world, with an estimated production of 19.07 million metric tons. During harvest, rice grains are separated from the rest of the plant, and the residual plant materials are piled up or spread out in the field. According to the International Rice Research Institute (IRRI), roughly 6 tons of rice straw is made for every 4 tons of grain, and in Asia alone, 600–800 million tons of rice straw are produced each year. Aside from straw, the agricultural residues generated from rice also include rice husk and rice bran. Sorghum is a cereal grain popularly used in the production of ethanol and livestock feed, but unlike corn and rice, it is not a widely known crop in the Philippines. However, with the launch of the National Sorghum Development Programme in the county, it is expected that sorghum production will reach 1 million hectares of planting area within the next three years. Hence, it is also to be expected that the production of agricultural wastes from various sources will also increase. Farmers tend to burn these wastes, resulting in the immense release of carbon dioxide into the atmosphere, the depletion of soil nutrients and organic matter and the destruction of natural and beneficial microbiota (Mandal et al., 2004). Burning agricultural waste is a long-term tradition to effectively deal with the massive amounts of biomass generated from farming. Such a method is very successful in controlling weeds, pests and pathogens. However, the benefits of burning agricultural waste far offset the gains derived from it. Thus, a paradigm shift in dealing with agricultural waste products is imperative.

A multitude of approaches exist to manage massive amounts of agricultural waste. For instance, the waste can be incorporated into the soil, but this requires careful management since improper incorporation leads to a decrease in production efficiency (Mandal et al., 2004). Moreover, cellulosic tissues require prolonged periods of decomposition, and farmers have a very short time interval between cropping seasons. Thus, rapid composting technology is proposed to resolve the situation. Rapid composting technology offers alternative yet efficient bioconversion of agro-industrial wastes into useful products. Farm and agro-industrial wastes, if utilized properly, could be a rich source of natural fertilizer, known as bio-organic fertilizer. Moreover, rice farming byproducts can be used in the synthesis of nanomaterials. For instance, silica nanoparticles can be harvested from rice husk, which can then be used as filler with a polyether-polyamide co-polymer to synthesize polymeric membrane nanocomposites (Bhattacharya and Mandal, 2018). Another application of agricultural waste is its usage in mushroom farming. Cellulosic materials such as rice bran and straw are used predominantly as a substrate in mushroom farming. In some areas in the Philippines, large-scale mushroom farmers often buy straw and bran to use in this endeavour. The spent mushroom substrate (SMS), on the other hand, has been shown to be a rich source of protein and nutrients, which can then be used as feed for ruminants and fish (Kakkar et al., 1990). Consequently, the waste of the ruminants can be combined with agricultural wastes and subjected to rapid composting and nitrogen enrichment, which can then be used as a fertilizer for farming.

To alleviate the adverse effects of the exponential increase in carbon dioxide concentration in the atmosphere and the rapid degradation of the soil health of farmlands, the sustainable management of agricultural waste is imperative. Moreover, sustainable waste management should be complemented with the use of agricultural inputs with minimal ecological risk such as biofertilizers (Brown et al., 2016), biopesticides (Nepomuceno et al., 2019) and microbial inoculants that promote micronutrient uptake (Nepomuceno et al., 2020). Although the traditional combustion of agricultural waste has long been prohibited, this is ineffective as long as no alternative methods of waste management are presented to small-scale and underprivileged farmers. In this chapter, we explore the different alternative and sustainable routes in managing agricultural waste in conjunction with the introduction of various products than can aid in facilitating these alternative routes (Figure 9.1).

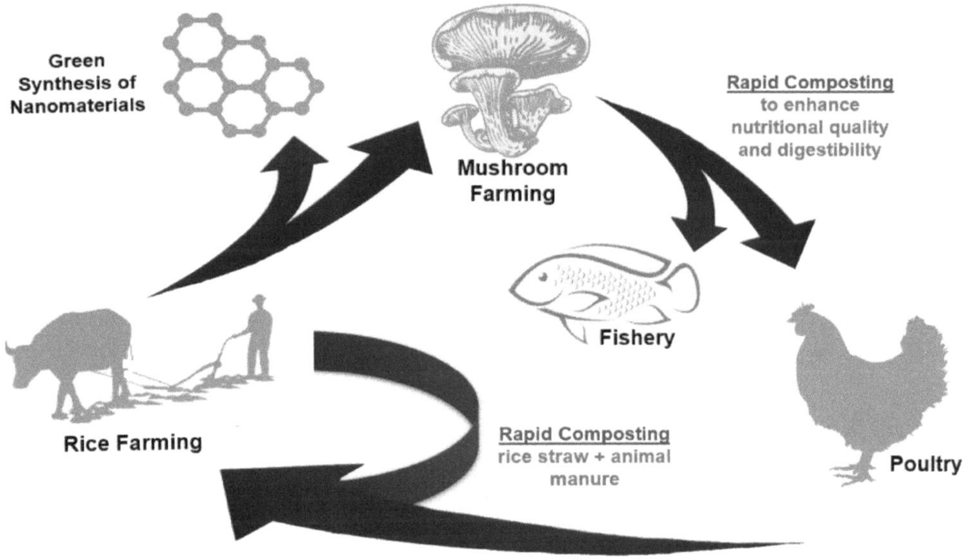

FIGURE 9.1 Sustainable management of agricultural waste through cyclical usage of waste products in the various agricultural and industrial sectors.

9.2 Green Synthesis of Nanomaterials

Nanotechnology is a multidisciplinary field that aims to engineer and reconstruct functional structures to the scale of molecules and atoms. This field of study has resulted in revolutionary and radical changes in the areas of pharmacology, food, medicine and agriculture. In particular, nanotechnology has brought a significant development to the field of material science through the innovative application of nanomaterials (Lopez-Lorente and Valcárcel, 2016). The technology has also been harnessed in the field of agriculture by developing nanotech-based devices used for the rapid detection of diseases, enhanced crop nutrient uptake and plant disease treatment (Huang et al., 2001). There are various ways to construct or synthesize nanoparticles, but the most popular methods include physical, chemical and biological processes (Khatoon et al., 2017). For the physical approach, the most used techniques are laser ablation, condensation and evaporation, but these methods use a tremendous amount of energy (Rajasekharreddy et al., 2010). Moreover, these methods are neither eco-friendly nor economical (Lopez-Lorente and Valcárcel, 2016). Green nanotechnology offers an alternative way to nanoparticle production with minimal cost and reduced environmental risks through the use of plant material. As compared to the chemical synthesis, nanoparticles synthesized with this method are more stable, safer and pose minimal ecological risk (Pandiyan et al., 2018).

9.2.1 Chestnut Shell

The extract of food byproducts such as the chestnut shell is now being utilized for the synthesis of nanoparticles. In India, coconut shells can also be used to synthesize silver nanomaterials through green technology (Altintig and Kirkil, 2016). Nanoparticles such as platinum, zinc oxide, gold, magnetite, cobalt, palladium and copper can now be successfully synthesized using plant parts.

9.2.2 Rice Husks

Lignocellulosic biomass (i.e. rice husk and walnut shell) has recently been shown to be a promising renewable source for the production of fuels, various chemicals and nanomaterials (Zamani et al., 2019). Rice husks are the hard coverings of grains of rice and comprise about 20–22% of the total yield of rice. It is one of the most abundant agricultural wastes containing 15.4–20.0% lignin, 28.7–35.6% cellulose and 12.0–29.3% hemicellulose prepared by the rice mill industry (Isikgor and Becer, 2015). It has been shown that the synthesis of 80–85 nm forsterite (Mg_2SiO_4) nanoparticles is possible by using the silica and magnesium oxide extracted from rice husk (Mathur et al., 2017). Silica nanoparticles obtained from rice husk were also used as filler with a polyether-polyamide co-polymer to synthesize polymeric membrane nanocomposites (Bhattacharya and Mandal, 2018). Rice husk ash was used to prepare silica nanowires and rice straw for the synthesis of 15–20 nm-sized SiO_2NPs (Bathla et al., 2018; Kauldhar and Yadav, 2018).

9.2.3 Walnut Shell

Walnut shell can be used in the synthesis of 11 nm copper nanoparticles (Saadati, 2017). Ceria (CeO_2), magnesia (MgO) and high surface alumina (Al_2O_3) nanoparticles were recently synthesized using walnut shells.

9.2.4 Sugarcane Leaves and Bagasse

The agricultural waste biomass of the sugar and ethanol industry is also abundant and widely available. It has been found that 50–200 nm magnetic iron oxide nanoparticles can be synthesized through the use of sugar cane bagasse and ammonia (Jabasingh et al., 2018). Cellulosic sugarcane leaves can also be used with magnesium acetate to synthesize magnesium oxide hybrid sponge-like carbonaceous composite (Li et al., 2018).

9.2.5 Egg Shells

Eggshell waste can also be used for nanomaterial synthesis. Calcined waste eggshell and rice husk were used to prepare Dy^{3+}-doped calcium silicate (Ca_2SiO_4) nanoparticles through a solid-state reaction technique (Devi et al., 2017). Calcium oxide nanoparticles have also been successfully synthesized using hen eggshell (Pandit and Fulekar, 2017).

9.2.6 Wheat Straw

Calcinated and acid-washed wheat straw can be used in the preparation of 100–200 nm silica nanoparticles (Patel et al., 2018). KOH-containing wheat straw and $Fe(NO_3)_3{\cdot}9H_2O$, on the other hand, was calcinated in the preparation of porous carbon-supported Fe_2O_3 ultrathin film (Fang et al., 2018).

9.2.7 Coconut Shell

Cazetta et al. (2016) carbonized coconut shell in the presence of ferric acid to prepare magnetic-activated carbon. Sebastian et al. (2018), on the other hand, used coconut husk extract and ferric chloride to synthesize 10–100 nm iron oxide particles, which is used in removing calcium and cadmium from aqueous media.

9.2.8 Banana Peel

In 2014, Yan et al. used banana peel extract in the synthesis of 20–50 nm Mn_3O_4 nanoparticles. Banana peel extract reduces $KMnO_4$ to form Mn_3O_4 while preventing nanoparticle agglomeration. Ali (2018), on the other hand, was able to synthesize 20 nm SiO_2NPs by adding banana peel extract to tetraethylorthosilicate in ethanol, followed by calcination of the precipitate.

9.3 Mushroom Production

Another profitable endeavour that makes use of rice or wheat straw to produce an alternative food source that is rich in protein and consequently facilitates the disposal of agricultural waste in an eco-friendly and sustainable manner is mushroom farming. Edible mushrooms, belonging to the division

Basidiomycota, possess varying metabolic profiles, which significantly influence the required substrate for its cultivation. Generally, these mushrooms can be categorized into (i) primary decomposers, which include oyster (*Pleurotus* spp.) and shiitake (*Lentinula edodes*) mushrooms, are capable of breaking down complex organic plant resources such as lignin, cellulose and hemicellulose; (ii) secondary decomposers, such as button mushrooms (*Agaricus bisporus*), prefer composted substrate; and (iii) tertiary decomposers such as *Agrocybe* spp. thrive in soil relying mainly on the degraded nutrients coming from primary and secondary decomposers (Grimm and Wösten, 2018). According to Panthapulakkal and Sain (2015), rice straw, corn stover and sorghum stalks are among the major lignocellulosic agricultural residues. Lignin and cellulose are insoluble and extremely complex compounds that normally become humic substances in soil, making it more difficult for several microorganisms to exploit them. The life cycle of mushrooms is composed of two phases: the vegetative and reproductive phases. Extensive production of hydrolytic and oxidative enzymes occurs during the vegetative phase wherein the mycelia actively colonize the substrate and degrade the organic matter present. Dissolved nutrients are then used to support the reproductive phase, where fruiting body formation is initiated (Julian et al., 2018). Mushroom cultivation is reported to yield 5–10% of the total straw input, which translates to 50–100 kg of mushrooms for every ton of waste straw (Zhang et al., 2002). *Volvariella volvacea*, known as the paddy straw mushroom, is one of the easiest mushrooms to cultivate due to its relatively short incubation time of 14 days (Reyes, 2000), while *Pleurotus* sp. or oyster mushroom, on the other hand, can be cultivated in poor quality rice straw.

With the inherent ability of mushrooms to take advantage of available nutrients in most common agricultural residues, several studies on mushroom cultivation have already been made to assess the yield, substrate combination, growth condition required, etc. In a study conducted by Adjapong et al. (2015), *Pleurotus ostreatus* grown in maize stalk and husk produced yield characteristics comparable to those grown in the widely accepted sawdust substrate. In a similar study by Zhang et al. (2002), rice straw was used in the cultivation of oyster mushroom, wherein grounded straw resulted in a higher mushroom growth rate as compared to its chopped straw counterpart. Also, rice straw generally produced 10% more fruiting bodies when compared to wheat straw under similar environmental conditions. The cellulose to lignin ratio reportedly had a positive correlation with the mycelia growth and yield of oyster mushroom grown in various types of agricultural residue, whereas *P. eryngii* and *A. aegerita* mushroom yield was directly correlated with the carbon to nitrogen ratio of the substrate (Kamthan and Tiwari, 2017).

The importance of mushrooms in the food and pharmaceutical industries cannot be overemphasized. Moreover, owing to their ubiquity and sustainability, mushrooms are currently being tapped as a substitute for less environmentally friendly materials such as plastic, rubber, wood and leather. They are currently regarded as the "material of the future". Mushrooms can also be a rich source of pharmaceutical compounds. Macrofungi, particularly nutraceutical and edible macrofungi (NEM), have long been used in East Asian countries and are proven to possess a myriad of health benefits. Many of these reported NEM grow abundantly in the Philippines, and harnessing their potential as a cheaper alternative and complementary source of medicine and food in the future could significantly ease the burden of healthcare and food security in the country.

In the Philippines, only 20 out of 500 species of edible mushrooms known worldwide are used as food, and only five of them are grown commercially (*Volvariella*, *Auricularia*, *Pleurotus*, *Agaricus* and *Lentinus*). These are traditionally used for culinary purposes, but due to their health benefits, their uses as nutraceuticals are now being recognized. A number of these nutraceutical fungi, such as *Ganoderma*, *Agaricus* and *Polyporus*, are already commercially available in tea form, capsules, tablets, decoctions and many other formulations. Research on the antimicrobial properties, phytochemical evaluation and antioxidant, hypoglycemic, antitumor and immunomodulating properties/activities of some of the indigenous macrofungi was initiated in the early 2000s. Research activities, however, were conducted in the laboratory and in vitro conditions. Initial attempts were made to produce the mycelia of some of the local nutraceutical fungi in liquid form.

The National Institute of Biotechnology and Molecular Biology (BIOTECH), which was established in 1980, played an active role in initiating and promoting NEM or mushroom research in the country. A mycelium bank was established, and initial funds were provided by the Food and Agricultural Organization (FAO) to supply quality cultures to growers and researchers not only in the Philippines

FIGURE 9.2 Small-scale mushroom farmer utilizing rice straw as a substrate in Guimaras, Philippines.

but also in many tropical countries. Moreover, the institute, in collaboration with the Department of Agriculture, has been training farmers in various provinces on mushroom farming (Figure 9.2).

9.4 Rapid Composting

To fully utilize agricultural waste and incorporate it into the soil as an amendment, rapid composting facilitated by beneficial microbes is recommended. Composting is the natural process of organic matter degradation facilitated by microorganisms. Rapid composting technology, on the other hand, involves the inoculation of microorganisms such as cultures of *Trichoderma harzianum*, which is a cellulose-decomposing fungus. The use of organic compost helps in restoring soil fertility and in replenishing organic matter in the soil. Moreover, organic compost can also be used to eliminate several polluting substances in the soil (de Bertoldi et al., 1996). Composting can be done in conjunction with nitrogen enrichment of the compost. Nitrogen is considered a critical limiting element for plant growth and reproduction. It is a major component of chlorophyll and proteins, and an absolute necessity in virtually all metabolic processes (Wagner, 2011). Compost enrichment can be conducted through the inoculation of nitrogen-fixing bacteria, specifically the Rhizobia group of bacteria, which exhibits a mutualistic relationship with terrestrial plants and has proven to be exceedingly crucial in various agricultural endeavours.

Different inoculants and methodologies enrich agricultural organic waste into useful soil amendments. Some notable inoculants for rapid composting are Bio-Quick and Bio-Fix, which were developed and produced at the National Institute of Molecular Biology and Biotechnology at the University of the Philippines—Los Baños (UPLB) in Laguna. Bio-Quick is a composting inoculum containing fungal strains of *Trichoderma* sp. that is used to increase the rate of decomposition of household and agro-industrial wastes. Bio-Fix, on the other hand, is an enrichment inoculum which contains a nitrogen-fixing bacteria used to enhance the nitrogen content of the compost. Together Bio-Quick and Bio-Fix produce an effective, reliable and enriched bio-organic fertilizer (sold as "Bio-Green") (Figure 9.3).

9.5 Utility of Spent Mushroom Substrate

Mushrooms serve as a good source of protein and other valuable nutrients with a low caloric index, while some species have medical applications. The global production of mushrooms has reached 25 billion

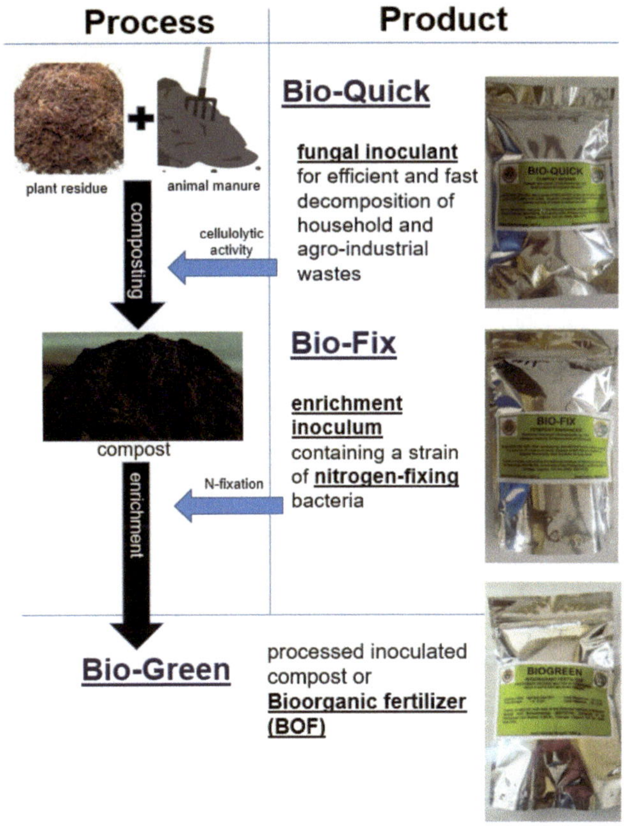

FIGURE 9.3 Rapid composting and compost enrichment process using Bio-Quick and Bio-Fix inoculants.

kilograms (Yang et al., 2015). Correspondingly, mushroom industries generate huge amounts of biomass waste known as spent mushroom substrate after every harvesting period (Moon, 2012). SMS refers to the composted organic growth media that is left unutilized after a complete period of mushroom cultivation (Phan and Sabaratnam, 2012). It is usually derived from renewable agricultural residues. These materials include sawdust, sugarcane bagasse, oil palm empty fruit bunch, wheat straw–bedded horse manure, hay, poultry manure, ground corncobs, cottonseed meal, cocoa shells, gypsum and other substances (Jordan et al., 2008). Approximately 5 million tons of SMS are generated annually as the trend for mushroom cultivation increases (Chiu et al., 1998). The production rate of some major mushroom-producing countries such as China, South Korea and Europe is already at a critical level, which can pose harmful effects to the environment (Hanafi et al., 2018).

The lack of waste disposal supervision has become one of the challenges in handling SMS (Magette et al., 1998). Mushroom industries are facing difficulties in waste management, mainly due to its abundance (Rasib, 2015). The majority of SMS is either spread on land as fertilizer, disposed in situ and/or incinerated (Williams, 2001). The composition of SMS plays a significant role in determining its suitable application (Hanafi et al., 2018). Previous studies have shown that SMS can be a potential source of animal feedstock (Oh et al., 2010).

9.5.1 Livestock Feed

SMS is generated in large volumes from the cultivation of mushrooms using agro-industrial wastes as a component of the growing medium. Due to its high organic matter content, SMS is utilized as a cheap source of animal feed and fertilizer in crop production. It contains fibres (cellulose, hemicellulose and lignin), carbohydrates, protein, fat, calcium, nitrogen (N), phosphorus (P) and potassium (K) in varying amounts depending on the composition of the substrate used, the mushroom species and other factors (Hanafi et al., 2018). Medina and colleagues (2009) also reported that SMS generated by *A. bisporus* has significantly higher macronutrients than that of *P. ostreatus*. The nutrients available in SMS can provide the requirements of livestock animals and even commercial fishes (Wu, 2018).

Numerous studies were published suggesting SMS can be used as a rich source of organic matter, enzymes and energy feedstocks to supply other industrials requirements. Hanafi (2018) enumerated a few applications of SMS on other industries, namely: crop production, livestock, energy feedstocks, pest management, wastewater treatment, reuse for the cultivation of mushrooms, along with bioremediation, enzymes and the promotion of fishery growth and health responses (Chirapongsatonkul, 2019)

Farmers find some constraints on their accessibility to commercial inputs such as fertilizers and pesticides due to their high cost. Compared to commercial fertilizers, SMS is cheaper since it is considered a waste material from mushroom production. SMS has been used in several flowering and vegetable crops in greenhouse and field experiments. SMS, as such, has also been evaluated as a general soil amendment (Sagar et al., 2009) and as compost with other agricultural wastes such as winery sludge (Gonzalez-Marcos et al., 2015), manure and rice husk (Meng et al., 2018) and so forth.

Utilizing SMS as ruminant feedstock can be done using *P. ostreatus* and *L. edodes*. Other species of spent substrates were also examined specifically for other types of livestock. These included rice straw and banana leaves–based *V. volvacea* SMS for sheep (Sevilla et al., 1989) and rice and oat-based *Coprinus fimetarius* SMS for goats (Mann et al., 1994). Additionally, SMS can also be a source of feedstock in aquaculture. For instance, *A. bisporus* spent substrate was investigated as an additive in the diet of carp (*Cirrhina mirigala*) (Sehgal et al., 1993).

However, several factors for formulating the diet of animal feeds should be considered. These include the species of the animal, mushroom strains, nutritive level of SMS, cell wall component, digestibility and voluntary intake. There is also a need to evaluate and guarantee the consistency in the performance of SMS in order to recognize the application of SMS and its related products in the future (Phan and Sabaratnam, 2012).

9.5.2 Fish Feed

Since tilapia farmers face major constraints in sustaining feeds (72% of the total production cost) (Asian Development Bank, 2004), the utilization of SMS as a feed supplement offers a more affordable alternative. The use of SMS as fish feed enhanced tilapia growth and immune responses (Doan et al., 2017). Doan and colleagues (2017) noted a significant increase in growth and immune response parameters when tilapia were fed with 1% SMS. Also, SMS as a feed component for carp, *C. mirigala*, has been reported (Sehgal et al., 1993). Sehgal and colleagues demonstrated that incorporating 5% SMS and other agricultural and animal wastes with rice bran at a certain formulation enhanced the growth rate of carp. However, a higher amount of SMS (12%) resulted in decreased growth.

It is also reported that higher mortality rates occur during the warm cropping season due to a shortage of dissolved oxygen, water pollution and higher water temperatures (Asian Development Bank, 2004). Many human activities and natural processes usually end up in water systems. Common pollutants come from industrial discharges, surface run-off, agricultural residues and sewage, leaching of the soil and accidental spills. Reports demonstrated the biodegrading ability of *A. bisporus* substrate and its associated microorganisms against toxic pollutants of soil, such as polycyclic aromatic hydrocarbons (PAHs), heavy metals and agricultural pesticides (García-Delgado et al., 2015). PAHs are persistent chemical pollutants that are released from industries and through natural processes by the combustion of organic and petroleum-derived materials and are found in soil, water and air.

9.6 Conclusion

Organic plant materials are considered of lower value during harvesting, and the processing of crops usually ends up as agricultural residue. The most common agricultural residues are maize stalks, rice straws, wheat straws and sugarcane bagasse, which are often left to decompose in the field as a method of improving the soil structure and fertility. Although these residues have found their function in amending soil fertility and are used in several household activities such as cooking and heating, and for household building materials and livestock feed, a substantial amount of agricultural residue is still left neglected and managed improperly. A common practice in the Philippines is rice straw burning, which is documented to occur in 76% of all rice farming areas in the country. According to Mendoza (2015), about 7.08 million tons of rice residue are burned per year, which would have generated PhP 18.41 billion had it been composted. A similar approach to field burning is also commonly applied in managing corn and sorghum residues. This practice is attributed to several environmental and health concerns, mainly due to its emission of greenhouse gases and particulate matters, such as dioxins and furans (Yodkhum et al., 2018). Hence, over the years, this area of concern has attracted diverse feasibility studies on how to properly utilize the bioenergy potential of agricultural residues ensuring sustainable practices. In this chapter, we discussed the suitability of using agricultural waste in various endeavours. Rice straw and other cellulosic materials have been shown to be a promising material for the biosynthesis of nanomaterials. This is relevant since the nanotechnology industry has been making great strides in multiple fields of study. Moreover, this is a promising venture, since the process of nanomaterial production makes use of agricultural waste, poses minimal ecological risk and produces a more stable nanomaterial product. Rice straw can also be utilized for other agricultural endeavours such as mushroom farming. Mushrooms are an affordable and protein-rich food that could potentially be an alternative to meat for underprivileged households. In addition, mushrooms have a longer shelf life and require less demanding storage conditions. The waste product of mushroom farming, on the other hand, has been shown to be as nutritious as the mushroom itself. It has been used and extensively studied as an alternative to commercial fish and livestock feed. Considering that more than 70% of the production cost is allotted to feeds in fish farming, the use of spent mushroom substrate as feed offers a lucrative alternative. Moreover, it was observed that the crude protein, fat and ash content of SMS generally increases over time, whereas the cell wall components degrade. This is due to the enzymatic process that occurs during the growth and cultivation of mushrooms, thereby increasing their digestibility by ruminants. Alternatively, the use of rapid composting technology, owing to its cellulolytic activity, can hasten this process of enrichment. The waste materials of livestock and poultry farming, which are mainly composed of manure, can be combined with cellulosic waste products and subjected to rapid composting. The manure, in conjunction with the nitrogen-fixing inoculants, makes the compost a rich source of nitrogen. The compost can then be used by farmers as a fertilizer in their next cropping season.

REFERENCES

Adjapong A, Ansah K, Angfaarabung F and Sintim H. 2015. Maize residue as a viable substrate for farm scale cultivation of oyster mushroom (*Pleurotus ostreatus*). *Advances in Agriculture.* 2015: 1–6. doi:10.1155/2015/213251.

Ali SM. 2018. Fabrication of a nano composite from an agricultural waste and its application as a biosorbent for organic pollutants. *Journal of Environmental Science and Technology.* 15: 1169–1178.

Altintig E and Kirkil S. 2016. Preparation and properties of Ag-coated activated carbon nano composites produced from wild chestnut shell by $ZnCl_2$ activation. *Journal of the Taiwan Institute of Chemical Engineers.* 63: 180–188.

Asian Development Bank (ADB) 2004. *Case Study 5: Farming Tilapia in Ponds in Central Luzon, Philippines.* Accessed Oct. 23, 2019 from www.adb.org/sites/default/files/evaluation-document/35937/files/farming-tilapia-phi.pdf.

Bathla A, Narula C and Chauhan RP. 2018. Hydrothermal synthesis and characterization of silica nanowires using rice husk ash: An agricultural waste. *Journal of Materials Science: Materials in Electronics.* 29: 6225–6231.

Bhattacharya M and Mandal MK. 2018. Synthesis of rice straw extracted nano-silica-composite membrane for CO_2 separation. *Journal of Cleaner Production.* 186: 241–252.

Brown MB, Brown CMB and Nepomuceno R (2016) Regulatory requirements and registration of biopesticides in the Philippines. In: Singh H, Sarma B and Keswani C (eds) *Agriculturally Important Microorganisms: Commercialization and Regulatory Requirements in Asia.* Springer, Singapore.

Cazetta AL, Pezoti O, Bedin KC, Silva TL, Junior AP, Asefa T and Almeida VC. 2016. Magnetic activated carbon derived from biomass waste by concurrent synthesis: Efficient adsorbent for toxic dyes. *ACS Sustainable Chemistry & Engineering*. 4: 1058–1068.

Chirapongsatonkul N, Mueangkan N, Wattitum S and U-Taynapun K. 2019. Comparative evaluation of the immune responses and disease resistance of Nile tilapia (*Oreochromis niloticus*) induced by yeast β-glucan and crude glucan derived from mycelium in the spent mushroom substrate of *Schizophyllum commune*. *Aquaculture Reports*. 15: 100205.

Chiu SW, Ching ML, Fong KL and Moore D. 1998. Spent oyster mushroom substrate performs better than many mushroom mycelia in removing the biocide pentachlorophenol. *Mycological Research*. 102(12): 1553–1562.

De Bertoldi M, Sequi P, Lemmes B and Papi T. 1996. The role of composting in sustainable agriculture. In: *The Science of Composting* (first edition). Springer Netherlands, 23–29. doi:10.1007/978-94-009-1569-5.

Devi LL, Basavapoornima CH, Venkatramu V, Babu P and Jayasankar CK. 2017. Synthesis of $Ca_2SiO_4Dy^{3+}$ phosphors from agricultural waste for solid state lighting applications. *Ceramics International*. 43: 16622–16627.

Dhugga KS. 2007. Maize biomass yield and composition for biofuels. *Crop Science*. 47(6), 2211. doi:10.2135/cropsci2007.05.0299.

Doan HV, Hoseinifar SH, Tapingkae W, Chitmanat C and Mekchay S. 2017. Effects of *Cordyceps militaris* spent mushroom substrate on mucosal and serum immune parameters, disease resistance and growth performance of Nile tilapia (*Oreochromis niloticus*). *Fish & Shellfish Immunology*. 67: 78–85.

Dominguez-Escriba L and Porcar M. 2010. Rice straw management: The big waste. *Biofuels, Bioproducts and Biorefining*. 4(2): 154–159. doi:10.1002/bbb.196.

Fang K, Chen J, Zhou X, Mei C, Tian Q, Xu J and Wong CP. 2018. Decorating biomass-derived porous carbon with Fe_2O_3 ultrathin film for high-performance supercapacitors. *Electrochimica Acta*. 261: 198–205.

Food and Agricultural Organization. 2013. Agricultural statistical database. Retrieved May 20, 2016 from http://faostat.org.

Garcia-Delgado C, Yunta F and Eymar E. 2015. Bioremediation of multi-polluted soil by spent mushroom (*Agaricus bisporus*) substrate: Polycyclic aromatic hydrocarbons degradation and Pb availability. *Journal of Hazardous Materials*. 300: 281–288.

Gonzalez-Marcos A, Alba-Elias F, Martinez-De-Pison FJ, Alfonso-Cendon J and Castejon-Limas M. 2015. Composting of spent mushroom substrate and winery sludge. *Compost Science & Utilization*. 23: 58–65.

Grimm D and Wösten H. 2018. Mushroom cultivation in the circular economy. *Applied Microbiology and Biotechnology*. 102(18): 7795–7803. doi:10.1007/s00253-018-9226-8.

Hanafi FM, Rezania S, Taib SM, Din MFM, Yamauchi M, Sakamoto M, Hara H, Park J and Ebrahimi SS. 2018. Environmentally sustainable applications of agri-based spent mushroom spent (SMS): An overview. *Journal of Material Cycles and Waste Management*. 20(3): 1383–1396.

Huang E, Satjapipat M, Han S and Zhou F. 2001. Surface structure and coverage of an oligonucleotide probe tethered onto a gold substrate and its hybridization efficiency for a polynucleotide target. *Langmuir*. 17: 1215–1224.

Isikgor FH and Becer CR. 2015. Lignocellulosic biomass: A sustainable platform for the production of bio-based chemicals and polymers. *Polymer Chemistry*. 6: 4497–4559.

Jabasingh SA, Belachew H and Yimam A. 2018. Iron oxide induced bagasse nanoparticles for the sequestration of Cr^{6+} ions from tannery effluent using a modified batch reactor. *Journal of Applied Polymer Science*. 135: 46683.

Jordan SN, Mullen GJ and Murphy MC. 2008. Composition variability of spent mushroom compost in Ireland. *Bioresource Technology*. 99: 411–418.

Julian AV, Reyes RG and Eguchi F. 2018. Agro-industrial waste conversion into medicinal mushroom cultivation. *Reference Module in Earth Systems and Environmental Sciences*. doi:10.1016/b978-0-12-409548-9.11042-5.

Kakkar VK, Garach HS, Dhanda S, Makkar GS. 1990. Mushroom harvested spent straw as feed for buffaloes. *Indian Journal of Animal Nutrition*. 7: 267–272.

Kamthan R and Tiwari I. 2017. Agricultural wastes: Potential substrates for mushroom cultivation. *European Journal of Experimental Biology*. 7(5). doi:10.21767/2248-9215.100031

Kauldhar BS and Yadav SK. 2018. Turning waste to wealth: A direct process for recovery of nano-silica and lignin from paddy straw agro-waste. *Journal of Cleaner Production*. 194, 158–166.

Khatoon N, Mazumder JA and Sarda M. 2017. Biotechnological applications of green synthesized silver nanoparticles. *Journal of Nanosciences: Current Research*. 2: 1.

Li R, Liang W, Wang J, Gaston LA, Huang D, Huang H, Shuang L, Awasthi MK, Zhou B, Xiao R and Zhang Z. 2018. Facilitative capture of As(V), Pb(II) and methylene blue from aqueous solutions with MgO hybrid sponge-like carbonaceous composite derived from sugarcane leafy trash. *Journal of Environmental Management*. 212: 77–78.

Lopez-Lorente AI and Valcárcel M. 2016. The third way in analytical nanoscience and nanotechnology: Involvement of nanotools and nanoanalytes in the same analytical process. *Trends in Analytical Chemistry*. 75: 1–9.

Magette W, Smyth S and Dodd V. 1998. Logistical considerations for spent mushroom compost utilization. In: *8th International Conference on Management Strategies for Organic Waste in Agriculture*. Rennes, France, 23–30.

Mandal KG, Misra AK, Hati KM, Bandyopadhyay KK, Ghosh PK and Mohanty M. 2004. Rice residue management options and effects on soil properties and crop productivity. *Journal of Food, Agriculture and Environment*. 2: 224–231.

Mann NS, Kumar N, Dahiya DVS, Khatta VK. 1994. Effect of Coprinus fimetarius on urea treated crop residues and its utilization by goats. *Indian Journal of Animal Sciences* 64(10): 1086–1091.

Mathur L, Hossain SKS, Majhi MR and Roy PK. 2017. Synthesis of nano-crystalline forsterite (Mg_2SiO_4) powder from biomass rice husk silica by solid-state route. *Boletin de la Sociedad Espanola de Ceramica y Vidrio*. 57: 112–118.

Medina E, Paredes C, Perez-Murcia M, Bustamante M and Moral R. 2009. Spent mushroom substrates as component of growing media for germination and growth of horticultural. *Bioresource Technology*. 100: 4227–4232.

Mendoza T. 2015. *Enhancing Crop Residues Recycling in the Philippine Landscape*. doi:10.1007/978-981-287-643-0_4.

Meng X, Liua B, Xi C, Luo X, Yuan X, Wang X, Zhu W, Wang H and Cui Z. 2018. Effect of pig manure on the chemical composition and microbial diversity during co-composting with spent mushroom substrate and rice husks. *Bioresource Technology*. 251: 22–30.

Moon YH, Shin PG and Cho SJ. 2012. Feeding value of spent mushroom (*Pleurotus eryngii*) substrate. *Journal of Mushroom*. 10(4): 236–243.

Nepomuceno RA, Brown CMB, Gargarino AMP, Pedro MS and Brown MB. 2020. Growth enhancement of rice (*Oryza sativa* L.) by zinc-solubilizing bacteria isolated from Vesicular-Arbuscular Mycorrhizal Root Inoculant (VAMRI). *Philippine Journal of Crop Science (PJCS)*. 45 (1): 34–40.

Nepomuceno RA, Brown CMB, Mojica PN and Brown MB. 2019. Biological control potential of Vesicular Arbuscular Mycorrhizal Root Inoculant (VAMRI) and associated phosphate solubilizing bacteria, pseudochrobactrum asaccharolyticum against soilborne phytopathogens of onion (*Allium cepa* L. var. Red Creole). *Archives of Phytopathology and Plant Protection*. 52: 7–8, 714–732, doi:10.1080/03235408.2019.1644058.

Oh YK, Lee, WM, Choi CW, Kim KH, Hong SK, Lee SC, Seol YJ, Kwak WS and Choi NJ. 2010. Effects of spent mushroom substrates supplementation on rumen fermentation and blood metabolites in hanwoo steers. *Asian-Australian Journal of Animal Science*. 23(12): 1608–1613.

Pandit PR and Fulekar MH. 2017. Egg shell waste as heterogeneous nanocatalyst for biodiesel production: Optimized by response surface methodology. *Journal of Environmental Management*. 198: 319–329.

Pandiyan N, Murugesan B, Sonamuthu J, Samayanan S and Mahalingam S. 2018. Facile biological synthetic strategy to morphologically aligned CeO_2/ZrO_2 core nanoparticles using *Justicia adhatoda* extract and ionic liquid: Enhancement of its bio-medical properties. *Journal of Photochemistry and Photobiology*. 178: 481–488.

Panthapulakkal S and Sain M. 2015. The use of wheat straw fibres as reinforcements in composites. In: Faruk O and Sain M (eds) *Biofiber Reinforcements in Composite Materials*. Woodhead Publishing, 423–453, ISBN 9781782421221, https://doi.org/10.1533/9781782421276.4.423.

Patel KG, Misra NM, Vekariya R and Shettigar R. 2018. One-pot multicomponent synthesis in aqueous medium of 1,4-dihydropyrano[2,3-c]pyrazole-5-carbonitrile and derivatives using a green and reusable nano-SiO_2 catalyst from agricultural waste. *Research on Chemical Intermediates*. 44: 289–304.

Phan CW and Sabaratnam V. 2012. Potential uses of spent mushroom and its associated lignocellulosic enzymes. *Applied Microbiology and Biotechnology.* 96: 863–873.

Rajasekharreddy P, Rani PU and Sreedhar B. 2010 Qualitative assessment of silver and gold 319 nanoparticle synthesis in various plants: A photobiological approach. *Journal of Nanoparticle Research.* 12: 1711–1721.

Ranum P, Peña-Rosas JP and Garcia-Casal MN. 2014. Global maize production, utilization, and consumption. *Annals of the New York Academy of Sciences.* 1312(1), 105–112.

Rasib NA, Zakaria Z, Tompang MF, Rahman RA and Othman H. 2015. Characterization of biochemical composition for different types of spent mushroom substrate in Malaysia. *Malaysian Journal of Analytical Sciences.* 19(1): 41–45.

Reyes RG. 2000. Indoor cultivation of paddy straw mushroom, *Volvariella volvacea*, in crates. Center for Tropical Mushroom Research and Development, Dept. of Biological Sciences, College of Arts and Sciences, CLSU, Nueva Ecija, Philippines. *Mycologist* 14(4): 174–176.

Saadati F, Leghaei V and Zamani A. 2017. Environmentally benign copper nanoparticles supported on walnut shell as a highly durable nanocatalyst for the synthesis of propargylamines. *Journal of the Serbian Chemical Society.* 82: 1–12.

Sagar MP, Ahlawat OP, Raj D, Vijay B and Indurani C. 2009. Indigenous technical knowledge about the use of spent mushroom substrate. *Indian Journal of Traditional Knowledge.* 8(2): 242–248.

Sebastian A, Nangia A and Prasad MNV. 2018. A green synthetic route to phenolics fabricated magnetite nanoparticles from coconut husk extract: Implications to treat metal contaminated water and heavy metal stress in *Oryza sativa* L. *Journal of Cleaner Production.* 174: 355–366.

Sehgal HS, Simmi S and Sharma S. 1993. A note on evaluation of some wastes and by products from agriculture and animal husbandry as feed ingredients for *Cirrhina mirigala* (Ham.) *Bioresource Technology.* 44(1): 9–11.

Sevilla, CC, Mojica RE, Quimio TH. 1989. Feeding value of spent mushroom (Volvariella volvacea Bull.) substrate-based diet in sheep. Recent advances in animal nutrition in Australia-1989. 22A.

Yan D, Zhang H, Chen L, Zhu G, Wang Z, Xu H and Yu A. 2014. Supercapacitive properties of Mn_3O_4 nanoparticles biosynthesized from banana peel extract. *RSC Advances.* 4: 23649–23652.

Yang Y, Huang C and Hu K. 2015. Purification and further characterizations of the cellulose from the spent mushroom compost of *Hypsizgus marmoreus. Research Journal of Biotechnology.* 120(2): 52–57.

Yodkhum S, Sampattagul S and Gheewala SH. 2018. Energy and environmental impact analysis of rice cultivation and straw management in northern Thailand. *Environmental Science and Pollution Research.* 25(18): 17654–17664. doi:10.1007/s11356-018-1961-y

Wagner SC. 2011. Biological nitrogen fixation. *Nature Education Knowledge.* 3(10): 15.

Williams BC, McMullan JT and McCahey S. 2001. An initial assessment of spent mushroom compost as a potential energy feedstock. *Bioresource Technology.* 79(3): 227–230.

Wu G. 2018. *Principles of Animal Nutrition.* Taylor & Francis Group, L.L.C, Florida.

Zamani A, Marjani AP and Mousavi Z. 2019. Agricultural waste biomass-assisted nanostructures: Synthesis and application. *Green Processing and Synthesis.* 8(1): 421–429. doi:10.1515/gps-2019-0010.

Zhang R, Li X and Fadel JG. 2002. Oyster mushroom cultivation with rice and wheat straw. *Bioresource Technology.* 82: 277–284.

10

Compost and Compost Tea from On-Farm Composted Agro-Wastes Improve the Sustainability of Horticultural Organic Cropping Systems

Ugo De Corato

CONTENTS

10.1 Introduction

Agricultural activities generate a significant amount of biomass waste in the form of cultivated green residues, non-marketable products and several agro-wastes from crop cultivation and food processing (Table 10.1). In the absence of productive uses, these unused or underused biomasses are considered waste and are destined for disposal with additional costs for companies. Furthermore, biomass waste such as vegetable waste should be avoided or reduced, especially waste from greenhouse cultivation and warehouse processing. This waste is usually stacked and left to decompose outdoors with consequent environmental problems. The agricultural biomass waste management through its recovery and further reusing in new production cycles can determine an interesting change of paradigm from the perspective of an agri-based circular economy by converting such biomass into more co/by-products useful in organic agriculture. This vision of a circular economy system can have a positive impact on the sustainability of organic cropping systems and their ability to recover agro-waste and turn it into a profitable resource (De Corato et al. 2018a). The production of quality composts and/or their derivative liquid (teas) and solid (humic substances) products represents one possibility of exploiting a rich and valuable source of organic molecules and microbial consortia from agro-wastes and plant residues produced on the farm. Recycling biomass waste using the composting process can lead either to rationalizing and reducing biomass waste or to guaranteeing the supply of beneficial microbial consortia associated with the organic matter that agricultural soils and crops need for good soil quality on the basis of its role as a suppressive agent, in plant nutrition and vigour and in plant health and productivity (De Corato 2020a).

TABLE 10.1

Classification of Residual Biomass from Agro-Industrial Activities Suitable to be Recycled for Producing a Range of High Value–Added Products Useful in Sustainable Horticulture

Typology	Feedstock	Agricultural Tool
Crop residues	Straws, stover, husks, cobs, press cakes from rape and soybean seeds, olive marc and lees, grapes, olives and other fruit, bagasse, sugar beet pulps, palm kernel meals, palm oil mill effluents, empty fruit bunches and nutshells.	**Green compost** and **compost tea**, biogas and **digestate**. Biodegradable-compostable plastic (only by flour or corn starch, wheat or other cereal grains).
Forestry residues	Treetops, branches, stumps, leaves, sawdust, cutter shavings, scrap wood, wood pulp and wood chip.	**Green compost** and **compost tea**, biogas and **digestate**.
Food processing residues	Oil-less seed cakes/meals, molasses, defatted olive marc, spent coffee-grounds, tea bags, exhausted beet pulps, tall oil pitch and animal fats.	**Green compost**, biogas and **digestate**, second-generation biodiesel.
Aquaculture and fisheries residues	Brown, red and green seaweeds, fish scales, viscera and scraps.	Biostimulant for plant growth and improver of crop quality (seaweeds). Phytosanitary drugs in fruit postharvest application against spoilage fungi and plant pathogens in preventing postharvest losses (seaweeds). Biogas and **digestate** (others).
Agro-wastes	Used cooking oils, exhausted engine oils, biomass fraction of mixed municipal solid waste (MSW), animal manure, biomass fraction of industrial waste and sewage sludge.	Second-generation biodiesel (oil), biogas and **digestate** (others).
Wet municipal solid waste (European Waste Code: 200108)	Unmarketable products such as vegetable and fruit byproducts; food wastes such as meat, fish, bread, eggs, cheeses, pastries, coffee grounds, tea bags; cropping residues such as grass, leaves, flowers, twigs, extinguished wood ashes.	**Compost**, biogas and **digestate**.
Green municipal solid waste (European Waste Code: 200201)	Vegetable processing leftovers, large pruning, cuttings and garden waste.	**Green compost** and **compost tea**.

Note: Organic products as soil improvers, suppressive organic amendments and plant growth promoters (including composts, teas and digestates) are in bold.

Currently, compost is commercially available to farmers and growers either as a conditioner to improve the physical structure and biological fertility of the soil or as an organic amendment to enhance the natural suppressiveness of the soil against plant pathogens (Albiach et al. 2000; Zaccardelli et al. 2011). Compost can be a suitable source of nutrition for crops, partially replacing mineral fertilizers, or a peat substitute in nursery potting mix substrates for seedling growth (Zaccardelli et al. 2006). Moreover, compost can also be employed as a bioremediation product for the recovery and clean up of marginal soils (Hickman and Reid 2008). Recently, emerging technologies have created more concrete possibilities for new preparations made of organic compounds derived from compost, including humic substances (mixtures of humic and fulvic acids) and humin extracted from compost using a chemical–physical process. Such soluble organic molecules seem to have a direct effect on some plant metabolic processes, thanks to their particular molecular structure (Ayuso et al. 1996; Diver and Greer 2001; Nardi et al. 2002; Eyheraguibel et al. 2008; Morard et al. 2011; Khaled and Fawy 2011). For example, due to its high content of soluble nutrients, the use of leachate recovered from piles during the composting phase was proposed as a foliar biofertilizer. But, this is a co-product that require disinfection, making it unsuitable for foliar applications without any pre-treatment (Diver 2002).

This chapter highlights the main advantages of bio-composting agro-wastes from different feedstocks for improving the sustainability of horticultural cropping systems managed under organic cultivation systems. The purpose is to focus the knowledge acquired about the use of suppressive compost (SC) and compost tea (CT) in horticulture, both in phytosanitary defence and plant growth promotion. This contribution is mainly addressed to farmers and growers by describing and comparing the most promising technologies to significantly reduce the amounts of either synthetic fungicides for controlling soil-borne plant pathogens of horticultural crops using suppressive composts, or mineral fertilizers for enhancing productivity and the quality of horticultural products using compost teas.

10.2 Compost

Compost can be defined as a matured and microbiologically stabilized organic material that has been naturally enriched during the composting process with hydrophobic carbonaceous molecules and humic substances, such as humic and fulvic acids, which make it a more recalcitrant organic media for microbial degradation (De Corato 2020a). Compost derives from a biological oxidative process through which biodegradable organic substances are transformed into a more stable and highly humified product, which is widely used in agriculture for its positive effects on plant growth and crop protection (ISO/IEC 17025:2005).

10.2.1 Bio-Composting

Bio-composting is defined as a natural and sustainable biological process that, thanks to the action of endogenous microorganisms that colonize highly biodegradable organic biomasses during the curing phase, allows them to be transformed into composted biomasses. This particular approach of bio-oxidative transformation in the solid phase, to which the biomass is submitted if arranged in oxygenated piles using forced air, is subjected to complex interactions between substrate and microbiota (Figure 10.1). The development of heat within the pile is the key point of the whole composting process that defines the two main phases of the active process (De Corato 2020a). The first phase, the so-called "thermophilic" phase, is characterized by elevated temperatures above 60°C to 75–78°C for at least 5–6 consecutive days. This allows the sanitation and pasteurization of the mass by eliminating pathogenic microbial species both for humans and animals and for plants. The second phase, the so-called "mesophilic" phase, is characterized by a progressive drop in temperature that takes place when most of the

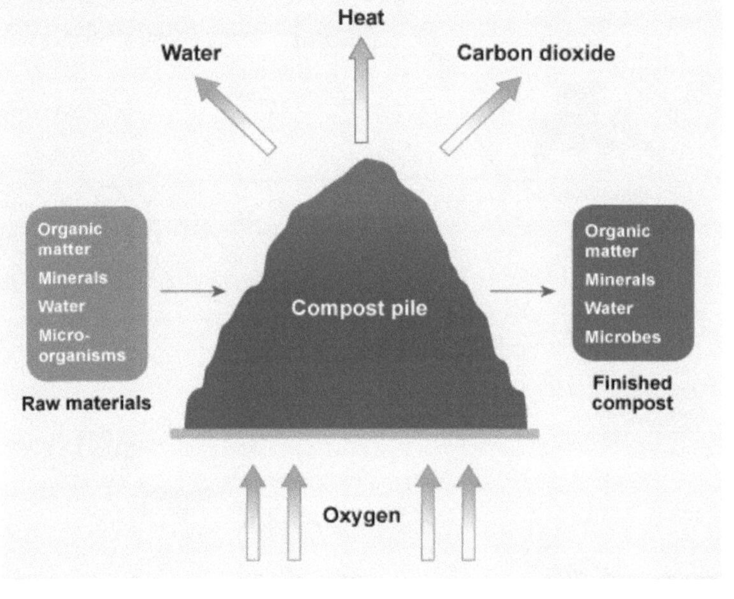

FIGURE 10.1 Scheme of a composting process based on the reuse of agro-wastes suitable for on-farm recycling into green compost.

carbon molecules are easily attacked, which are metabolized by the microorganisms in the pile. Before turning into compost, however, the pile will undergo a third and final phase of composting, the so-called "curing" or "maturation" phase, providing adequate stabilization and humification of the organic matter. According to European guidelines (ISO/IEC 17025:2005) and European (Regalement UE No. 2003/03) and Italian (Italian Legislative Decree No. 217/06; Italian Legislative Decree No. 75/10) legislation, it takes about 90 days from the start of the composting process to achieve a highly marketable mature compost. Obviously, the curing time and the maturation conditions strongly depend on the composted feedstock typology and environmental conditions (De Corato et al. 2016).

Typically, the bio-composting process is carried out in industrial biomass processing plants on a small, medium and large scale for processing the organic fraction that is a result of the separate collection of municipal solid organic waste (MSW), which adopt very different ventilation systems such as mechanical forced ventilation in bioreactors or insufflation chambers. These plants meet most of the demand for compost in agriculture. However, regarding the development of the organic agriculture sector, the transport cost factor becomes more decisive and restrictive, such as limiting the supply capacity of compost in the farm from outside. This seemingly insurmountable obstacle could be easily bypassed through a particular management practice of organic agro-wastes that is spreading to various agricultural sectors, "in situ composting", otherwise known as "on-farm composting".

10.2.2 On-Farm Composting

In addition to industrial composting systems, among the safer and more innovative ones suitable for producing high-quality compost, we can include those based on "on-farm technologies", which open the way to the most effective reuse of different types of agricultural wastes (Zaccardelli et al. 2010). The self-production of compost by on-farm techniques, thanks to the greater availability of a wide range of agricultural residues, allows the recycling of biomass waste on the same farm that produces it, solving the dual problem of agricultural waste disposal by allowing the agricultural players to get supplies of humified organic substance at lower costs by applying compost to depleted soil (Scotti et al. 2016). On-farm composting requires simple technologies based on suitably sized systems, possibly made with equipment already present on the farm or, in any case, easily accessible to the farmer at lower costs (Figure 10.2) than those required for industrial composting. The systems that can be adopted depend on

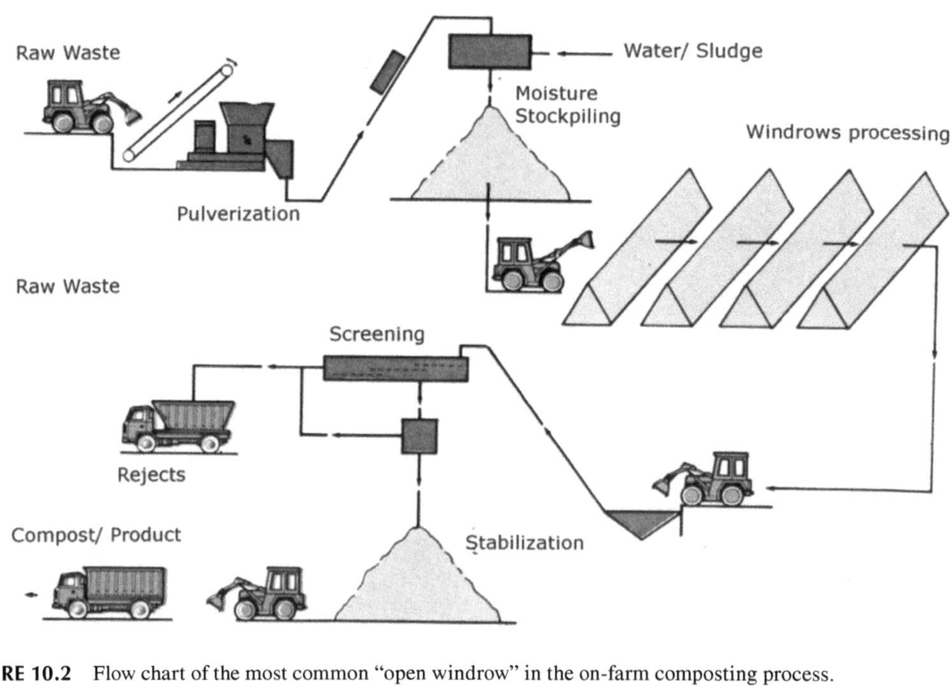

FIGURE 10.2 Flow chart of the most common "open windrow" in the on-farm composting process.

the quantity and type of biomass to be treated, the frequency of its production and the available equipment. The simplest technology of on-farm composting is the "pile static" method in which the biomasses, once accumulated, are not marketable until the end of the process. This system adapts to a very structured type of biomass, such as a lignocellulosic type (wood) of biomass that, mixed with herbaceous material (cellulose), gives sufficient porosity to the mass and allows efficient air diffusion inside. On the other hand, the "periodic turning" method of the pile represents an expensive system that depends on the availability of staff, machines and tools to perform the composting function in a timely manner. Finally, the "static accumulation with forced ventilation" method represents the most advanced technology that can be conveniently adopted in an on-farm system (Pergola et al. 2020). Periodic mass ventilation is provided by a blower, activated by a timer, which diffuses air through perforated pipes placed at the base of the pile. A balanced mixture of woody material (structuring) and green-grassy material (nutritional) is necessary in order to guarantee the most adequate porosity that allows air diffusion and the exchange of heat with the outside. The dynamics of decomposition are based on the fact that, at the end of the process, the unmatured biomass loses its natural phytotoxicity and acquires a specific and higher microbiological, chemical and molecular maturation, providing the compost with the necessary quality to be safely used in agricultural applications (Pergola et al. 2020).

10.2.3 Utilization

Intensive horticultural cropping systems, with frequent soil processing, are the main cause of soil depletion in terms of organic matter loss from arid and semi-arid soil (Costantini and Lorenzetti 2013). In addition to this, is the failure to return green waste to the soil which, for phytosanitary or other management reasons, is removed from the soil surfaces when preparing fields for mono-cropped crops. Moreover, the widespread use of monoculture systems in advanced economies and the rapid succession of "short-cycle crops" in the same year, amplifies the soil decline with the onset of typical manifestations of soil fatigue. For instance, all these issues are more pronounced in protected cultivation under greenhouse conditions due to higher soil temperatures that are compatible with the acceleration of mineralization processes. The organic matter loss, due to agricultural activities and environmental factors, leads inexorably to soil fertility degradation. Therefore, external inputs of organic material by compost represent an agricultural practice not only desirable, but also increasingly necessary (Scotti et al. 2015).

The use of compost in unbalanced soils contributes organic matter and nutrients and aids in the general recovery of the soil's fertility. Compost improves the growth of plants, increases carbon reserves and overall stimulates microbial activity and its biodiversity linked to the nutrient cycle of the soil (De Corato 2020b). The addition of compost to soil represents the main mode of application of organic agriculture with well known benefits in contrasting soil decline. In fact, the external input of organic matter into the soil plays a crucial role not only in maintaining the fertility and productivity of agricultural systems performing nutritional functions, but also in stimulating microbial activity, maintaining the soil structure, ventilating, retaining moisture and increasing buffer capacity (De Corato 2020a). Moreover, the use of disease-suppressive compost in horticulture for crop protection and plant growth promotion represents a technological advancement to solve the dual problem of the disposal of agro-wastes and the production of high value–added products in a circular economy system (De Corato et al. 2018a).

10.2.4 Disease-Suppressive Composts

Certain composts can create an environment adverse to the development of plant diseases caused by soil-borne plant pathogens especially in horticulture (Hoitink and Fahy 1986; Hoitink et al. 1991, 1997, 2001; Bonilla et al. 2012). The capacity of a compost to suppress plant diseases has been widely studied, documented and reviewed (Lazarovits 2001; Noble and Coventry 2005; Bonanomi et al. 2007; Avilés et al. 2011; Noble 2011). Compost amendment has been proposed to control *Pythium ultimum* and *Pythium irregulare* (Pascual et al. 2000; Scheuerell et al. 2005), *Phytophthora nicotianae* and *Phytophthora cinnamomi* (Hardy and Sivasithamparam 1991; Widmer et al. 1999), *Rhizoctonia solani* (Tuitert et al. 1998; Termorshuizen et al. 2006) and *Fusarium oxysporum* (Cotxarrera et al. 2002). All these issues have been critically revised by De Corato (2020b) under the perspective of an agri-based bioeconomy. Composts sourced from MSW have been tested for controlling *P. ultimum* damping-off of cucumber, *F. oxysporum* f. sp. *basilici* wilt of basil and *Sclerotinia sclerotiorum* root rot of lettuce (Pugliese et al. 2010). Composts resulting from MSW and cow manure enhanced peat suppressiveness towards damping-off caused by *P. ultimum*, *R. solani* and *Sclerotinia minor* (Pane et al. 2011). Compost suppression is related not only to the antagonistic microbial communities sourced from the organic matter used as feedstock (Lumsden et al. 1983; Conrath et al. 2002; Manici et al. 2004; Hadar and Papadopoulou 2012), but also to its phytotoxicity level (Aslam and Van der Gheynst 2008). For all these reasons, suppressive composts belong to a special compost category suitable for improving the sustainability of the horticultural cropping systems managed under organic cultivation conditions (De Corato 2020b).

The suppressive property of compost can be explained through different anti-pathogen mechanisms exerted by some molecules of the compost, or through the actions of microbial antagonism of the microflora residing in the soil conditioner. In particular, the mechanisms underlying the suppressive effects of compost are primarily associated with the biological activity of its specific microbiota which are able to interact with the soil and host plant (Shi et al. 2017). Several mechanisms explaining compost suppressivess have been proposed: (i) increased microbial activity (Melero et al. 2006) and the release of mineral nutrients during organic matter decomposition (Berry et al. 2002); (ii) activation of competition between microbes for available nutrients (Noble and Coventry 2005); (iii) elicitation of microbiostasis and hyperparasitism (Hoitink et al. 1993); (iv) release of diffusible antibiotic-like compounds (Weller et al. 2002); and (v) activation of systemic disease resistance in host plants, where the protection exercised by ground treatment with suppressive composts can also extend to the aerial part through the mechanism of the induction of resistance and the improvement of the general conditions of plant development (Bulluck et al. 2002). A "general suppression effect" is normally attributed to a microbial biomass affecting more than one pathogen simultaneously, whenever diversified mechanisms are activated, offering basal protection against a broad range of pathogens (Berendsen et al. 2012; Cha et al. 2016). In contrast, a "specific suppression effect" is usually attributed to a specific biocontrol-based microbiota that is active against one or a few pathogens (Termorshuizen et al. 2006; Termorshuizen and Jeger 2008). Nevertheless, the evidence suggests that the ability of compost to suppress plant pathogens depends on the complex interactions among beneficial microbes and pathogens, and the physicochemical properties of plant growth media (Hadar and Papadopoulou 2012).

Its suppressive characteristics make compost one of the most suitable tools for the biological control of soil-borne plant pathogens. Numerous examples based on the use of compost in suppressing the pathogenic capacity of plant fungi can be found in literature. The addition of suppressive compost in appropriate rates into plant growth media is a consolidated strategy for the effective biological control of many soil-borne pathogens in organic cropping systems (Avilés et al. 2011; Pane et al. 2013; Angelopoulou et al. 2014; Chilosi et al. 2017; Kefalogianni et al. 2017). Banned synthetic chemicals as soil fumigants (methyl bromide, 1,3-dichloropropene, chloropicrin, etc.) and their inability to totally control a wide range of soil-borne diseases in different agro-ecosystems have prompted the exploration and evaluation of a wider range of feedstocks derived from agro-wastes and agro-industrial co/byproducts for making the so-called "tailored multi-suppressive composts" a reliable and effective mean for controlling many soil-borne phytopathogens in the greenhouse (Tables 10.2, 10.3 and 10.4). Composts derived from green agro-wastes and several agro-industrial co/byproducts (green composts) provided the most varied microbiomes related either to the suppression of *Rhizoctonia* damping-off of bean and *Verticillium* wilt of eggplant or to the control of *Pythium* damping-off of cucumber and *Phytophthora* root rot of tomato (De Corato et al. 2019). On the other hand, composted municipal solid wastes and co-composted cow manure with household waste provided a very specific microbiota related to the suppression of *Fusarium* wilt of melon when compared to green composts (De Corato et al. 2019). In this regard, the composition, diversity and variability of microbiomes were recently investigated for the effective biological control of soil-borne phytopathogens in a collection of suppressive composts with microbiome-assisted strategies using next-generation amplicon sequencing (De Corato 2020c). Amplicon sequencing resulted in a reliable and faster approach to characterizing fungal and bacterial microbiomes in complex microbial communities of composts derived from feedstock of various compositions, origins and geographical provenances, where differences in taxonomic structure were primarily related to the origin of the feedstock (De Corato 2019).

TABLE 10.2

Classification of 10 Disease-Suppressive Composts into 3 Main Typologies on the Basis of the Feedstock Used during Composting Process

Typology	Compost	Feedstock #1: Agricultural and Agro-Industrial Residues	Feedstock #2: Plant Green-Waste	Feedstock #3: Organic Fraction of Municipal-Solid-Waste, Cattle Manure and Other Bio-Waste	Composting Starter (2%)	Geographical Origin
Group-AC (green composts)	Com-A	Defatted olive marc (50%)	Fennel (48%)	—	Yes	Bitonto (Bari, Italy)
	Com-B	Un-defatted olive marc (50%)	Artichoke (48%)	—	Yes	Bitonto (Bari)
	Com-C	Coffee ground (50%)	Celery (28%) + carrot (20%)	—	Yes	Foggia (Italy)
	Com-D	Tea bag (50%)	Tomato (28%) + lettuce (20%)	—	Yes	Monopoli (Bari)
Group-MC (green composts)	Com-E	Wood chip (50%)	Tomato (28%) + escarole (20%)	—	Yes	Policoro (Matera, Italy)
	Com-F	Aspen chip (50%)	Artichoke (28%) + fennel (20%)	—	Yes	Policoro (Matera)
	Com-G	Vineyard pruning wastes (21%) + vinery residues (21%) + straw (8%)	Potato (24%) + pepper (24%)	—	Yes	Turi (Bari)
Group-LC (traditional composts)	Com-H	—	—	Urban bio-waste (100%)	No	Modugno (Bari)
	Com-I	—	—	Wet MSW (100%)	No	Modugno (Bari)
	Com-L	—	—	Cow manure (70%) + household waste (30%)	No	Lavello (Potenza, Italy)

TABLE 10.3

Classification of 10 Disease-Suppressive Composts into 3 Main Typologies on the Basis of the Different Composting Techniques

Typology	Composting Cycle Length and Plant Type	Biomass Composted as Dry Weight (kg)	Pile Size (m^3)	Maximum Temperature (°C) Reached for At Least 7 Consecutive Days	Thermophilic Stage	Curing Stage	References
Group-AC	Accelerated cycle into a semi-industrial plant	500	5	68	65 days	2 months	De Corato et al. (2019)
Group-MC	Middle cycle into an on-farm organic farming system	500	5	70	90 days	6 months	Pane et al. (2011, 2013)
Group-LC	Long cycle into an accredited industrial plant	500	5	75	3 months	2 years	ISO/IEC 17025:2005

TABLE 10.4

Classification of 10 Disease-Suppressive Composts into 3 Main Typologies on the Basis of the Different Suppressive Models Studied

Typology	Target Pathogen	Susceptible Host	Suppression Model	Disease Suppression	Greenhouse	Open Field	References
Group-AC	*Verticillium dahliae*, *Rhizoctonia solani*, *Phytophthora nicotianae*, *Pythium ultimum*	Eggplant, bean, tomato, cucumber	Multiple range (general and specific)	From 60 to 80% in glasshouse and between 45 and 64% in open field	Yes	Yes	De Corato et al. (2019)
Group-MC	*Verticillium dahliae*, *Phytophthora nicotianae*, *Pythium ultimum*	Eggplant, tomato, cucumber	General	From 48 to 57%	Yes	No	De Corato et al. (2019)
Group-LC	*Fusarium oxysporum* f. sp. *melonis*	Melon	Specific	From 57 to 70%	Yes	No	De Corato et al. (2019)

10.3 Compost Teas

Two important compost derivatives suitable for improving the sustainability of the horticultural cropping systems managed under organic cultivation are (i) aqueous extracts (WEs) and (ii) compost teas. These two products differ essentially in their production processes. WEs are obtained from a water extraction procedure of soluble components of compost that occurs through an almost instantaneous process carried out over a few hours (Joshi et al. 2009). In this case, the composition of the liquid formulation depends on the composition of the starting compost. Instead, CT is obtained by a complex process of fermentation that changes the main chemical and microbiological characteristics of it, when compared to the compost from which it is derived (Ingham 1999). In recent years, increasing interest has been shown in WEs from next-generation green composts for enhancing the suppressive properties of composted MSW-based biomass, using WEs recruited from compost during the composting process (De Corato et al. 2018b), as well as next-generation CTs for their effects on plant growth promotion and disease suppression (Zaccardelli et al. 2012). In particular, growers can benefit from CTs both from a phytopathological point of view for controlling plant diseases and agronomically for promoting plant growth and other physiological properties related to the productivity of crops. The effectiveness of these products is appreciated in the improvement in the yield and quality of the marketable products in horticulture.

CTs can be defined as organic liquid formulations produced by a compost extraction-oxygenation process of immersion in a liquid, generally water, for a period of time varying from a few hours to two weeks, with or without the addition of additives. The presence of organic molecules and soluble inorganic substances, such as humic acids and mineral elements carried in suspension, increases in CTs with respect to the compost from which they derive. Likely, bacteria, fungi and yeasts have a decisive effect on the ability of the formulation to suppress diseases and/or promote plant growth. The biostimulant capacity of CTs is exercised on the plant through direct or indirect mechanisms on nutrition and hormone-like activity by inducing its photosynthetic capacity. Plant growth promotion is also exercised by useful microbiota that reside in CT which induce the maintenance of normal phytosanitary conditions and/or promote plant growth (De Corato 2020a). This action is reflected in the different morphological, physiological and productive characteristics that contribute to the agronomic performances of crops.

CTs can produce many beneficial effects when directly applied to plants, showing interesting disease-suppressive abilities towards foliar and fruit pathogens by stimulating the physiology of the plant and making more efficient use of nutrients with positive consequences on the quantitative and qualitative aspects of the products (Reeve et al. 2010). More than composts, the formulations of CTs present an abundance and diversity of bacteria, fungi, protozoa and nematodes that are potentially useful for both their disease-suppressive properties and as promoters of plant growth and plant health.

10.3.1 Production

CTs are organic formulations usually produced inside a bioreactor and extracted from high-quality composts under fermentation or oxidative conditions in the liquid phase carried out over a period of a few hours up to two weeks. Fermentation takes place without ventilation in the production of non-aerated CTs, while aerated CTs are produced by oxygenation at the liquid phase. As previously mentioned, CT consists of an aqueous suspension made of useful microorganisms and water-soluble organic and inorganic molecules extracted from the starting compost and further rearranged, which can have positive effects either on plant growth and development or in suppressing plant diseases. The bioreactor can vary from the most complex layout to a simple bench bioreactor, with little or no impact on the effectiveness of the product. Therefore, it is possible to encounter an artisanal fermenter system for the on-farm production of CTs (Figure 10.3) up to more sophisticated industrial processing plants (bio-fermenters). Since this technology should ensure the complete oxygenation of the liquid phase, the uniform distribution of the liquid mass and the efficient extraction, overall in the case of aerated CTs, the starting compost should be carefully placed in a permeable bag inside the bio-fermenter. Oxygenation of the liquid mass can be achieved by continuous recirculation of the liquid extraction or through direct ventilation by active insufflation. A particularly efficient innovation process for producing CTs is the use of additives that partially or can totally replace water in the

FIGURE 10.3 Liquid-based multi-extractor system for on-farm aerated compost tea production under oxygenation conditions.

liquid phase during the fermentation process. The use of co-products other than water, such as wastewaters from the agro-food industry, e.g. whey, buttermilk, "borlanda" and molasses (Pane et al. 2010, 2012b), has been proposed as additives. In this case, the list of additives is further enriched with other materials, such as caseins and fish flours (Al-Mughrabi 2006). The purpose of these replacements is to improve the quality of the biotic and abiotic component in determining the efficacy of the final products.

10.3.2 Composition

CTs contain a larger quantity of organic and inorganic substances than compost, which are dissolved in aqueous solutions. Nitrogen, phosphorus and potassium and an abundance of humic and fulvic acids are present in CTs. In general, the salinity of a CT does not reach levels that are toxic to crops. If the compost of origin is not of good quality, in particular due to heavy metals content beyond the legal limit, elements such as lead, cadmium, copper and iron can easily be released into CTs in bioavailable forms for crops during extraction. Thus, repeated treatments could increase the concentration of heavy metals in the drenched soil and could have a negative long-term impact on the environment and on human health. If the compost of origin is of good quality, the concentration of heavy metals in the resulting CT is very low and the impact on the soil and the environment is negligible, if compared to the direct supply of compost (Baldantoni et al. 2010).

The microbiological composition of CTs influences their use in particular areas. A suspension of CTs contains high microbial diversity that includes bacteria, actinomycetes, yeasts and fungi, both saprophytic and mycorrhizal, which represent a valuable microbiota source. In this regard, Palmer et al. (2010) have clarified that the microbiota diversity and structure, rather than its abundance, is considered the main contributing factor to the microbiological quality of CTs. Such microorganisms are capable of conferring specific properties on CTs, such as biological control of the main foliar and fruit diseases and/or plant growth promotion of vegetative organs and the productivity of the crops. The different technological variables of the fermentation/oxidation process affect the microbial component of CTs. A study reported that the increased efficiency in the extraction of nutrients and the increased microbial population are related to the adoption of conditions specific to CT production (Shrestha et al. 2011). In particular, the optimal compost/water ratio (1:10, v/v) ensures a continuous exchange between the two phases of the mixes, maintaining the optimal oxygen level higher than 6 ppm as the concentration limit. Furthermore, particular emphasis is placed on the use of additives in the fermentation/oxidation process for increasing the microbial biomass (Scheuerell 2006). However, this last aspect is still much debated because, on the one hand, additives can affect the microbial community of CT increasing its biodiversity, and, on the other hand, the development of human bacteria such as *Escherichia coli*, which is dangerous to human health, could be stimulated (Scheuerell 2003).

The abiotic component most important for CTs is represented by the humic fraction that originates from the chemical breakdown of the organic matter, the so-called humic substances including humin. They can also be present in CTs with tips of 0.66 g L^{-1} using an acidifying additive (Ciancio 2010). Such processes, due to the high variability of the starting and elevated materials and the number of reactions involved, lead to the formation of highly complex heterogeneous organic structures with a higher molecular weight and a higher carbon content that are more stable than the starting materials and, thus, are able to remain longer in the soil. Generally, the humic fraction can be divided into three sub-fractions: fulvic and humic acids and humin (Stevenson 1994).

10.3.3 Formulation and Application

Among the most important parameters of the CT production process, the quality of the compost and the use of additives seem to influence the effectiveness of the preparation more than other variables. Composts with disease-suppressive properties have the potential to generate high-quality CTs that are particularly effective in the biocontrol activity of phytopathogens (Joshi et al. 2009). In addition to the features intrinsic to starting composts, the degree of maturity can determine the quality of the final CTs. Recent research has shown that aerated CTs produced by immature composts, tested on the bean/grey mould (*Botrytis cinerea*) pathosystem, were more suppressive compared to those derived from stabilized composts (Palmer et al. 2010). The addition of additives that promote microbial and/or antifungal growth substances such as humic acids (Ingham 2005) and/or antagonists in the production of bioactive formulations (Siddiqui et al. 2008), generally enhanced the suppressive potential of CTs, thereby increasing their efficiency and broadening their spectrum of action. The addition of additives, even at lower doses, provides macroscopic effects on disease suppression. For instance, a significant reduction in mildew symptoms (*Phytophthora infestans*) on potatoes occurred by adding a final concentration of just 1% of additive at the time of treatment (Al-Mughrabi 2007). On the other hand, oxygenation does not seem to be decisive in the effectiveness of CTs. Direct comparisons of aerated and non-aerated CTs did not show significant differences in the suppression of *P. ultimum* on cucumber (Scheuerell and Mahaffee 2004) or *Xanthomonas vesicatoria* on tomato (Al-Dahamani et al. 2003).

The application of organic formulations based on liquid (teas) and solid (humic substances) compost derivatives for crop protection is generally carried out in the same way as traditional fungicides through wetting leaves or drenching soils and/or roots, and is facilitated by sprayers and/or mini-sprinklers. In the case of CT application, it is necessary to pay attention to working pressures which must not exceed certain limits in order to preserve the integrity of the foliar resident microbiota. CTs can be applied by spraying the leaves both in the preventive and curative phases. Research conducted on several CTs indicated that the preventive application of tea is more effective than the curative application, as a result of the possibility of the epiphytic development of tea microbes and the greater possibilities of these microbes interacting with the infectious pathogens (Koné et al. 2010). CTs are commonly employed as undiluted liquid formulates or, alternatively, they can be further diluted with water in the ratio 1:5 to 1:10 (v/v). The doses used in the open field are calibrated in relation to the targeted pathogen/crop system. Generally, doses range from 50 to 140 L ha^{-1} in the case of *P. infestans* on potatoes and up to 900 L ha^{-1} for the control of *Septoria lycopersici* on tomatoes (Al-Mughrabi 2007). Unlike CTs, humic substances are already present in the phytosanitary market. This makes the application methods less arbitrary for farmers, if they carefully follow the indications on the manufacturer's label.

10.3.4 Effects on Plant Growth

In several cases, organic formulations have the effect of plant growth promotion when directly applied to plants. These effects are represented not only by increased biomass production, but also by increased physiological activity. Different mechanisms related to the CT activity of plant growth promotion have been studied. In particular, direct and/or indirect effects on plant nutrition (Virgine Tenshia and Singaram

2005; Aguirre et al. 2009), effects on hormone-like activities (Bernal-Vicente et al. 2008; Mora et al. 2010), improved photosynthetic efficiency (Siddiqui et al. 2008), effects on the inoculation of useful microorganisms for plant nutrition (Bess 2000; Scheuerell 2003; Lanthier 2007) and growth promoters (Baldotto et al. 2010) and improved phytosanitary conditions (Scheuerell and Mahaffee 2002) have been described. There are numerous references in the literature on the use of CTs as organic biostimulants in horticulture. This is the case, for example, of *Centella asiatica* in which the application of CTs increased the vegetative growth, dry biomass and antioxidants content (such as asiaticoside, madecassoside and Asian acid) (Siddiqui et al. 2011). Similarly, foliar applications of CTs showed significant effects on plant growth promotion of *Abelmoschus esculentus* with respect to its morphological (height of the aerial part, root length, number of leaves and leaf area) and physiological (chlorophyll content and photosynthetic ratio) characteristics and its productive performance (Siddiqui et al. 2008). Other authors reported interesting results on tomatoes sprayed with CT obtained from a compost of vegetable origin every 7–10 days (Pane et al. 2012a). These treatments determined a very high increase in the production of berries (on average 46%) overall in the absence of fungicide treatment. This result was also confirmed during the following years of experiments (unpublished data) and highlights how CTs can perform the dual action of plant growth promotion and protection. The root application on melon of CTs with composts from citrus fruit residues showed a significant increase in total biomass due to the biostimulation effect of molecules with auxino-like and cytokine-like activities (Bernal-Vicente et al. 2008).

The abiotic component of CTs, in particular the humic acids content, seems to play the main role in plant growth promotion. Indeed, it has been reported that the humic fraction can induce an increase in production by direct action on the plant physiology and nutrition, even if applied in purified form (Virgine Tenshia and Singaram 2005; Selim et al. 2009, 2010; Hartz and Bottoms 2010; Mora et al. 2010; Varanini and Piton 2001). For example, humic acids derived from various composts stimulated vegetative growth in chicory (Valdrighi et al. 1996), an increase in the growth of the root system in pepper (Arancon et al. 2003) and increments of fresh and dry biomass in cucumber and tomato (Atiyeh et al. 2002). Such actions carried out by humic acids at the physiological level of the crop, can be under gene control (Aguirre et al. 2009). Table 10.5 summarizes the principal biostimulation effects observed on plant growth in horticulture after the application of compost teas and their derivatives.

10.3.5 Effects on Plant Disease Suppression

Crop protection is widely performed through the consistent use of synthetic fungicides that, in view of the practical and economic advantages they can offer, concurs with consumers' requests to reduce chemical inputs in agriculture to ensure greater sustainability for the production supply chain and greater safety for human health. The best way to study new natural formulations against phytopathogens is to research eco-friendly alternatives to synthetic chemicals for crop protection. CTs are enjoying success and a great deal of interest in the agricultural sector thanks to their ability to partially replace traditional fungicides in the control of many foliar and fruit diseases of horticultural crops (Scheuerell and Mahaffee 2002). Several mechanisms have been studied to explain the suppressiveness of CTs against plant pathogens.

As with SCs, numerous studies are in agreement that the main contribution to suppressiveness comes from the microbiota that can act as an antagonist (El-Masry et al. 2002). CTs can be considered as a suitable mean in the biological control of certain horticultural crop diseases. The microbiota contained in CTs can develop their own antagonistic action through mycoparasitization and antibiosis (Bernal-Vicente et al. 2008), and competition for space, nutrients and infection sites (Segarra et al. 2009). However, the microbial community of CTs is not always essential in the control of diseases. A study has shown that the effectiveness of formulated CTs does not depend on the presence of antagonist microflora only (Al-Dahamani et al. 2003). In this case, it has been hypothesized that the abiotic component of CT may also play a significant role in determining the suppressive effects. For instance, nutrients and organic molecules, such as humic acids and phenolic compounds, might significantly contribute to the protection of the plant through toxicity effects by inhibiting and/or limiting the activity of the pathogen. Humic fractions (humic and fulvic acids) and humus-like fractions (humin) are composed of organic natural bioactive substances that can directly and/or indirectly influence the plant and/or pathogen, thereby inducing disease suppression. In this case, the mechanisms of action are not fully known, although recent studies have shown significant

TABLE 10.5

Plant Growth Promotion due to Biostimulation Effects of Different Compost Teas (Including Their Derivatives) on Some Horticultural Crops

Plant Growth Promoter	Horticultural Crop	Biostimulant Effect	References
Humic acid	Tomato	Increase in plant height, leaf area and dry weight of shoots and roots. Increase in the absorption of macro- and micro-elements (increased efficiency in use of nutrients).	Atiyeh et al. (2002)
Humic acid from vermicompost	Pepper and strawberry	Increase in root growth. Increase in production per square metre (on average 23% for pepper).	Arancon et al. (2003)
Humic acid from vermicompost	Cucumber	Increase in plant height, leaf area and dry weight of shoots and roots.	Atiyeh et al. (2002)
Compost + humic acid + amino acid	Beans	Productivity increases.	Shehata e El-Helaly (2010)
Humic acid	Chicory	Biostimulation of total fresh biomass.	Valdrighi et al. (1996)
Compost tea + NPK mineral fertilizer at half dosage	*Centella asiatica*	Increase in plant vegetative growth, productivity and antioxidant substances content in the tissue.	Siddiqui et al. (2011)
Compost tea from green compost	*Abelmoschus esculentus*	Increase in plant vegetative growth, photosynthesis efficiency and productivity.	Siddiqui et al. (2008)
Compost tea from green compost	Oilseed rape and cabbage-turnip	Increase in plant vegetative growth. Increase in production per square metre (on average 35% for cabbage-turnip).	Keeling et al. (2003)
Compost tea from composted oranges and lemons co-products	Melon	Increase in total fresh biomass.	Bernal-Vicente et al. (2008)
Compost tea from plant green waste and agro-industrial co-product	Tomatoes	Increase in production per square metre (on average 46%).	Pane et al. (2012a)

relationships between the chemical and functional properties of such groups of substances and their ability to inhibit phytopathogenic fungi (Moliszewska and Pisarek 1996; Loffredo et al. 2008; Loffredo and Senesi 2009). Moreover, the effects of such molecules can determine improvements in the nutritional and physiological stages of the plants which can exhibit less susceptibility to diseases. Using soil drenching, mixtures of humic and fulvic acids applied to soil naturally infested with *Fusarium* spp. have determined a significant increase in the radical biomass resulting from a reduction in the parasitic rot severity on tomato (Yigit and Dikiliatas 2008). Some studies have shown the possibility of using humic-based substances derived from various types of composts in the in vitro and in vivo control of plant pathogens. For example, humic fractions have been reported to be abundant in SCs having suppressive effects towards *P. ultimum* on cucumber and cress (Pascual et al. 2002; Pane et al. 2011).

In most of the known cases, the CT's biotic and abiotic components perform complementary functions and synergic action that reflect on the suppressive properties of the CT. The selective pressure exerted from the abiotic component of CTs toward the resident microbiota helps to define the structure and the degree of biodiversity of the same microbiome. Moreover, the role that secondary metabolites produced by microorganisms during fermentation can play in suppression should not be overlooked. The biotic and abiotic components of CTs can induce systemic resistance in the host by revealing another mechanism of action for such organic formulations. In analyzing the enzymatic activities in plant tissue such as polyphenol oxidase, peroxidase and phenylalanine-ammonium lyase commonly related to the induction of resistance, it was possible to find this type of induced response to *Choanephora cucurbitarum* in *Cucurbita cucurbitarum* treated with two CTs, sterilized and non-sterilized by Millipore filtration (Siddiqui et al. 2009). A similar phenomenon was previously described in tomato and onion plants treated with non-aerated CTs against *Alternaria solani* and *A. porri*, respectively (Hagag and Saber 2007).

CTs have been studied in many horticultural cropping systems where the formulates seem to act against a broad range of phytopathological case studies such as powdery mildew of tomato and artichoke caused by *Oidium neolycopersici*, *Leveillula taurica* and *Erisiphe polygoni*; downy mildew of tomato and potato by *P. infestans*; gray mould (*B. cinerea*) and alternariosis (*Alternaria* spp.) of tomato; septoriosis (*S. lycopersici*) of celery; corky root (*Pyrenochaeta lycopersici*) of tomato; and damping-off of cucumber and bean (*P. ultimum* and *R. solani*) (Elad and Shtienberg 1994; Pane et al. 2007, 2012b; Segarra et al. 2009; Koné et al. 2010). Authors report that root treatments with CTs have suppressed rot root in Iceberg lettuce caused by *S. sclerotiorum* and *S. minor* (Zaccardelli et al. 2012). Moreover, CT applications on tomatoes have been reported to suppress some bacterial foliar diseases caused by *Pseudomonas siringae* pv. *tomato* (Pane et al. 2007) and mottling bacterial tomato by *X. vesicatoria* (Al-Dahmani et al. 2003). Full-field CT applications have occurred for controlling several pathogens of potato including *Helminthosporium solani*, *A. solani*, *Fusarium* spp., *R. solani* (Al-Mughrabi 2006) and *P. infestans* (Al-Mughrabi 2007; Al-Mughrabi et al. 2008).

10.4 Conclusion

This chapter highlights one of the most common derivative products from green composts with the greatest potential for dissemination in horticultural cropping systems. A first consideration is the numerous benefits that can be gained through the direct application of CTs to plants. From the quantitative-qualitative point of views, the two main benefits, either as a plant biostimulant or a protective agent against plant diseases enclosed in a single intervention, represents an ambitious challenge for the formulates market in organic agriculture. Additionally, the simple CTs production technology that can be realized by on-farm systems makes possible a self-supply of green compost at very low cost to farmers. The use of organic formulations in plant growth promotion and crop protection represents one of the most effective strategies for the production of environmentally low-impact products with significant health benefits, where the use of chemical inputs are banned or heavily reduced. In light of all this, CTs are currently among the most innovative products available for sustainable agriculture. However, the benefits described here are subject to new developments and more attention by research groups dedicated to exploiting the maximum favourable and promising properties of CTs. In the future, we will progress towards better and more efficient CTs with enriched components, both biotic (selected microbiota) and abiotic (purified humic fractions), and with greater responsibility for plant growth promotion and crop protection. Finally, we must not overlook how the consolidated microbial balance at the telluric level and the protective thrust towards plant production achieved by the use of SCs can reduce the need for synthetic fungicides in crop protection, where the sustainability of the horticultural cropping systems will be more pronounced in the next years.

REFERENCES

Aguirre E, Lemènager D, Bacaicoa E, Fuentes M, Baigorrir ZA, Garcìa-Mina J (2009) The root application of a purified leonardite humic acid modifies the transcriptional regulation of the main physiological root responses to Fe deficiency in Fe-sufficient cucumber plants. *Plant Physiology and Biochemistry* 47:215–223

Albiach R, Canet R, Pomares F, Ingelmo F (2000) Microbial biomass content and enzymatic activities after the application of organic amendments to a horticultural soil. *Bioresource Technology* 75:43–48

Al-Dahamani JH, Abbasi PA, Miller SA, Hoitink HA (2003) Suppression of bacterial spot of tomato with foliar sprays of compost extracts under green house and field conditions. *Plant Disease* 87:913–919

Al-Mughrabi KI (2006) Antibiosis ability of aerobic compost tea against foliar and tuber potato diseases. *Biotechnology* 5:69–74

Al-Mughrabi KI (2007) Suppression of *Phytophthora infestans* in potatoes by foliar application of food nutrients and compost tea. *Aust J Basic Appl Sci* 1:785–792

Al-Mughrabi KI, Berthélémé C, Livingston T, Burgoynea PR, Vikram A (2008) Aerobic compost tea, compost and combination of both reduce the severity of common scab (*Streptomyces scabiei*) on potato. *J Plant Sci* 3:168–175

Angelopoulou DJ, Naska EJ, Paplomatas EJ, Tjamos SE (2014) Biological control agents (BCAs) of *V. dahliae* wilt: influence of application rates and delivery method on plant protection, triggering of host defense mechanisms and rhizosphere populations of BCAs. *Plant Pathol* 63(5):1062–1069

Arancon NQ, Lee S, Edwards CA, Atiyeh R (2003) Effects of humic acids derived from cattle, food and paper-waste vermicomposts on growth of greenhouse plants. *Pedobiologia* 47:741–744

Aslam DN, Van Der Gheynst JS (2008) Predicting phytotoxicity of compost-amended soil from compost stability measurements. *Environ Eng Sci* 25:72–81

Atiyeh RM, Lee S, Edwards CA, Arancon NQ, METZGER (2002) The influence of humic acids derived from earthworm processed organic wastes on plant growth. *Bioresource Technology* 84:7–14

Avilés M, Borrero C, Trillas MI (2011) Review on compost as an inducer of disease suppression in plants grown in soilless culture. *Dyn Soil Dyn Plant* 5(2):1–11

Ayuso M, Hernandez T, Carcia C, Pascual JA (1996) Stimulation of barley growth and nutrient absorption by humic substances originating from various organic materials. *Bioresource Technol* 57:251–257

Baldantoni D, Leone A, Iovieno P, Morra L, Zaccardellim AA (2010) Total and available soil trace element concentrations in two Mediterranean agricultural systems treated with municipal waste compost or conventional mineral fertilizers. *Chemosphere* 80:1006–1013

Baldotto Leb BMA, Canellas LP, Bressansmithr OF (2010) Growth promotion of pineapple"Vitória" by humic acids and *Burkholderia* spp. during acclimatization. *Revista Brasileira de Ciencia do Solo* 34:1593–1600

Berendsen RL, Pieterse CMJ, Bakker PAHM (2012) The rhizosphere microbiome and plant health. *Trends Plant Sci* 17(8):478–486

Bernal-Vicente A, Ros M, Tittarelli F, Intrigliolo F, Pascual JA (2008) Citrus compost and its water extract for cultivation of melon plants in green house nurseries. Evaluation of nutriactive and biocontrol effects. *Bioresource Technol* 99:8722–8728

Berry BM, Sylvester-Bradley R, Philipp L, Hatch DJ, Cuttle SP, Rayns FW, Gosling P (2002) Is the productivity of organic farms restricted by the supply of available nitrogen? *Soil Use Manage* 18:248–255

Bess VH (2000) Understanding compost tea. *BioCycle* 41:71–72

Bonanomi G, Antignani V, Pane C, Scala F (2007) Suppression of soil-borne fungal diseases with organic amendments. *J Plant Pathol* 89:311–324

Bonilla N, Gutiérrez-Barranquero JA, De Vicente A, Cazorla FM (2012) Enhancing soil quality and plant health through suppressive organic amendments. *Diversity* 4:475–491

Bulluck LR, Brosius M, Evanylo GK, Ristaino JB (2002) Organic and synthetic fertility amendments influence soil microbial, physical and chemical properties on organic and conventional farms. *Appl Soil Ecol* 19(2):147–160

Cha J-Y, Han S, Hong H-J, Cho H, Kim D, Kwon Y, Kwon S-K, Crusemann M, Bok Lee Y, Kim JF, et al. (2016) Microbial and biochemical basis of a *Fusarium* wilt-suppressive soil. *ISME J* 10(1):119–129

Chilosi G, Aleandri MP, Bruni N, Tomassini A, Torresi V, Muganu M, Paolocci M, Vettraino A, Vannini A (2017) Assessment of suitability and suppressiveness of on-farm green compost as a substitute of peat in the production of lavender plants. *Biocontrol Sci Technol* 27(4):539–555

Ciancio D (2010) Valutazione in laboratorio di diversi compost tea. Tesi di Laurea Magistrale in Scienze e Tecnologie Agrarie, Università della Basilicata, A.A. 2009–2010 (In Italian)

Conrath U, Pieterse CMJ, Mauch-Mani B (2002) Priming in plant–pathogen interactions. *Trends Plant Sci* 7:210–216

Costantini EAC, Lorenzetti M (2013) Soil degradation processes in the Italian agricultural and forest ecosystems. *Ital J Agron* 8, 233–243. https://doi.org/10.4081/ija.2013.e28.

Cotxarrera L, Trillas MI, Steinberg C, Alabouvette C (2002) Use of sewage sludge compost and *Trichoderma asperellum* isolates to suppress *Fusarium* wilt of tomato. *Soil Biol Biochem* 34:467–476

De Corato U (2019) Use of omic approaches for characterizing microbiota from suppressive compost to control soil-borne plant pathogens. *Archives of Phytopathology and Plant Protection Special Issue "Applications of Agriculturally Important Microorganisms for Management of Soilborne Phytopathogens"* 52(7–8):757–775. https://doi.org/10.1080/03235408.2018.1554199

De Corato U (2020a) Agricultural waste recycling in horticultural intensive farming systems by on-farm composting and compost-based tea application improves soil quality and plant health: a review under the perspective of a circular economy. *Sci Total Environ* 738:139840. https://doi.org/10.1016/j.scitotenv.2020.139840

De Corato U (2020b) Disease-suppressive compost enhances natural soil suppressiveness against soil-borne plant pathogens: a critical review. *Rhizosphere* 13:100192. https://doi.org/10.1016/j.rhisph.2020.100192

De Corato U (2020c) Soil microbiota manipulation and its role in suppressing soil-borne plant pathogens in organic farming systems under the light of microbiome-assisted strategies. *Chem Biol Technol Agriculture*, 7:17. https://doi.org/10.1186/s40538-020-00183-7

De Corato U, Viola E, Arcieri G, Valerio V, Zimbardi F (2016) Use of composted agro energy co-products and agricultural residues against soil-borne pathogens in horticultural soil-less systems. *Scientia Horticulturae* 210:166–179. http://dx.doi.org/10.1016/j.scienta.2016.07.027

De Corato U, De Bari I, Viola E, Pugliese M (2018a) Assessing the main opportunities of integrated biorefining from agro-bioenergy co/by-products and agroindustrial residues into high-value added products associated to some emerging markets: a review. *Renew Sustain Energy Rev* 88:326–346. https://doi.org/10.1016/j.rser.2018.02.041

De Corato U, Salimbeni R, De Pretis A, Patruno L, Avella N, Lacolla G, Cucci G (2018b) Microbiota from "next-generation green compost" improves suppressiveness of composted municipal-solid-waste to soil-borne plant pathogens. *Biol Control* 124:1–17. https://doi.org/10.1016/j.biocontrol.2018.05.020

De Corato U, Patruno L, Avella N, Lacolla G, Cucci G (2019) Composts from green sources show an increased suppressiveness to soilborne plant pathogenic fungi: relationships between physicochemical properties, disease suppression, and the microbiome. *Crop Prot* 124:104870. https://doi.org/10.1016/j.cropro.2019.104870

Diver S (2002) *Compost Tea for Plant Disease Control*. ATTRA. http://www.attra.org/attra-pub/comptea.htm

Diver S, Greer L (2001) *Sustainable Small-scale Nursery Production. Horticulturae System Guide*. ATTRA. http://www.http://attra.org/attra-pub/PDF/nursery.pdf

Elad Y, Shtienberg D (1994) Effect of compost water extracts on grey mould (*Botrytis cinerea*). *Crop Prot* 13:109–114

El-Masry MH, Khalil AI, Hassouna MS, Ibrahimhah (2002) *In situ* and *in vitro* suppressive effect of agricultural composts and their water extracts on some phytopathogenic fungi. *World J Microbiol Biotechnol* 18:551–558

Eyheraguibel B, Silvestre J, Morard P (2008) Effects of humic substances derived from organic waste enhancement on the growth and mineral nutrition of maize. *Bioresource Technol* 99:4206–4212

Hadar Y, Papadopoulou KK (2012) Suppressive composts: microbial ecology links between abiotic environments and healthy plants. *Annu Rev Phytopathol* 50:133–153

Hagag WM, Saber MSM (2007) Suppression of early blight on tomato and purple blight on onion by foliar sprays of aerated and non-aerated compost teas. *J Food, Agri Environ* 5:302–309

Hardy GE, Sivasithamparam K (1991) Suppression of *Phytophthora* root rot by a composted eucalyptus bark mix. *Aust J Bot* 39:153–159

Hartz TK, Bottoms TG (2010) Humic substances generally ineffective in improving vegetable crop nutrient uptake or productivity. *Scientia Horticulturae* 45:906–910

Hickman ZA, Reid BJ (2008) The co-application of earthworms (*Dendrobaena veneta*) and compost to increase hydrocarbon losses from diesel contaminated soils. *Environ Int* 34:1016–1022

Hoitink HAJ, Fahy PC (1986) Basis for the control of soil-borne plant pathogens with composts. *Annu Rev Phytopathol* 24:93–114

Hoitink HAJ, Inbar Y, Boehm MJ (1991) Status of compost-amended potting mixes naturally suppressive to soil-borne diseases of horticultural crops. *Plant Dis* 75:869–873

Hoitink HAJ, Boehm MJ, Hadar Y (1993) Mechanisms of suppression of soil borne plant diseases in composted-amended substrates. In: Hoitink HAJ, Keener HM (Eds.), *Science and Engineering on Composting: Design, Environment, Microbiological and Utilization Aspects*. Ohio State University, Ohio, pp. 601–621

Hoitink HAJ, Stone AG, Han DY (1997) Suppression of plant disease by composts. *Scientia Horticulturae* 32:184–187

Hoitink HAJ, Krause MS, Han DY (2001) Spectrum and mechanisms of plant disease control with composts. In: Stoffella PJ, Kahn BA (Eds.), *Compost Utilization in Horticultural Cropping Systems*. Lewis Publishers, Boca Raton, pp. 263–274

Ingham ER (1999) What is compost tea? Part 1. *BioCycle* 40:74–75

Ingham ER (2005) *The Compost Tea Brewing Manual*, 5th Edition. Soil Foodweb, Incorporated, Corvallis, OR.

Joshi D, Hooda KS, Bhatt JC, Mina BL, Gupta HS (2009) Suppressive effects of composts on soil-borne and foliar diseases of French bean in the field in the western Indian Himalayas. *Crop Prot* 28:608–615

Keeling AA, Mckallum KR, Beckwith CP (2003) Mature green waste compost enhances growth and nitrogen uptake in wheat (*Triticum aestivum* L.) and oilseed rape (*Brassica napus* L.) through the action of water extractable factors. *Bioresource Technol* 90:127–132

Kefalogianni I, Gkizi D, Pappa E, Dulaj L, Tjamos SE, Chatzipavlidis I (2017) Combined use of biocontrol agents and zeolite as a management strategy against *F. oxysporum* f. sp. *melonis* and *V. dahliae* wilt. *BioControl* 62(2):139–150

Khaled H, Fawy HA (2011) Effect of different levels of humic acids on the nutrient content, plant growth, and soil properties under conditions of salinity. *Soil Water Res* 6:21–29

Koné SB, Dionne A, Tweddell RJ, Antoun H, Avis TJ (2010) Suppressive effect of non-aerated compost teas on foliar fungal pathogens of tomato. *Biol Control* 52:167–173

Lanthier M (2007) Compost tea and its impact on plant diseases. *BC Organic Grower* 10:7–11

Lazarovits G (2001) Management of soil-borne plant pathogens with organic soil amendments: a disease control strategy salvaged from the past. *Can J Plant Pathol* 23:1–7

Loffredo E, Senesi N (2009) *In vitro* and *in vivo* assessment of the potential of compost and its humic acid fraction to protect ornamental plants from soil-borne pathogenic fungi. *Scientia Horticulturae* 122: 432–439

Loffredo E, Berloco M, Senesi N (2008) The role of humic fractions from soil and compost in controlling the growth *in vitro* of phytopathogenic and antagonistic soil-borne fungi. *Ecotoxicol Environ Safety* 69:350–357

Lumsden RD, Lewis JA, Millner PD (1983) Effect of composted sewage sludge on several soilborne pathogens and diseases. *Phytopathology* 73:1543–1548

Manici LM, Caputo F, Babini V (2004) Effect of green manure on *Pythium* spp. population and microbial communities in intensive cropping systems. *Plant Soil* 263:133–142

Melero S, Porras JCR, Herencia JF, Madejon E (2006) Chemical and biochemical properties in a salty loam soil under conventional and organic management. *Soil Till Res* 90(1–2):162–170

Moliszewska E, Pisarek I (1996) Influence of humic substances on the growth of two phytopathogenic soil fungi. *Environ Int* 22:579–584

Mora V, Bacaicoa E, Zamarreno AM, Aguirre E, Garnicam FM, Garcìa-Mina J (2010) Action of humic acid on promotion of cucumber shoot growth involves nitrate related changes associated with the root-to-shoot distribution of cytokinins, polyamines and mineral nutrients. *J Plant Physiol* 167:633–642

Morard P, Eyheraguibel B, Morard M, Silvestre J (2011) Direct effects of humic-like substance on growth, water, and mineral nutrition of various species. *J Plant Nutr* 34:46–59

Nardi S, Pizzeghello D, Muscolo A, Vianello A (2002) Physiological effects of humic substances on higher plants. *Soil Biol Biochem* 34:1527–1536

Noble R (2011) Risks and benefits of soil amendment with composts in relation to plant pathogens. *Australasian Plant Pathol* 40:157–167

Noble R, Coventry E (2005) Suppression of soil-borne plant diseases with composts. *Biocontrol Sci Technol* 15:3–20

Palmer AK, Evans KJ, Metcalf DA (2010) Characters of aerated compost tea from immature compost that limit colonization of bean leaflets by *Botrytis cinerea*. *J Appl Microbiol* 109:1619–1631

Pane C, Valentini F, Bonanomi G, Cozzolino L, Antignaniv PG, Zoina A, Scala F (2007) Control of plant pathogens by using a compost tea. *J Plant Pathol* 89:52

Pane C, Celano G, Villecco D, Zaccardelli M (2010) Suppressive effect of aerated compost teas produced in water and in whey on plant fungal pathogens. *J Plant Pathol* 92:93

Pane C, Spaccini R, Piccolo A, Scala F, Bonanomi G (2011) Compost amendments enhance peat suppressiveness to *Pythium ultimum*, *Rhizoctonia solani* and *Sclerotinia minor*. *Biol Control* 56:115–124

Pane C, Villecco D, Zaccardelli M, Ronga D, Celano G (2012a) Il compost-tea su pomodoro dà più resa e migliorequalità. *L'Informatore Agrario* 7:43 (In Italian)

Pane C, Celano G, Villecco D, Zaccardelli M (2012b) Control of *Botrytis cinerea*, *Alternaria alternata* and *Pyrenochaeta lycopersici* on tomato with compost-tea applications. *Crop Prot* 38:80–86

Pane C, Piccolo A, Spaccini R, Celano G, Villecco D, Zaccardelli M (2013) Agricultural waste-based composts exhibiting suppressivity to diseases caused by the phytopathogenic soil-borne fungi *Rhizoctonia solani* and *Sclerotinia minor*. *Appl Soil Ecol* 65:43–51

Pascual JA, Hernandez T, Garcia C, De Leij FAAM, Lynch JM (2000) Long-term suppression of *Pythium ultimum* in arid soils using fresh and composted municipal wastes. *Biol Fertil Soils* 30:478–484

Pascual JA, Garcia C, Hernandez T, Lerma S, Lynch JM (2002) Effectiveness of municipal waste compost and its humic fraction in suppressing *Pythium ultimum*. *Microb Ecol* 44:59–68

Pergola M, Persiani A, Pastore V, Palese AM, D'adamo C, De Falco E, Celano G (2020) Sustainability assessment of the green compost production chain from agricultural waste: a case study in southern Italy. *Agronomy* 10, 230. https://doi.org/10.3390/agronomy10020230.

Pugliese M, Gullino ML, Garibaldi A (2010) Efficacy of microorganisms selected from compost to control soil borne pathogens. *Commun Agric Appl Biol Sci* 75(4):665–669

Reeve JR, Carpenter-Boggs L, Reganold JP, York AL, Brinton WF (2010) Influence of biodynamic preparations on composts development and resultant compost extracts on wheat seedling growth. *Bioresource Technol* 101:5658–5666

Shehata SA, El-Helaly MA (2010) Effect of compost, humic acid and amino acids on yield of snap beans. *JHSOP* 2:107–110

Scheuerell SJ (2003) Understanding how compost tea can control disease. *BioCycle* 44:20–25

Scheuerell SJ (2006) Variability associated with suppression of gray mold (*Botrytis cinerea*) on geranium by foliar applications of no aerated and aerated compost teas. *Plant Disease* 90:1201–1208

Scheuerell SJ, Mahaffee WF (2002) Compost tea: principles and prospects for plant disease control. *Compost Sci Util* 10:313–338

Scheuerell SJ, Mahaffee WF (2004) Compost tea as container medium drench for suppressing seedling damping-off caused by *Pythium ultimum*. *Phytopathology* 94:1156–1163

Scheuerell SJ, Sullivan DM, Mahaffee WF (2005) Suppression of seedling damping-off caused by *Pythium ultimum*, *P. irregulare*, and *Rhizoctonia solani* in container media amended with a diverse range of Pacific Northwest compost sources. *Phytopathology* 95(3):306–315

Scotti R, Bonanomi G, Scelza R, Zoina A, Rao M (2015) Organic amendments as sustainable tool to recovery fertility in intensive agricultural systems. *J. Soil Sci Plant Nutr* 15, 333–352. https://doi.org/10.4067/S0718-95162015005000031.

Scotti R, Pane C, Spaccini R, Palese AM, Piccolo A, Celano G, Zaccardelli M (2016) On-farm compost: a useful tool to improve soil quality under intensive farming systems. *Appl Soil Ecol* 107, 13–23. https://doi.org/10.1016/j.apsoil.2016.05.004.

Segarra G, Reis M, Casanova E, Trillas MI (2009) Control of powdery mildew (*Erysiphe polygoni*) in tomato by foliar applications of compost tea. *J Plant Pathol* 91:683–689

Selim EM, Mosa AA, El Ghamry AM (2009) Evaluation of humic substances fertigation through surface and subsurface drip irrigation systems on potato grown under Egyptian sandy soil conditions. *Agri Water Manag* 96:1218–1222

Selim EM, El-Neklawy AS, El-Ashry SM (2010) Beneficial effect of humic substances on soil fertility to fertirrigated potato grown on sandy soil. *Libyan Agri Res Center J Int* 1:255–262

Shi L, Du N, Shu S, Sun J, Li S, Guo S (2017) *Paenibacillus polymyxa* NSY50 suppresses *Fusarium* wilt in cucumbers by regulating the rhizospheric microbial community. *Sci Rep* 7:412–434

Shrestha K, Shrestha P, Walsh KB, Harrower KM, Midmore DJ (2011) Microbial enhancement of compost extracts based on cattle rumen content compost: characterization of a system. *Bioresource Technol* 102:8027–8034

Siddiqui Y, Meon S, Ismail MR, Ali A (2008) *Trichoderma* fortified compost extracts for the control of *Choanephora* wet rot in okra production. *Crop Prot* 27:385–390

Siddiqui Y, Meon S, Ismail MR, Rahmani M (2009) Bio-potential of compost tea from agro-waste to suppress *Choanephora cucurbitarum* L. the causal pathogen of wet rot of okra. *Biol Control* 49:38–44

Siddiqui Y, Islam TM, Naidu Y, Meon S (2011) The conjunctive use of compost tea and inorganic fertiliser on the growth, yield and terpenoid content of *Centella asiatica* (L.) urban. *Scientia Horticulturae* 130: 289–295

Stevenson FJ (1994) *Humus Chemistry: Genesis, Composition, and Reactions*. 2nd Edition, Wiley, New York

Termorshuizen AJ, Jeger MJ (2008) Strategies of soil-borne plant pathogenic fungi in relation to disease suppression. *Fungal Ecology* 1(4):108–114

Termorshuizen AJ, Van Rijn E, Van Der Gaag DJ, Alabouvette C, Chen Y, Lagerlöf J, Malandrakis AA, Paplomatas EJ, Rämert B, Ryckeboer J, Steinberg C, Zmore-Nahum S (2006) Suppressiveness of 18 composts against 7 pathosystems: variability in pathogen response. *Soil Biol Biochem* 38:2461–2477

Tuitert G, Szczech M, Bollen GJ (1998) Suppression of *Rhizoctonia solani* in potting mixtures amended with compost made from organic household waste. *Phytopathology* 88:764–773

Valdrighi MM, Pera A, Agnolucci M, Frassinetti S, Lunardi D, Vallini G (1996) Effects of composts-derived humic acids on vegetable biomass production and microbial growth within a plant (*Cichorium intybus*)-soil system: a comparative study. *Agri Ecosys Environ* 58:133–144

undefinedVaranini Z, Pinton R (2001) Direct versus indirect effects of soil humic substances on plant growth and nutrition. In: Pinton R, Varanini Z, Nannipieri P (Eds.), *The Rizosphere*, pp. 141–158

Virgine Tenshia JS, Singaram P (2005) Influence of humic acid application on yield, nutrient availability and uptake in tomato. *Madras Agri J* 92:670–676

Weller DM, Raaijmakers JM, Gardener BM, Thomashow LS (2002) Microbial populations responsible for specific soil suppressiveness to plant pathogens. *Annu Rev Phytopathol* 40:309–348

Widmer TL, Graham JH, Mitchell DJ (1999) Composted municipal solid wastes promote growth of young citrus trees infested with *Phytophthora nicotianae*. *Compost Sci Util* 7:6–16

Yigit F, Dikilitas M (2008) Effect of humic acid applications on the root-rot diseases caused by *Fusarium* spp. on tomato plants. *Plant Pathol* 7:179–182

Zaccardelli M, Perrone D, Del Galdo A, Giordano I, Villari G, Bianco M (2006) Multi disciplinary approach to validate compost use in vegetable crop systems in Campania Region (Italy): effect of compost fertilization on processing tomato in field cultivation. *Acta Horticulturae* 700:285–288

Zaccardelli M, Villecco D, Pane C, Ragosta G, Paleseam CG (2010) Realizzazione di un sistema di compostaggioon farm dei residui di pomodoro. *Biol It* 8:63–67 (In Italian)

Zaccardelli M, Perrone D, Pane C, Pucci N, Infantino A (2011) Control of corky root of tomato with compost and role of spore-forming bacteria to inhibit *Pyrenochaeta lycopersici*. *Acta Horticulturae* 914:393–396

Zaccardelli M, Pane C, Scotti R, Palese AM, Celano G (2012) Use of compost-teas as biopesticides and biostimulants in horticulture. *Italushortus* 19:17–28 (In Italian)

11

Cultural, Economic and Environmental Impact of Conservation and Sustainable Exploitation Management Areas (UMA) of Freshwater Turtles in Tabasco, Mexico

Claudia Elena Zenteno Ruiz, Casiano Alberto Méndez Sánchez, Diana Ivette Triana Ramírez and Angel Sol Sánchez

CONTENTS

11.1 Introduction

Mexico's land turtles and freshwater turtles comprise 56 species, many of which are in some level of vulnerability status. In states where turtles are distributed, threats are common, for example, overexploitation, illegal commerce and habitat destruction. However, an alternative to an increase in the turtle population is through ex situ management in zoo hatcheries, which allows sustainable exploitation and cooperates to conserve the resource while observing Mexico's and international normativity.

In 1978, freshwater turtle ex situ group management started successfully in Tabasco through the Programa de Inversion para el Desarrollo Rural (PIDER) with "Granja de Tortugas" located in Tucta, Nacajuca. Its initial objective was to protect the freshwater turtle species and to develop necessary zootechnics for cultivation. This farm was the first in the country to use the semi-artificial nest incubation methods of commonly known species such as Hicotea (*Trachemys venusta*) and Guao (*Staurotypus triporcatus*) (Méndez-Sánchez & Zenteno-Ruíz, 2017). Nowadays, there is management experience of seven freshwater turtle species (*Kinosternum leucostomun*, *T. venusta*, *Rhinoc lemysareolata*, *Chelydra rosignoni*, *Claudius angustatus*, *S. triporcatus* and *Dermatemys mawii*) (Zenteno-Ruíz et al., 2019).

In 1997, Mexico's government created the national System of Management Units for the Conservation of Wildlife (SUMA) which groups management units (UMAs) as a promotion space for alternative production schemes compatible with the conservation of wildlife. Since its creation, the scheme has gone through a consolidation process, facing multiple challenges on implementation and efficiency evaluation (Méndez-Sánchez & Zenteno-Ruíz, 2017). However, freshwater turtles, especially in Tabasco, are an example of the integration of the population into this environmental policy, which had both economic and social impacts. In this chapter, we analyze the cultural, economic and environmental impacts of Tabasco's freshwater turtles' UMAs.

11.2 Context of Traditional Turtle Exploitation in Tabasco

Traditionally, all nine species of reported freshwater turtles in Tabasco are used, regardless if their status is vulnerable as specified by the national normativity (NOM-059-SEMARNAT-2010) or that some are on the International Union for the Conservation of Nature's (IUCN) red list. Exploitation of freshwater turtles is mainly local with illegal points of sale near roads and highways in the region (Arriaga, 1987) or they are taken to Villahermosa city, the state's most important market (Lazcano et al., 1988).

Poaching with little respect for closed seasons, habitat deterioration caused by wetland desiccation and the clearance of vast tracts of land for agriculture and cattle are the main sources of population decline and, in some cases, the extinction of species (Reyes, 1981). In 1985, closed seasons were established for turtle fishing. However, today it is common to find sellers on highways and even in Villahermosa city who do not comply with the law.

Formerly, Tenosique municipality was a strategic place between Petén peninsula and Tabasco's plains, and many agricultural products, hunting and fishing (like turtles) products were concentrated in this area for their commercialization and/or distribution by river to other ports (Reyes, 1981). Nowadays, Tenosique is no longer an important distribution centre due to land communication. However, it is now a point for endangered or threatened species coming from Guatemala to Mexico, as it is with the Tortuga Blanca (*D. mawii*), which is highly priced.

In field work on turtle consumption in the river region (Tenosique municipality), Méndez-Sánchez and Barragán-Vázquez (2010) found that the most consumed species in Tabasco are *T. venusta, D. mawii and S. tripocartus*. The frequency of turtle consumption ranges from once every two weeks to four times a year. Within this range, the highest frequency was at least once a year. On the other hand, Carrillo-Torres (2004) found that near the swamp region, the main type of exploitation is for self-consumption, followed by commercialization and that turtles are captured all year round. Prices may vary depending on the species and the season. Likewise, 67% of the interviewed population mentioned that the turtles they found or captured were for their consumption.

11.3 Sustainable Use of Freshwater Turtles in Management Units for the Conservation of Wildlife (UMA) in Tabasco

The state of Tabasco has registered 114 UMAs (52 intensive and 62 extensive), of which 31 intensive UMAs are registered with some species of turtle. Among these are units for showing, research and use. UMAs for commercial harvesting purposes have increased markedly in the last decade, mainly due to the pet market's need for pups for aquariums coupled with the import restriction on the turtle *Tachemys scripta*, a species that was massively sold but its entry into the country was prohibited because it was classified as an invasive species by the National Commission for the Knowledge and Use of Biodiversity (CONABIO) (Born-Schmidt et al., 2017; Koleff et al., 2010).

These units present different phases of development and consolidation, both operational and productive. Despite the fact that there are successful examples of units in relation to production, there are great challenges in ensuring the permanence of this sustainable use scheme and reducing the pressure on the wild populations. However, these spaces have accumulated basic knowledge about species, acting as centres for environmental awareness and planning conservation actions under an adaptive management scheme.

There are strengths and opportunities within the system:

Strengths	**Opportunities**
There are experiences of zoo technical transfers between UMAs	Growing demand at the national and international level for the products and byproducts of UMAs
UMAs retain a biological collection of species in the vulnerable status	Spaces for monitoring the effects of climate change on biodiversity, production processes and infrastructure

Strengths	Opportunities
Organism production and the management of extractive use rates of products and sub-products for commercialization	Compatible strategy to achieve goals in national and international agreements and conventions on biodiversity conservation and management (e.g. CITES and AICHI)
Grouping in the legally constituted organization of UMAs	Areas of opportunity for the development of basic and applied research for the generation of public policies on wildlife

11.4 Contribution of the Management Areas in Tabasco

11.4.1 Financial Contributions

In relation to the profitability of the UMAs, no assessment has been made that allows the economic contributions of the UMAs to be fully appreciated. Some evaluations carried out indicate that the UMAs do not represent a secure and continuos source of income. In fact, operating costs generated in the early years often had to be met by owners or participants or by government subsidies (Pineda-Vázquez et al., 2019).

Currently, the production of turtles generates an economic return through the sale of young turtles, which are acquired by traders and distributed throughout the country. Tabasco produces a high percentage of turtles that are authorized through the use rate granted by the authorities. The type of product that is mainly generated are calves between 30 and 90 days after hatching, of good quality in terms of appearance and liveliness, followed by adults that are sold as reproducers to other areas. Indirectly, UMAs also generate economic resources through exhibitions and tourism, attracting interested parties and offering food services, tours and sales of handicrafts.

The turtle hatchling trade in Tabasco is mainly concerned with two species: *T. venusta* and *D. mawii* (Zenteno et al., 2019). *T. venusta* is produced in high volumes under a relatively simple zootechnics because the species is quite resistant to handling. The market for this species is well established, with large buyers purchasing in the capital of the country and distributing to other states. In the case of *D. mawii*, trade in this species is low volume. The production of offspring requires more specialized management and particular captive conditions, which increases its cost; each specimen costs up to 20 times more than *Trachemys*. The baby market is preponderantly international, and the UMAs in Tabasco mostly satisfy international demand.

Turtles continue to occupy a place in the informal trade, since a considerable proportion of the inhabitants of rural Mexico continue to take advantage of some species of wild fauna as a food source, while others commercialize it locally (Naranjo et al., 2010). The cultural importance of turtles in Tabasco lies in the fact that they have formed part of the local diet, their traditional celebrations and even the worldview and self-identification of the original and current inhabitants (Guevara-Chumacero et al., 2017).

However, the valuation only of production (direct use value) leaves valuation as an environmental asset that is vital to increase the efficiency of government intervention, as it is a way to justify and prioritize decisions management on the use of resources and also provides information on their own areas and an acknowledgment of their importance for society (Aznar-Bellver & Estruch-Guitart, 2015).

In this sense, the values assigned to the turtle UMAs should include not only their market value but also those derived from their indirect use, including the protection of the territory, carbon sinks, habitat construction and the protection of biodiversity; in this case, a species at risk. It is admissible to add partial values that lead to a correct rough estimate of the real value. This type of asset also has a social value, which is the benefit or well-being it provides (Aznar-Bellver & Guijarro-Martínez, 2012). The value sought is called the total economic value (VET), which is made up of a set of values that may contain UMAs for environmental services. Among them, the option value (VO) is obtained by guaranteeing that in the future the benefits of an asset will be available and enjoyed and it is also the value generated by the uncertainty of knowing the future uses and benefits of an asset. This is an area of opportunity for creating new knowledge and technology packages on turtle farming and generating a manual on establishing and maintaining costs for the production activities that increase its profitability.

11.4.2 Environmental Contributions of UMAs

The UMAs in Tabasco have allowed lands that were previously destined for agriculture and livestock but became unproductive, to be used for the conservation of wildlife and the management of habitats. With this, they have generated alternative sources of employment and greater well-being for families in the rural sector. Perhaps the greatest importance of this conservation scheme is that it fosters social awareness of the value of biodiversity and ecosystems and the environmental services they provide (Ortega-Argueta et al., 2016).

Regarding the protection of the territory, in Tabasco the UMAs (intensive and extensive) protect approximately 1% of the state territory (Pineda-Vázquez et al., 2019). A strategy for recovering populations under in situ conditions depends on the comprehensive work of managers of natural protected areas and population managers in UMAs, since a central strategy to achieve this conservation objective is to evaluate the genetic diversity of the individuals to be reintroduced to avoid problems such as inbreeding, the fixation of harmful alleles or the loss of allelic diversity due to genetic drift. The research carried out by Gallardo et al. (2019) found that the inbreeding coefficient was higher for the wild populations studied than for the UMAs, some with better value probably as a consequence of the diverse geographic origins of the founding individuals of the UMAs. Therefore, like botanical gardens that represent leading institutions in the fields of conservation, research, education and recreation on a global scale (Borsch & Löhne, 2014; Chen & Sun, 2018), UMAs also play an important role in providing a scientific basis for the sustainable use and conservation of biodiversity.

Regarding the contribution of the UMAs to the scientific knowledge of the species, Zenteno et al. (2019) indicate that basic research in management has been obtained working with turtle colonies in captivity. In this sense, it is worth highlighting two UMAs: the UMA of the Government of Tabasco (called U OtotAk) has the largest number of studies related to turtles, and has dedicated 42 years to the conservation of this group of animals. The UMA CICEA (Research Centre for the Conservation of Endangered Species) located at the DACBiol–UJAT is also positioned with the participation of its students and teachers, to carry out studies on various topics.

REFERENCES

Arriaga, WS (1987). *Fauna de vertebrados de la región Usumacinta.* Usumacinta No. 2. Investigación Científica en la Cuenca del Usumacinta. SECUR-DESIC, 127–143.

Aznar-Bellver, J, & Estruch-Guitart, AV (2015). *2da Ed. Valoración de activos ambientales: teoría y casos.* Valencia: Universidad Politécnica de Valencia. ISBN:978-84-9048-307-7

Aznar-Bellver, J, & Guijarro-Martínez, F (2012). *Nuevos métodos de valoración.* Modelos multicriterio. Editorial Universitat Politècnica de València. ISBN:978-84-8363-982-5

Born-Schmidt, G, de Alba, F, Parpal, J, & Koleff, P (2017). *Principales retos que enfrenta México ante las especies exóticas invasoras.* Centro de estudios sociales y de opinión pública. ISBN:978-607-8501-58-8

Borsch, T, & Löhne, C (2014). Botanic gardens for the future: Integrating research, conservation, environmental education and public recreation. *Ethiopian Journal of Biological Sciences*, 13: 115–133.

Carrillo-Torres, DM (2004). *Aprovechamiento actual y comercialización de las tortugas de agua dulce en la zona oeste de la Reserva de la Biosfera Pantanos de Centla, Tabasco. (Tesis de licenciatura).* Tabasco: Universidad Juárez Autónoma de Tabasco.

Chen, G, & Sun, W (2018). The role of botanical gardens in scientific research, conservation, and citizen science. *Plant Diversity*, 40: 181–188. https://doi.org/10.1016/j.pld.2018.07.006

Gallardo-Álvarez, MI, Lesher-Gordillo, JM, Machkour-M'Rabet, S, Zenteno-Ruiz, CE, Olivera-Gómez, LD, Barragán-Vázquez, MR, Ríos-Rodas, L, Valdés-Marín, A, Vázquez-López, HG, & Arriaga-Weiss, SL (2019). Genetic diversity and population structure of founders from wildlife conservation management units and wild populations of critically endangered *Dermatemys mawii. Global Ecology and Conservation*, 19: 1–14. https://doi.org/10.1016/j.gecco.2019.e00616

Guevara-Chumacero, M, Pichardo-Fragoso, A, & Martínez-Cornelio, M (2017). The turtle in Tabasco: Food, identity and representation. *Estudios de Cultura Maya*, XLIX: 97–122. doi:10.19130/iifl.ecm.2017.49.758

Koleff, P, González, AI, & Born-Schmidt, G (2010). *Estrategia nacional sobre especies invasoras en México, prevención, control y erradicación. Comisión Nacional para el Conocimiento y Uso de la Biodiversidad.* México: Comisión Nacional de Áreas Protegidas, Secretaría de Medio Ambiente y Recursos Naturales.

Lazcano, BMA, Flores, OA, Bonabid, M, Hernández, JA, Chávez, MP, & Cabrera A (1988). Estudio y Conservación de los anfibios y reptiles de México: una propuesta. Cuadernos de Divulgación INIREB. *Xalapa, Veracruz. México* 25: 27–213.

Méndez-Sánchez, CA, & Barragán-Vázquez, MR (2010). *Aspectos generales sobre la utilización de las tortugas de agua dulce, en la ciudad de Tenosique.* Tabasco: Mexico (Datos no publicados).

Méndez-Sánchez, CA, & Zenteno-Ruiz, CE (2017). Unidades de manejo para la conservación de la vida silvestre (uma). En: Mata-Zayas, EE, & Palma-López, DJ (Eds.) *La biodiversidad en Tabasco. Estudio de Estado*, Vol. iii. México: CONABIO, pp. 251–258.

Naranjo, EJ, López-Acosta, JC, & Dirzo, R (2010). La cacería en México. *Biodiversitas*, 91(1): 6–10.

Ortega-Argueta, A, González-Zamora, A, & Contreras-Hernández, A (2016). A framework and indicators for evaluating policies for conservation and development: The case of wildlife management units in Mexico. *Environmental Science & Policy*, 63: 91–100. http://dx.doi.org/10.1016/j.envsci.2016.05.003

Pineda-Vázquez, M, Ortega-Argueta, A, Mesa-Juradob, MA, & Escalona-Segura, G (2019). Evaluating the sustainability of conservation and development strategies: The case of management units for wildlife conservation in Tabasco, Mexico. *Journal of Environmental Management*, 248: 1–11. https://doi.org/10.1016/j.jenvman.2019.109260

Reyes, CP (1981). *La fauna silvestre en el plan Balancán – Tenosique.* Xalapa: INIREB, 45 p.

Zenteno-Ruíz, CE, Rangel-Mendoza, JA, Triana-Ramírez, DI, & Méndez-Sánchez, CA (2019). Aplicaciones al conocimiento sobre tortugas dulceacuícolas desde la División Académica de Ciencias Biológicas: una revisión retrospectiva. *Kusulkab*, 25(53): 13–25, https://doi.org/10.19136/kuxulkab.a25n53.3400

Zenteno Ruíz, C.E., C.A Méndez Sanchez, C.A., Hernández V. y C.M. Burelo-Jiménez 2019. Unidades de manejo para la conservación de la vida silvestre(UMA) en: la biodiversidad de Tabasco. Vol III CONABIO, México. pp 251–257.

12

Innovation to Produce Basic Food in Recycled Containers in Saline Soils of the Coast of Tabasco, Mexico

Ángel Sol Sánchez and Gloria I. Hernández Melchor

CONTENTS

12.1 Introduction

The current status of climate change has had significant impacts on various parts of the world, mainly the coasts (Bardají et al., 2009), where the tides, tropical storms and others have caused coasts and roads to disappear, and exposed human settlements to inland salinization, negatively impacting agricultural and livestock production areas.

This phenomenon is very obvious on the coast of Tabasco since a large part of the coastal road has disappeared due to the impact of sea waves leading to an expansion of the saline surface (Gama et al., 2008). In Mexico, the coast of the state of Tabasco is an area that is most vulnerable to climate change due to its extensive flat surface; however, it is also one of the most productive economic zones as it is located within the 10 states with the highest gross national product (GDP) (INEGI, 2018, 2020).

With 192 km of coastline distributed among the municipalities of Huimanguillo, Cárdenas, Paraíso and Centla, many economic activities are carried out, such as the extraction of crude oil, commercial and artisanal fishing, agriculture and livestock and, recently, the construction of the Dos Bocas refinery.

These preponderant activities have been so important that other activities, such as food production, have been relegated because saline soils do not allow the growth of produce of culinary interest, such as vegetables, fruits and other short-cycle products (Sol et al., 2018). This is due to the fact that poorly planned infrastructure has led to the encroachment of seawater on the land, the growth of the saline surface and the mangrove surface and the abandonment of some areas by the population. This situation has made life more expensive since the population cannot produce even the most essential foods, such as vegetables.

Taking into account that the coast road is a watershed, the population that lives next to the mangrove swamp has been more affected than their counterpart living on the other side of the road, since salt levels are higher in the former site. This has led to the community capturing species that are not for direct consumption but can be transformed for animal consumption. In this sense, the present investigation was developed with the objective of evaluating the production of vegetables for self-consumption.

12.2 Location

To develop the research, a coastal community with the greatest problems due to salinity and no food production was taken as a reference. The community is the Ejido las Coloradas which is located on the coast of Tabasco. Geographically, the community is located between 18°19'46.44" LN and 93°32'59.35" LW and it has a high degree of marginalization. The relief of the area is irregular, composed of flat terrain with areas of depression, with a variable altitude of up to 2 m in the project area. The project area is a low zone influenced by the saltwater of the Machona Lagoon and the soils are clay type with poor drainage. The most important hydrographic resource is the lagoon system, which is formed by the Carmen, Pajonal and Machona lagoons, separated from the Gulf of Mexico by the Santana bar. Likewise, in the municipality there is the Chicozapote River which borders the town of Benito Juárez and empties into the Tonalá River near the Gulf of Mexico (Palma et al., 2007).

A meeting was held with the community to expose this project and evaluate the feasibility of using 20 L recycled containers to produce vegetables for daily consumption. This activity began with 39 interested people and 30 g of saladette tomato seeds, 35 g of serrano pepper seeds, 30 g of white cucumber seeds, 30 g of Italian pumpkin seeds, 1,100 disposable 20 L cans of purified water, 28 m^3 of garden soil, 14 m^3 of sand, 7 m^3 of oyster shells, 250 g of coriander seeds and 30 chayote fruits.

12.3 Process

The neck of the water bottle was removed, leaving the handle in the boat. Each boat was painted with oil paint to protect it from corrosion. An average of 14 kg of nursery soil was deposited on each boat and mixed with sand; 750 g of crushed oyster shell was deposited on the surface of each boat as a source of calcium, which was obtained from the community. Tomato and pepper seeds began to germinate in beds prepared for this purpose and those of Italian squash and chayote fruit were put directly into containers. These seeds were commercial and thus had already been treated against soil insects.

The containers were arranged in rows and by species. At 26 and 30 days after being planted in seedlings, the peppers and tomatoes were transplanted into the containers and placed in a half-shaded position (Torres, 2017). Evaluations were subsequently made every week to assess their development. The sites used for these activities were unused family gardens in the community.

12.4 Information Obtained

The growth achieved by the average plant was 80 cm in the case of pepper and 1.25 m in the case of tomato; the growth of chayote was difficult to measure because as a vine, it climbs to any source of light. This was also the case for cucumber although it was developed on a wood mesh. No case of pests or diseases was observed. Coriander, which has the shortest growth cycle of all the planted species, was harvested at 30 days, without waiting for flowering or fruiting. All species were followed agronomically with continuous irrigation every two days to ensure the availability of water to the plant. Tables 12.1 and 12.2 detail the species planted and their growth.

Considering the precarious conditions of the site and of the community in general, the production of short-cycle vegetables in containers is a viable option for the community; however, it should be borne in mind that in each case the seeds must be transported from the city to the communities since the coast lacks such seeds because the communities are accustomed to bringing the produce from the city. In general terms, 1,100 kg of salad tomato were obtained. Some plants produced 3.65 kg of tomato and some produced less (1.64 kg) with an average of 2.2 kg per plant. In all plants the first harvest was ripe tomato and the subsequent harvest was green tomato for various types of sauces and these were weighed in general terms.

At the fatal level, the tomato had problems because the relative humidity generated drops of saltwater that adhered to the tomato fronds causing regressive death in terminal leaves; however, tomato

TABLE 12.1

Species Planted and Physiological Details

Species	Number of Plants	Transplanted (days after planting)	Flowering	Days to Fruiting
Tomato	500	30	60	105
Serrano pepper	450	26	59	90
Chayote	30	–	311	330
Cucumber	50	–	45	60
Coriander	6 m^2	–	–	30

TABLE 12.2

Production and Savings by Producing Their Own Vegetables

Species	Production per Plant (kg)	Total Production	Market Price	Savings (Mexican Pesos)
Tomato	2.2 kg p/p	1,100	30	33,000
Serrano pepper	1.6 kg p/p	720	42	30,240
Chayote	2.45 kg p/p	73.5	18	1,323
Cucumber	3.2 kg p/p	160	5	800
Coriander	2.8 kg/m^2	18.8	80	1,504
Income				66,667

production is considered of importance given that it is produced locally. It is necessary to mention that not all tomatoes reached their total physiological maturity. Some plants lost much of their foliage and the fruits became yellowish before physiological maturity, at which point they were harvested to take advantage of this. The tomato and pepper plants suffered stress and did not present a second bloom as is common in these species, rather they died.

In the case of serrano pepper, problems occurred during the development of the chili plants not only due to the abortion of flowers caused by the excess salt in the environment, but also due to the malnutrition of the plant since the soil used presented characteristics from sand to soil, due to the fact that age and size were appropriate, but physiologically they were not (Pino et al., 2009). Also, after fertilizing the first flowers, the plants began to wilt due to the intense relative humidity of the area, resulting in many flowers unable to withstand the heat of the site.

Pepper production was 720 kg; all the peppers were harvested before reaching maturity. Like the tomato, harvesting was irregular, that is, when people required the peppers.

Coriander was the most effective crop because after 30 days, the harvest was gathered, there were no pests or diseases and it was used for local consumption. The average amount of coriander obtained was 2.8 kg/m^2. This was not uprooted from the ground, but cut to ground level so that a second crop was obtained.

The production of cucumber was not as significant. Although the main root was on the prepared substrate, the guides were in contact with the local saline soil, which impacted on its development. A total of 160 kg was obtained. The most important fact about this backyard activity is that the population stopped spending USD 66,667, which is the amount that would have been spent if the population bought the same products.

12.5 Conclusion

Although the project was carried out effectively, the relative humidity was high and loaded with salt particles, which impacted the development and production of the plants. Taking this experience as a base, it is advisable to work in a greenhouse to prevent saltwater from reaching the plants, thus ensuring that the foliage production are not affected. It is also advisable to plant at the beginning of rain to avoid evapotranspiration.

The produce obtained was for self-consumption. The production system was replicated in non-saline sites far from the mangrove swamp where the results were more satisfactory, thus suggesting that it is possible to produce short-cycle vegetables in agricultural areas using the method described. The total number of beneficiaries was 139, among workers, their wives and children. Nursery soil was the most difficult product to obtain because it had to be transported over 25 km, which increased its cost.

Under these conditions, a producer could not pay the cost represented by the substrate used as it is too expensive, but it has been proven that, using 3 L plastic bottles of waste, the same activity could be carried out in this marginalized area with equally satisfactory results. If we consider that with only five species, by integrating others such as purple chives and leaf chives, this activity could be an alternative for small-scale local income.

REFERENCES

Andrea TP 2017. *Manual de cultivo del tomate al aire libre.* Santiago de Chile: Instituto Nacional de investigaciones Agropecuarias num. 11. 93 p. ISSN 0717 – 4829.

Bardají T, Zazo C, Cabero A, Dabrio CJ, Goy JL, Lario J, Silva PG 2009. Impacto del cambio climático en el litoral. climate change impacts on coastal áreas. *Enseñanza de las Ciencias de la Tierra*, 17(2): 141–154. I.S.S.N.: 1132–9157. Revista de la Asociación Española para la Enseñanza de las Ciencias de la Tierra.

Gama, LM, Villanueva C, Macias ME, Moguel EJ, Benítez, B 2008. *Inundaciones y lluvias extremas en Tabasco.* En Semana de Divulgación y Video Científico 2008, pp. 160–166.

INEGI 2018. *México en cifras. Información Nacional, por entidad federativa y municipios.* www.inegi.org.mx/app/areasgeograficas/.09/032020

INEGI 2020. *Aportación al Producto Interno Bruto (PIB) nacional.* www.inegi.org.mx/temas/pib/09/032020

Palma, DJ, Cisneros J, Moreno E, Rincón JA 2007. *Suelos de Tabasco: su uso y manejo sustentable, Colegio de Postgraduados-ISPROTAB-FUPROTAB.* México: Villahermosa, TAB, 184 p.

Pino M, delos A, Terry E, León A, Marrero P, Soto F 2009. Respuesta de las plantas de tomate a la modificación de algunas variables del microclima en un sistema protegido con sombra natural. *Cultivos Tropicales*, 21(2): 33–36.

Sol Sánchez Á, Melchor GIH, Cruz JMZ, González CAZ, Galarza JLS 2018. Mangrove restoration an economical alternative for generating incomes. In: Leal Filho W, Pociovălişteanu D, Borges de Brito P, Borges de Lima I (Eds) *Towards a Sustainable Bioeconomy: Principles, Challenges and Perspectives. World Sustainability Series.* Springer, Cham.

13

A Comprehensive Review of Commercial Production of Microbial Fertilizers

Arash Hemati, Nasser Delangiz, Behnam Asgari Lajayer and Mansour Ghorbanpour

CONTENTS

13.1 Introduction

Although the precise history of the emergence of agriculture is not well understood, evidence indicates that cultivation and production operations are at least 10,000 years old. Continuous cultivation, on the one hand, has gradually reduced soil fertility, and the gradual increase in the population, on the other hand, has meant that crop production through traditional farming is insufficient, which was the only way to fertilize farm soils. The role of organic matter and plant and animal residues in increasing soil fertility is indisputable in traditional agriculture. The European Industrial Revolution followed by the introduction of industrial tools into agriculture resulted in further exploitation of soils. Certainly, when Dutch scientist Begerink was studying soil microorganisms, and when his experiences led to the discovery of rhizobia and tuberculosis, he could not have imagined that his discoveries would have a huge influence on agricultural productivity in the following years. The first biofertilizer was introduced into the United States in 1895 by Nobbe and Hiltner (Jones and Lewis, 1993). By isolating *Rhizobium* from the root nodes of the broad bean, these researchers purified and amplified it and delivered the final product in glass containers to the market. In the following years, other countries such as Russia, Canada, Sweden and Australia turned to fertile fertilizers (Jones and Lewis, 1993). The range of soil microorganisms used in the production of biofertilizers has expanded over the 20th century, and today a wide range of soil bacteria are identified as *Rhizobium*, *Pseudomonas*, *Azotobacter* and *Azocyloris*, *Bacillus*, *Azocylorum*. Different mechanisms use algae to produce biofertilizers, but bacteria are the most useable organisms in the production of biofertilizers. Nowadays, biofertilizers are used in numerous formulations for various agricultural products such as cereals, herbs and vegetables, and in orchards and greenhouses. In this chapter, we introduce different types of biofertilizers and their production methods. Some manufacturing costs as well as the main production tool, the fermenter, are explained.

13.2 Effectiveness of Biological Fertilizers

The most controversial issue concerning biofertilizers is their effectiveness, which today affects research, production, government procurement and other policymaking related to this area. In general, a well-known feature of biological fertilizers is the inconsistency of the results of their application. This means that the results of applying a particular biofertilizer on a particular plant at two different points will not be the same. The reason for this is the living nature of these fertilizers and the influence of the environment, soil, plant species and native soil microorganisms on their effectiveness. This is often overlooked in considerations about biofertilizers, especially during executive meetings. Obviously, the ineffectiveness of a biofertilizer in certain conditions cannot be a sign of its ineffectiveness in other conditions. One of the reasons for this confusion could be the failure by experts in the field to provide specific and measurable criteria for the evaluation of biological fertilizers. Another problem in this area is the comprehensive reliance on a statistical analysis of experiments. For example, in an experiment, the application of a specific biofertilizer increased the yield of wheat by 12% (based on the yield of 5 ton/ha) compared with the control sample. Statistical analysis indicated that there was no significant difference between the two treatments; therefore, in the final consideration, the inefficiency of this fertilizer was emphasized. However, from the economic point of view of this experiment, one finds that the use of this biofertilizer is significant and justified. It does not matter what the ingredients are. But the effectiveness of fertilizer depends on the quality of the ingredients. In other words, if separation, screening, bioassay and effectiveness tests are not performed properly in the case of the microorganism used in the biofertilizer production, it cannot be effective. Therefore, based on these examinations, the unauthorization of biofertilizers can disrupt the biofertilizer market in some parts of the world.

13.3 Biofertilizers

13.3.1 Bacterial Biofertilizers

Rhizobacteria affect plant growth, plant production and health through various mechanisms, including dissolution and enhancement of nutrient bioavailability, stimulation of root growth and control of root diseases (Zehnder et al., 2001). The main goal of developing biotechnology for plant growth–promoting bacteria is to increase the population of effective soil bacteria that can contribute to sustainable agricultural development and reduce the need for chemical fertilizers and poisons (Adesemoye and Kloepper, 2009). Beneficial extracellular bacteria are defined by three inherent characteristics: (1) the ability to colonize the root; (2) the ability to survive, grow and reproduce at the root surface microenvironment enabling them to compete with other microorganisms; and (3) the ability to exhibit growth-promoting activities and enhance plant growth (Ahemad and Khan, 2011). *Rhizobium* bacteria have also been shown to contribute to the growth of non-legume plants. There is evidence that *Rhizobium* isolates are capable of enhancing rice plant growth and yield through changes in root morphology and physiology (Hafeez et al., 2004; Hussain et al., 2009; Gamalero and Glick, 2011). In order for a plant growth–promoting rhizobacteria (PGPR) to have a positive effect on plant growth, a close relationship must be established between the bacterium and the host plant. The relationship between PGPR and the host plant is divided into two levels: (1) extracellular and (2) endophytic (Vessey, 2003). The role of 1-aminocyclopropane-1-carboxylic acid (ACC)-deaminase-producing rhizobacteria to modulate the adverse effects of ethylene stress on wheat was investigated. The results showed that inoculation with these isolates significantly increased stem length, root length, number of early roots and root and shoot weight compared to the control. Due to inoculation with superior isolates, the biological yield and nutrient uptake of phosphorus, iron and zinc in these plants were significantly increased compared to the control (Mayak et al., 2004; Li et al., 2005; Nadeem et al., 2007; Swain et al., 2007).

13.3.2 Mycorrhizal Biofertilizers

Mycorrhizal fungi are among the most important microorganisms present in most undestroyed soils. About 70% of the biomass of the microbial community is composed of the mycelium of these fungi (Mukerji et al., 2002). Increasing crop yields is essential, especially in low fertility soils. Mycelium can penetrate into the deep soils and increase the uptake of nutrients through plants inoculated with related mycorrhiza (Swift, 2004). Mycorrhizal symbiosis with fungi by improving the spread of fungal hyphae in the soil pores has physically increased the phosphorus uptake of the vegetative parts and the dry weight of the plant (Nagahashi and Douds, 2003). Increasing the percentage of root mycorrhizal colonization increases plant and root growth and shoot dry weight through better absorption of water and minerals (Mirzakhani et al., 2009). The inoculation of onion seedlings with mycorrhiza increases phosphorus, sodium, chlorine, zinc, potassium, copper and total nitrogen concentrations (Schnepf et al., 2011; Viswanathan et al., 2011). Increasing biomass is accomplished by increasing the absorption of water and minerals and the production of growth phytohormones. The production of growth phytohormones such as auxin, cytokinin and gibberellin has been proven by these fungi (Swift, 2004).

13.4 Types of Microbial Fertilizers

Nitrogen fertilizers (biological nitrogen fixation [BNF]): The molecular nitrogen fixation process is performed under three general conditions: free living, symbiosis and associative symbiosis. In the free-living condition, the diazotrophic bacterium also stabilizes the molecular nitrogen during its free and independent life. The distinctive feature of the symbiotic state is the very specific relationship between the bacterium and the host plant. During symbiosis, the continuous and often compulsive communication between two symbionts results in the formation of a specific structure or organ called the node, which is the major site of molecular nitrogen fixation. In symbiotic systems, the nitrogen-fixing bacterium is the one side of the symbiotic relationship, which is referred to as microsymbiont, and usually endophyte, which is within the other symbiotic component. The other side of the symbiosis, called macrosymbiont, involves a wide range of organisms, including angiosperms, gymnosperms, fungi, ferns and algae. Known symbiotic systems include three general groups: cyanobacterial, actinorhizal and rhizobial symbiosis. There is also a specific state of coexistence in which a specific structure does not form as a main site for the nitrogen fixation process. This is termed associative symbiosis. Known types of nitrogenous microbial fertilizers include: *Cyanobacteria*, which are phototrophic bacteria and have oxygenic photosynthesis. Actinorhizal symbiosis: Symbiosis between actinomycetes of the family Frankiaceae and tree plants such as *Casuarina* and *Elaeagnus*. Rhizobial bacteria: Symbionts of the legume family plants. *Azotobacter* genus bacteria: *Azospirillum*. *Thiobacillus* genus bacteria: Includes chemo-organotrophic bacteria and *Beggiatoa* bacteria.

13.4.1 Mineral and Organic Phosphate-Solubilizing Microorganisms (PSM)

Microorganisms, whether prokaryotes or eukaryotes, play a key role in the phosphorus cycle. The phosphorus cycle is achieved through the sequential reduction and oxidation processes of phosphorus compounds. The electron transfer reactions are between the two ranges from the redox state of –3 (phosphines) to 5+ (phosphates). The ability of microorganisms to dissolve insoluble phosphates in soil and convert them into absorbable forms for plants was first demonstrated by Gerretsen (1948). His experiments showed that rhizosphere microorganisms can affect the rate of phosphorus dissolution and its uptake by the plant. These microorganisms can be found in most soils, but their numbers and types vary in different soils. Now, more than 50 years after its discovery, many researchers from different countries have isolated different types of phosphate-solubilizing microorganisms and studied their properties. Goldstein (1995) suggested mineral phosphate solubilization (MPS) for microbial phenotypes capable of solubilizing insoluble inorganic phosphates (mineral phosphate). Also, the phosphate-solubilizing microorganism abbreviation (PSM) refers to different types of microbes capable of solubilising mineral phosphate. PSMs include bacteria, actinomycetes and fungi, but actinomycetes are not efficient in solubilizing phosphates.

13.4.2 Phosphate-Solubilizing Fungi (PSF) (*Penecellium, Aspergillus*)

Numerous reports show the ability of different bacterial strains to dissolve insoluble inorganic phosphates, such as tricalcium phosphate, dicalcium phosphate, hydroxyapatite and soil phosphate. Among the bacterial species that have been shown to have phosphate-solubilizing ability are *Pseudomonas*,

Bacillus, *Acrobacterium*, *Agrobacterium*, *Micrococcus*, *Enterobacter*, *Flavobacterium* and *Arvinia*, and different genera of *Rhizobium* bacteria can also be named (Gizaw et al., 2017).

13.4.3 Biological Potassium Fertilizer (BPF)

Since 1988, Chinese scientists have been extensively researching, isolating and evaluating the efficacy of these bacteria. Using these bacteria, known as silicate soluble bacteria (SSB), they have produced biological fertilizers called biological potassium fertilizer. In recent years, research in this field has successfully mass-produced biological fertilizers, which are now used on an area of $10^6 \times 10^4$ ha. Using these fertilizers on more than 20 product types such as wheat, rice, corn, millet, peanuts, soybeans, potatoes, tomatoes, watermelons, cabbage, green peppers, cucumbers, red beans, grapes, apples and fruit trees has shown positive results. Some microorganisms (bacteria, fungi, yeasts, algae, lichens, etc.) are able to decompose silicates and release elements such as potassium, iron, zinc, silica and phosphorus, among which bacteria are the most important. Silicate dissolving bacteria (SDB) produce a biological fertilizer called biological potassium fertilizer. Bacteria in this biological fertilizer release minerals such as biotite, muscovite, goethite, orthoclase and microcline. Potassium is the most important of these elements. These bacteria are divided into orthoclase, biotite and albite bacteria based on the type of minerals they grow on. *Bacillus circulans*, *Arthrobacter*, *B. megaterium*, *Azotobacter*, *B. mucilagenose* and *B. polymixa* are some of these bacteria. SDB dissolve silicates by secreting polysaccharides, organic acids (oxalic, citric, malonic, lactic, ketogluconic) and inorganic acids (carbonic, sulphuric and nitric). SDB efficacy depends on soil type (especially absorbable K), plant type, environmental conditions and bacterial type. Increased plant yield was reported between 0 and 30% through BPF application (Basak et al., 2015).

13.4.4 Silicate Bacteria (SDB)

These bacteria are named for their ability to decompose silicates and release elements such as potassium, silicon, aluminium, iron, zinc and molybdenum. The most important role of these bacteria is in the release of potassium; however, there are also reports of phosphorus release. The effect of these bacteria on the release of elements from minerals such as biotite, muscovite, chitosite, orthoclase and microcline has been reported (Rogers and Bennett, 2004).

13.4.5 Sulphur Fertilizers

This biological fertilizer helps plants take up sulphur, phosphorus, iron and zinc for better nutrition through sulphur oxidation in saline-sodium and sodium soils, subsequently enhancing plant yield. A wide range of soil microorganisms are able to oxidize sulphur-reducing compounds, the most important of which are *Thiobacillus* bacteria, which are a rod-shaped, Gram-negative, chemolithotroph, sulphur-oxidizing bacteria. More than 20 species of this genus have been identified and introduced. The species of this genus are divided into two groups: acidophile and neutrophile. Acidophiles are predominant in acidic soils (optimum pH about 3) and neutrophils are dominant in alkaline and calcareous soils. These bacteria oxidize sulphur to produce energy and some sulphuric acid. The acid produced at the microsite points around the roots increases the solubility of insoluble compounds (insoluble phosphates, Ca, etc.) by decreasing the local pH of the soil. *Thiobacillus* bacteria are also used in bioleaching. In this process, bacteria oxidize sulphur (sulphide) to release metals such as zinc, copper, gold and uranium. A group of sulphur-oxidizing microorganisms include a wide range of bacteria, e.g. photosynthetic sulphur bacteria, heterotrophic bacteria, microtrophic bacteria and sulphur bacteria. The group of non-oxygenated photosynthetic sulphur bacteria use H_2S as the electron acceptor and oxidize sulphur, for instance, green sulphur bacteria such as chlorobium and purple sulphur bacteria such as chromate (Aleem, 1975; Bdgander, 1980; Bremner and Steele, 1978). The sulphur oxidation reaction in this group is as follows:

$$CO_2 + 2H_2S \rightarrow (CH_2O)_n + 2S\downarrow + H_2O$$

Heterotrophic bacteria use organic matter as a carbon substrate to provide the energy and carbon source they need and oxidize sulphur as a secondary reaction. The sulphur oxidation reaction in this group is as follows:

$$4Na_2S_2O_3 + 2H_2O + O_2 \rightarrow 2Na_2S_4O_6 + 4NaOH$$

Among this group are bacteria of the genus *Pseudomonas*, *Arthrobacter* and *Bacillus*. Other sulphur-oxidizing heterotrophic microorganisms include *Aspergillus niger*, *Mucor flavus*, *Trichoderma harzianum* and *Fusarium solani*. The mixotroph *Beggiatoa* is a group of bacteria with sliding motions and is classified in the α-proteobacteria branch. It is a microaerobic bacterium that lives in the root of rice plants and in semi-aerobic conditions reduces sulphur oxidation and H_2S toxicity. Colourless sulphur bacteria are a group of aerobic chemolithotrophic bacteria whose colourless term distinguishes them from pigmented photolithotrophic bacteria. An important feature of these bacteria is that they do not contain any photosynthetic pigments and use nitrate or oxygen as the final electron acceptor. The most important of these bacteria are *Thiobacillus*, which are mandatory or optional chemoattractive bacteria that include neutral and acid-friendly neutral types, as well as thermophilic and mesophilic types (Wainwright, 1984; Aleem, 1975; Kuenen, 1975; Bdgander, 1980).

13.5 Conclusion

The production of biofertilizers is basically explained in three stages. Firstly, the main challenge in producing effective biofertilizers is that much research is needed to select the most efficient bacterial or fungal strain. Secondly, equipment is very important. The type of fermenter used is important for the speed of production as well as for contamination in the production process. A third factor that is important for the viability of biofertilizers is the carrier of biofertilizers. The choice of carrier should be commensurate with the type of microbe, so an appropriate carrier should be selected for each microbe. The type of biofertilizer is also important. Three types of biofertilizers have different production stages. For example, fermenter is not required during the mycorrhizal fertilizer production process. But in the other two types this step is necessary. The use of biofertilizers is justified if they are economically viable compared to chemical fertilizers. Biofertilizers are also a good investment for employment and profitability in the future.

REFERENCES

Adesemoye AO, Kloepper JW 2009. Plant microbes' interactions in enhanced fertilizer-use efficiency. *Appl. Microbiol. Biotechnol.* 85: 1–12.

Ahemad M, Khan MS 2011. Functional aspects of plant growth promoting rhizobacteria: Recent advancements. *Insight Microbiol.* 1(3): 39–54.

Aleem MIH 1975. Biochemical reaction mechanisms in sulfur oxidation by chemosynthetic bacteria. *Plant Soil.* 43: 587–607.

Basak BB, Meena HM, Meena MS, Dotaniya ML 2015. Potassium bio-fertilizers. Viable technology for sustainable agriculture. *Ind. Farm.* 64(11): 2–4.

Bdgander LE 1980. Bacterial cycling of sulfur in a Baltic sediment: An *in situ* study in closed systems. *Geomicrobiol. J.* 2: 141–159.

Bremner JM, Steele CG 1978. Role of microorganisms in the atmospheric sulfur cycle. *Adv. Microb. Ecol.* 2: 155–201.

Gamalero E, Glick BR 2011. Mechanisms used by plant growth-promoting bacteria. *Bacteria Agrobiol.: Plant. Nutr. Manage.* 17–46.

Gerretsen FC 1948. The influence of microorganisms on the phosphate intake by the plant. *Plant Soil* 1: 51–81.

Gizaw B, Tsegay Z, Tefera G, Aynalem E, Wassie M 2017. Phosphate solubilizing fungi isolated and characterized from teff rhizosphere soil collected from North Showa and Gojam, Ethiopia. *J. Fertil. Pestic.* 8: 180.

Goldstein AH 1995. Recent progress in understanding the molecular genetics and biochemistry of calcium phosphate solubilization by Gram-negative bacteria. *Biol. Agri. Hortic.* 12: 185–193.

Hafeez FY, Safdar ME, Chaudhry AU, Malik KA 2004. Rhizobial inoculation improves seedling emergence, nutrient uptake and growth of cotton. *Aust. J. Exp. Agr.* 44: 617–622.

Hussain MB, Mehboob I, Zahir ZA, Naveed M, Asghar, HN 2009. Potential of *Rhizobium* spp. for improving growth and yield of rice (*Oryza sativa* L.). *Soil Environ.* 28(1): 49–55.

Jones DG, Lewis DM 1993. *Rhizobium* inoculation of crop plants. In: *Exploitation of Microorganisms*, Jones DG (Ed.), Chapman and Hall, London. 197–224.

Kuenen JG 1975. Colorless sulfur bacteria and their role in the sulfur cycle. *Plant Soil.* 43: 49–76.

Li Q, Saleh-Lakha S, Glick BR 2005. The effect of native and ACC deaminase-containing *Azospirillum brasilense* Cd1843 on the rooting of carnation cuttings. *Can. J. Microbiol.* 51: 511–514.

Mayak S, Tirosh T, Glick BR 2004. Plant growth-promoting that confer resistance to water stress in tomatoes and peppers. *Plant Sci.* 166: 525–530.

Mirzakhani M, Ardakani MR, Aeeneband A, Shiranirad AH, Rejali F 2009. *Effects of Co-Inoculation of Azotobacter and Mycorrhiza under Nitrogen and Phosphorus Levels on Nutrients Absorbtion Efficiency in Safflower (Carthamustinctorius L.).* Ph.D. Thesis of Agricultural on Agronomy. Islamic Azad University Science and Research Branch-Khouzestan, Iran. (In Persian with English Summary).

Mukerji KG, Manoharachary C, Chamola BP 2002. *Techniques in Mycorrhizal Studies.* Kluwer Academic Publishers.

Nadeem SM, Zahir ZA, Naveed M, Arshad M 2007. Preliminary investigations on inducing salt tolerance in maize through inoculation with rhizobacteria containing ACC deaminase activity. *Can. J. Microbiol.* 53: 1141–1149.

Nagahashi G, Douds D 2003. Action spectrum for the induction of hyphal branches of an arbuscular mycorrhizal fungus exposure sites versus branching sites. *Mycol. Res.* 107: 1075–1082.

Rogers JR, Bennett PC 2004. Mineral stimulation of subsurface microorganisms: Release of limiting nutrients from silicates. *Chem. Geol.* 203: 91–108.

Schnepf A, Jones D, Roose T 2011. Modelling nutrient uptake by individual hyphae of arbuscular mycorrhizal fungi: Temporal and spatial scales for an experimental design. *Bull. Math. Biol.* 73: 2175–2200.

Swain MR, Naskar SK, Ray RC 2007. Indole-3-acetic acid production and effect on sprouting of yam (*Dioscorea rotundata* L.) minisetts by *Bacillus subtilis* isolated from culturable cow dung microflora. *Pol. J. Microbiol.* 56: 103–110.

Swift CE 2004. Mycorrhiza and soil phosphorus levels. Area extension agent. www.colostate.edu/Depts/Coop Ext/TRA/Plants/mycorrhiza

Vessey JK 2003. Plant growth promoting rhizobacteria as biofertilizers. *Plant Soil.* 255: 571–586.

Viswanathan G, Panchaksharam T, Kurusangu V 2011. Influence of arbuscular mycorrhizal fungi and *Azospirillum* co-inoculation on the growth characteristics, nutritional content and yield of tomato crops grown in south India. *Ind. J. Fund. Appl. Life. Sci.* 1: 84–92.

Wainwright M 1984. Sulfur oxidation in soil. *Adv. Agron.* 37: 349–396.

Zehnder GW, Murphy JF, Sikora EJ, Kloepper JW 2001. Application of rhizobacteria for induced resistance. *Eur. J. Plant Pathol.* 107: 39–50.

14

Microbial Solubilizers of Rock-Forming Minerals: Opportunities and Challenges

Julio Alves Cardoso Filho

CONTENTS

14.1 Introduction

One of the major global challenges of the 21st century is the probability that the world's population will exceed nine billion by 2050 (FAO 2009; Godfray et al. 2010; Shelden and Roessner 2013; Godfray and Garnett 2014), an increase of 2.25 billion over the next 30 years (Bruinsma 2011). By 2025, global agricultural production may need to be increased by ~40% to meet these increasing demands (Pennisi 2008; Verma et al. 2015a,b), as well as to provide food security for ~870 million now chronically undernourished people in several world regions (e.g. sub-Saharan Africa and South Asia) due to a combination of low per capita food consumption (in terms of kcal/person/day) and high uncontrolled population growth, according to the Organization for Economic Co-operation and Development (OECD) and the Food and Agriculture Organization of the United Nations (FAO) [2012]. In the 1940s, the "Green Revolution" (or "Third Agricultural Revolution") began in Mexico. It is attributed to Norman Ernest Borlaug ("The Father of Crop Revolution"), an American agronomist (winner of the Nobel Peace Prize in 1970). Between 1950 and the late 1960s, the Green Revolution spread global agricultural practices and renewing technologies (Pearce and Turner 1990). The Green Revolution was the result of an international development program supported by US philanthropy (the Rockefeller Foundation) to equate and solve the problems of global hunger after the Second World War (FAO 2019). The Green Revolution combined high-yielding cultivars (from hybrid seeds), inorganic chemical fertilizers and the use of pesticides (synthetic or non-synthetic) to foster agri-food industrial production (Neethirajan et al. 2018). However, the dark side of the Green Revolution entailed heavy negative impacts on environmental health, such as the rapid depletion of earth's natural resources (Lindenmayer and Sato 2018), ecological imbalance (Tydecks et al. 2018), persistent organic pollutants (POPs) (Thompson and Darwish 2019), herbicide resistance in weeds (Schulz and Segobye 2016; Westwood et al. 2018) and insecticide resistance in pests (Pose-Juan et al. 2015), and converted modern hybrid crops into agrochemical-dependent synthetic input (Bomgardner 2016). Moreover, the Green Revolution inflicted severe health issues on humans and animals (e.g. cancers, chronic neurodegenerative disorders, infertility, sperm chromosomal abnormalities, liver and kidney damage, congenital birth defects, disruption of the endocrine system, epigenetic alterations and mental and physical disabilities), mainly from contamination by synthetic pollutants (e.g. in air, water, soil, livestock manure and food due to high and repeated occupational and environmental exposure to POPs) (Zacharia 2019). Fortunately, there are solutions to the problems of environmental unsustainability arising from age-old agricultural practices. Thus, due to the immediate necessity to reduce chemical synthetic or non-synthetic product consumption with the aim of sustainable agriculture and to protect environmental resources, many eco-friendly and sustainable approaches have been implemented across the globe (Patibanda and Ranganathswamy 2018). Among these, the use of commercial microbial bio-inoculants (e.g. microbial phosphorus [P], potassium [K], silicon [Si], sulphur [S], zinc [Zn], solubilizers, nitrogen [N_2] symbiotic and non-symbiotic fixers, plant growth-promoting bacteria [PGPB], plant growth-promoting fungi [PGPF] and bio-inoculants for induced systemic resistance [ISR]) can be highlighted (Rouphael and Colla 2018; Bulgari et al. 2019). The use of commercial microbial bio-inoculants can be considered an alternative and innovative solution to ensure food security and to restore and guarantee human, animal and environmental health (Alori and Babalola 2018; Luziatelli et al. 2019), due to the total or partial replacement of synthetic or non-synthetic agrochemicals used in traditional agro-food industrial production systems (Bradácová et al. 2020). Consequently, this chapter summarizes the advances in microbial solubilizing approaches and describes the strategies used to explore this natural and biological resource for the sustainable maintenance of agroecosystems, ecosystems and human, plant and animal health.

14.2 Phosphate (P) Solubilizers (P-S)

14.2.1 P in Soils

Phosphorus (a non-metal belonging to the fifteenth group of the periodic table) is a biologically essential element for all life forms (e.g. cell constituents such as nucleic acids, nucleotides, phosphoproteins and phospholipids) on earth (prokaryotic, eukaryotic cells and virions) (Krishnaraj and Dahale 2014; Alori and Babalola 2018). The earth's lithosphere has ~1,200 mg P kg^{-1} (mostly found in sedimentary rocks with a few of igneous origin) (Porder and Ramachandran 2013), making it the eleventh most abundant element (Krishnaraj and Dahale 2014; Daneshgar et al. 2018). P in soil is usually found associated with minerals, such as the apatite group (e.g. hydroxyapatite [$Ca_5(PO_4)_3(OH)$], fluorapatite [$Ca_5(PO_4)3F$] or chlorapatite [($Ca_5(PO_4)3Cl$)]), and oxides and hydroxides of aluminium (Al), calcium (Ca) and iron (Fe) (e.g. goethite, magnetite, gibbsite) (Goss 1987). Globally, the highest reserves of phosphate rock (PR) are found in Morocco, the Western Sahara (71.5%), China (4.6%) and Algeria (3%) (U.S. Geological Survey 2019). Considering the finite natural reserves of P (~67 Gt. of P rock: Sverdrup et al. 2009), a global P crisis is very near (Fink et al. 2016). However, none of these studies focused on the Presidual in soil (Sattari et al. 2012). Considering P uses in agriculture, ~85% comes from ground rock phosphate (GRP) (especially a hydroxyapatite[$Ca_5(PO_4)_3OH$](Baya et al. 2013). Globally, the chemical P fertilizers applied to soil include mono-ammonium phosphate [MAP$(NH_4)H_2PO_4$~48–50% of P_2O_5], di-ammonium phosphate [DAP$(NH_4)_2HPO_4$], single superphosphate [SSPCa$(H_2PO_4)_2$ + $CaSO_4$~18–22% of P_2O_5] and triple superphosphate [TPPCa$(H_2PO_4)_2$~46–47% of P_2O_5] (Chagas et al. 2016). Worldwide, P deficiency occurs over ~5.7 billion hectares (e.g. Histosols, Oxisols, Andisols, Vertisols). Soil P deficiency induces limitations on global crop production (Molin and Tavares 2019), especially in tropical soils with variable charge clays (e.g. Ferralsols, Acrisols, Oxisols, Ultisols, Latosols, Lateritic, Krasnozems, Ferrisols and Ferrallitic) (Uehara and Gillman 1981), or soil that is high in calcium carbonate (e.g. calcixerepts with petrocalcic horizon [$CaCO_3$]) and gypsum (e.g. xerorthents with palygorskite horizon [$CaSO_4.2H_2O$]) (Silva et al. 2016). These factors reduce the availability of P in soil (e.g. in both inorganic [Pi] and organic [Po] forms) (Tipping et al. 2016). Both soil acidity (Holland et al. 2019) and soil alkalinity (Bai et al. 2018) induce fixation of P applied by oxides of Fe (e.g. hematite α-Fe_2O_3) and Al (e.g. corundum α-Al_2O_3), sesquioxides of Fe (e.g. goethite α-FeO(OH)) and Al (e.g. gibbsite $Al(OH)_3$). In non-calcareous soils, P fixation is a ligand exchange reaction, in which an OH^- molecule is released from a hydroxylated mineral surface (Field et al. 1985). These authors use the generalized reaction to describe the ligand exchange reaction as follows:

$$\mathbf{aSOH_{(s)} + HbPO_4^{b-3}{}_{(aq)} + cH^+{}_{(aq)} \rightleftharpoons SaHcPO_{4(s)} + bH_2O_{(l)} + (a-b)OH^-{}_{(aq)}}$$

where:

$SOH_{(s)}$ = reactive surface hydroxyl with a metal ion

b = degree of protonation of orthophosphate ion (ranging from 0 to 3)

The effects of the concentration of electrolytes on the adsorption of phosphate by goethite (colloid with variable charge surface) can be correlated with the model of Bowden, which describes the adsorption of phosphate, citrate and selenite on a variable charge mineral surface (Tan 2011).

P fixation in alkaline soil conditions is illustrated as follows (Tan 1982):

$$\mathbf{3Ca^{2+} + 2PO_4^{3-} \rightarrow Ca_3\left(PO_4\right)_2 \downarrow \left(insoluble\right)}$$

$$\mathbf{3CaCO_3 + 2PO_4{}^{3-} \rightarrow Ca_3(PO_4)_2 + 3CO_2 - \left(insoluble\right)}$$

Calcite (calcium carbonate [$CaCO_3$]) and dolomite (calcium magnesium carbonate ($CaMg(CO_3)_2$) limestones are used in tropical soils to increase the pH, to neutralize Al^{3+}, to raise the soil's capacity by adsorption and exchange cations (cation exchange capacity [CEC.∴Tan 2011: CE=ΣmEq exchangeable cations per 100 g soil], and to provide P in soil solution (Tan 1982). Calcitic (or $CaCO_3$) is insoluble in H_2O. Added to acid soil (Al-soil), $CaCO_3$ reacts with H_2O containing CO_2 as follows (Tan 2011):

$$\mathbf{CaCO_3 + H_2CO_3 \rightarrow Ca\left(HCO_3\right)_2}$$

After limestone is added to acid soil, $CaCO_3$ reacts with H_2O and the dissociated $Ca^{2+}$is adsorbed by Al^{3+}, reducing soluble Al by two reactions (Tan 2011):

(1) $\mathbf{CaCO_3 + H_2O \rightarrow Ca^{2+} + 2OH^- + CO_2}$

(2) $\mathbf{Al^{3+}{}_{[soluble]} + 3OH^- \rightarrow Al(OH)_{3[insoluble]}}$

The process of P precipitation in soil by Al^{3+} is described as follows (Olsen and Khasawneh 1980):

$$Al^{3+} + H_2PO^-_{4(aq)} + 2H_2O \leftrightarrow AlPO_4 \cdot 2H_2O_{(s)}\ (\text{variscite}) + 2H^+\ \text{Log}\,K = 2.50$$

Th organic soil P (Po) fractions have been characterized as inositol phosphates (e.g. homocyclicsugar–like compound, $C_6H_{12}O_6$, myoinositol hexa-phosphoric acid), phospholipids (e.g. phosphatidylcholine and phosphatidylethanolamine), nucleic acids (e.g. RNA and DNA), nucleotides (e.g. purine and pyrimidine) and sugar phosphates (e.g. ribose or deoxyribose) (Chardon et al. 1997). The mineralization of Po is related to the C/N/P ratio in soil (Chardon et al. 1997). The C/N/P ratio of 100:10:1 is the theoretical ideal composition for soil organic matter (SOM), but real values ranging from 229:10:0.39 to 71:10:3.05 have been found (Brookes et al. 1984). It is suggested that if C/Pi = 200:1, Pi mineralization (Po → Pi) will occur, whereas if C/Pi = 300:1, immobilization (Pi → Po) will occur (Beck and Sanchez 1994). Plant growth and yield is adversely affected by the lack of P (Marschner and Marschner 2012; Caporale and Violante 2016). P in the soil-solution-plant system is modeled from three main soil phosphate fractions: non-labile P, labile P and solution P (Barrow 1983; Bolan et al. 1997; MacDonald et al. 2011). The phosphate fractions (non-labile P, labile P and P solution) are illustrated as follows (Bolan et al. 1997):

Soil solution $\rightleftharpoons$ labile P $\rightleftharpoons$ non-labile P

Phosphate "adsorption" (or phosphate retention) is used to characterize the reaction of phosphate ions in solution with atoms on the surface of soil colloids (Barrow 1978). The main forces that governs adsorption include physical and chemical forces, hydrogen bonding, hydrophobic bonding, electrostatic bonding, coordinations reactions, and ligand exchange (Tan 1982). Usually, the quantity of solid-phase P in soils acts as a reserve; this can be summarized by plotting the calculated adsorption of phosphate versus the observed solution concentration; these plots are called Q/I plots (quantity/intensity) (Barrow 1983). Soil P maximum adsorption capacity (PMAC) can be determined by Freundlich's Equation (14.1) and Langmuir's Equation (14.2), and can be written as follows:

$$\frac{x}{m} = kC^{1/n} \tag{14.1}$$

where:

x = amount of material adsorbed
m = amount of adsorbens
C = concentration of the equilibrium solution
k = Freundlich adsorption constant
$\frac{1}{n}$ = is the Freundlich exponent

Taking the logarithm, Equation (14.1) changes into linear form: $\log(x/m) = \log k + (1/n) \log C$, when $\tan \alpha = 1/n$ (the slope of the curve) or the regression coefficient and the value of the intersection is $\log k$.∴; then the log equation gives a "straight-line" curve, or linear, curve (Tan 1982).

$\frac{1}{n}$ = range between 0.2 and 0.7 (Tan 1982).

The Langmuir equation (originally) is formulated as follows:

$$\theta = \frac{(bP)}{(1 + bP)} \tag{14.2}$$

Where
θ = the amount of gas adsorbed per unit of area,
b = constant related to free energy of adsorption,
P = the equilibrium gas pressure.

In soil solutions P can be replaced by C, and the Equation (14.2) assumes the following formatting:

$$\frac{x}{m} = \frac{k_1 C}{1 + k_2 C} \tag{14.3}$$

where:
x = amount adsorbed
m = amount of adsorbens
k_1, k_2 = constants
C = concentration of equilibrium solution

When the concentration k_2C reaches a value equal to 1, Equation (14.3) becomes

$$\frac{x}{m} = \frac{k_1}{k_2}\frac{x}{m} = \frac{k_1}{k_2}$$

In other words, at a high C value, the adsorption is raised to the maximum, and the surface of the adsorbents reaches saturation. Equation (14.3) is called the Langmuir equation to most scientists. However, this equation was updated to the power of β, α, or x, as follows:

$$\frac{x}{m} = \frac{(k_1 C)^{\alpha}}{(1 + (k_1 C)^{\alpha}}$$

and called Langmuir–Freunlich equation (Tan 2011).

These equations can be useful for summarizing data about a specific soil, and to describe the effects of a particular treatment (e.g. hydroxylated mineral surfaces, anaerobically digested poultry manure) (Field et al. 1985).

14.2.2 P in Plants

Plants take up either $H_2PO_4^-$ or $H_2PO_4^{2-}$ from soil solution (Bieleski 1973; Shen et al. 2011). Due to low P solubility, the mobility of P and the high absorption rate by root uptake, a P depletion zone (~0.2–1.0 mm around the root) is created in the rhizosphere (Tinker 1975), which restricts the $H_2PO_4^-$ or $H_2PO_4^{2-}$ uptake from the soil solution to the roots (Bhat et al. 1976). The P concentration is ~1 μM, 400 μM in the xylem and 10,000 μM in the cytoplasm (Ahmad et al. 2018). Plants use high-affinity active transport for Pi uptake against a chemical gradient across the apoplast (plasma membrane of root epidermal and cortical cells) (Fageria 2009), mediated by high-affinity Pi/H^+ symporters of *PHT1*gene family membrane proteins (e.g. *PHT1*, *PHT2*, *PHT3* and *PHT4*) (Ceasar et al. 2014; Ahmad et al. 2018). The P from the xylem to the cytoplasm and vacuole is an energy-dependent process (Ullrich and Novacky 1990). Inside the cell, Pi is used to generate adenosine triphosphate (ATP; phosphate carrier) by the substrate as per the following reaction sequence (Marschner and Marschner 2012):

Triosephosphate dehydrogenase
3-Phosphateglyceraldehyde + NAD^+ + Pi → 1,3 bisphosphos glyceraldehyde + NADH + H^+
ADP kinase
1,3 Bisphosphosglyceraldehyde + AD → 3-phosphate glyceraldehyde + ATP

Alternatively, ATP can be generated by electron transport phosphorylation or photophosphorylation by membrane-bound ATP synthase at the expense of proton (H^+) or sodium (Na^+) ion fluxes across the cytoplasmic membrane (Marschner and Marschner 2012), as follows:

$$ADP + Pi \rightarrow ATP$$

Then, inorganic pyrophosphate (PPi) is generated (as a side metabolite) by energy-expensive reactions and hydrolyzed to 2Pi as follows (Marschner and Marschner 2012):

$$PP \rightarrow 2Pi$$

P occurs in the phosphorylated intermediates of many metabolic pathways (e.g. the Calvin cycle; in the photorespiratory pathway; in glycolysis; in the pentose phosphate pathway; in nitrogen and sulphur assimilation; in the amino acid and nucleotide metabolism pathways; and in pathways leading to the synthesis of polyphenols and lignin) (Chouhan et al.2017). The plant phosphate reserve (especially in seeds) is phytin (Meis et al. 2003). Phytic acid is indigestible by monogastric animals (Dersjant-Li et al. 2015). However, it is an essential compound in the phosphate cycle due to its capacity to hinder the uptake of a range of minerals (Daneshgar). P deficiency in plants occurs in older leaves (general symptoms are a darkish green, reddish color, due to the increase in anthocyanins), with a decrease in the biosynthesis of phospholipids and an increase in the biosynthesis of non-P lipids (Kuppusamy et al. 2014). P-deficient plants are characterized by delayed maturity, sparse flowering, poor seed quality and premature senescence of the leaves (Marschner and Marschner 2012; Malhotra et al. 2018). It is considered an adaptive and evolutive response to increase the acquisition of Pi and its translocation, to improve the utilization of stored P by adjusting the Pi recycling internally (e.g. restricting Pi consumption) and by reallocating Pi from old tissues to actively growing tissues (Pi released from glycosylation) (Selle et al. 2006). Therefore, matching the exploration of the molecular mechanisms of P with the physical, chemical and biological properties of soil holds promise for nutrient use efficiency (NUE) and breeding in crop varieties (Ahmad et al. 2018). Organisms of P-solubilizers (*Bacillus megaterium*, *B. subtilis*, *B. malabarensis*, *Serratia* sp., *Proteus* spp., *Arthrobacter* spp., *Streptomyces* spp., and fungi such as *Aspergillus* sp., *Penicillium* sp., *Rhizopus* sp. and *Cunninghamella*sp.) secrete or, upon death, release phosphatases (e.g. phytase) with greater or lesser substrate (e.g. phytin) specificity (Taga and Kobori 1978). The liberation of phosphate from phytin needs the enzyme phytase:

$$\text{Phytin} + 6H_2O \rightarrow \text{inositol} + 6P_i \ \text{Phytin} + 6H_2O \rightarrow \text{inositol} + 6P_i$$

On the other hand, the release of phosphate from nucleic acid (DNA or RNA) needs the action of nucleases, which yield nucleotides, nucleosides and inorganic phosphate, as follows (Taga and Kobori 1978):

$$\text{Nucleic acid} \xrightarrow[+H_2O]{\text{nuclease}} \text{nucleotides} \xrightarrow[+H_2O]{\text{nucleotidase}} \text{nucleosides} + P_i$$

14.2.3 Mechanisms of Phosphate Solubilization

14.2.3.1 Solubilization from P-inorganic (Pi) Compounds

The most common mineral form found on the earth's lithosphere is apatite, usually from igneous, sedimentary and metamorphic rocks. Apatitesare primarily phosphates and are represented as the isomorphous end-members series flour-apatite ($Ca_5(PO_4)_3F$), chlorapatite ($Ca_5(PO_4)_3Cl$), hydroxyapatite ($CA_5(PO_4)_3OH$) and carbonate-apatite ($Ca_5(PO_4, CO_3, OH)_3(F,OH)$) (Deer et al. 1996). In soil, the primary phosphate forms of apatites, such as francolite (Ca, Mg, Sr, Na)10(PO_4, SO_4, CO_3)6$F_{2\text{-}3}$ are more solubilized than the pure forms of fluorapatite (McArthur 1985), as shown in the following reaction (Schuffert et al. 1990):

$$\mathbf{Ca_{10}F_2(PO_4)_6 + 12H^+ \rightleftharpoons 10Ca^{2+} + 6H_2PO_4^- + 2F}$$

Fluorapatite orthophosphate ion

Insoluble secondary phosphate minerals are aluminium salts such as variscite ($AlPO_4 \times 2H_2O$), iron salts vivianite ($Fe_3(PO_4)2 \times 8H_2O$) and strengite ($FePO_4 \times 2H_2O$) (Deer et al. 1996). Pi-solubilizing microorganisms can use Ca–P (e.g. brushite [dicalciumphosphatedihydrate] [$CaHPO \bullet 2H_2O$]) andmonetite[dicalciumphosphate] [$CaHPO_4$]), Al–P (e.g. amorphous variscite [$Al(OH)_2H_2PO_4$]) and Fe–P (e.g. strengite [$FePO_4 \cdot 2H_2O$]) compounds to release Pi (Deer et al. 1996). The equilibrium reactions for brushite, monetite and amorphous variscite are described as follows (Nathan 1984):

$$\mathbf{CaHPO_4 \bullet 2H_2O + H^+ \rightleftharpoons Ca^{2+} + H_2PO_4^- + 2H_2O \therefore Log\ K^o = 0.6}$$

Dicalcium orthophosphate

Phosphate ion dihydrate

$$\mathbf{Ca\left(H_2PO_4\right)_2 \bullet H_2O \rightleftharpoons Ca^{2+} + 2H_2PO_4^- + H_2O \therefore Log\ K^o = -1.15}$$

Monocalcium orthophosphate

phosphate ion

$$\mathbf{Al\left(OH\right)_2 H_2PO_4 \rightleftharpoons A1^{3+} + 2OH^- + H_2PO_4^- \therefore Log\ K^o = -27\ to\ -29}$$

Amorphous variscite

orthophosphate ion

According to the literature data, the mechanisms of Pi solubilization are (1) the production of inorganic or organic acids (e.g. oxalic acid and CO_2) that degrade the insoluble phosphates; (2) the production of the agent's chelators (e.g. siderophores); (3) the reduction of Fe in ferric phosphate mineral (e.g. strengite) to ferrous iron (or iron[II]sulphate [FeO_4S]) via the extracellular electron transfer (EET) process; (4) the production of hydrogen sulphide (H_2S), which reduces the iron and precipitates it as iron sulphide (or iron sulphide [FeS_2], pyrite), thereby mobilizing phosphate; and (5) the mineralization of biochemical Po (Chatterjee and Nandi 1964; Cosgrove 1977; Bonting et al. 1992; Dyhrman et al. 2009; Boge et al. 2012; Lever et al. 2015).

14.2.3.1.1 Production of Inorganic or Organic Acids

Microbial cells can alter their metabolism (e.g. due to lack of C) to produce organic acids such as gluconic and citric acid (Cunningham and Kuiack 1992; Illmer and Schinner 1992; Kim et al. 1997), and 2-ketoglutaric acid (KGA) (Duff and Webley 1959; Kang et al. 2002) in the medium (Schink et al. 2002; Khan et al. 2017), and to release a free orthophosphate form (HPO_4^- and HPO_4^{2-}) from insoluble inorganic phosphates (e.g. vivianite [$Fe_3(PO_4)_2.8H_2O$], strengite[$FePO_4.2H_2O$] and variscite [$AlPO_4.2H_2O$]) (Khan et al. 2017). In fermentations (organic acids production), the ATP is produced by substrate-level phosphorylation, and in oxidative phosphorylation (inorganic acids production) by a proton (H^+) motive force (Atlas and Bartha 1997). The role of organic ligands (of organic acids) on the availability of Pi was demonstrated in yellow-brown soil and showed the order: oxalate > citrate > malate > tartrate, and in paddy soils, the order was tartrate > citrate ≈ oxalate > malate (Zhang et al. 2014). Organic acids can produce carboxyl (–COOH) and hydroxyl ($^-$OH) ions (chelate agent of cations) or reduce the pH (Slate 1967: by extrusion of H^+ according to the chemiosmotic hypothesis of oxidative phosphorylation or Mitchell's hypothesis) (Illmer and Schinner 1992). These carboxyl and hydroxyl groups can act as donors for protons (H^+) under specific pH conditions such as NH_4^+ assimilation (Beever and Burns 1980; Nahas 1996; Baya et al. 2013) or from respiratory $H_2CO_3^-$production (Jurinak et al. 1986). The origin of the H^+ occurs according to the following equation (Illmer et al. 1995):

$$\mathbf{AlPO_{4(so)} + 2H^+_{(aq)} \rightarrow Al^{3+}_{(aq)} + H_2PO_4^-{}_{(aq)}}$$

The selective extrusion of H^+ ions across a proton-impermeable membrane generates a proton motive force (localized in the plasmalemma), and this energy is used (with the help of H^+ translocating ATPase) for ATP synthesis in respiration and photosynthesis (Beever and Burns 1980).

14.2.3.1.2 Production of Agents Chelators

14.2.3.1.2.1 Siderophores

Siderophores, which are peptides biosynthesized by non-ribosomal peptide synthetase (NRPS), are metal chelating (200–2,000 Da) with a high affinity for Fe, and are produced by fungi (Saha et al. 2016), hydroxamate-type siderophores (e.g. ferrioxamine B), bacteria (Glick et al. 2010), catecholate-type siderophores (e.g. enterobactin) and plants (Takemoto et al. 1978) and, phytosiderophore (e.g. mugineic acid), under Fe-starvation conditions (Sugiura et al. 1981). In general, acquisition starts with Fe-III [$FePO_4$] that binds with the excreted siderophores forming a ferri–siderophore complex and this complex binds to a specific receptor protein (e.g. FepA, FecA and FhuA) present on the microbial outer membrane surface (Khan et al. 2018). The ferri–siderophore complex is then translocated by active transport (e.g. proton gradient) into the cell and released (Khan et al. 2018). Because of their relevancy (e.g. bioweathering and soil formation), the role of siderophores in the dissolution of Fe-binding minerals has been extensively studied (Bou-Abdallah 2010; Qi and Zhao 2013; Ahmed and Holmström 2014; Saha et al. 2016; Khan et al. 2018).

14.2.3.1.2.2 Extracellular Electron Transfer (or Direct Oxidation Pathway)

Under specific environmental conditions (e.g. hydrosphere rich in Fe (III) hydroxides, and lack of molecular oxygen [O_2] or other terminal electron acceptors), bacteria (e.g. *Geobactermetallireducens* GS-15) use the Fe (III) hydroxides as terminal electron acceptors, and exchange electrons between the redox proteins (e.g. quinol/quinone pools) in the cytoplasmic or inner membrane and extracellular substrates (e.g. metal ions associated with minerals, such as strengite), and microbial cells of the same or different Gram-negative and Gram-positive bacteria, as well as archaea species, via an extracellular electron transfer process (reviewed by Lovley 2012; Hori et al. 2015). The porin cytochrome (PpcA and PpcD in the periplasm) EET pathways mediate electron transfer across the cell envelope (OmaB, OmaC, OmcB and OmcC in the outer membrane) in *G. sulfurreducens* (Shi et al. 2016).

14.2.3.1.3 The Production of Hydrogen Sulphide

In nature, organisms that are able to use inorganic chemicals (e.g. hydrogen sulphide [H_2S], hydrogen gas [H_2], ferrous iron [Fe^{2+}] and ammonia [NH_3]) as electron donors are called chemolithotrophs (*Thiobacillus*spp., *Acidithio bacillus* spp., green sulphur bacterium [e.g. Chlorobiaceae] and green gliding bacteria [Chloroflexaceae]) (Simeonova et al. 2010). On the other hand, those that use organic chemicals (e.g. glucose) are called chemoorganotrophs (Simeonova et al. 2010). The H_2S reacts with ferric phosphate (or ferric orthophosphate, iron (III) phosphate [$FePO_4$]) and is insoluble in H_2O, but soluble in hydrochloric acid (HCl), yields ferrous sulfate (or iron (II) sulfate [FeO_4S]) and releases phosphate (Swaby and Sperber 1958), making it available for plant uptake (Menezes-Blackburn et al. 2013). Then, due to microbial activity (sulphur oxidation), nitrate (NO^{3-}) production and CO_2 formation occur (Sperber 1958), as follows:

$$\mathbf{2FePO_4 + 3H_2S \rightarrow 2FeS + 2HPO_4{}^{2-} + 4H^+ + S^0}$$

Some industrial activities release H_2S, which is toxic at high concentrations and odorous at even very low levels (Gerrity et al. 2016). The use of chemolithotrophic sulfur-oxidizing bacteria (*Gluconobacteroxydans* 621H and *Zunongwangia profunda* SM-A87) in H_2S bioremediation (under aerobic conditions), due to the chemical oxidation of H_2S by 2FePO4 or Fe_3O_4 was recently reported (Chen and Jiao 2016).

14.2.3.2 Biochemical Po Mineralization

Organic P (Po) solubilization plays an essential role in P cycling in agroecosystems (Stamenković et al. 2018). Po constitutes ~4–90% of the total soil P (Khan et al. 2017, Mangaset al. 2016). Pesticides [(e.g.

synthetic compounds Organochlorine Pesticides (OCPs), especially the synthetic pyrethroid] are associated with conventional agriculture inputs (~50,000 tonnes per year in the United States) in use since the Green Revolution after the end of the Second World War, and modern lifestyles (petroleum additives and plasticizers) (Mangas et al. 2016). Synthetic OCPs is highly toxic to animals (e.g. mammals) due to their capacity to block acetylcholinesterase (AChE, choline hydrolase, EC 3.1.1.7) activity, acetylcholine receptors in the synapses of the nervous systems (autonomic and central) and neuromuscular junctions, inducing death in animals and insects due to respiratory failure (Galloway and Handy 2003; Belzunces et al. 2012). The OCP pool composition is complex and can be divided into P esters (C–O–P bonds) and phosphonates (C–P bonds) (Singh and Walker 2006). The P uptake of plants needs the mineralization of the Po-soil component to orthophosphate. The microbial degradation of natural and synthetic organo phosphonates is associated with enzymes, collectively named phosphatases, such as phosphono pyruvate hydrolase, phosphonoacetate hydrolase, phosphono acetaldehyde hydrolase and carbon–phosphorus (C–P) lyase (Nannipieri et al. 2011). The most common, isolated from soil organophosphorus-degrading bacteria, and their associated phosphatases include *Flavobacterium* sp. ATCC 27551 (OP hydrolase, encoding genes*opd A*) (Horne et al. 2002), *Pseudomonas diminuta* (OP hydrolase, encoding genes *opd*) (Serdar and Gibson 1985), *Alteromonasundina* (OPDA, OP-degrading enzyme, encoding genes *opa*) (Cheng et al. 1993) and *Pseudomona s*sp. WBC-3 (MPH, methyl parathion hydrolase, encoding genes *mpd*) (Singh and Walker 2006). Our knowledge and understanding of natural OP or xenobiotic-degrading enzymes (e.g. *Flavobacterium* sp. ATCC27551 [organophosphate Parathion (insecticide)], *Sphingobium chlorophenolicum* ATCC39723 [Pentachlorophenol (fungicide)], *Geobacilluscaldoxylosilyticus* T20 and *Flavobacterium* sp. [Glyphosate (herbicide)], *Sphingomonas paucimobilis* UT26 [Lindane (insecticide)]), has advanced in recent years (Siddavattam et al. 2019), and this information is being exploited in several environmental, agro-industrial and medical applications (Sarwar et al. 2019).

14.3 Potassium (K) Solubilizers (K-S)

14.3.1 K in Soils

The origin of soil K (~90–98%) (Zörb et al. 2014) is related to the desintegration and decomposition of metamorphic, sedimentary and igneous rocks (Deer et al. 1996; U.S. Geological Survey 2019), forming minerals that include sheet silicates related to the mica group (non-expanding 2:1 silicate minerals) (e.g. muscovite–$K_2A_{14}[Si_6AlO_{20}](OH,F)_4$, paragonite–$NA_2Al_4[Si_6Al_2O_{20}](OH)_4$, glauconite–$(K,Ca,Na)_{\sim1.6}(Fe^{3+}, AlMg,Fe^{2+})_{4.0}Si_{7\cdot3}A_{10\cdot7}O_{20}(OH)_4$, phlogopite-biotite–$K_2(MgFe^{2+})_{6-4}(Fe^{3+}), AlTi_{0-2}[Si_{6-5}Al_{2-3}O_{20}](OH,F)_4$, biotite–$[K(Mg, Fe)_3AlSi_3O_{10}(OH_2)]$, lepidolite–$K_2(LiAl)_{5-6}[Si_{6-7}Al_{2-1}O_{20}](OH,F)_4$, zinnwaldite–$K_2Fe^{2+}{}_{3-1}Li_{1-3}(Al,Fe^{3+})_2[Si_{5-7}Al_{3-1}O_{20}](F,OH)_4$, margarite–$Ca_2Al_4[Si_4O_{20}](OH)_4$ and clintomite–$Ca_2(MgFe)_4(Al_2[Si_2Al_6O_{20}](OH)_4)$); clay minerals of the kaolinite group (non-expanding 1:1 silicate minerals) $[Al_4[SiO_{10}](OH)_8]$, illite $[K_{1.5-1.0}A_{14}[Si_{6.5-7.0}Al_{1.5-1.0}O_{20}](OH_4)]$, the smectite group (expanding 2:1 silicate minerals) [dioctahedral $(1/2Ca, Na)_{0.7}(AlMg, Fe)_4(Si,Al)_8O_{20}](OH)_4.nH_2O]$ and vermiculite (expanding 2:1 silicate minerals) $[(Mg,Ca)_{0.6-0.9}(Mg,Fe^{3+}), Al_{6.0}[(Si,Al)_8 0_{20}](OH)_4.nH_2O]$; and framework silicates of the feldspar group [alkali feldspar $[(K,Na)[AlSi_3O_8]]$, plagioclase $[Na[AlSi_3O_8]–Ca[Al_2SiO_8]]$, barium feldspar (e.g. celsian–$Ba[Al_2SiO_8]$ and hyalophane–$(K,Na,Ba)[(Al,Si)_4O_8]$) (Deer et al. 1996; USGS 2019). The formation of K-feldspar (in high-grade metamorphic rocks) is due to the instability of micas, and the following reactions (14.1, 14.2 and 14.3, respectively) may operate:

$$K(Mg,Fe)_3 AlSi_3O_{10}(OH)_2 + 3SiO_2 \rightarrow KAlSi_3O_8 + 3(Mg,Fe)SiO_3 + H_2O \tag{14.1}$$

biotite orthoclase orthopyroxene

$$2K(Mg, Fe)_{1.5} Al_2Si_3O_{10}(OH)_2 + 3SiO_2 \rightarrow 2KAlSi_3O_8 + (Mg, Fe)_3 Al_2SiO_{12} + H_2O \tag{14.2}$$

siderophyllite orthoclase pyrope-almadine

$$\underset{\text{Muscovite}}{KAl_3Si_3O_{10}(OH)_2} + SiO_2 \rightarrow \underset{\text{orthoclase}}{KAlSi_3O_8} + \underset{\text{silimanite}}{Al_2SiO_5} + H_2O \tag{14.3}$$

The soil K exists in four fractions (forms and equilibria) based on their availability to plants: K-mineral (or structural~5,000 to 25,000 mg of K per gram of soil); non-exchangeable K (fixed or more or less easily reversible, ~50–750 mg of K per gram of soil); exchangeable K (held by electrostaction attraction on the clay surface or organic matter of soil, rapid "release" in case of need, ~40–600 mg of K per gram of soil); and K dissolved in a soil solution (~1–10 mg of K per gram of soil) (Fageriaet al. 2011). In general, the relative impact of these K groupings is associated with the soil mineralogy composition (Inda et al. 2010; Šolc et al. 2011). Like others cations, the charged soil colloids (e.g. illite, vermiculite and chlorite) can adsorb the exchangeable K^+ ion by electrostatic attraction (Balan et al. 2014). This phenomenon affects the availability of K^+ ion in the soil solution and consequently its availability to plants, and is recognized as K fixation (Tazaki 2005). Soil orders such as Spodosols, Entisols, Mollissols, Aridisols and Alfisols/Ultisols (poorly drained) have a deficiency related to K^+ ion availability, which is associated with K fixation by soil 2:1 clay minerals (e.g. weathered mica and smectite) (Cygan et al. 2004). In addition to 2:1 clay minerals, other edaphic factors may affect soil K^+ ion fixation, such as soil reaction (pH) (Cygan et al. 2004), the concentration of K added to the soil (Lee et al. 2013), drying and rewetting (Haria et al. 2013) and freezing and thawing (Tazaki 2013). In strongly acid soils (pH <5.5), the K^+ ion fixation by clay minerals can be reduced by the action of H^+ and hydroxy aluminium ions (Tazaki 2005). On the other hand, high calcium (Ca^{2+}) and magnesium (Mg^{2+}) levels in the soil solution may reduce K^+ absorption by the plant because these cations (Ca^{2+} and Mg^{2+}) have a tendency to compete against one another for uptake by roots in the soil solution. This relationship can be calculated using the following ratio:

$$\text{Activity ratio}\left(AR_e^k\right) = \frac{\left[K^+\right]}{\sqrt{\left[Ca^{2+}\right]+\left[Mg^{2+}\right]}}$$

This ratio is a measure of the "intensity" of the K-labile in the soil solution (Tisdale et al. 1985). $\left(AR_e^k\right)$ is a single measurement only momentarily available (Beckett 1964), and not the current status of the K-labile pool (Bertsch and Thomas 1985). Barnhisel and Bertsch (1989) described the relations between exchangeable and soluble K, called K quantity–intensity (Q/I) relations, shown in Figure 14.1. The static equilibria for soil K may be represented as follows (adapted from Tisdale et al. 1985):

$$\textbf{Nonexchangeable K} \rightleftharpoons \textbf{Exchangeable K} \rightleftharpoons \textbf{Soil solution K}$$

K^+ deficiency also occurs in calcareous soils (e.g. calcixereptspH >7.5) because K^+ fixation as well as cations ratios (as above) contributes to these negative effects on calcium carbonate ($CaCO_3$)-rich soil (Marschner and Marschner 2012). Usually sandy soils (e.g. soil orders Alfisols, Entisols, Inceptisols, Spodosols and Ultisols) have low cation exchange capacity and are poorly buffered with respect to K^+ ion (Darunsontaya et al. 2012). However, because of the continuous K^+ uptake and leaching by plants, static equilibria never occurs (Britzke et al. 2012). For a long time, interest has been shown in the correlation between exchangeable K (Q for quantity) and K ion activity in soil solution (I for intensity) (Moterle et al. 2019). The quantity–intensity (Q/I) concept (introduced by Becket in 1964) can be used to describe the availability of K to plants (Sparks and Liebhardt 1981). A typical K quantity/intensity (Q/I) concept curve is shown in Figure 14.1.

For potassium, the intensity factor ARk is plotted from measured concentrations of Ca, magnesium (Mg), potassium (K) and sodium (Na) adjusted to specific activities by application of the Debye–Hückel (DH) theory of electrolytes (reviewed by Liuaand Li 2019). According to the Q/I concept, higher values of labile (ΔK^o)indicate that more K is released into the soil solution. Moreover, a low value of the potential buffering capacity (PBCK) signifies a necessity for K fertilization (Rosolemet al. 1993). In general, the Q/I concept is a good predictor for the correlation between exchangeable K^+ and the K^+ available in

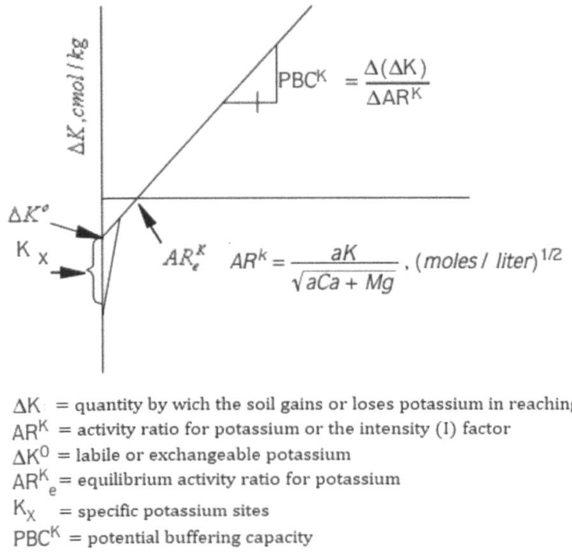

FIGURE 14.1 A typical K quantity/intensity (Q/I) concept curve graphic plot. (Redrawn from Sparks and Liebhardt, 1981.)

the soil solution (Sparks and Liebhardt 1981; Darunsontaya et al. 2012). Usually, the K^+ flux in soils is related to its diffusibility in the medium and it is correlated with the soil moisture (mass flux) and the K^+ concentration in the soil solution, as shown in the following formula (Barber 1962):

$$J = D_1\left(\frac{dc_1}{dx}\right) + D_2\left(\frac{dc_2}{dx}\right) + c_3 v + a \text{ replacement factor}$$

where:

J = total quantity of ions reaching the root per unit time per area of root surface
D_1 = diffusion coefficient of the ions in the soil
D_2 = diffusion coefficient for the movement on the particle surfaces
c_1 = concentration of the ions in the soil solution not associated with the particle surfaces
c_2 = concentration of the ions moving on the solid surfaces
c_3 = concentration of the ions flowing in the soil solution with velocity v towards the root surface

According to this model, when the amount of K^+that reaches the roots is equal to or exceeds the absorbed quantity (factor Q), the mass flow (which depends on the amount of water used by plant) is the main mechanism of K^+ transport to roots. However, when the plant demand exceeds the amount of K^+ carried by the mass flow, then diffusion (movement of K in response to a concentration gradient, high to lower concentration) mechanisms become most predominant (Schimitz and Pratt 1953; Wilson 1999). K transportation in the soil solution is achieved by mass flow and diffusion and both are controlled by K intensity (Marschner and Marschner 2012). Chemical fertilizers containing potassium that are usually applied to the soil include potassium chloride (KC~50–52% of K_2O), potassium sulphate (K_2SO50–53% of K_2O), potassium magnesium sulphate (K_2SO_4, $MgSO_4$~22% of K_2O), potassium nitrate (KNO_{3z}~44% of K_2O) and potassium phosphates (KPO_3, $K_4P_2O_7$, KH_2PO_4, K_2HPO_4~40% of K_2O) (Hurtado et al. 2008).

14.3.2 K in Plants

Potassium ions (K^+) are taken up by soil solution via a proteic system called transporters (K^+high-affinity components) and channels (K^+low-affinity components, voltage-dependent K^+ channel [*TPK1*, *TPK2*, *TPK3* and *TPK5*]) found in the plasma membrane of the root pericycle and stelar parenchyma cells (Ahn

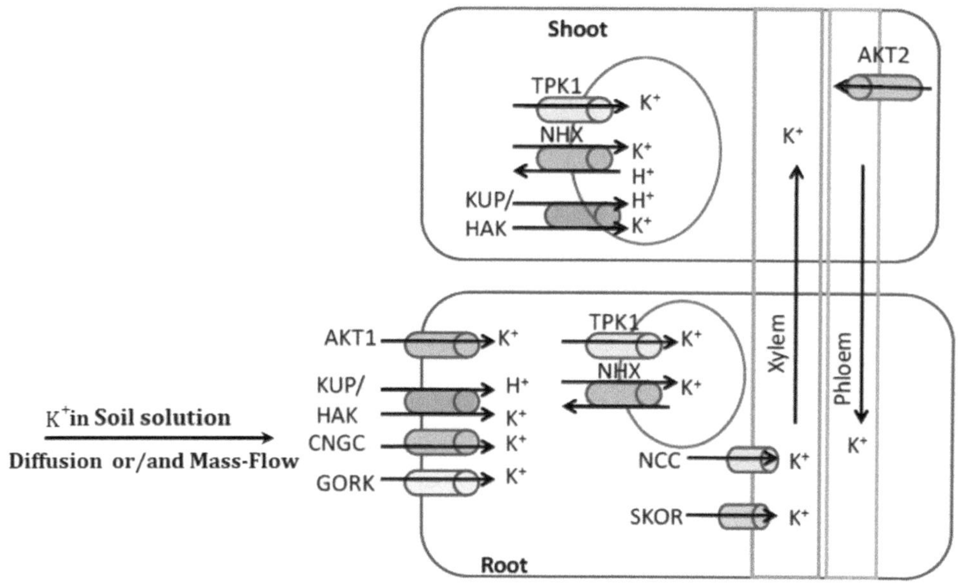

FIGURE 14.2 GeneralK$^+$uptake and transport in plants. AKT: *Arabidopsis* (Shaker-type) K$^+$ channel: CNGC: cyclic nucleotide-gated channel; GORK: guard cell outward rectifying K$^+$channel; HAK/KUP: high affinity K$^+$ transporters; KAT: *Arabidopsis* (Shaker type) K$^+$ channel; NCC:non-selective cation channels; NHX: Na$^+$ proton exchanger; SKOR: stellar outward rectifying K$^+$ channel; TPK: tonoplast two-pore K$^+$ channel. (Modified from Ahmad and Maathuis 2014.)

et al. 2004). In roots, the K$^+$ uptake from soil solution is mediated by two proteins, AKT1 and HAK5 (Amtmann and Beilby 2010). The K$^+$ long-distance transport (from the root cortex to the xylem) is mediated by channels (also called stellar outward rectifying channels) (Figure 14.2).

The stellar outward rectifying K$^+$(SKOR) channel is related to the K$^+$ content in the xylem sap (Nguyen et al. 2017). Shutdown of the stomata is mediated by the guard cell outward rectifying K$^+$(GORK) channel (Hosy et al. 2003). In *Arabidopsis*, cell turgor and stomatal function are regulated by ion exchangers NHX1 and NHX2, which control the active K$^+$ uptake in vacuoles (Barragán et al. 2012). The cyclic nucleotide-gated ion channel (CNGC) mediates the uptake of cations (e.g. Na$^+$, K$^+$ and Ca^{2+}), and can provide biotic and abiotic stress tolerance (Jha et al. 2016). The passive fluxes of K$^+$ (and other cations) via plant membranes are catalyzed by non-selective cation channels (NSCCs) and the perception of external stimuli (e.g. signal transducers for reactive oxygen species [ROS]) (Sewelam et al. 2016). *OsKAT1* (KAT members of rice) can confer salt stress tolerance in yeast (*Saccharomyces cerevisiae*) and rice (*Oryza sativa*) suspension cells (Gierth et al. 2005).

14.3.3 Mechanisms of K-Compounds Solubilization

Solubilizers, microorganisms and clay minerals (e.g. feldspar group minerals, abundant constituents of igneous rocks) have many similarities such as micrometric dimensions, a high adsorption–specific surface and an electric-charged surface (negative or positive) (Basak and Biswas 2010; Bhattacharya et al. 2016).

14.3.3.1 Feldspar Bioweathering

Various models (considering their structure, chemistry, optical and physical properties and paragenesis) are proposed to describe rock-forming clay minerals within soils (e.g. feldspar bioweathering) (Ahmad et al. 2016). More than any other sheet silicate mineral, studies have focused their attention on the mica (especially the kaolinite, illite, smectite and vermiculite groups) and the feldspar (especially alkali feldspar and silica minerals) groups' mechanism of dissolution and their rates (Crundwell 2014). During the dissolution of the K-feldspar crystal lattice, a variable quantity of aluminium and silicais leached from the crystal to release the residual clay phase (see Reactions 14.4 and 14.5) (Tan 2011):

$$2KAl\,Si_3O_8\left(K-\text{feldspar}\right)+2H^+ +9H_2O\rightarrow Al_2Si_2O_5\left(OH\right)_4\left(\text{kaolinite}\right)+4Si\left(OH\right)_4+2K^+ \quad (14.4)$$

$$2KAlSi_3O_8+2H^+ +14H_2O\rightarrow 2Al(OH)_3+6Si\left(OH\right)_4+2K^+ \quad (14.5)$$

Under anoxic (oxygen-free) environmental conditions, chemolithotrophs and anoxygenic phototrophic bacteria use ferrous iron (Fe^{2+}) (or other inorganic compounds such as SiO_2, TiO_2, MnO, MgO, CaO and K_2O) as an acceptor terminal of electrons, and C-organic as the energy source (Ceci et al. 2019). This process, called chemolithotrophy, can lead to the solubilization and precipitation of rock-forming clay mineral groups (e.g. kaolinite, illites, smectites and vermiculites) (Badr et al. 2006), with commercial biohydrometallurgy ("biomining") applications for the recovery of metals (e.g. via biosorption or bioaccumulation) from primary and secondary resources, soil bioremediation (e.g. immobilization or detoxification of metallic and radionuclide pollutants) and water treatment (clean-up of organic and inorganic pollutants) (Igiri et al. 2018).

14.3.3.2 Organic Acids in Feldspar Bioweathering

Using biomineralization processes (the formation of inorganic minerals in organisms), kaolinite interacts with microorganisms and can nucleate and precipitate on cell walls (Kirtzel et al. 2018). Studies indicate that organic acids such as oxalic and citric at concentrations that are likely to occur in soils may increase the dissolution rates of feldspars in soils (Burford et al. 2003). Calcium-rich plagioclase feldspars (a framework silicate from a series of tectosilicate) are more susceptible to attack by organic acids (e.g. oxalic acid—$(COOH)_2$) induced by lichens than K-feldspars due to their higher structural Al content (Song and Huang 1988; Ganor and Lasaga 1994). Lichen is considered an association between a fungus (mainly an ascomycete or, less frequently, a basidiomycete or deuteromycete, called mycobiont lichen) and green algae or cyanobacteria (also called photobionts lichen) (Kappen 1993). The bioweathering of basalt ([labradorite-bytownite] a Ca-rich plagioclase feldspar) and ferromagnesian minerals by *Pertusariacorallina* (a crustose lichen) yields a thin ochreous crust of poorly ordered ferruginous (ferrihydrite) and alumino-silicate materials, because the mycobiont excretion of oxalic acid (Willmann and Fakoussa 1997) corroborates with other experimental observations with *Fuscideacyathoides*, *Ochrolechiatartarea*, *Ophioparmaventosa* and *P. corallina* (Moses and Smith 1993; McCarroll and Viles 1995). It has been well documented in the literature that fungi (plant symbionts or plant pathogens) can acidify their microenvironment (Kirtzel et al. 2018) by excreting metabolites with metal-chelating properties, such as carboxylic acids, amino acids, siderophores and phenolic compounds (Scheerer et al. 2009). Ectomycorrhizal fungi in boreal forest soils induce the dissolution of silicate and phosphate minerals (e.g. apatite, quartz, potassium-feldspar) (Burford et al. 2003). In European coniferous forests, the release of oxalic, citric, succinic, formic and malic acids by ectomycorrhizal hyphae contributes to the bioweathering of hornblendes, feldspars and granitic bedrock (Fomina et al. 2005). Mycogenic precipitated calcium oxalate crystals on the hyphae of *Piloderma* (ectomycorrhizal fungi) were detected as a result of their nutritional requirements for K and/or Mg during the dissolution of biotite, microcline and chlorite (Gadd 2007). The oxalic acid biosynthesis (in a tricarboxylic acid [TCA] cycle) of *Aspergillus niger* occurs through the conversion of oxaloacetate to oxalate and acetate by cytoplasmic Mn^{2+}-dependent oxaloacetate acetylhydrolase (OAH) (Belén et al. 2010). Oxalate biosynthesis via OAH can also occur in wood-rot fungi (*Fomitopsispalustres*) via the glyoxylate (GLOX) cycle by the action of glyoxysomal glyoxylate dehydrogenase on glyoxylate (Lammers et al. 2001). Calcium oxalate precipitation can be illustrated using the following equation (Gadd et al. 2014):

$$Ca^{2+}+G_2O_4{}^{2-}+xH_2O\rightarrow Ca\left(G_2O4\right)\bullet xH_2O_{(s)}\downarrow$$

Oxalic acid can induce the acidolysis (from the release of H^+), ligand-promoted dissolution (arising from metal complexation by oxalate) and reductive mobilization of metals from minerals. The dissolution of quartz (stable solid phase of silica) only occurs at pH <2 or pH >8 where the H^+ extrusion to induce the deprotonation of surface Si–OH–H bonds occurs (Gadd and Raven 2011). Freshwater has a higher concentration of dissolved silica (~mg L^{-1} at 2°C)that illustrates feldspar hydrolysis, as follows:

$$SiO_2 + 2H_2O \rightarrow Si(OH)_4$$

Despite the recalcitrance of quartz, studies have shown the desilication of tropical soils relative to aluminium or iron, due to the increase in the soluble silica chelates dissolved by organic compounds (Hallbauer and Jahns 1977). Extracellular polysaccharides (EPS) were detected in weathering plagioclase feldspar (bytownite), in a batch reactor by the action of bacteria (Barker et al. 1998). Similar results were shown by Welch and Ullman (1993) in plagioclase feldspar in solutions of acid polysaccharide (e.g. alginates) with the inhibition of Si and Al production at the neutral pH phase. In this case, the rate of mineral solubilization increased due to H^+activity related to the carboxyl acid functional groups (Seifert et al. 2011).

14.3.3.3 Organic Ligands in Feldspar Bioweathering

It is well documented in the literature that organic ligands (e.g. metabolic byproducts, extracellular enzymes, chelates and both simple and complex organic acids) produced by microorganisms can increase the feldspar dissolution rates by decreasing the pH by forming framework destabilizing surface complexes or by complexing metals in solution (Burford et al. 2003). In particular, the knowledge that the bioweathering of feldspar minerals releases limiting inorganic nutrients is of tremendous importance for subsurface microbial ecology (Rhee et al. 2014), and may change concepts regarding the mineral weathering sequence ("Goldich's Weathering Sequence") in the rock record (Wieler et al. 2019). Nowadays, studies of subsurface geomicrobiology are considered a new frontier in microbiology (Ceci et al. 2019).

14.4 Silicon (Si) Solubilizers (Si-S)

14.4.1 Si in Soils

Si is commonly deposited (~27.7%) in the earth's lithosphere in the form of Si-bearing minerals in crystalline, poorly crystalline such as nesosilicates or orthosilicates (e.g. olivine $(Mg,Fe)_2[SiO_4]$), inosilicates or chain silicates (e.g. augite $(Ca,Mg,Fe^{2+},Al)_2(Si,Al)_2O_6$), diopside ($CaMgSi_2O_6$), phyllosilicates or sheet silicates (e.g. muscovite $[K_2Al_4[Si_6Al_2O_{12}](OH,F)_4]$) or sorosilicates or pyrosilicates (e.g. epidote group $(Ca_2(Fe^{3+},Al)Al_2O.OH.Si_2O_7.SiO_4)$) (Deer et al. 1996; U.S. Geological Survey 2019). In terms of geological timescales, silicate weathering is the most important net carbon sink for atmospheric CO_2, due to removing CO_2 from the atmosphere and to increasing dissolved bicarbonate (e.g. $[HCO_3^-]$), regulating the amount of CO_2 in the atmosphere and global climate change (Frondini et al. 2019). Si in a solution at pH 2–9 exists as monosilicic acid (H_4SiO_4); however, at pH 9 and above, it occurs as silicate anions ($H_3SiO_4^-$) (Hall 1972). H_4SiO_4 polymerizes in acid aqueous solutions supersaturated with amorphous silica (e.g. the biogenic phytoliths) forming oligomers of polysilicic acids (Iler 1979), or amorphous biosilica, such as silicateins, found in spicules of the siliceous skeletal of sponges, and diatoms fossilized skeletons (also referred to as diatomaceous earth, diatomite)with many biotechnological and medical applications (Harriss 1972; Barker et al. 1997; Tobler et al. 2008).

14.4.2 Si in Plants

Si is considered a quasi-essential nutrient to the life cycle of plants (Epstein and Bloom 2005; Coskun et al. 2016), but Si is essential for *silicophile* species (e.g. Poaceae, Equisetaceae and Cyperaceae) which show high Si accumulation (e.g. *Equisetum arvense* with ~4% Si) (Epstein 1999). In highly weathered tropical soils, the anion form of monosilicic acid adsorption is more dominant (Schultze-Lam et al. 1996). The Si concentration in soil solution is correlated with the pH solution (below pH 9.0) (Epstein 2009). Some anthropogenic activities, such as mining, conventional agricultural practices and industrialization, can enhance the heavy metal, metals or metalloids concentration in soil, water and air (Detmann et al. 2013). Si is taken up by plants in an uncharged form $[Si(OH)_4]$ (monosilicic acid) or $Si(OH)_3^-$, by a low affinity transporter (*Lsi1*) found on the lateral roots (Epstein 2009), or rejective mechanisms (Chen et al. 2011). Si present in plant tissues is often found as hydrogen-bound Si–organic complexes (Epstein 1999). Si is related to cell wall rigidity and cell wall elasticity (Epstein 2009). After the removal of H_2O, Si accumulates in the leaf in $[Si(OH)_4]$ form and becomes condensed in a polymerized silica gel $(SiO_2.nH_2O)$, the phytoliths or silica bodies (Fleck et al. 2015). After plant death, the phytoliths take part in the global silicon cycle (Haynes 2014) or can be incorporated into the soil sediments (DasSarma and Schwieterman 2018; Katz 2018), and can be used to reconstruct the past earth's vegetation, under global climatic conditions, with specific geochemical and morphological signatures (Cornelis and Delvaux 2016). Recent reviews have shown the beneficial effects of Si plant fertilization under environmentally adverse growing conditions (e.g. heavy metal toxicity, salinity and osmotic stress) for Si hyperaccumulating plant species (e.g. rice [*O. sativa* L.], soybean [*Glycine max* L. Merr.], sugarcane [*Saccharum officinarum* L.], maize [*Zea mays* L.] and wheat [*Triticum aestivum* L.]) (Liang et al. 2007, 2015; Yin et al. 2016).

14.4.3 Mechanisms of Si-Compounds Solubilization

Si solubilization is related to the cleavage of Si–O–Si bonds (siloxane) or Al–O framework bonds or the removal of cations from the natural crystal lattice of silicate, entailing a mineral neoformation (Schultze-Lam et al. 1996), due to the collapse of the silicate lattice structure, or the physical distortion of the primary mineral lattice structure (Bonneville et al. 2009).

14.4.3.1 Si Solubilization by Microbial Ligands

Some soil *Erwinia* spp., *Erwinia herbicola*, *Pseudomonas* strains (in vitro conditions) can produce 2-ketogluconic acid from glucose by the dissolution of synthetic silicates of Ca, Zn and Mg, and silicated minerals, such as wollastonite $(CaSiO_3)$, apophyllite $(KCa_4Si_8O_{20}(F,OH)\bullet 8H_2O)$ and olivine $(Mg,Fe)_2SiO_4$ (Webley et al. 1960; Duff et al. 1963). The dissolution of silicates, such as wollastonite $(CaSiO_3)$, can induce the complexation of the cationic components of the silicates by 2-ketogluconate, according to the following example (Jongmans et al. 1997):

$$\mathbf{CaSiO_3 \Leftrightarrow Ca^{2+} + SiO_3{}^{2-} \text{ and } Ca^{2+} + 2 - ketogluconate \rightarrow Ca\left(2 - ketogluconate\right)}$$

Gluconic acid produced from glucose by bacteria (e.g. *P. fluorescens*) has been shown to solubilize bytownite $(Ca_{0.7-0.9}, Na_{0.3-0.1})$ $(Al(Al, Si)Si_2O_8)$, albite $(Na(AlSi_3O_8))$ and quartz (SiO_4, Si–O tetrahedra form, with oxygen shared between two tetrahedra, giving two SiO_2) (Eckhardt 1980; Vandevivere et al. 1994), suggesting that the mechanism of action (MOA) of gluconate is related to chelation (Vandevivere et al. 1994), due to an electron donor–acceptor (hydrolysis and protonation) system (Bennett et al. 2001), as seen in the following reaction:

$$\mathbf{4\ NaAlSi_3O_8(s) + 4H^+ + 18\ H_2O\left(l\right) = Si_4A_{14}O_{10}(OH)_8 + 8\ Si(OH)^0{}_4 + 4\ Na^+}$$

$$\mathbf{\left(albite\right)\left(kaolinite\right)}$$

14.4.3.2 Si Solubilization by Microbial Acids Compounds

Studies of the solubilization of silicates by microbial acids were first conducted by Waksman and Starkey (1931) and Etesami et al. (2017). These showed the MOA of CO_2 (a product of respiration or fermentation) on orthoclase ($2KAlSi_3O_8$, a feldspar [alkali] group, tectosilicate mineral), according to the following chemical reaction:

$$\mathbf{2KAlSi_3O_8 + 2H_2O + CO_2 \rightarrow H_4Al_2Si_2O_9 + K_2CO_3 + 4SiO_2}$$

The MOA of CO_2 on orthoclase was reviewed by Etesami et al. (2017). Karavaiko et al. (1979) showed the MOA of spodumene ($LiAlSi_2O_6$), a lithium aluminium inosilicate (pyroxene mineral), as illustrated by the following chemical reaction:

$$\mathbf{4Li\ AlSi_2O_6 + 6H_2O + 2H_2SO_4 \rightarrow 2Li_2SO_4 + Al_4\left(Si_4O_{10}\right)\left(OH\right)_8 + 4H_2SiO_3},$$

In this reaction, the final product is kaolinite ($Si_4Al_4O_{10}(OH)_8$), a hydrated alumino-silicate mineral (Barker et al. 1997). Non-spore-forming heterotrophic bacteria (e.g. *Arthrobacter pascens*, *A. globiformis*, *A. simplex*, *Nocardia globerula*, *P. fluorescens*, *P. putida* and *P. testeronii*) and fungi (e.g. *Trichoderma lignorum*, *Cephalosporium atrium* and *Penicillium decumbens*) were found to predominate in weathered spodumene (Karavaiko et al. 1979; Avakyan et al. 1986), to smectite in a lateritic environment in southwestern Australia. The Si mobilization from halloysite ($Al_2Si_2O_5(OH)_4{\cdot}2H_2O$), serpentine (Mg, Fe, Ni, Mn, $Zn)_{2-3}$(Si, Al, $Fe)_2O_5(OH)_4$ and muscovite ($KAl_2(AlSi_3O_{10})(OH)_2$) by citric and oxalic acids secreted by filamentous mycelium (e.g. *A. niger and P. chrysosporium*) and yeasts (e.g. *S. cerevisiae*), at the expense of glucose, was demonstrated (Balland et al. 2010; Gadd et al. 2014). Oxalic acid results from the hydrolysis of oxaloacetate to oxalate and acetate, whereas on citric acid, oxalate production occurs due to glyoxylate oxidation (Gadd 1999, 2007).

14.4.3.3 Si Solubilization by Microbial Alkali Compounds

Understanding the role of microorganisms (e.g. *Alteromonadales*, *Bdellovibrionales*, *Caulobacterales*, *Desulphovibrionales*, *Ferrovales*, *Hydrogenophilales*, *Nitrosomonadales*, *Oceanospirillales*, *Rhodobacterales* and *Rhodocyclales*) in the minerogenesis of quartz-sandstones and quartzites or quartz and silicates (Basak et al. 2017) will help elucidate a paradox, the silica mobilization from quartz in low-temperature and neutral-acidic pH conditions (Rimstidt 1997). This is attributable to the lability of the Al–O and Si–O bonds under these conditions because both types of bonds are susceptible to microbial nucleophilic attack (Wray and Sauro 2017). The general reaction for the breaking of a >Si–OSi< surface structure is represented by the following equation:

$$> \mathbf{Si - O + H_2O} = (> \mathbf{Si - O})'' \cdot \left(\mathbf{OH_2}\right)^{\ddagger},$$

Where > or < represents a surface-bound species;‡ indicates an activated complex that reacts with H_2O and decays to the stable product, H_4SiO_4; and the=sign indicates the equilibrium nature of this rate-limiting reaction, which is in accordance with the transition state theory (TST) that assumes quasi-equilibrium between reactants and their activated transition state complexes (reviewed and discussed by Dove and Nix [1997]). *Sarcinaureae* (bacteria genus *Sporosarcina*) release SiO_2 (in vitro in media peptone and urea) from nepheline, plagioclase and quartz (Aristovskaya and Kutuzova 1968). The interaction of NH_3 with H_2O causes alkalinolysis (pH ≥10), which can be achieved by the transfer of urea by hydrolysis (catalyzed by the enzyme urease) and urea cleavage, by the following equation:

(*Sarcina ureae*)

$$\left(NH_2\right)_2 CO + 2H_2O \rightarrow \left(NH_4\right)_2 CO_3$$

urea water urease ammonium carbonate

(or carbamide)

$$\left(NH_4\right)_2 CO_3 + 2H^+ \rightarrow 2\left(NH_4\right)^+ + CO - + H_2O$$

Ammonium carbonate ion ammonium carbondioxide water

$$NH_4{}^+ + OH^- \rightarrow NH - + H_2O$$

Ammonium

In this instance, NH_3 resulting from the hydrolysis of urea was the source of the alkalinolysis (Aristovskaya and Kutuzova 1968). Orthoquartzite caves are rich in quartz-rich lithologies (orthoquartzites, meta-quartzites and granites) (Basak et al. 2017). In 2018, Wray and Sauro showed that the mobilization of silica (SiO_2) does not occur only in hydrothermal conditions. This discovery created new perspectives on the relationship between silica and microbial-mediated silica solubilization in quartz-rich lithologies deposits, and provided new insights for understanding the SiO_2 cycle (e.g. quartz weathering and silica mobility) in the earth's crust and the subsurface (e.g. in orthoquartzite caves) (Sauro et al. 2018).

14.4.3.4 Si Solubilization by Microbial Extracellular Polysaccharide

EPS has an important role in silica release (e.g. adhesion, cell protection, ligand binding and as organic carbon sources), for metabolism in diatoms (Ahmed and Holmstrom 2015), in extreme environmental conditions (e.g. Antarctic and Arctic sea ice), especially from quartz (Sauro et al. 2018). EPS can react with siloxanes to form organic siloxanes (Banfield et al. 1999). EPS has an untapped industrial market, suitable for applications in the bio-nanotechnology, food, pharmaceutical, cosmetics, petroleum, civil construction and environmental sectors (Hyde et al. 2019).

14.5 Sulphur (S) Solubilizers (S-S)

14.5.1 S in Soils

Sulphur (S) is considered the fourteenth most abundant element (a non-metallic chemical element) in the earth's crust (~0.06 and 0.10% of S content), and S-inorganic forms occur in redox states from 2^- to 6^+(Deer et al. 1996). The surface sulphur reservoir is in contact with the mantle, and the volcanic outgassing, hydrothermal input and ocean crust weathering processes carry sulphur to the earth's surface (from the mantle) (Hofer 2018). Sulphates occur naturally in numerous minerals, including barite ($BaSO_4$), celestine ($SrSO_4$), epsomite ($MgSO_4{\cdot}7H_2O$), gypsum ($CaSO_4{\cdot}2H_2O$), mirabilite ($Na_2SO_4{\cdot}10H_2O$), pyrite (FeS_2), sphalerite (ZnS), chalcopyrite ($CuFeS_2$), cobaltite (CoAsS), anhydrite ($CaSO_4$) (Deer et al. 1996) and other sulphides such as pyrrhotite ($Fe_{11}S_{12}$), galena (PbS), arsenopyrite ($FeS_2{\cdot}FeAs_2$) and pentlandite (Fe,Ni)9 S_8 (Tisdale et al. 1985). In elemental form (So), sulphur is found in salt dome deposits, volcanic deposits and deposits associated with calcite, gypsum and anhydrite (Bowles et al. 2014; Jørgensen et al. 2019). S deficiency in soils can be induced by emissions of S dioxide (SO_2) from fossil fuel burning to the atmosphere (Maynard et al. 1985). In specific global areas, such as the Brazilian savanna (also called Cerrado biome), S deficiency in soils is a natural process (Ribeiro et al. 1991). S-organic (>95%) in the soil is bound to organic molecules (e.g. organic-S directly bound to C [C–S], primarily of sulphate esters [C–O–S]) and other organic fractions such as sulphonates, sulphoxides and heterocyclic S (Fike et al. 2015). S-organic can be separated (based on the oxidation state) into two S reduced fractions (poly, di and monosulphides, thiols and thiophenes), two S reduced intermediate oxidation state fractions and one S highly oxidized state organic fraction (Wasmund et al. 2017). The S-organic soil fractions can be converted to sulphate by biological mineralization (Visser et al. 1997; Tang et al. 2009; Luther et al. 2011). Sulphur mineralization is the conversion of S-organic to inorganic (Klotz et al. 2011) and depends on the mineral content of the organic matter or the C/N/S ratio (Tisdale et al. 1985: ideal C/S ratio = 200:1); temperature (from 20 to 5°C; ~Q10 = 1.9); moisture (<15 and<40%) or at 60% of field moisture-holding capacity; wetting and drying (to induce gradual release by microbial decomposition); soil pH (at pH >7.5, to increase mineralization); the effect of rhizosphere (>plant presence;<plant absence); time and

cultivation (at first cultivation < mineralization, when at equilibrium level) (Marschner and Marschner 2012). The sulphates hydrolyze organic sulphate esters according to the following reaction (Schinner et al. 1996):

$$ROSO_3^- + H_2O + \text{sulphatase} \rightleftharpoons ROH + HSO_4^-$$

Chemical fertilizers (Mengel and Kirby 1987; Marschner and Marschner 2012) are low in sulphur release, and the ongoing decrease in atmospheric sulphur deposition appears to be one of the primary causes of sulphur deficiency in crop plants (Lucheta and Lambais 2012). Sulphur volatilization compounds, such as dimethyl sulphide, are produced by soil microorganisms from the decomposition of animal manure (Malavolta et al. 1997). In high-pH calcareous soils (e.g. in rice yields in Arkansas), applications of sulphuric acid increase the availability of plants (Malavolta et al. 1997). When added to soil, aluminium sulphate $Al_2(SO_4)_3$ in soil solution Al replaces any exchangeable hydrogen ion (H^+) on the surface of the soil colloid and drops the pH:

$$Al_2\left(SO_4\right)_3 + \left[\text{clay}\right]\frac{4H}{Ca} \rightarrow \frac{Al}{Al}\left[\text{clay}\right] + CaSO_4 + H_2SO_4$$

Iron sulphate ($FeSO_4$), when applied, has the same effect on aluminium sulphate (Nor 1981). Soil has different capacities to adsorb SO_4^{2-} and is influenced by the clay content and the type of clay mineral (e.g. kaolinite>ill it>bentonite), hydrous oxides (e.g. iron hydrous oxides such as goethite [α-FeOOH]), soil horizon (SO_4^{2-} >adsorption in subsoils), pH effect (SO_4^{2-} > adsorption at pH > 6.5), sulphate (adsorbed [SO_4^{2-}] is in kinetic equilibrium with [SO_4^{2-}] in soil solution), time effect (SO_4^{2-} >adsorption with>time), anions competition (< SO_4^{2-} retention in hydroxyl phosphate nitrate), cations effects ([SO_4^{2-}]> H^+>Ca^{2+} > Mg^{2+} >K^+> NH_4^+ $Na^+$$Li^+$) and soil organic matter >(SOM>retention of SO_4^{2-}) (Tisdale et al. 1985). S is applied to the soil as the elemental sulphur (S^o), gypsum hydrated ($CaSO \bullet H_2O$~15% of S), ammonium phosphate (MAP crude~24% of S), ammonium sulphate ($(NH_4)_2SO_4$~24.2% of S), magnesium sulphate ($MgSO_4$~30% of S), single superphosphate (SSP [$Ca(H_2PO_4)_2$]+$CaSO_4$~12% of S) and triple superphosphate (TPP [$Ca(H_2PO_4)_2$]~1.5% of S), potassium sulphate (K_2SO~17% of S), zinc sulphate ($ZnSO_4$~18% of S) and sulphate of potassium-magnesia ($K_2SO \bullet 2MgSO$~22% of S) (Boswell and Friesen 1993).

14.5.2 S in Plants

Plants require sulphur in the same quantities as phosphorus (P) (Mengel and Kirby 1987; Marschner and Marschner 2012). Usually, S uptake by plant roots occurs in the form of sulphate (Malavolta and Kliemann 1985). S can also be absorbed as thiosulphate (Droux 2004). S is critical for the synthesis of coenzyme A, which is crucial for fatty acid biosynthesis and oxidation, amino acid uptake, oxidation of intermediates of the citric acid cycle and ferredoxin oxidation, which is essential in photosynthesis, biological N fixation and detoxification/protection mechanisms (Marschner and Marschner 2012). Sulphur is the structural component in two aminoacids: cysteine (Cys) and methionine (Met) (Droux 2004). Cys occurs as carbon/nitrogen reduced in the sulphate assimilation pathway (Takahashi et al. 2011). Recent findings have shown the regulatory processes of sulphate transport and metabolism and their connection with transcriptional and posttranscriptional mechanisms (Bohrer and Takahashi 2016), which will help our understanding of sulphur regulatory networks (e.g. sulphate transporters and assimilatory enzymes) (Cotrim et al. 2019).

14.5.3 Mechanisms of S-Compounds Solubilization

14.5.3.1 Microbial Sulphate Reduction

Some species of sulphate-reducing bacteria (e.g. *Desulfovibrio sulfodismutans*) can catalyze unusual reactions (e.g. disproportionation, phosphite oxidation and sulphur reduction), which are not characteristic of all species (Rabus et al. 2013).

$$S - SO_3^2 + H_2O \rightarrow S_4^{2-} + H_2S\ \Delta G^{o'} = -21.9\ \text{kj/reaction}$$

Sulphur-reducing prokaryotes in the anoxic mineralization of organic matter in sulphate-rich environments (e.g. subsurface environments), which use sulphate as terminal electron acceptors, carry out dissimilatory sulphate reduction, and couple anaerobic electron transport to ATP synthesis (Eckert et al. 2011). The mining industries release great waste streams which are an environmental pollution challenge (Stanley and Southam 2018). Bioremediation with sulphate-reducing microorganisms is an environmental alternative in the treatment of acid mine waste (Anawar 2015). Currently, Bio Sulphide R (Biote Q Environmental Technologies Inc., Canada) and Thiopaq R (Paques Technologies Inc., Netherlands) are bioremediation technologies with sulphur-reducing prokaryotes to remove sulphate and heavy metals from acid mine drainage and produce high-quality and marketable metal sulphides (Ayangbenro and Babalola 2017).

14.5.3.2 *Pyrite and Chalcopyrite Oxidation*

Pyrite is the most abundant sulphide mineral, with origins as a primary and secondary mineral in igneous and sedimentary rocks (especially argillaceous and carbonaceous) (Deer et al. 1996). Chalcopyrite ($CuFeS_2$) is similar to sphalerite (ZnS) and occurs with copper (Cu)-bearing minerals (Bai et al. 2013). Considering metals sulphides (e.g. MoS_2 [molybdenite] and WS_2 [tungstenite]), the oxidation of pyrite (FeS_2) by Gram-negative mesophiles *Leptospirillum ferrooxidans* and *Thiobacillus ferrooxidans* is the most studied, because of its importance in acid mine drainage formation, but it has a negative environmental impact (Barton and Fauque 2009; Hallberg 2010; Johnson 2012). Pyrite (considered "fool's gold") is present in mining impurities (e.g. copper [Cu] in chalcopyrite, lead [Pb] in galena [PbS], arsenic [As] in arsenopyrite [FeAsS] and zinc [Zn] in sphalerite [(Zn, Fe)S]) (Barahona et al. 2014). Metal sulphide oxidation is performed by two pathways: the thiosulphate intermediate pathway and the polysulphide intermediate pathway (Stanley and Southam 2018). The thiosulphate pathway (pyrite) can be described by Reactions 14.6–14.8 (Schippers and Sand 1999):

The thiosulphate pathway (pyrite)

$$FeS_2 + \tfrac{7}{2}O_2(aq) + H_2O \rightarrow Fe^{2+} + 2S_4^{2-} + 2H^+ \tag{14.6}$$

$$Fe^{2+} + \tfrac{1}{4}(aq) + \mathbf{H}^+ \rightarrow Fe^{3+} + \tfrac{1}{2}H_2O \tag{14.7}$$

$$FeS_2 + 14Fe^{3+} + 8H_2O \rightarrow 15Fe^{2+} + 2SO_4^{2-} + 16H^+ \tag{14.8}$$

The polysulphide pathway (chalcopyrite)

$$CuFeS_2 \rightarrow Cu_{1-x}Fe_{1-y}S_{2-z} + xCu^{2+} + yFe^{2+} + zS + 2(x+y)e$$

$$Cu \rightarrow Cu^{2+} + 2^e$$

$$Cu_{1-x}Fe_{1-y}S_{2-z} \rightarrow (1-x)Cu^{2+} + (1-y)Fe^{2+} + (2-z)S + 2(2-x-y)e$$

The leaching rate of chalcopyrite ($CuFeS_2$) is very slow and tends to be depressed with time (Li et al. 2013). This is a major limitation to its commercial use in biohydrometallurgy (Zhu et al. 2011) when compared to pyrite bioleaching (Peng et al. 2019). The pyrite and chalcopyrite biomining methodologies (e.g. BIOX® process) have been studied and applied for several decades (Stanley and Southam 2018). Nowadays, new electrostatic attachment and the planktonic process, related to pyrite and chalcopyrite based on "bioleaching" and "bio-oxidation" mechanisms, are gradually being applied in biohydrometallurgy with promising results (Vera et al. 2013). Both processes are attractive eco-friendly alternatives to conventional biohydrometallurgy processes, which induce environmental contamination due to the discharge of toxic materials (e.g. arsenic) (Pollmann et al. 2016). In well-drained upland soils, sulphides and polysulphides do not occur (Johnson 2013). Calcium polysulphide (CaS_x) can occur spontaneously as gypsum ($CaSO_4{\cdot}2H_2O$) or plaster of Paris ($CaSO_4{\cdot}H_2O$) (Stanley and Southam 2018) in the environment.

CaS_x can be used in agriculture industries to remove potentially toxic elements (PTEs) from the environment (Dahlawi and Siddiqui 2016). Polysulphides (such as ammonium polysulphide $(NH_4)_2(S_3)$) are used as S fertilizers and soil conditioners and in water irrigation treatment (Dahlawi and Siddiqui 2016). Ammonium polysulphide and calcium polysulphide are changed to colloidal sulphur and sulphide when added to soil or irrigation water, as the following reaction for CaS_x shows:

$$CaS_5 + H_2O + CO_2 \rightarrow H_2S + CaCO_3 + 4S$$

Ammonium polysulphide $(NH_4)_2(S_3)$ and CaS_x adsorbed to the soil colloid can be rapidly converted into sulphates (Dahlawi and Siddiqui 2016).

The kinetics of CaS_x (wastewater) in the presence of heavy metals (Hg^{2+}, Cd^{2+} and Cr^{6+}) provides the following reactions:

$$CaS_x \frac{3}{2O_2} \rightarrow CaS_2O_3 + (x-1)S \tag{14.9}$$

$$CaS_x + CO_2 + H_2O \rightarrow CaCO_3 + H_2S + (x-1)S \tag{14.10}$$

CaS_x decomposes rapidly with CO_2 and produces H_2S, $CaCO_3$ and $CaSO_4$ (Dahlawi and Siddiqui 2016). CaS_x precipitation (in wastewater) can be shown by the following reactions (Dahlawi and Siddiqui 2016):

$$M^{2+} + CaS_2O_3 + H_2O \rightarrow MS + CaSO_4 + 2H^+ \tag{14.11}$$

$$M^{2+} + H_2S \rightarrow MS + 2H^+ \tag{14.12}$$

where:

M = heavy metal cations

MS = precipitated metal sulphides

14.5.3.2.2 Paracoccus Sulphur Oxidation or "PSO Pathway"

Another mechanism for S-compounds oxidation is the PSO pathway (also called the Kelly–Friederick pathway), a metabolic pathway found in the Alphaproteobacteria *Paracoccus pantotrophus* which use the sox operon (comprises of at least two transcriptional units with 15 genes) (Dahlawi and Siddiqui 2016). Thiosulphate is oxidized by Equation (14.13) and sulphite is oxidized by Equation (14.14):

$$\begin{array}{l} S - SO_3 + 5H_2O \rightarrow SO_4^{2-} + 8e^- + 10H^+ \\ \text{sulphur-oxidizing enzyme system} \end{array} \tag{14.13}$$

$$\begin{array}{l} SO_3^{2-} + H_2O \rightarrow SO_4^{2-} + 2e^{2-} + 2H^+ \\ \text{Sox system, sulphite oxidoreductase} \end{array} \tag{14.14}$$

There are other sulphur-oxidizing bacteria (e.g. *Acidianus brierleyi* and *A. ambivalens*) and different enzymes can catalyze the oxidation or hydrolysis of inorganic sulphur compounds, as illustrated by the following equations:

$$\begin{array}{l} 2\,{}^-SO_3 \rightarrow {}^-O_3S - S - SO_3 + 2e^{2-} \\ \text{Thiosulphate dehydrogenase} \end{array} \tag{14.15}$$

$$^{-}O_3S-S-SO_3^- + H_2O \rightarrow {}^{-}O_3S-S-S^- + SO_4^{2-} + 2H^+ \quad (14.16)$$

Tetrathionate hydrolase

$$^{23}O_3S-S-SO_3^- + H_2O \rightarrow {}^{-}S-SO_3^- + SO_4^{2-} + 2H^+ \quad (14.17)$$

Trithionate hydrolase

$$5S^O + O_2 + 4OH^- \rightarrow HSO_3^- + {}^{-}S-SO_3^- + 2HS^- + H^+ \quad (14.18)$$

Sulphur oxygenase – reductase

Recently, with the advances in molecular biology techniques, and omics studies, we have started to learn more about the biology of sulfate-reducing bacteria and sulfur-oxidizing bacteria on the sulfur cycle (Berben et al. 2019).

14.6 Zinc (Zn) Solubilizers (Zn-S)

14.6.1 Zn in Soils

In the earth's lithosphere, the transition metal Zn occurs in unavailable forms in minerals such as smithsonite (a zinc carbonate mineral [$ZnCO_3$]), sphalerite (a zinc sulphidemineral [ZnS]), zincite (a zinc oxide crystal [ZnO]), franklinite (an iron-rich zinc mineral [$ZnFe_2O_4$]), willemite (a zinc silicate mineral[Zn_2SiO_4]) and hopeite ($Zn_3(PO_4)_2.4H_2O$) (Lindsay 1979), derived from weathering of the parental rock (metamorphic, igneous and sedimentary rocks) and from other inputs such as atmospheric deposition (e.g. volcanic activity, forests fires) and anthropogenic activities (e.g. mining and chemical fertilizers applications) (Alloway 2013). The most important global reserves of Zn (~1.9 billion tons) are located in China, Australia, Peru, Canada, the United States and Iran (U.S. Geological Survey 2019). Zn is considered an essential micronutrient for all forms of organisms in the earth (Andrades and Martínez 2014). Zn can also be found on clay minerals exchange sites and soil organic matter or adsorbed as Zn^{2+}, $ZnOH^+$ or $ZnCl^+$on the solid surfaces of oxides (e.g. goethite) (Mengel and Kirkby 1987). The forms of zinc in soil are water-soluble Zn^{2+}; exchangeable Zn^{2+}; adsorbed Zn^{2+} (on clay surfaces, soil organic matter, carbonates and oxides minerals), Zn^{2-} organically complexed; and Zn^{2+} exchangeable with Mg^{2+} in the crystal lattices of clay minerals (Tisdale et al. 1985; Andrades and Martínez 2014). Zn deficiency in soils (approximately one-third of worldwide soils) affects vast global areas of cultivated soil, particularly for cereal production (Andrea et al. 2019), and is a widespread problem across the world causing Zn nutritional disorders in humans (FAO 2019). Zn in soil solution is available as Zn^{2+}, and its concentration and speciation are usually pH-dependent, competing ions, precipitation-dissolution, sorption and complexation reactions (Lindsay 1979). Globally, sources applied to soil as chemical fertilizers include zinc sulphate monohydrate ([$ZnSO_4 \cdot H_2O$] ~36% of Zn), zinc oxysulphate ([$xZnSO_4 \cdot xZnO$] ~2–50% of Zn), zinc oxide ([ZnO] ~5–80% of Zn), zinc carbonate ([$ZnCO_3$]~5–56% of Zn), zinc chloride ([$ZnCl_2$]~50% of Zn), zinc nitrate ([$Zn(NO_3).3H_2O$] ~23% of Zn), chelates ([$Na_2ZnEDTA$] ~14% of Zn; [NaZnNTA] ~13% of Zn; [$Zn3(C_6HsO_7)z.2H_2O$]~1–18% of Zn) and ammoniated zinc sulphate solution ([$Zn(NH_3)_4SO_4$]~10% of Zn) (Tisdale et al. 1985; Mengel and Kirkby 1987).

14.6.2 Zn in Plants

Zn is uptake by root plants as a divalent cation (Zn^{+2}), or as a monovalent cation ($ZnOH^+$), by a reversibly dissociable ion-carrier (into plasmalemma) or a membrane-bound carrier (Hefferon 2019). In xylem transport, Zn is linked with organic acids or exists as Zn^{+2}, whereas phloem transport (e.g. from old to new leaves) is more selective (Gupta et al. 2016). The ZIP (ZRT, IRT-like protein) (Milner et al. 2013) family is involved in transport and cellular Zn homeostasis (Caldelas and Weiss 2017), and its diffusion across the cell membrane to the cytoplasm (Bashir et al. 2012, 2016). Zn participates in numerous biochemical reactions (Tsonev and Lidon 2012) in plant cellular metabolism (e.g. carbohydrate and auxin metabolism) (Sadeghzadeh 2013), enzymatic cofactor (e.g. carbonic anhydrase [CA][EC: 4.2.1.1]), catalase (CAT) (EC 1.11.1.6), peroxidase (POX) (EC 1.11.1.7) and RNA polymerase (EC 2.7.7.6), minimizing the generation of reactive oxygen species induced by abiotic stress (Janků et al. 2019). Plant zinc deficiency reduces the copper/zinc superoxide dismutase (CuZnSOD) activity and the photosynthetic rate and limits protein concentration (Sadeghzadeh 2013).

14.6.3 Mechanisms of Zn-Compounds Solubilization

Many methods have been used to minimize the global deficiency of Zn in soil, such as Zn agronomic crop biofortification by inorganic and organic fertilizers inputs, traditional breeding, crop rotation and intercropping, genetic engineering and transgenic biotechnological approaches (Borrill et al. 2014; Manwaring et al. 2016). However, these are expensive, laborious and time-consuming methodologies (Ludwig and Slamet-Loedin 2019). A viable and ecological alternative is the use of Zn-solubilizing soil microorganisms (e.g. bacteria, fungi and algae) (Glick 2012; Deepaket al. 2013; Kamran et al. 2017), to improve socio-economic viability, soil health and agro-environmental sustainability (Khan et al. 2017). Bacteria that can solubilize Zn include *Micro bacteriumsaperdae, P. monteilli, Enterobacter cancerogens, P. fluorescens, Panto eaagglomerans, P. aeruginosa, Azotobacter chroococcum, B. megaterium, B. edaphicus, Gluconacetobacter diaztorophicus* and *Burkholderia susongensis* (Gadd 2007; Saravananet al. 2007; Selvakumaret al. 2009; Sharma et al. 2012; Ramesh et al. 2014; Majumdar and Chakraborty 2015; Oteino et al. 2015). Similar to the mechanisms associated with the solubilization of phosphate and silicate minerals, zinc solubilization mechanisms and other rock minerals might be a consequence of proton extrusion and the production of organic acids (e.g. gluconic acid, oxalic acid and citric acid) (Gadd et al. 2014) and other chelating metabolites (e.g. citrate, gluconate, lactate, 2-ketogluconate and oxalate) (Gadd 2007). The mechanism of dissolution by acids is a process associated with proton activity and the solubilization of ionized forms of minerals from rocks, and can be illustrated as follows (Berthelin 1983):

$$\mathbf{(Mineral)^{-}\,M^{+} + H^{+}R^{-} \rightarrow H^{+}(Mineral)^{-} + M^{+}R^{-}}$$

$$\mathbf{\left(R = NO_3^{\,-},\ SO_4^{\,2-},\ HCO^{3-},\ R_1COO^{-}\right)}$$

$$\mathbf{(Mineral)^{-}\,M^{+} = silicates,\ phosphates,\ carbonates,\ oxides\ \ldots\ from\ rocks}$$

$$\mathbf{M^{+} = metal\ ions}$$

The general microbial complexation dissolution of an insoluble mineral is an equilibrium reaction (metal ion and a complexing agent) and characterized by the formation of a metal ion in a ring structure, illustrated as follows (Berthelin 1983; Berthelin et al. 1991):

$$\mathbf{(Mineral)^{-}\,M^{+} + H^{+}L^{-} \rightarrow H^{+}(Mineral)^{-} + ML}$$

Such models provide information on complex (e.g. polycations or polyanions) formations in natural environmental conditions (Berthelin 1983; Berthelin et al. 1991).

14.7 Opportunities and Challenges

14.7.1 Opportunities

High dependence and exorbitant expenditure on chemical fertilizers has necessitated the development of eco-friendly, sustainable and viable economic mechanisms to fertilize the soil using rock-forming minerals with timed solubility of its nutrients in environmental conditions that are suitable for biological solubilization (Mahanty et al. 2016), ensuring food safety and soil health (Sharma et al. 2013; Etesami et al. 2017). In this context, the microbial solubilizing of rock-forming minerals (e.g. basalt, granite, granodiorite, rhyolite, andesite, peridotite, dunite and quartzite) (Setiawati and Mutmainnah 2016) has emerged as a potential environmentally friendly input with a vital role in the integrated nutrient management (INM) system (Bargaz et al. 2018). By definition, any formulations of substances (e.g. humic and fulvic acids) or compounds (e.g. phytohormones) of natural origin, their mixtures (e.g. essential oils), bio-extracts (e.g. seaweed extracts) and microorganisms (e.g. plant growth–promoting microorganisms [PGPMs]) applied to plants aiming to improve nutrient use efficiency, abiotic stress tolerance (e.g. soil acidity) and/or crop quality traits and minimize the dependency on chemical fertilizers, especially on off-farm chemical inputs, are considered plant biostimulants (PBs) or biofertilizer (Calvo et al. 2014; Bulgari et al. 2015; De Pascale et al. 2017; Caradonia et al. 2018, 2019; Rouphael and Colla 2018; Rouphael et al. 2018; Ricci et al. 2019). The global biofertilizer market (also known as the agricultural biologicals market, see Figure 14.3) is expected to achieve~US$ 5.23 billion by 2026 growing at a compound annual growth rate (CAGR) of ~12% between 2018 and 2026 (https://www.mordorintelligence.com/industry-reports/global-biofertilizers-market-industry). Given the above, and considering the diversity and the mode of their action, PB scan be used to overcome NUE limitations in organic farming systems, thereby minimizing the gap between organic and conventional agricultural systems production (Traon et al. 2014). The global biofertilizers market is segmented by microorganism (e.g. *Rhizobium*, *Azotobacter*, *Azospirillum*, blue-green algae), solubilizers bacteria (e.g. phosphate, potassium, silicon, sulphur and zinc) and mycorrhizal fungi (e.g. *Paxillus involutus*, *Piloderma croceum*, *Hebeloma longicaudum*, *Suillus bovinus* and *Rhizopogon luteolus*); by technology (e.g. type into carrier-enriched biofertilizers, liquid biofertilizers and other technology types); by application (seed treatment and soil treatment); and by crop (e.g. cereals, legumes, fruits and vegetables and forest plantations) (https://mordorintelligence.com/industry-reports/global-biofertilizers-market-industry). An analysis of the global biofertilizers market has shown that North America (United States and Canada) is the largest market for biofertilizers with~28% of the global market in 2017 (Yakhin et al. 2017; Willer and Lernoud 2019). Recent research indicates that the increase in organic farming practices (e.g. biofertilizers) helps to improve the soil and human health, due to the extensive use of organic products (Bulgari et al. 2019). In the global context, the size of the organic market in the United States (~US$ 45.2 billion) is the most promising market for organic food, followed by Germany (~US$ 11.3 billion) and France (~US$ 8.9 billion) (Willer and Lernoud 2019). The microorganism types used in the US biological market as biofertilizers include diazotrophic bacteria (e.g. *Rhizobium*, *Azotobacter* and *Azospirillum*), blue-green algae (also called cyanobacteria), potassiumv-solubilizer (e.g. *Frateuria aurantia*), phosphate solubilizer (e.g. *Achromobacter*, *Arthrobacter*, *Azotobacter*, *Azospirillum*, *Bacillus*, *Burkholderia*, *Enterobacter*, *Klebsiella*, *Microbacterium*, *Paenibacillus*, *Pantoea*, *Pseudomonas*, *Serratia*, *Streptomyces*), silicate solubilizer (e.g. *Thiobacillus thiooxidans*) and mycorrhiza (endomycorrhiza and ectomycorrhiza) (Dudaš et al. 2016; Olanrewaju et al. 2017; Alori and Babalola 2018).

Globally, the demand for organic products is increasing (partly due to the increase in conscientious consumers) and this is driving the necessity for biofertilizers (Caradonia et al. 2019). Undoubtedly, this has further triggered the search for biofertilizer products (Alori and Babalola 2018), as they are organic and help in maintaining the ecological balance and sustainable agriculture (Olanrewaju et al. 2017). Biofertilizer products are considered an innovative agronomic and eco-friendly tool as demonstrated by the increase in scientific publications (Yakhin et al. 2017; Caradonia et al. 2018; Rouphael et al. 2018), and the constant expansion of their market (Ricci et al. 2019). The European Committee for

Standardization (CEN) has started working on a range of processes and methods that will provide the basis for a system of European standards protocols (Caradonia et al. 2019).

14.7.2 Challenges

The current global marketplace for bio-inoculants has no international standards for bioformulations, giving rise to the commercialization of bioproducts without quality assurance (reviewed by Bona et al. 2016). In practical terms every country is looking out for their own best interest, with bioformulations development and commercialization under governmental regulations of each country, for example the Netherlands, India, China, Thailand, Russia, Canada, France, Australia (voluntarily) and Argentina (since 2016) (reviewed by Keswani et al. 2016). Another aspect to be considered is the final consumers' negative perceptions (e.g. non-standardized packaging, labelling and pricing of biofertilizers) of bioproducts. Moreover, the commercialization of biofertilizers is problematic because the final product has a limited shelf life due to the microbial life contained within the package (Caradonia et al. 2019). Taking this background into consideration, successful bioproduct commercialization is directly related to interdisciplinary research, large-scale production, viable formulation methods, product marketing and education of farmers. The biological industry needs to work together with academia, under appropriate government regulations, to get its bioproducts registered and credibility ensured (Yakhin et al. 2017). The new Fertilizing Products Regulation (FPR) 2019/1009 was passed by the European Union (EU), recognizing PBs or phyto stimulators as a distinct category of agricultural inputs (e.g. pesticides or fertilizers) (Caradonia et al. 2019), encouraging innovation in the EU and giving European farmers equal opportunities to access these new technologies and to contribute to the development of the bioeconomy by farming in according to the more stringent global consumer demands (Yakhin et al. 2017; Caradonia et al. 2018; Rouphael et al. 2018; Ricci et al. 2019). The European Biostimulant Industry Council (EBIC) promotes the plant biostimulants market, which gained legal identity in 2013 (www.biostimulants.eu). In Belgium, France, Denmark, Hungary and Austria, some substances that are used as plants biostimulants and agronomic fertilizer additives (PB&AFA) are registered under other EU regulatory frameworks (e.g. Fertilizers [Regulation EC No 2003/2003]) (reviewed by Caradonia et al. 2019).

In the United States, the Environmental Protection Agency (EPA) and the United States Department of Agriculture (USDA) regulate the registration of biostimulant products (Alori and Babalola 2018), with a lower level of bureaucracy, thereby stimulating the market (De Pascale et al. 2017). To date, Europe's inconsistency and regionalism of laws at the national level as relate to biofertilizers/biostimulants, places them in categories with different names based on regional councils, such as *Corroboranti* (invigorators) in Italy, *Pflanzenstärkungsmittel* (plant strengtheners) in Germany and *Substances Naturelles à usage*

FIGURE 14.3 General vision of global biofertilizer market segmented by geography region (2018). (Source: https://mordorintelligence.com)

Biostimulant (natural substances for use as biostimulants) in France (Caradonia et al. 2018), leading to uncertainty among manufacturers (e.g. BASF SE, Bayer AG, Italpollina Spa, Valagro, Biovert S.A., Arysta Life Science Ltd, Agrinos, Biostadt India Ltd, Koppert B.V. and Omex Agrifluids Ltd) (Ricci et al. 2019), and restricting the expansion of this market in the EU (Caradonia et al. 2018, 2019). Thus, it is necessary to change the current scenario and equalize EU legislation (Willer and Lernoud 2019).

14.8 Conclusion and Future Prospects

Nowadays, knowledge is a driver for economic growth, requiring high investment in education (e.g. academia), research and development (R&D) and training, and harmonizing relations between governments and industry. The content of biofertilizers, the promotion of simplified registration, and the standardization of regulations governing these products under national law (e.g. European Union regulatory frameworks) would help accelerate the registration process and commercialization. In bioproduct development is necessary to recognize the bottlenecks in the roadmap, which are usually related to the process of bioformulation development of biofertilizers (e.g. from isolation, screening of microbes to bioproduct development and its implementation). Regardless of governmental regulatory frameworks, more research is required (e.g. concept of bioformulation), and suitable management conditions need to be identified that can contribute to stringent global consumer demands.

REFERENCES

Ahmad I, Maathuis FJ (2014) Cellular and tissue distribution of potassium: physiological relevance, mechanisms and regulation. *J Plant Physiol* 171:708–714

Ahmad I, Mian A, Maathuis FJ (2016) Overexpression of the rice AKT1 potassium channel affects potassium nutrition and rice drought tolerance. *J Exp Bot* 67:2689–2698. https://doi.org/10.1093/jxb/erw103.

Ahmad M, Nadeem SM, Naveed M, Zahir ZA (2016) Potassium-solubilizing bacteria and their application in agriculture. In: Meena VS, Maurya BR, Verma JP, Meena RS (eds.) *Potassium Solubilizing Microorganisms for Sustainable Agriculture*. Springer India, New Delhi, pp. 293–313

Ahmad Z, Nadeem F, Wang R, Diao X, Han Y, Wang X, Li X (2018) A larger root system is coupled with contrasting expression patterns of phosphate and nitrate transporters in Foxtail millet [*Setariaitalica* (L.) Beauv.] under phosphate limitation. *Front Plant Sci* 9:1367. https://doi.org/10.3389/fpls.2018.01367.

Ahmed E, Holmstrom SJM (2014) Siderophores in environmental research: roles and applications. *Microb Biotechnol* 7:196–208

Ahmed E, Holmstrom SJM (2015) Microbe-mineral interactions: the impact of surface attachment on mineral weathering and element selectivity by microorganisms. *Chem Geol* 403:13–23

Ahn SJ, Shin R, Schachtman DP (2004) Expression of KT/KUP genes in *Arabidopsis* and the role of root hairs in K+ uptake. *Plant Physiol* 134:1135–1145

Alloway BJ (2013) Sources of heavy metals and metalloids in soils. Heavy metals in soils. In: Alloway BJ (ed.) *Trace Metals and Metalloids in Soils and Their Bioavailability*, 3th edn. Springer, vol 22. Springer, Dordrecht,11–50

Alori ET, Babalola OO (2018) Microbial inoculants for improving crop quality and human health in Africa. *Front Microbiol* 9:2213. https://doi.org/10.3389/fmicb.2018.02213

Amtmann A, Beilby MJ (2010) The role of ion channels in plant salt tolerance. In: Demidchik V, Maathuis F (eds.). *Ion Channels and Plant Stress Responses*. Springer, Berlin, pp. 23–46

Anawar HM (2015) Sustainable rehabilitation of mining waste and acid mine drainage using geochemistry, mine type, mineralogy, texture, ore extraction and climate knowledge. *J Environ Manage* 158:111–121. https://doi.org/10.1016/j.jenvman.2015.04.045

Andrades M, Martínez E (2014) *Fertilidad del Suelo y Parâmetros que la Definen, 34. Tercera Edicion*. Universidad de la Rioja, Logroño, espana

Andrea EMM, Carolina ATE, Tito JCB, Luis JMN, Carlos LGM (2019) Evaluation of contaminants in agricultural soils in an irrigation district in Colombia. *Heliyon* 5. e02217. https://doi.org/10.1016/j.heliyon.2019.e02217

Aristovskaya TV, Kutuzova RS (1968) Microbiological factors in the mobilization of silicon from poorly soluble natural compounds. *Pochvovedenie* 12:59–66.

Atlas RM, Bartha R (1997) *Microbial Ecology: Fundamentals and Applications*. 4th edn. Benjamin/Cummings Publishing Company, Redwood City, California

Avakyan ZA, Pivovarova TA, Karavaiko GI (1986) Properties of a new species, *Bacillus mucilaginosus*. *Mikrobiologiya* 55:477–482 (English translation pp. 477–482)

Ayangbenro AS, Babalola OO (2017) A new strategy for heavy metal polluted environments: a review of microbial biosorbents. *Int J Environ Res Public Health* 14:94. https://doi.org/10.3390/ijerph14010094

Badr MA, Shafei AM, Sharaf El-Deen SH (2006) The dissolution of K and P-bearing minerals by silicate dissolving bacteria and their effect on sorghum growth. *Res J Agri Biol Sci* 2:5–11

Bai H, Kang Y, Quan H, Han Y, Sun J, Feng Y (2013) Treatment of acid mine drainage by sulfate reducing bacteria with iron in bench scale runs. *Bioresour Technol* 128, 818–822. https://doi.org/10.1016/j.biortech.2012.10.070

Bai L, Wang C, Zang S, Wu C, Luo J, Wu Y (2018) Mapping soil alkalinity and salinity in Northern Songnen Plain, China with the HJ-1 hyperspectral imager data and partial least squares regression. *Sensors* 18: 3855.https://doi.org/10.3390/s18113855

Balan E, Calas G, Bish DL (2014) Kaolin group minerals: from hydrogen-bonded layers to environmental recorders. *Elements* 10:183–188

Balland C, Poszwa A, Leyval C, Mustin C (2010) Dissolution rates of phyllosilicates as a function of bacterial metabolic diversity. *Geochim Cosmochim Acta* 74:5478–5493.

Banfield JF, Barker WW, Welch SA, Taunton A (1999) Biological impact on mineral dissolution: application of the lichen model to understanding mineral weathering in the rhizosphere. *Proc Natl Acad Sci U S A* 96:3404–3411

Barahona S, Dorador C, Zhang R, Aguilar P, Sand W, Vera M, Remonsellez F (2014) Isolation and characterization of a novel *Acidithiobacillus ferrivorans* strain from the Chilean Altiplano: attachment and biofilm formation on pyrite at low temperature. *Res Microbiol* 165:782–793

Barber SA (1962) A diffusion and mass flow concept of soil nutrient availability. *Soil Sci* 93:39–49

Bargaz A, Lyamlouli K, Chtouki M, Zeroual Y, Dhiba D (2018) Soil microbial resources for improving fertilizers efficiency in an integrated plant nutrient management system. *Front Microbiol* 9:1606. https://doi.org/10.3389/fmicb.2018.01606

Barker WW, Welch SA, Banfield JF (1997) Biogeochemical weathering of silicate minerals. In: Banlleld JF, Nealson KH (eds) *Geomicrobiology: Interactions Between Microbes and Minerals*. Mineralogical Society of America, Washington DC, pp 391–428

Barker WW, Welch SA, Chu S, Banfield JF (1998) Experimental observations of the effects of bacteria on aluminosilicate weathering. *Am Mineral* 83:1551–1563

Barnhisel RI, Bertsch PM (1989) Chlorites and hydroxy-interlayered vermiculite and smectite. In: Dixon JB, Weed SB (ed.) *Minerals in Soil Environments*. 2nd ed. Soil Science Society of America, Madison, pp. 729–788.

Barragán V, Leidi EO, Andrés Z, Rubio L, De Luca A, Fernández JA, … Pardo JM (2012) Ion exchangers NHX1 and NHX2 mediate active potassium uptake into vacuoles to regulate cell turgor and stomatal function in *Arabidopsis*. *Plant Cell* 24:1127–1142. https://doi.org/10.1105/tpc.111.095273

Barrow NJ (1978) The description of phosphate adsorption curves. *J Soil Sci* 29:447–462

Barrow NJ (1983) A mechanistic model for describing the sorption and desorption of phosphate by soil. *J Soil Sci* 34:733–750

Barton LL, Fauque GD (2009) Biochemistry, physiology and biotechnology of sulfate-reducing bacteria. *Adv Appl Microbiol* 68:41–98. https://doi.org/10.1016/S0065-2164(09)01202-7

Basak BB, Biswas DR (2010) Co-inoculation of potassium solubilizing and nitrogen fixing bacteria on solubilization of waste mica and their effect on growth promotion and nutrient acquisition by a forage crop. *Biol Fertil Soils* 46:641–648

Basak BB, Sarkar B, Biswas DR et al. (2017) Bio-intervention of naturally occurring silicate minerals for alternative source of potassium: challenges and opportunities. *Adv Agron* 141:115–145. https://doi.org/10.1016/bs.agron.2016.10.016

Bashir K, Ishimaru Y, Nishizawa N (2012) Molecular mechanisms of zinc uptake and translocation in rice. *Plant Soil* 36:189–201. https://doi.org/10.1007/s11104-012-1240-5

Bashir K, Rasheed S, Kobayashi T, Seki M, Nishizawa NK (2016) Regulating subcellular metal homeostasis: the key to crop improvement. *Front Plant Sci* 7:1192. https://doi.org/10.3389/fpls.2016.01192.

Baya DT, Effebi KR, Tangou TT, Keffala C, Vasel JL (2013) Effect of hydroxyapatite, octacalcium phosphate and calcium phosphate on the auto-flocculation of the microalgae in a high-rate algal pond. *Environ Tech* 34:2407–2414. https://doi.org/10.1080/09593330.2013.770563

Beck MA, Sánchez PA (1994) Soil phosphorus fraction dynamics during 18 years of cultivation on a Typic Paleudult. *Soil Sci Soc Am J* 58:1424–1431

Beckett PHT (1964) Studies of soil potassium: II the immediate q/I relations of labile potassium in the soil. *J Soil Sci* 15:9

Beever RE, Burns DJW (1980) Phosphorus uptake, storage and utilization by fungi. *Adv Bot Res* 8:127–219

Belén M, Manuel SJ, Noelia R, Sandra C, Attilio C, Domínguez J (2010) Biotechnological production of citric acid. *Braz J Microbiol* 41:862–875 https://doi.org/10.1590/S1517-83822010000400005

Belzunces LP, Tchamitchian S, Brunet JL (2012) Neural effects of insecticides in the honey bee. *Apidologie* 43:348–370. https://doi.org/10.1007/s13592-012-0134-0

Bennett PC, Rogers JR, Choi WJ, Hiebert FK (2001) Silicates, silicate weathering, and microbial ecology. *Geomicrobiol J* 18:3–19

Berben T, Overmars L, Sorokin DY, Muyzer G (2019) Diversity and distribution of sulfur oxidation-related genes in *Thioalkalivibrio*, a genus of chemolithoautotrophic and haloalkaliphilic sulfur-oxidizing bacteria. *Front Microbiol* 10:160. https://doi.org/10.3389/fmicb.2019.00160

Berthelin J (1983) Microbial weathering process. In: Krumbein WE (ed.) *Microbial Geochemistry*. Blackwell, Oxford, pp. 223–263

Berthelin J, Leyval C, Laheurte F, de Giudici P (1991) Involvements of roots and rhizosphere microflora in the chemical weathering of soil minerals. In: Atkinson D (ed.) *Plant Root Growth: An Ecological Perspective*, Blackwell Scientific Publications, Oxford, pp. 187–200 (Special Publications Series of the British Ecological Society, No, 10)

Bertsch PM, Thomas GW (1985) *Munson RD Potassium in Agriculture*. Soil Science Society of America, Madison, WI, pp. 131–162

Bhat KKS, Nye PH, Baldwin JP (1976) Diffusion of phosphate to plant roots in soil. IV. The concentration distance profile in the rhizosphere of roots with root hairs in a low-P soil. *Plant Soil* 44:63–72

Bhattacharya S, Bachani P, Jain D, Patidar SK, Mishra S (2016) Extraction of potassium from K-feldspar through potassium solubilization in the halophilic *Acinetobacter* soli (MTCC 5918) isolated from the experimental salt farm. *Int J Miner Process* 152:53–57

Bieleski RL (1973) Phosphate pools, phosphate transport, and phosphate availability. *Annu Rev Plant Physiol* 24:225–252

Boge G, Lespilette M, Jamet D, Jamet JL (2012) Role of sea water DIP and DOP in controlling bulk alkaline phosphatase activity in NW Mediterranean Sea (Toulon, France). *Mar Pollut Bull* 64:1989–1996

Bolan NS, Elliott J, Greeg PEH, Weil S (1997) Enhanced dissolution of phosphate rocks in the rhizosphere. *Biol Fertil Soils* 24:169–174

Bohrer AS, Takahashi H (2016) Compartmentalization and regulation of sulfate assimilation pathways in plants. *Int Rev Cell Mol Biol* 326:1–31. https://doi.org/10.1016/bs.ircmb.2016.03.001

Bomgardner MM (2016) Transforming agriculture, again. *Chem Eng News* 94:32–38

Bona E, Lingua G, Todeschini V (2016) Effect of bioinoculants on the quality of crops. In: Arora N, Mehnaz S, Balestrini R (eds.) *Bioformulations: for Sustainable Agriculture*. Springer, New Delhi, pp. 93–124

Bonneville S, Smits MM, Brown A, Harrington J, Leake JR, Brydson R, Benning LG (2009) Plant-driven fungal weathering: early stages of mineral alteration at the nanometer scale. *Geology* 37:615–618

Bonting CFC, van Veen HW, Taverne A, Kortstee GJJ, Zehnder AJB (1992) Regulation of polyphosphate metabolism in *Acinetobacter* strain 210A grown in carbon limited and phosphate limited continuous cultures. *Arch Microbiol* 158:139–144

Borrill P, Connorton JM, Balk J, Miller AJ, Sanders D, Uauy C (2014) Biofortification of wheat grain with iron and zinc: integrating novel genomic resources and knowledge from model crops. *Front Plant Sci* 5:53. https://doi.org/10.3389/fpls.2014.00053. eCollection2014

Boswell CC, Friesen DK (1993) Elemental sulfur fertilizers and their use on crops and pastures. *Fert Res* 35:127–149

Bou-Abdallah F (2010) The iron redox and hydrolysis chemistry of the ferritins. *Biochim Biophys Acta (BBA) Gen Subj* 1800:719–731

Bowles MW, Mogollón JM, Kasten S, Zabel M, Hinrichs KU (2014) Global rates of marine sulfate reduction and implications for sub-sea-floor metabolic activities. *Science* 344:889–891. https://doi.org/10.1126/science.1249213

Britzke D, Silva LS, Moterle DF, Rheinheimer DS, Bortoluzzi EC (2012) A study of potassium dynamics and mineralogy in soils from subtropical Brazilian lowlands. *J Soils Sediments* 2:185–197. https://doi.org/10.1007/s11368-011-0431-7

Bradáčová K, Kandeler E, Berger N, Ludewig U, Neumann G (2020) Microbial consortia inoculants stimulate early growth of maize depending on nitrogen and phosphorus supply. *Plant Soil Environ* 66:105–112. https://doi.org/10.17221/382/2019-PSE

Bruinsma J (2011) The resource outlook to 2050: by how much do land, water use and crop yields need to increase by 2050? In: Conforti P (ed.) *Looking Ahead in World Food and Agriculture: Perspectives to 2050, Food and Agricultural Organization of the United Nations (FAO)*, Chapter 6, Rome. https://www.fao.org/docrep/014/i2280e/i2280e06.pdf

Bulgari R, Cocetta G, Trivellini A, Vernieri P, Ferrante A (2015) Biostimulants and crop responses: a review. *Biol Agric Hortic* 31:1–17. https://doi.org/10.1080/01448765.2014.964649

Bulgari R, FranzoniG, Antonio Ferrante A (2019) mBiostimulants application in horticultural crops under abiotic stress conditions. *Agronomy* 9:306. https://doi.org/10.3390/agronomy9060306

Burford EP, Fomina M, Gadd GM (2003) Fungal involvement in bioweathering and biotransformation of rocks and minerals. *Mineral Mag* 67:1127–1155. https://doi.org/10.1180/0026461036760154

Caldelas C, Weiss DJ (2017) Zinc Homeostasis and isotopic fractionation in plants: a review. *Plant Soil* 411:17–46. https://doi.org/10.1007/s11104-016-3146-0

Calvo P, Nelson L, Kloepper JW (2014) Agricultural uses of plant biostimulants. *Plant Soil* 383:3–41. https://doi.org/10.1007/s11104-014-2131-8

Caporale AG, Violante A (2016) chemical processes affecting the mobility of heavy metals and metalloids in soil environments. *Curr Pollution Rep* 2:15–27 https://doi.org/10.1007/s40726-015-0024-y

Caradonia F, Battaglia V, Righi L et al. (2019) Plant biostimulant regulatory framework: prospects in Europe and current situation at international level. *J Plant Growth Regul* 38:438–448. https://doi.org/10.1007/s00344-018-9853-4

Caradonia F, Battaglia V, Righi L, Pascali G, La Torre A (2018) Plant biostimulant regulatory framework: prospects in Europe and current situation at international level. *J Plant Growth Regul* 1–11. https://doi.org/10.1007/s00344-018- 9853-4

Ceasar SA, Hodge A, Baker A, Baldwin SA (2014) Phosphate concentration and arbuscular mycorrhizal colonization influence the growth, yield and expression of twelve PHT1 family phosphate transporters in foxtail millet (*Setariaitalica*). *PLoS One* 9:e108459. https://doi.org/10.1371/journal.pone.0108459

Ceci A, Pinzari F, Russo F, Persiani AM, Gadd GM (2019) Roles of saprotrophic fungi in biodegradation or transformation of organic and inorganic pollutants in co-contaminated sites. *Appl Microbiol Biotechnol* 103:10353–10368

Chagas WFT, Guelfi DR, Emrich EB, Silva AL, Faquin V (2016) Agronomic efficiency of polymer-coated triple superphosphate in onion cultivated in contrasting texture soils. *Rev Ci Agron* 47:439–446. https://doi.org/10.5935/1806-6690.20160053

Chardon WJ, Oenema O, Del Castilho P, Vriesema R, Japenga J, Blaauw D (1997) Organic from phosphorus in solution and leachates from soils treated with animal slurries. *J Environ Qual* 26:372–378

Chatterjee AK, Nandi P (1964) Solubilization of insoluble phosphates by rhizosphere fungi of legumes Arachis, Cyamopsis, Desmodium. *Transl Bose Res Inst (Calcutta)* 27:115–120

Chen JF, Jiao C (2016) Hydrogen sulfide alleviates chlorobenzene toxicity in soybean seedlings. *Ciênc Rural* 46:1743–1749. http://doi.org/10.1590/0103-8478cr20151503

Chen W, Yao X, Cai K, Chen J (2011) Silicon alleviates drought stress of rice plants by improving plant water status, photosynthesis and mineral nutrient absorption. *Biol Trace Elem Res* 142:67–76. https://doi.org/10.1007/s12011-010- 8742-x

Cheng TC, Harvey SP, Stroup AN (1993) Purification and properties of a highly active organophosphorus acid anhydrolase from *Alteromonasundina*. *Appl Environ Microbiol* 59:3138–3140

Chouhan S, Sharma K, Zha J, Guleria S, Koffas MAG (2017) Recent advances in the recombinant biosynthesis of polyphenols. *Front Microbiol* 8:2259. https://doi.org. 10.3389/fmicb.2017.02259.

Cornelis JT, Delvaux B (2016) Soil processes drive the biological silicon feedback loop. *Funct Ecol* 30:1298–1310. https://doi.org/10.1111/1365-2435.12704

Cosgrove DJ (1977) Microbial transformations in the phosphorus cycle. In: Alexander M (ed.) *Advances in Microbial Ecology*, vol. 1. Plenum Press, Springer, New York, 95–134

Coskun D, Britto DT, Huynh WQ, Kronzucker HJ (2016) The role of silicon in higher plants under salinity and drought stress. *Front Plant Sci* 7:1072. https://doi.org/10.3389/fpls.2016.01072

Cotrim MF, Silva JBD, Lourenço FMDS, Teixeira AV, Gava R, Alves CZ, Candido ACDS, Campos CNS, Pereira MD, Torres SB, Bacchetta G, Teodoro PE (2019) Studying the link between physiological performance of *Crotalaria ochroleuca* and the distribution of Ca, P, K and S in seeds with X-ray fluorescence. *PLoS One* 14:e0222987. https://doi.org/10.1371/journal.pone.0222987

Crundwell FK (2014) The mechanism of dissolution of minerals in acidic and alkaline solutions: part II application of a new theory to silicates, aluminosilicates and quartz. *Hydrometallurgy* 149:265–275. https://doi.org/10.1016/j.hydromet.2014.07.003

Cunningham JE, Kuiack C (1992) Production of citric and oxalic acids and solubilization of calcium phosphate by *Penicillium bilaii*. *Appl Environ Microbiol* 58:1451–58

Cygan RT, Liang J-J, Kalinichev AG (2004) Molecular models of hydroxide, oxyhydroxide, and clay phases and the development of a general force field. *J Phys Chem B* 108:1255–1266

Dahlawi SM, Siddiqui S (2016) Calcium polysulphide, its applications and emerging risk of environmental pollution—a review. *Environ Sci Pollut Res* 24(1). https://doi.org/10.1007/s11356-016-7842-3

Daneshgar S, Buttafava A, Capsoni D, Callegari A, Capodaglio A (2018) Impact of pH and ionic molar ratios on phosphorous forms precipitation and recovery from different wastewater sludges. *Resources* 7:71. https://doi.org/10.3390/resources7040071

Darunsontaya T, Suddhiprakarn A, Kheoruenromne I, Prakongkep N, Gilkes RJ (2012) The forms and availability to plants of soil potassium as related to mineralogy for upland Oxisols and Ultisols from Thailand. *Geoderma* 170:11–24. https://doi.org/10.1016/j.geoderma.2011.10.0

Dassarma S, Schwieterman EW (2018) Early evolution of purple retinal pigments on Earth and implications for exoplanet biosignatures. *Int J Astrobiol* 1–10. https://doi.org/10.1017/S1473550418000423

De Pascale S, Rouphael Y, Colla G (2017) Plant biostimulants: innovative tool for enhancing plant nutrition in organic farming. *Eur J Hortic Sci* 82:277–285. https://doi.org/10.17660/eJHS.2017/82.6.2

Deepak J, Geeta N, Sachin V, Anita S (2013) Enhancement of wheat growth and Zn content in grains by zinc solubilizing bacteria. *Int J Agr Environ Biotechnol* 6:363–370. https://doi.org/10.5958/j.2230-732X.6.3.004

Deer WA, Howie RA, Zussaman J (1996) *An Introduction to the Rock-forming Minerals*. 2nd ed. Pearson-Prentice Hall, London

Dersjant-Li Y, Awati A, Schulze H, Partridge G (2015) Phytase in non-ruminant animal nutrition: a critical review on phytase activities in the gastrointestinal tract and influencing factors. *J Sci Food Agric* 95:878–896. https://doi.org/10.1002/jsfa.6998

Detmann K, Araújo W, Martins S, Fernie AR, DaMatta F (2013) Metabolic alterations triggered by silicon nutrition: is there a signaling role for silicon? *Plant Signal Behav* 8:e22523. https://doi.org/10.4161/psb.22523

Droux M (2004) Sulfur assimilation and the role of sulfur in plant metabolism: a survey. *Photosynth Res* 79:331–48

Dove PM, Nix CJ (1997) The influence of the alkaline earth cations, magnesium, calcium, and barium on the dissolution kinetics of quartz. *Geochim Cosmochim Acta* 61: 3329–3340

Dudaš S, Šola I, Sladonja B, Erhatic R, Ban D, Poljuha D (2016) The effect of biostimulant and fertilizer on "low input" lettuce production. *Acta Bot Croat* 75:253–259. https://doi.org/10.1515/botcro-2016-0023

Duff RB, Webley DM (1959) 2-ketogluconic acid as a natural chelator produced by soil bacteria. *Chim Ind* 44:1376–1377

Duff RB, Webley DM, Scott RO (1963) Solubilization of minerals and related minerals by 2-ketogluconic acid-producing bacteria. *Soil Sci* 95:105–114

Dyhrman ST, Benitez-Nelson CR, Orchard ED, Haley ST, Pellechia PJ (2009) A microbial source of phosphonates in oligotrophic marine systems. *Nat Geosci* 2:696–699
Eckert T, Brunner B, Edwards EA, Wortmann UG (2011) Microbially mediated re-oxidation of sulfide during dissimilatory sulfate reduction by *Desulfobacter latus. Geochim Cosmochm Acta* 75:3469–3485
Eckhardt FEW (1980) Microbial degradation of silicates. Release of cations from aluminosilicate minerals by yeast and filamentous fungi. In: Oxley TA, Becker G, Allsopp D (eds.) *Biodeterioration. Proceedings of the 4th International Biodeterioration Symposium.* London, pp. 107–116
Epstein E (1999) Silicon. *Annu Rev Plant Physiol Plant Mol Biol* 50:641–664. https://doi.org/10.1146/annurev.arplant.50.1.641
Epstein E (2009) Silicon: its manifold roles in plants. *Ann Appl Biol* 155:155–160. https://doi.org/10.1111/j.1744-7348.2009.00343.x
Epstein E, Bloom AJ (2005) *Mineral Nutrition of Plants: Principles and Perspectives.* 2nd ed. Sinauer Associates Inc, Sunderland
Etesami H, Emami S, Alikhani HA (2017) Potassium solubilizing bacteria (KSB): mechanisms, promotion of plant growth, and future prospects: a review. *J Soil Sci Plant Nutr* 17:897–911. https://doi.org/10.4067/s0718-95162017000400005
Fageria NK (2009) *The Use of Nutrients in Crop Plants.* CRC Press, Boca Raton, FL
Fageria NK, Baligar VC, Jones CA (2011) *Growth and Mineral Nutrition of Field Crops.* 3th ed. CRC Press, Taylor and Francis Group, Boca Raton, FL
FAO (2009) High Level Expert Forum—how to Feed the World in 2050, Economic and Social Development. Food and Agricultural Organization of the United Nations, Rome
FAO (2012) *State of Food and Agriculture 2012.* Food and Agricultural Organization of the United Nations, Rome. https://www.fao.org/3/i3028e/i3028e00.htm
FAO (2019) *FAO: Challenges and Opportunities in a Global World.* Food and Agricultural Organization of the United Nations (FAO), Rome. https://www.fao.org/3/ca4305en/ca4305en.pdf
Field JA, Reneau RB, Kroontje W (1985) Effects of anaerobically digested poultry manure on soil phosphorus adsorption and extractability. *J Environ Qual* 14:105–107
Fike AD, Bradley AS, Rose CV (2015) Rethinking the ancient sulfur cycle. *Annu Rev Earth Planet Sci* 43:593–622. https://doi.org/10.1146/annurev-earth-060313-054802
Fink JR, Inda AV, Tiecher T, Barrón V (2016) Iron oxides and organic matter on soil phosphorus availability. *Ciênc Agrotec* 40:369–379. https://doi.org/10.1590/1413-70542016404023016
Fleck AT, Schulze S, Hinrichs M, Specht A, Wassmann F, Schreiber L, et al (2015) Silicon promotes exodermal Casparian band formation in Si-accumulating and Si-excluding species by forming phenol complexes. *PLoS ONE* 10:e0138555. https://doi:.org/10.1371/journal.pone.0138555
Fomina M, Alexander IJ, Colpaert JV, Gadd GM (2005) Solubilization of toxic metal minerals and metal tolerance of mycorrhizal fungi. *Soil Biol Biochem* 37:851–866
Frondini F, Vaselli O, Vetuschi Zuccolini M (2019) Consumption of atmospheric carbon dioxide through weathering of ultramafic rocks in the Voltri Massif (Italy): quantification of the process and global implications. *Geosciences* 9:258. doi:10.3390/geosciences9060258
Gadd GM (1999) Fungal production of citric and oxalic acid: importance in metal speciation, physiology and biogeochemical processes. *Adv Microb Physiol* 41:47–92. http://doi.org/10.1016/S0065-2911(08)60165-4
Gadd GM (2007) Geomycology: biogeochemical transformations of rocks, minerals, metals and radionuclides by fungi, bioweathering and bioremediation. *Mycol Res* 111:3–49. http://dx.doi.org/10.1016/j.mycres.2006.12.001
Gadd GM, Bahri-Esfahani J, Li Q, Rhee YJ, Wei Z, Fomina M, et al. (2014) Oxalate production by fungi: significance in geomycology, biodeterioration and bioremediation. *Fungal Biol Rev* 28:36–55.
Gadd GM, Raven JA (2011) Geomicrobiology of eukaryotic microorganisms. *Geomicrobiol J* 27:491e519
Galloway T, Handy R (2003) Immunotoxicity of organophosphorous pesticides. *Ecotoxicology* 12:345–363
Ganor J, Lasaga AC (1994) The effects of oxalic acid on kaolinite dissolution rate. *Min Mag* 58A:315–316
Gerrity S, Kennelly C, Clifford E, Collins G (2016) Hydrogen sulfide oxidation in novel horizontal-flow biofilm reactors dominated by an *Acidithiobacillus* and a *Thiobacillus* species. *Environ Technol.* https://doi.org/10.1080/09593330.2016.1147609

Gierth M, Mäser P, Schroeder JI (2005) The potassium transporter AtHAK5 functions in K+ deprivation-induced high-affinity K+ uptake and AKT1 K+ channel contribution to K+ uptake kinetics in *Arabidopsis* roots. *Plant Physiol* 137:1105–1114

Glick B (2012) *Plant Growth-Promoting Bacteria: Mechanisms and Applications*. Hindawi Publishing, New York

Glick R, Gilmour C, Tremblay J, Satanower S, Avidan O, Deziel E, Greenberg EP, Poole K, Banin E (2010) Increase in rhamnolipid synthesis under iron-limiting conditions influences surface motility and biofilm formation in *Pseudomonas aeruginosa*. *J Bacteriol* 192:2973–2980

Godfray HCJ, Beddington JR, Crute IR, Haddad L, Lawrence D, Muir JF, Pretty J, Robinson S, Thomas SM, Toulmin C (2010) Food security: the challenge of feeding 9 billion people. *Science* 327:812–818

Godfray HCJ, Garnett T (2014) Food security and sustainable intensification. *Philos Trans R Soc Lond B* 369:20120273. https://doi.org/10.1098/rstb.2012.0273

Goldberg S, Sposito G (1985) On the mechanisms of specific phosphate adsorption by hydroxylated mineral surfaces. A review. *Commun Soil Sci Plant Anal* 16:801–821

Goss CJ (1987) The kinetics and reaction mechanisms of the goethite to hematite transformation. *Min Mag* 51:437–451. https://doi.org/10.1180/minmag.1987.051.361.11

Gupta N, Ram H, Kumar B (2016) Mechanism of zinc absorption in plants: uptake, transport, translocation and accumulation. *Rev Environ Sci Biotechnol* 15:89–109. https://doi.org/10.1007/s11157-016-9390-1

Hall FR (1972) Silica cycle. In: Fairbridge RW (ed.) *The Encyclopedia of Geochemistry and Environmental Sciences. Encyclopedia of Earth Science Series*, vol. IVA.Van Nostrand Reinhold, New York, pp. 1082–1085

Hallbauer DK, Jahns HM (1977) Attack of lichens on quartzitic rock surfaces. *Lichenologist* 9:119–122

Hallberg K (2010) New perspectives in acid mine drainage microbiology. *Hydrometallurgy* 104:448–453. Https ://doi.org/10.1016/j.hydromet.2009.12.013

Haria NR, Grest GS, Lorenz CD (2013) Viscosity of nanoconfined water between hydroxyl basal surfaces of kaolinite: classical simulation results. *J Phys Chem C* 117:6096–6104

Harriss RC (1972) Silica–biogeochemical cycle. In: Fairbridge RW (ed.) *The Encyclopedia of Geochemistry and Environmental Sciences. Encyclopedia of Earth Science Series*. vol. IVA.Van Nostrand Reinhold, New York, pp. 1080–1082

Haynes RJ (2014) A contemporary overview of silicon availability in agricultural soils. *J Plant Nutr Soil Sci* 177:831–844. https://doi.org/10.1002/jpln.201400202

Hefferon K (2019) Biotechnological approaches for generating zinc-enriched crops to combat malnutrition. *Nutrients* 11. 253. https://doi.org/10.3390/nu11020253

Hofer U (2018) New diversity in the sulfur cycle. *Nat Rev Microbiol* 16:260–261. https://doi.org/10.1038/nrmicro.2018.32

Holland JE, White PJ, Glendining MJ, Goulding KWT, McGrath SP (2019) Yield responses of arable crops to liming: an evaluation of relationships between yields and soil pH from a long-term liming experiment. *Eur J Agron* 105:176–188. https://doi.org/10.1016/j.eja.2019.02.016

Hori T, Aoyagi T, Itoh H, Narihiro T, Oikawa A, Suzuki K, Ogata A, Friedrich MW, Conrad R, Kamagata Y (2015) Isolation of microorganisms involved in reduction of crystalline iron (III) oxides in natural environments. *Front Microbiol* 6:386. https://doi.org/10.3389/fmicb.2015.00386

Horne I, Sutherland TD, Harcourt RL, Russell RJ, Oakeshott JG (2002) Identification of an opd (organophosphate degradation) gene in an *Agrobacterium* isolate. *Appl Environ Microbiol* 68:3371–3376. https://doi .org/10.1128/aem.68.7.3371-3376.2002

Hosy E, Vavasseur A, Mouline K, Dreyer I, Gaymard F, Porée F, Boucherez J, Lebaudy A, Bouchez D, Very AA, Simonneau T, Thibaud JB, Sentenac H (2003) The *Arabidopsis* outward K+ channel GORK is involved in regulation of stomatal movements and plant transpiration. *Proc Natl Acad Sci U S A* 100:5549–5554. https://doi.org/10.1073/pnas.0733970100.

Hurtado SMC, Resende AV, Corazza EJ, Shiratsuchi LS, Higashikawa FS (2008) Otimização da adubaçãoem-lavouras com altos teores de potássio no solo. *Rev Plantio Direto* 106:28–31

Hyde KD, Xu J, Rapior S et al. (2019) The amazing potential of fungi: 50 ways we can exploit fungi industrially. *Fungal Diversity*. 97:1–136 https://doi.org/10.1007/s13225-019-00430-9

Igiri BE, Okoduwa SIR, Idoko GO, Akabuogu EP, Adeyi AO, Ejiogu IK (2018) Toxicity and bioremediation of heavy metals contaminated ecosystem from tannery wastewater: a review. *J Toxicol* 2018:2568038. https://doi.org/10.1155/2018/2568038. PMID:30363677

Iler R (1979) *Chemistry of Silica*. Wiley, New York

Illmer P, Barbato A, Schinner F (1995) Solubilization of hardly-soluble AlPO4, with P-solubilizing microorganisms. *Soil Biol Biochem* 27:265–270

Illmer P, Schinner F (1992) Solubilization of inorganic phosphates by microorganisms isolated from forest soil. *Soil Biol Biochem* 24:389–395

Inda AV, Torrent J, Barrón V, Bayer C (2010) Aluminum hydroxy-interlayered minerals and chemical properties of a subtropical Brazilian Oxisol under no-tillage and conventional tillage. *Rev Bras Cienc Solo* 34:33–41. https://doi.org/10.1590/S0100-06832010000100004

Janků M, Luhová L, Petřivalský M (2019) On the origin and fate of reactive oxygen species in plant cell compartments. *Antioxidants (Basel)* 8:105. https://doi.org/10.3390/antiox8040105.

Jha SK, Sharma M, Pandey GK (2016) Role of cyclic nucleotide-gated channels in stress management in plants. *Curr Genomics* 17:315–329. https://doi.org/10.2174/1389202917666160331202125

Johnson DB (2012) Geomicrobiology of extremely acidic subsurface environments. *FEMS Microbiol Ecol* 81:2–12. https://doi.org/10.1111/j.1574-6941.2011.01293.x

Johnson DB (2013) Development and application of biotechnologies in the metal mining industry. *Environ Sci Pollut Res* 20:7768–7776

Jongmans AG, van Breemen N, Lundstrom U, van Hees PAW, Finlay RD, Srinivasan M, Unestam T, Giesler R, Melkerud PA, Olsson M (1997) Rock-eating fungi. *Nature* 389:682–683

Jørgensen BB, Findlay AJ, Pellerin A (2019) The biogeochemical sulfur cycle of marine sediments. *Front Microbiol* 10:849. https://doi.org/10.3389/fmicb.2019.00849

Jurinak JJ, Dudley LM, Allen MF, Knight WG (1986) The role of calcium oxalate in the availability of phosphorus in soils of semiarid regions: a thermodynamic study. *Soil Sci* 142:255–261

Kamran S, Shahid I, Baig DN, Rizwan M, Malik KA, Mehnaz S (2017) Contribution of zinc solubilizing bacteria in growth promotion and zinc content of wheat. *Front Microbiol* 8:2593. https://doi.org/10.3389/fmicb.2017.02593

Kang SC, Ha GC, Lee TG, Maheshwari DK (2002) Solubilization of insoluble inorganic phosphates by a soil inhabiting fungus sp. Ps 102. *Curr Sci* 79:439–442

Kappen L (1993) Lichens in the Antarctic region. In: Friedmann EI (ed.) *Antarctic Microbiology*. Wiley-Liss, New York, pp. 433–490

Karavaiko GI, Avakyan ZA, Krutsko VS, Mel'nikova EO, Zhdanov V, Piskunov VP (1979) Microbiological investigations on a spodumene deposit. *Mikrobiologiya* 48:502–508 (English translation pp. 383–398)

Katz O (2018) Plant silicon and phytolith research and the earth-life super discipline. *Front Plant Sci* 9:1281. https://doi.org/10.3389/fpls.2018.01281

Keswani C, Sarma BK, Singh HB (2016) Synthesis of policy support, quality control, and regulatory management of biopesticides in sustainable agriculture. In: Singh H, Sarma B, Keswani C (eds.) *Agriculturally Important Microorganisms*. Springer, Singapore, pp. 3–12

Khan A, Singh P, Srivastava A (2018) Synthesis, nature and utility of universal iron chelator – Siderophore: a review. *Microbiol Res* 213:103–111.

Khan MS, Rizvi A, Saif S, Zaidi A (2017) Phosphate-solubilizing microorganisms in sustainable production of wheat: current perspective. In: Kumar V, Kumar M, Sharma S, Prasad R (eds.) *Probiotics in Agroecosystem*. Springer, Singapore. pp. 51–81. http://link.springer.com/10.1007/978-981-10-4059-7_3

Kim KY, McDonald GA, Jordan D (1997) Solubilization of hydroxyapatite by *Enterobacter agglomerans* and cloned *Escherichia coli* in culture medium. *Biol Fertil Soils* 24:347–352. https://doi.org/10.1007/s003740 050256

Kirtzel J, Madhavan S, Wielsch N, Blinne A, Hupfer Y, Linde J, Krause K, Svatoš A, Kothe E (2018) Enzymatic bioweathering and metal mobilization from Black Slate by the *Basidiomycete Schizophyllum* commune. *Front Microbiol* 9:2545. https://doi.org/10.3389/fmicb.2018.02545

Klotz MG, Bryant DA, Hanson TE (2011) The microbial sulfur cycle. *Front Microbiol* 2:241. https://doi.org/10.3389/fmicb.2011.00241

Krishnaraj PU, Dahale S (2014) Mineral phosphate solubilization: concepts and prospects in sustainable agriculture. *Proc Indian Natl Sci Acad* 80: 389. https://doi/org/10.16943/ptinsa/2014/v80i2/55116

Kuppusamy T, Giavalisco P, Arvidsson S, Sulpice R, Stitt M, Finnegan PM, … Jost R (2014) Lipid biosynthesis and protein concentration respond uniquely to phosphate supply during leaf development in highly phosphorus-efficient *Hakea prostrata. Plant Physiol* 166:1891–1911. https://doi.org/10.1104/pp.114.248930

Lammers PJ, Jun J, Abubaker J, Arreola R, Gopalan A, Bago B, Hernandez-Sebastia C, Allen JW, Douds DD, Pfeffer PE, Shachar-Hill Y (2001) The glyoxylate cycle in an arbuscular mycorrhizal fungus. carbon flux and gene expression. *Plant Physiol* 127:1287–1298

Lee SG, Choi JI, Koh W, Jang SS (2013) Adsorption of β-D-glucose and cellobiose on kaolinite surfaces: density functional theory (DFT) approach. *Appl Clay Sci* 71:73–81

Lever MA, Rogers KL, Lloyd KG, Overmann J, Schink B, Thauer RK, et al. (2015) Life under extreme energy limitation: a synthesis of laboratory- and field-based investigations. *FEMS Microbiol Rev* 39:688–728.

Li Y, Kawashima N, Li J, Chandra AP, Gerson AR (2013) A review of the structure, and fundamental mechanisms and kinetics of the leaching of chalcopyrite. *Adv Colloid Interface Sci* 197–198:1–32

Liang Y, Nikolic M, Bélanger R, Gong H, Song A (2015) *Silicon in Agriculture*. Dordrecht: Springer

Liang Y, Sun W, Zhu Y-G, Christie P (2007) Mechanisms of silicon mediated alleviation of abiotic stresses in higher plants: a review. *Environ Pollut* 147:422–428. https://doi:.org/10.1016/j.envpol.2006.06.008

Lindenmayer DB, Sato C (2018) Hidden collapse is driven by fire and logging in a socioecological forest ecosystem. *Proc Natl Acad Sci U S A* 115:5181–5186. https://doi.org/10.1073/pnas.1721738115

Lovley DR (2012) Electromicrobiology. *Annu Rev Microbiol* 66:391–409

Lucheta AR, Lambais MR (2012) Sulfur in agriculture. *Rev Bras Ciênc Solo* 35:1369–1379. https://doi.org/10.1590/S0100-06832012000500001

Ludwig Y and Slamet-Loedin IH (2019) Genetic biofortification to enrich rice and wheat grain iron: from genes to product. *Front Plant Sci* 10:833. https://doi.org/10.3389/fpls.2019.00833

Luther GW, Findlay AJ III, MacDonald DJ, Owings SM, Hanson TE, Beinart RA, Girguis PR (2011) Thermodynamics and kinetics of sulfide oxidation by oxygen: a look at inorganically controlled reactions and biologically mediated processes in the environment. *Front Microbiol* 2:62. https://doi.org/10.3389/fmicb.2011.00062

Luziatelli F, Ficca AG, Colla G, Baldassarre Švecová E, Ruzzi M (2019) Foliar application of vegetal-derived bioactive compounds stimulates the growth of beneficial bacteria and enhances microbiome biodiversity in lettuce. *Front Plant Sci* 10:60. https://doi.org/10.3389/fpls.2019.00060

MacDonald GK, Bennett EM, Potter PA, Ramankutty N (2011) Agronomic phosphorus imbalances across the world's croplands. *Proc Natl Acad Sci U S A* 108:3086–3091

Mahanty T, Bhattacharjee S, Goswami M, Bhattacharyya P, Das B, Ghosh A, et al. (2016) Biofertilizers: a potential approach for sustainable agriculture development. *Environ Sci Pollut Res* 24:3315–3335. https://doi.org/10.1007/s11356-016-8104-0

Majumdar S, Chakraborty U (2015) Phosphate solubilizing rhizospheric *Pantoea agglomerans* Acti-3 promotes growth in jute plants. *World J Agric Sci* 11:401–410. https://doi.org/10.5829/idosi.wjas.2015.11.6.1893

Malavolta E, Kliemann HJ (1985) *Desordens Nutricionais no Cerrado*. Associação Brasileira para Pesquisa da Potassa e do Fosfato, Piracicaba, SP

Malavolta E, Vitti GC, Oliveira SA (1997) *Avaliação do Estado Nutricional das Plantas: Princípios e Aplicações*. 2ndPotafos, Piracicaba, SP

Malhotra H, Vandana SS, Pandey R (2018) Phosphorus nutrition: plant growth in response to deficiency and excess. In: Hasanuzzaman M., Fujita M., Oku H., Nahar K., Hawrylak-Nowak B (eds.) *Plant Nutrients and Abiotic Stress Tolerance*. Springer, Singapore, pp. 171–190

Mangas I, Vilanova E, Estévez J, França TCC (2016) Neurotoxic effects associated with current uses of organophosphorus compounds. *J Braz Chem Soc* 27:809–825. http://doi.org/10.5935/0103-5053.20160084

Manwaring HR, Bligh HFJ, Yadav R (2016) The challenges and opportunities associated with biofortification of Pearl Millet (*Pennisetum glaucum*) with elevated levels of grain iron and zinc. *Front Plant Sci* 7:1944. doi:10.3389/fpls.2016.01944

Marschner H (1995) *Mineral Nutrition of Higher Plants.* 2nd ed. Academic, London
Marschner H, Marschner P (2012) *Marschner's Mineral Nutrition of Higher Plants.* Academic Press, New York
Maynard DG, Stewart JWB, Betanny JR (1985) The effects of plants on soil sulfur transformations. *Soil Biol Biochem* 17:99–103
McArthur JM (1985) Francolite geochemistry-compositional controls during formation, diagenesis, metamorphism and weathering. *Geochim Cosmochim Acta* 49:23–35
McCarroll D, Viles H (1995) Rock-weathering by the lichen *Lecidea auriculata* in an arctic alpine environment. *Earth Surf Processes Landforms* 20:199–206
Meis SJ, Fehr WR, Schnebly SR (2003) Seed source effect on field emergence of soybean lines with reduced phytate and raffinose saccharides. *Crop Sci* 43:1336–1339
Melton ED, Swanner ED, Behrens S, Schmidt C, Kappler A (2014) The interplay of microbially mediated and abiotic reactions in the biogeochemical Fe cycle. *Nat Rev Microbiol* 12:797–808. https://doi.org/10.1038/nrmicro3347
Menezes-Blackburn D, Jorquera MA, Greiner R, Gianfreda L, de la Luz Mora M (2013) Phytases and phytase-labile organic phosphorus in manures and soils. *Crit Rev Environ Sci Technol* 43:916–954
Mengel K, Kirbky EA (1987) *Principles of Plant Nutrition.* 4th ed. International Potash Institute, Worblufen-Bern, Switzerland
Milner MJ, Seamon J, Craft E, Kochian LV (2013) Transport properties of members of the ZIP family in plants and their role in Zn and Mn homeostasis. *J Exp Bot* 64:369–381.doi:10.1093/jxb/ers315
Mohammadi K (2012) Phosphorus solubilizing bacteria: occurrence, mechanisms and their role in crop production. *Resour Environ* 2:80–85 https://doi.org.10.5923/j.re.20120201.10
Molin JP, Tavares TR (2019) Sensor systems for mapping soil fertility attributes: challenges, advances, and perspectives in Brazilian tropical soils. *Eng Agríc* 39:126–147. https://doi.org/10.1590/1809-4430-Eng.Agric.v39nep126-147/2019
Moses CA, Smith BJ (1993) A note on the role of the lichen *Collemaauriforma* in solution basin development on a carboniferous limestone substrate. *Earth Surf Processes Landforms* 18:363–368
Moterle DF, Bortoluzzi EC, Kaminski J, Rheinheimer DS, Caner L (2019) Does ferralsol clay mineralogy maintain potassium long-term supply to plants? *Rev Bras Cienc Solo* 43:e0180166
Nahas E (1996) Factors determining rock phosphate solubilization by microorganism isolated from soil. *World J Microbiol Biotechnol* 12:18–23
Nannipieri P, Giagnoni L, Landi L, Renella G (2011) Role of phosphatase enzymes in soil. In Bunemann E, Oberson A, Frossard E (ed.) *Phosphorus in Action: Biological Processes in Soil Phosphorus Cycling. Soil Biology*, vol 26. Springer, Berlin, pp. 251–244
Nathan Y (1984) The mineralogy and geochemistry of phosphorites. In: Nriagu JO, Moore PB (eds.) *Phosphate Minerals.* Springer, Berlin, pp. 275–291
Neethirajan S, Ragavan V, Weng X, Chand R (2018) Biosensors for sustainable food engineering: challenges and perspectives. *Biosensors* 8:23. https://doi.org/10.3390/bios8010023
Nguyen TH, Huang S, Meynard D, Chaine C, Michel R, Roelfsema MRG, Guiderdoni E, Sentenac H, Véry AA (2017) A dual role for the OsK5.2 Ion channel in stomatal movements and K+loading into Xylem sap. *Plant Physiol* 174:2409–2418. https://doi.org/10.1104/pp.17.00691.
Nor YM (1981) Sulfur mineralization and adsorption in soil. *Plant Soil* 60:451–459
Olanrewaju OS, Glick BR, Babalola OO (2017) Mechanisms of action of plant growth promoting bacteria. *World J Microbiol Biotechnol* 33:197. https://doi.org/10.1007/s11274-017-2364-9
Olsen SR, Khasawneh FE (1980) Use and limitations of physical-chemical criteria for assessing the status of phosphorus in soils. In: Khasawneh FE, Sample EC, Kamprath EJ (eds.) *The Role of Phosphorus in Agriculture.* American Society of Agronomy, Madison, pp. 361–410
Oteino N, Lally RD, Kiwanuka S, Lloyd A, Ryan D, Germaine KJ, et al. (2015) Plant growth promotion induced by phosphate solubilizing endophytic *Pseudomonas* isolates. *Front Microbiol* 6:745. https://doi.org/10.3389/fmicb.2015.00745
Patibanda AK, Ranganathswamy M (2018) Effect of agrichemicals on biocontrol agents of plant disease control. In: Panpatte D, Jhala Y, Shelat H, Vyas R (eds.) *Microorganisms for Green Revolution. Microorganisms for Sustainability*, vol 7. Springer, Singapore, pp. 1–21

Pearce D, Turner RK (1990) *Economics of Natural Resources and the Environment*. Harvester, Wheatsheaf, New York

Peng T, Zhou D, Liu Y, Yu R, Qiu G, Zeng W (2019) Effects of pH value on the expression of key iron/sulfur oxidation genes during bioleaching of chalcopyrite on thermophilic condition. *Ann Microbiol* 69:627–635

Pennisi E (2008) Plant genetics. The blue revolution, drop by drop, gene by gene. *Science*. 320:171–173. https://doi.10.1126/science.320.5873.171

Pollmann K, Kutschke S, Matys S, Kostudis S, Hopfe S, Raff J (2016) Novel biotechnological approaches for the recovery of metals from primary and secondary resources. *Minerals* 6:54. https://doi.org/10.3390/min6020054

Porder S, Ramachandran S (2013) The phosphorus concentration of common rocks—a potential driver of ecosystem p status. *Plant Soil* 367:41–55. https://doi.org/10.1007/s11104-012-1490-2

Pose-Juan E, Sanchez-Martin MJ, Andrades MS, Rodriguez-Cruz MS, Herrero Hernandez E (2015) Pesticide residues in vineyard soils from Spain: spatial and temporal distributions. *Sci Total Environ* 514:351–358. https://doi.org/10.1016/j.scitotenv.2015.01.076

Qi W, Zhao L (2013) Study of the siderophore producing *Trichoderma asperellum* Q1 on cucumber growth promotion under salt stress. *J Basic Microbiol* 53(4):355–364

Rabus R, Hansen TA, Widdel F (2013) Dissimilatory sulfate- and sulfur-reducing prokaryotes. In: Rosenberg E, DeLong EF, Lory S, Stackebrandt E, Thompson F (eds.) *The Prokaryotes*. Springer, Berlin, pp. 309–404

Ramesh A, Sharma SK, Sharma MP, Yadav N, Joshi OP (2014) Inoculation of zinc solubilizing *Bacillus aryabhattai* strains for improved growth, mobilization and biofortification of zinc in soybean and wheat cultivated in vertisols of central India. *Agric Ecosyst Environ Appl Soil Ecol* 73:87–96. https://doi.org/10.1016/j.apsoil.2013.08.009

Rhee YJ, Hillier S, Pendlowski H, Gadd GM (2014) Pyromorphite formation in a fungal biofilm community growing on lead metal. *Environ Microbiol* 16:1441–1451

Ribeiro AC, Accioly LJO, Alvarez VH, Braga JM, Alves VMC (1991) Avaliação do enxofre disponível pelo método da resinatrocadora de ânions. *R Bras Ci Solo* 15:321–327

Ricci M, Tilbury L, Daridon B, Sukalac K (2019) General principles to justify plant biostimulant claims. *Front Plant Sci* doi:10.3389/fpls.2019.00494

Rimstidt JD (1997) Quartz solubility at low temperatures. *Geochim Cosmochim Acta* 61:2553–2558

Rosolem CA, Bessa AM, Pereira HFM (1993) Dinâmica do potássio no solo e nutriçãopotássica da soja. *Pesqui Agropec Bras* 28:1045–1054

Rouphael Y, Colla G (2018) Synergistic biostimulatory action: designing the next generation of plant biostimulants for sustainable agriculture. *Front Plant Sci* 9:1655. doi:10.3389/fpls.2018.01655

Rouphael Y, Spíchal L, Panzarová K, Casa R, Colla G (2018) High throughput plant phenotyping for developing novel biostimulants: from lab to field or from field to lab? *Front Plant Sci* https://doi.org/10.3389/fpls.2018.01197

Sadeghzadeh BA (2013) Review of zinc nutrition and plant breeding. *J Soil Sci Plant Nutr* 13:905–927 http://doi.org/10.4067/S0718-95162013005000072

Saha M, Sarkar S, Sarkar B, Sharma B, Bhattacharjee S, Tribedi P (2016) Microbial siderophores and their potential applications: a review. *Environ Sci Pollut Res* 23:3984–3999

Saravanan VS, Madhaiyan M, Thangaraju M (2007) Solubilization of zinc compounds by the diazotrophic, plant growth promoting bacterium *Gluconacetobacter diazotrophicus*. *Chemosphere* 66:1794–1798. https://doi.org/10.1016/j.chemosphere.2006.07.067

Sarwar A, Parthasarathy S, Singh C, Kumar S, Siddavattam D (2019) Genome organization and adaptive potential of archetypal organophosphate-degrading *Sphingobiumfuliginis* ATCC 27551. *Genome Biol Evol* 11:2557–2562. https://doi.org/10.1093/gbe/evz189

Sattari SZ, Bouwman AF, Giller KE, van Ittersum MK (2012) Residual soil phosphorus as the missing piece in the global phosphorus crisis puzzle. *Proc Natl Acad Sci U S A* 109:6348–6353. https://doi.org/10.1073/pnas.1113675109

Sauro F, Cappelletti M, Ghezzi D, Columbu A, Hong P-Y, Zowawi HM, et al. (2018) Microbial diversity and biosignatures of amorphous silica deposits in orthoquartzite caves. *Sci Rep* 8:17569. https://doi.org/10.1038/s41598-018-35532-y

Scheerer S, Ortega-Morales O, Gaylarde C (2009) Microbial deterioration of stone monuments: an updated overview. *Adv Appl Microbiol* 66:97–139. https://doi.org/10.1016/S0065-2164(08)00805-8

Schimitz GW, Pratt PF (1953) Exchangeable and nonexchangeable potassium as indexes to yield increases and potassium absorption by corn in the greenhouse. *Soil Sci* 76:345–354

Schink B, Thiemann V, Laue H, Friedrich MW (2002) *Desulfotignum phosphitoxidans* sp. nov., a new marine sulfate reducer that oxidizes phosphite to phosphate. *Arch Microbiol* 177:381–391

Schinner F, Öhlinger R, Kandler E, Margesin E (eds.) (1996) *Methods in Soil Biology*. Springer-Verlag, Berlin

Schippers A, Sand W (1999) Bacterial leaching of metal sulfides proceeds by two indirect mechanisms via thiosulfate or via polysulfides and sulfur. *Appl Environ Microbiol* 65:319–321

Schuffert J, Kastner M, Emanuele G, Jahnke R (1990) Carbonate-ion substitution in francolite: a new equation: *Geochim Cosmochim Acta* 54:2323–2328. https://doi.org/10.1016/0016-7037(90)90058-S

Schultze-Lam S, Fortin D, Davis BS, Beveridge TJ (1996) Mineralization of bacterial surfaces. *Chem Geol* 132:171–181

Schulz B, Segobye K (2016) 2,4-D transport and herbicide resistance in weeds. *J Exp Bot* 67:3177–3179. https://doi.org/10.1093/jxb/erw199

Seifert AG, Trumbore S, Xu X, Zhang D, Kothe E, Gleixner G (2011) Variable effects of labile carbon on the carbon use of different microbial groups in black slate degradation. *Geochim Cosmochim Acta* 75:2557–2570. https://doi.org/10. 1016/j.gca.2011.02.037

Selle PH, Ravindran V, Bryden WL, Scott T (2006) Influence of dietary phytate and exogenous phytase on amino acids digestibility in poultry: a review. *J Poult Sci* 43:89–103

Selvakumar G, Joshi P, Nazim S, Mishra PK, Bisht JK, Gupta HS (2009) Phosphate solubilization and growth promotion by *Pseudomonas fragi* CS11RH1 (MTCC 8984), a psychrotolerant bacterium isolated from a high altitude Himalayan rhizosphere. *Biologia* 64:239–245. https://doi.org/10.2478/s11756-009-0041-7

Serdar CM, Gibson DT (1985) Enzymatic hydrolysis of organophosphates: cloning and expression of a parathion hydrolase gene from *Pseudomonas diminuta*. *Biotechnology* 3:567–571

Setiawati TC, Mutmainnah L (2016) Solubilization of potassium containing mineral by microorganisms from sugarcane rhizosphere. *Agri Sci Procedia* 9:108–117. https://doi.org/10.1016/j.aaspro.2016.02.134

Sewelam N, Kazan K, Schenk PM (2016) Global plant stress signaling: reactive oxygen species at the crossroad. *Front Plant Sci* 23:187. https://doi.org/10.3389/fpls.2016.00187

Sharma SB, Sayyed RZ, Trivedi MH, Gobi TA (2013) Phosphate solubilizing microbes: sustainable approach for managing phosphorus deficiency in agricultural soils. *Springer Plus* 2:587. https://doi.org/10.1186/2193-1801-2-587

Sharma SK, Sharma MP, Ramesh A, Joshi OP (2012) Characterization of zinc solubilizing *Bacillus* isolates and their potential to influence zinc assimilation in soybean seeds. *J Microbiol Biotechnol* 22:352–359. doi:10.4014/jmb.1106.05063

Shelden MC, Roessner U (2013) Advances in functional genomics for investigating salinity stress tolerance mechanisms in cereals. *Front Plant Sci* 4:123. http://doi.org/10.3389/fpls.2013.00123

Shen J, Yuan L, Zhang J, Li H, Bai Z, Chen X, et al (2011) Phosphorus dynamics: from soil to plant. *Plant Physiol* 156:997–1005

Shi L, Dong H, Reguera G, Beyenal H, Lu A, Liu J, et al. (2016) Extracellular electron transfer mechanisms between microorganisms and minerals. *Nat Rev Microbiol* 14:651–662

Siddavattam D, Yakkala H, Samantarrai D (2019) Lateral transfer of organophosphate degradation (opd) genes among soil bacteria: mode of transfer and contributions to organismal fitness. *J Genet* 98:23. https://doi.org/10.1007/s12041-019-1068-3

Silva MP, do Santos MSF, Santos MRMC, Santos Júnior L de S, Fonseca MG da, Silva Filho EC (2016) Natural Palygorskite as an industrial dye remover in single and binary systems. *Mater Res* 19:1232–1240. https://doi.org/10.1590/1980-5373-MR-2015-0439

Simeonova DD, Wilson MM, Metcalf WW, Schink B (2010) Identification and heterologous expression of genes involved in anaerobic dissimilatory phosphite oxidation by Desulfotignum phosphitoxidans. *J Bacteriol* 192:5237–5244

Singh BK, Walker A (2006) Microbial degradation of organophosphorus compounds. *FEMS Microbiol Rev* 30:428–471. https://doi.org/10.1111/j.1574-6976.2006.00018.x

Šolc R, Gerzabek MH, Lischka H, Tunega D (2011) Wettability of kaolinite (001) surfaces—molecular dynamic study. *Geoderma* 169:47–54

Song SK, Huang PM (1988) Dynamics of potassium release from potassium-bearing minerals as influenced by oxalic and citric acids. *Soil Sci Soc Am J* 52:383–390

Sparks DL, Liebhardt WC (1981) Effect of long-term lime and potassium applications on quantity-intensity (Q/I) relationships in sandy soil. *Soil Sci Soc Am J* 45:786

Sperber JI (1958) Release of phosphates from soil minerals by hydrogen sulfide. *Nature* 181:934

Stamenković S, Beškoski V, Karabegović I, Lazić M, Nikolić N (2018) Microbial fertilizers: a comprehensive review of current findings and future perspectives. *Span J Agric Res* 16:e09R01. https://doi.org/10.5424/sjar/2018161-12117

Stanley W, Southam G (2018) The effect of gram-positive (*Desulfosporosinusorientis*) and gram-negative (*Desulfovibriodesulfuricans*) sulfate-reducing bacteria on iron sulfide mineral precipitation. *Can J Microbiol* 64:629–637

Sugiura Y, Tanaka H, Mino Y, Ishida T, Ota N, Inoue M (1981) Structure, properties, and transport mechanism of iron (III) complex of mugineic acid, a possible phytosiderophore. *J Am Chem Soc* 103:6979–6982

Sverdrup H (2009) Chemical weathering of soil minerals and the role of biological processes. *Fungal Biol Rev* 23:94–100. https://doi.org/10.1016/j.fbr.2009.12.001

Swaby RJ, Sperber J (1958) Phosphate-dissolving microorganisms in the rhizosphere of legumes. In: Hallsworth EG (ed.) *Nutrition of Legumes. Proc Univ Nottingham Easter School Agric Sci*, 5th ed. Academic Press, London, pp. 289–294

Taga N, Kobori H (1978) Phosphatase activity in eutrophic Tokyo Bay. *Marine Biology (Berlin)* 49:223–229

Takahashi H, Kopriva S, Giordano M, Saito K, Hell R (2011) Sulfur assimilation in photosynthetic organisms: molecular functions and regulations of transporters and assimilatory enzymes. *Annu Rev Plant Biol* 62:157–184. https://doi.org/10.1146/annurev-arplant-042110-103921

Takemoto T, Nomoto K, Fushiya S, Ouchi R, Kusano G, Hikino H, Takagi S, Matsuura Y, Kakudo M (1978) Structure of mugineic acid, a new amino acid possessing an iron-chelating activity from root washings of water-cultured *Hordeum vulgare* L. *Proc Jpn Acad* 54:469–473

Tan KH (1982) *Principles of Soil Chemistry*. Marcel Dekker, New York, pp. 179–194

Tan KH (2011) *Principles of Soil Chemistry*. 4th ed. CRC Press/Taylor and Francis, Boca Raton/Broken Sound Parkway, pp. 246–266

Tang K, Baskaran V, Nemati M (2009) Bacteria of the sulphur cycle: an overview of microbiology, biokinetics and their role in petroleum and mining industries. *Biochem Eng J* 44:73–94

Tazaki K (2005) Microbial formation of a halloysite-like mineral. *Clays Clay Miner* 53:224–233

Tazaki K (2013) Clays, micro-organisms, and biomineralization. In: Bergaya F, Lagaly G (eds.) *Handbook of Clay Science 5*, Elsevier, Amsterdam, 613–653

Thompson LA, Darwish WS (2019) Environmental chemical contaminants in food: review of a global problem. *J Toxicol* 1–14. https://doi.org/10.1155/2019/2345283

Tinker PB (1975) Soil chemistry of phosphorus and mycorrhizal effects on plant growth. In: Sanders FE, Mosse B, Tinker PB (eds.) *Endomycorhizas*. Academic Press, London, pp. 353–371

Tipping E, Somerville CJ, Luster J (2016) The C:N:P:S stoichiometry of soil organic matter. *Biogeochemistry* 130:117–131. https://doi.org/10.1007/s10533-016-0247-z

Tisdale SL, Nelson WL, Beaton JD (1985) *Soil Fertility and Fertilizers*. 4th ed. Macmillan Publishing Company, New York

Tobler DJ, Stefansson A, Benning LG (2008) In-situ grown silica sinters in Icelandic geothermal areas. *Geobiology* 6:481–502

Traon D, Amat L, Zotz F, du Jardin PA (2014) *Legal Framework for Plant Biostimulants and Agronomic Fertiliser Additives in the EU*. Report for the European Commission; Enterprise and Industry Directorate-General, Brussels, Belgium, p. 133

Tsonev T, Lidon FJC (2012) Zinc in plants: an overview. *Emir J Food Agric* 24:322–333

Tydecks L, Jeschke JM, Wolf M, Singer G, Tockner K (2018) Spatial and topical imbalances in biodiversity research. *PLoS ONE* 13(7):e0199327. https://doi.org/10.1371/journal.pone.0199327

U.S. Geological Survey (2019) *Mineral Commodity Summaries 2019*. U.S. Geological Survey, 200p, http://doi.org/10.3133/70202434

Uehara G, Gillman GP (1981) *The Mineralogy, Chemistry and Physics of Tropical Soils with Variable Charge Clays*. Boulder, Westview Press

Ullrich C, Novacky A (1990) Extra- and intracellular pH and membrane potential changes induced by K+, Cl– , H2PO– and NO– uptake and fusicoccin in root hairs of Limnobiumstoloniferum. *Plant Physiol* 94:1561–1567

Vandevivere P, Welch SA, Ullman WJ, Kirchman DL (1994) Enhanced dissolution of silicate minerals by bacteria at near-neutral pH. *Microb Ecol* 27:241–251

Vera M, Schippers A, Sand W (2013) Progress in bioleaching: fundamentals and mechanisms of bacterial metal sulfide oxidation: part A. *Appl Microbiol Biotechnol* 97:7529–7541

Verma JP, Jaiswa DK, Meena VS, Meena RS (2015a) Current need of organic farming for enhancing sustainable agriculture. *J Clean Prod* 102:545–547

Verma JP, Jaiswal DK, Meena VS, Kumar A, Meena RS (2015b) Issues and challenges about sustainable agriculture production for management of natural resources to sustain soil fertility and health. *J Clean Prod* 107:793–794

Visser JM, Robertson LA, van Verseveld HW, Kuenen G (1997) Sulfur production by obligately chemolitroautotrphic *Thiobacillus* species. *Appl Environ Microbiol* 63:2300–2305

Wasmund K, Mußmann M, Loy A (2017) The life sulfuric: microbial ecology of sulfur cycling in marine sediments. *Environ Microbiol Rep* 9:323–344. https://doi.org/10.1111/1758-2229.12538

Webley DM, Duff RB, Mitchell WA (1960) A plate method for studying the breakdown of synthetic and natural silicates by soil bacteria. *Nature* 188:766–767

Welch SA, Ullman WJ (1993) The effect of organic acids on plagioclase dissolution rates and stoichiometry. *Geochim Cosmochim Acta* 57:2725–2736

Westwood JH, Charudattan R, Duke SO, Fennimore SA, Marrone P, Slaughter DC, Swanton C, Zollinger R (2018) Weed management in 2050: perspectives on the future of weed science. *Weed Sci* 66:275–285. https://doi.org/10.1017/wsc.2017.78

Wieler N, Ginat H, Gillor O, Angel R (2019) The origin and role of biological rock crusts in rocky desert weathering. *Biogeo Sci* 16:1133–1145, https://doi.org/10.5194/bg-16-1133-2019

Willer H, Lernoud J (2019) *The World of Organic Agriculture. Statistics and Emerging Trends 2019*. FiBL, Frick, and, IFOAM – Organics International, Bonn

Willmann G, Fakoussa RM (1997) Extracellular oxidative enzymes of coalattacking fungi. *Fuel Process Technol* 52:27–41. https://doi.org/10.1016/S0378-3820(97)00013-1

Wilson MJ (1999) The origin and formation of clay minerals in soils: past, present and future perspectives. *Clay Miner* 34:7–25. https://doi.org/10.1180/000985599545957

Wray RA, Sauro F (2017) An updated global review of solutional weathering processes and forms in quartz sandstones and quartzites. *Earth-Science Reviews* 171:520–557

Yakhin OI, Lubyanov AA, Yakhin IA, Brown PH (2017) Biostimulants in plant science: a global perspective. *Front Plant Sci* 7:2049. https://doi.org/10.3389/fpls.2016.02049

Yin L, Wang S, Tanaka K, Fujihara S, Itai A, Den X, et al. (2016) Siliconmediated changes in polyamines participate in silicon-induced salt tolerance in *Sorghum bicolor* L. *Plant Cell Environ* 39:245–258. https://doi:.org/10.1111/pce.12521

Yost RS, Kamprath EJ, Lobato E, Naderman G (1979) Phosphorus response of corn on an Oxisol as influenced by rates and placement. *Soil Sci Soc Am J* 43:338–343. https://doi.org/10.2136/sssaj1979.03615995004300020020x

Zacharia JT (2019) *Degradation Pathways of Persistent Organic Pollutants (POPs) in the Environment, Persistent Organic Pollutants, Stephen Kudom Donyinah*. IntechOpen, doi:10.5772/intechopen.79645. Available from: https://www.intechopen.com/books/persistent-organic-pollutants/degradation-pathways-of-persistent-organic-pollutants-pops-in-the-environment

Zhang L, Wang MX, Li H, Yuan L, Huang JG, Penfold C (2014) Mobilization of inorganic phosphorus from soils by ectomycorrhizal fungi. *Pedosphere* 24:683–689

Zhang Y, Chen F-S, Wu X-Q, Luan F-G, Zhang L-P, Fang X-M, et al. (2018) Isolation and characterization of two phosphate-solubilizing fungi from rhizosphere soil of moso bamboo and their functional capacities when exposed to different phosphorus sources and pH environments. *PLoS ONE* 13(7):e0199625. https://doi.org/10.1371/journal.pone.0199625

Zhu W, Xia J, Yang Y, Nie Z, Zheng L, Ma C, Zhang R, Peng A, Tang L, Qiu G (2011) Sulfur oxidation activities of pure and mixed thermophiles and sulfur speciation in bioleaching of chalcopyrite. *Bioresour Technol* 102:3877–3882

Zörb C, Senbayram M, Peiter E (2014) Potassium in agriculture – status and perspectives. *J Plant Physiol* 171:656–669. http://doi.org/10.1016/j.jplph.2013.08.008

15

Advances in Creation and Development of Bacteriophage Collections: Viral Genomes Resources for Biotechnology and Agriculture

Galina Novik and Victoria Savich

CONTENTS

15.1 Introduction

Bacteriophages are viral agents that only infect bacterial species. Discovered in the late 19th and early 20th century, bacteriophages showing antagonistic action towards specific hosts attracted attention as potential antibacterial agents. However, a poor understanding of bacteriophage biology, controversial results of phage studies as well as the discovery of antibiotics significantly undermined interest in bacterial viruses as antimicrobials (Sulakvelidze and Kutter 2005).

Despite their limited application at that time, phages remained objects of intense research. The morphology, biochemistry and genetics of phages and their interaction with hosts were studied and attempts were made to systemize bacteriophages. However, the initial classifications were based on general features, especially the type of nucleic acid and the morphology of the virion. The development of molecular-genetic methods resulted in a revision of the classification schemes and a better understanding of the host–phage interaction process.

The prevalence of antibiotic-resistant strains triggered the search for new methods to eliminate bacterial pathogens, reviving the interest in bacteriophages. The phage market is steadily expanding with many studies now focusing on the application of bacterial viruses in different fields.

15.2 Biodiversity of Bacterial Viruses: Viral Resources in Nature and Environment

Viruses represent one of the most ubiquitous life forms. They can be found in any habitat and biome—aquatic and terrestrial media and other organisms. For example, the oceans contain about 10^{30} viral particles. Microorganisms constitute more than 90% of living marine biomass of which viruses kill approximately 20% per day. It appears likely therefore that the latter affect nutrient and energy cycles as well as the structure of marine microbial communities. However, viruses comprise only about 5% of the total biomass. Phages with small genomes and low burst sizes infect the most widespread, slow-growing members of the community, while phages with large burst sizes take advantage of microorganisms that are growing rapidly in response to transient favourable conditions (Suttle 2007). Global viral abundance in the top 1 m layer of fresh water and marine sediments is estimated at 0.5 and 28.72×10^{28} viruses, respectively. Although the whole surface of the oceans exceeds overall lake area, viral occurrence calculated per unit area of marine sediment is 2.5 times inferior to the corresponding lake value (Danovaro et al. 2008).

The terrestrial distribution of viruses ranges from 2.2×10^3 gdw^{-1} (per gram dry weight) in deserts to 5.8×10^9 gdw^{-1} in forest soil. Viral presence appears to be minimal in hot deserts; median in agricultural soils, fields and cold deserts; and highest in forest and wetland soils. Soil viral abundance is directly correlated with bacterial abundance and inversely correlated with soil pH. Temperature also influences viral appearance. Lower temperatures foster the persistence of viruses, while elevated temperatures promote thermal decay of particles (Williamson et al. 2017).

Bacteriophages demonstrate the ability to penetrate into sterile regions of the human body, such as the blood, lymph, internal organs and brain. Bacterial viruses are capable of crossing the epithelial cell layer, gaining access to both the vesicular and cytosolic compartments of the eukaryotic cell. It is suggested that phage transit is carried out through the Golgi apparatus via the endomembrane system. Around 3.1×10^{10} phage particles that are transferred every day across the epithelial cell layers of the gut are migrating within human bodies (Nguyen et al. 2017). Comparing the human gastrointestinal tract with other habitats, it was presumed that over 1×10^{12} virions were likely to occupy the former niche. However, the actual measurements have shown lower viral counts than might be expected. The inefficient elution of viral particles coupled with their binding and aggregation is expected to decrease the calculated amount (Shkoporov and Hill 2019).

Despite wide phage abundance, so far just over 6,000 bacterial viruses have been described morphologically. The majority of them are tailed phages (Ackermann and Prangishvili 2012). Global sampling surveys indicate that the morphological characteristics of marine bacterial viruses do not vary consistently with depth or oceanic region, in terms of parameters such as temperature, salinity and oxygen concentration. Non-tailed viruses numerically prevail in the upper ocean strata where they comprise 51–92% of the viral particles observed, but the majority of cultivated bacterial viruses are tailed (Brum et al. 2013).

About 2,500 complete phage genomes are listed in GenBank (NCBI 2019). They are characterized by high diversity. Phages infecting the same bacteria may show little or no relationship (Kwan et al. 2006). Vice versa, bacterial viruses may bear a closer resemblance to bacteriophages of other species than to their neighbours (Kotsonis et al. 2008).

15.3 Bacteriophage Culture Collections: Significance of Bacteriophage Conservation

Culture collections play an important role in microbiology. They are engaged in the identification, preservation, maintenance and distribution of microbial strains and cell cultures for scientific, medical and industrial goals. The World Federation for Culture Collections (WFCC) promotes and supports the establishment of culture collections and related services; provides liaison and establishes information networks between the depositories and their users; and arranges workshops, conferences, publications, newsletters and activities to ensure the stable performance of important collections (WFCC 2019). The World Data Centre for Microorganisms (WDCM) is the data bank of the WFCC, distributing information among various culture collections. The WDCM stockpiles data on 785 culture collections in 76 countries and regions. There are about 3 million microorganisms in the above-mentioned culture collections, of which only 38,622 are viral strains. At least 49 collections contain data about the preservation of bacteriophage strains (Table 15.1), but this information is incomplete and sometimes outdated. In most cases, bacteriophages are available in the collections with other cultures. Only a few depositories specializing exclusively in bacterial viruses are listed in the WDCM, for example the Bacteriophage Bank of Korea (Korea), the Collection of Bacteriophages and Microorganisms MicroWorld (Russia) and the Felix d'Herelle Reference Centre for Bacterial Viruses (Canada). The China Centre for Virus Culture Collection (China) works with viruses of various taxonomic groups, including bacteria (WDCM 2019).

The Belarusian collection of non-pathogenic microorganisms is registered as a member of the World Federation of Culture Collections (WFCC-MIRCEN World Data Centre for Microorganisms, WDCM, Global Catalogue of Microorganisms [GCM]; acronym BIM; collection number 909). The fund of microbial resources deposited in the Belarusian collection of non-pathogenic microorganisms was characterized in the sixth edition of the *World Directory of Collections of Cultures of Microorganisms*. Catalogues and databases of microbial cultures comprising entries of type, reference and industrial strains were compiled in Russian and English. The Belarusian collection of non-pathogenic microorganisms is considered a national asset of the Republic of Belarus (Figure 15.1).

Currently, 98 bacteriophage BIM strains are maintained at the Belarusian collection of non-pathogenic microorganisms. The main function of the bacteriophage collection is the organization and centralized recording of all bacteriophages. The principal research trends of the bacteriophage collection are

- Depositing and maintaining bacteriophage cultures valuable for different areas of microbiology and biotechnology in a specialized collection fund
- Selecting highly active bacteriophage strains to develop a novel generation of biopreparations for agriculture
- Identifying the molecular-genetic characteristics of bacteriophages and studying the physiology of bacteriophages representing diverse taxonomic groups (Figure 15.2)
- Optimizing the long-term preservation methods for bacteriophage cultures in the collection depository; elaborating and introducing new procedures of low-temperature conservation of bacteriophages; optimizing and defining regimens of refrigerating, defrosting and desiccation of bacteriophage suspensions and host bacteria concentrates
- Classifying information on deposited bacteriophage cultures, setting up data banks describing bacteriophage collection stocks, consulting on the identification and preservation of bacteriophages, depositing bacteriophage strains for patent procedures and providing cultures upon request

Bacteriophages are important biological objects demonstrating practical value for various industries, medicine and fundamental and applied research. The application of bacterial viruses requires the accumulation of data concerning their preservation in depositories. There are standard methods for the maintenance of bacteriophages, but these are not suitable for all cultures. Phage lysates stored at 4°C can retain stability for several months, even years, without significant loss of infectivity (Fortier and Moineau 2009). At the same time, lipid-containing phages are usually labile and should therefore be subcultured more frequently or other preservation methods should be used. The addition of chloroform caused the inactivation of lipid-containing phages as well as some tailed ones (Ackermann et al. 2004). Cryoconservation (preservation at low temperatures) and lyophilization/freeze-drying (process of water removal from the product after it is frozen and placed under vacuum) are long-term preservation methods. Both are standard techniques for bacteriophage maintenance, but they do not always suit certain viruses. Specifically, particle size and the presence of lipids may influence phage survival (Ackermann et al. 2004; Fortier and Moineau 2009). Another possible method of long-term bacteriophage preservation is based on freezing freshly infected susceptible bacteria at the early stages of bacteriophage development.

TABLE 15.1

Collections Containing Data about Preservation of Bacteriophage Strains

No.	Acronym	WDCM No.	Collection	Country
1.	ATCC	WDCM 1	American Type Culture Collection	United States
2.	BCRC	WDCM 59	Bioresource Collection and Research Centre	Chinese Taipei
3.	BDUZ	WDCM 17	Biological Sciences	Zimbabwe
4.	BGSC	WDCM 573	Bacillus Genetic Stock Centre	United States
5.	BIM	WDCM 909	Belarusian Collection of non-pathogenic microorganisms	Belarus
6.	BTCC	WDCM 66	Bulgarian Type Culture Collection	Bulgaria
7.	CBMW	WDCM 986	Collection of bacteriophages and microorganisms MicroWorld	Russian Federation
8.	CCGVCC	WDCM 613	China Centre for General Viruses Culture Collection	China
9.	CCTCC	WDCM 611	China Centre for Type Culture Collection	China
10.	CCVCC	WDCM 612	China Centre For Virus Culture Collection	China
11.	CFQ	WDCM 100	Cepario de la Facultad de Quimica	Mexico
12.	CICC	WDCM 582	China Centre of Industrial Culture Collection	China
13.	CMMC	WDCM 865	China Marine Microbe Collection	China
14.	CM-PUJ	WDCM 857	Coleccion Microorganismos Pontificia Universidad Javeriana	Colombia
15.	CNCM	WDCM 174	Collection Nationale de Cultures de Microorganismes	France
16.	CREA-CMVE	WDCM 1142	Council for Agricultural Research and Analysis of Agricultural Economics—Microbial Culture Collection of Oenological and Viticultural Environment	Italy
17.	CVCM	WDCM 815	Centro Venezolano de Colecciones de Microorganismos	Venezuela
18.	DBKKU3	WDCM 689	Department of Biology, Faculty of Science	Thailand
19.	DBUM;IPT	WDCM 765	Department of Biochemistry, Faculty of Medicine, University of Malaya	Malaysia
20.	DSMZ	WDCM 274	Leibniz-Institut DSMZ-Deutsche Sammlung von Mikroorganismen und Zellkulturen GmbH	Germany
21.	EMTC	WDCM 974	Collection for extremophile microorganisms and type cultures	Russian Federation
22.	GCL	WDCM 295	Central Laboratories	Israel
23.	HAMBI	WDCM 779	HAMBI Culture Collection	Finland
24.	HER	WDCM 6	Felix d'Herelle Reference Centre for Bacterial Viruses	Canada
25.	IFAM	WDCM 145	Institut fur Allgemeine Mikrobiologie	Germany
26.	IFM	WDCM 805	IFM Quality Services Pty Ltd	Australia

(*Continued*)

TABLE 15.1 (CONTINUED)

Collections Containing Data about Preservation of Bacteriophage Strains

No.	Acronym	WDCM No.	Collection	Country
27.	IFO	WDCM 191	Institute for Fermentation, Osaka	Japan
28.	ISS	WDCM 375	Collection of Bacteria	Serbia
29.	KFCC	WDCM 585	Korean Federation of Culture Collection	Republic of Korea
30.	LMS	WDCM 530	Carolina Biological Supply Company	United States
31.	MAFF	WDCM 637	NARO Genebank, Microorganism Section	Japan
32.	MRCKKU	WDCM 1130	Melioidosis Research Centre, Faculty of Medicine	Thailand
33.	MSCL	WDCM 854	Microbial Strain Collection of Latvia	Latvia
34.	NBRC	WDCM 825	NBRC Culture Collection	Japan
35.	NCC	WDCM 985	NIZO Culture Collection	Netherlands
36.	NCCB	WDCM 797	NCCB, the Netherlands Culture Collection of Bacteria	Netherlands
37.	NCIMB	WDCM 653	National Collections of Industrial Food and Marine Bacteria	UK
38.	NTCCI	WDCM 107	Culture Collection, Microbiology and Cell Biology Laboratory	India
39.	NZRD	WDCM 318	New Zealand Reference Culture Collection of Microorganisms, Dairy Section	New Zealand
40.	NZRM	WDCM 457	New Zealand Reference Culture Collection, Medical Section	New Zealand
41.	OCM	WDCM 592	Oregon Collection of Methanogens	United States
42.	Phagebank	WDCM 992	Bacteriophage Bank of Korea	Republic of Korea
43.	SAITP	WDCM 569	School of Pharmacy and Medical Sciences, University of South Australia	Australia
44.	SGSC	WDCM 338	Salmonella Genetic Stock Centre	Canada
45.	SMTWA	WDCM 90	School of Medical Technology Western Australia	Australia
46.	UCM	WDCM 1203	Ukrainian Collection of Microorganisms	Ukraine
47.	UPRM	WDCM 76	Rhizobium Culture Collection	United States
48.	USRCB	WDCM 855	Culture Collection for Science and Research.	Ukraine
49.	WCUM	WDCM 56	Working Collection	United States

The infected bacteria release mature bacteriophages upon melting. This method allows the preservation of bacteriophages susceptible to deep freezing in the form of lysate. It can serve as a high-throughput method of storing a large number of bacteriophages of unknown properties. The only limitation is the sensitivity of the bacteriophage host to deep-freezing (Łobocka et al. 2018). Encapsulation is the technology of packing solids, liquids or gaseous materials within sealed capsules that can release their contents

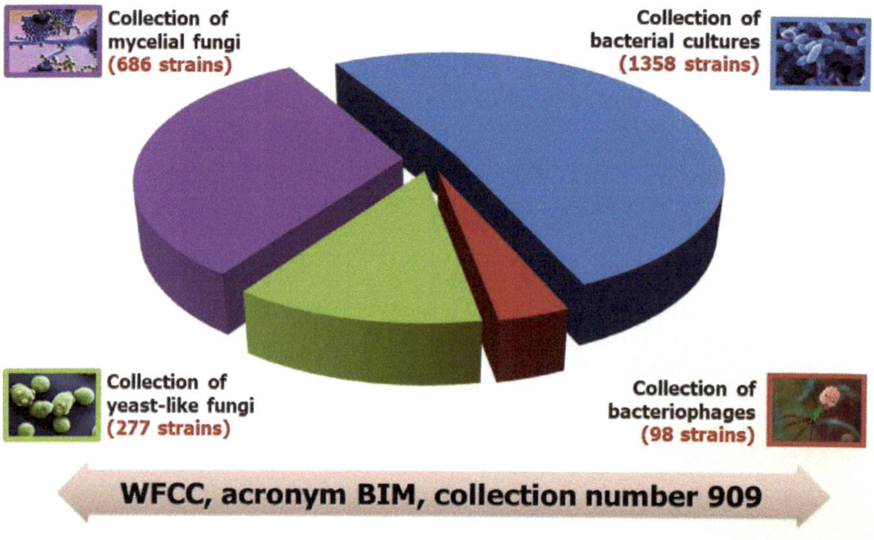

FIGURE 15.1 Main fund composition of Belarusian collection of microorganisms.

at controlled rates under the influence of specific conditions (Anal and Singh 2007). However, encapsulation can also be used as a storage method. *Staphylococcus* phage phiIPLA-RODI enclosed in alginate capsules retained high viability when stored at 4°C for 6 months and at 20°C for 1 month (González-Menéndez et al. 2018). *Listeria*, *Salmonella* and *Escherichia coli* bacteriophages encapsulated in sugar glasses composed of a mixture of pullulan and trehalose were able to retain infectivity for up to 3 months

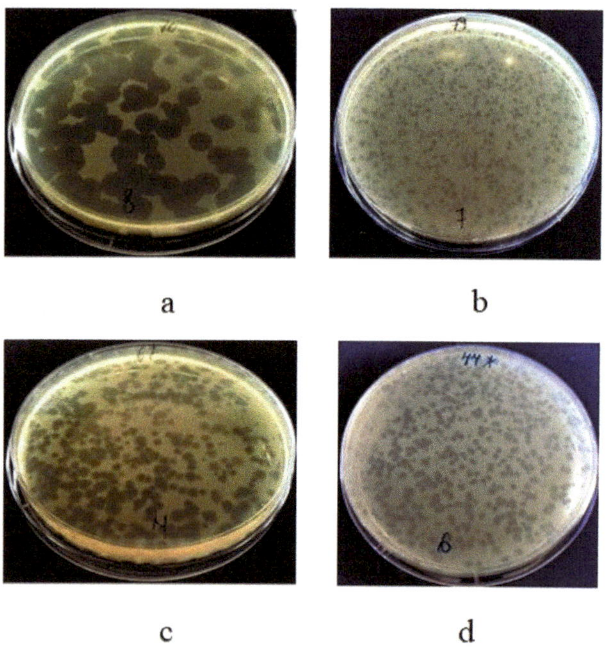

FIGURE 15.2 Bacteriophage cultures from the Belarus collection of non-pathogenic microorganisms: (a) *Pseudomonas fluorescens* phage BIM BV-70 plaques on host bacterial culture *P. fluorescens* BIM B-582; (b) *P. putida* phage BIM BV-79 plaques on host bacterial culture *P. putida* BIM B-1044; (c) *Escherichia coli* phage BIM BV-67 plaques on host bacterial culture *E. coli* BIM B-984; (d) *Escherichia coli* phage BIM BV-44 plaques on host bacterial culture *E. coli* BIM B-496. (Photo Galina Novik and Victoria Savich.)

at ambient storage conditions (Leung et al. 2017). Spray-drying is a method of producing a dry powder from a liquid by rapid desiccation with a hot gas. This method is sometimes used in bacterial preservation, but it is rarely used for bacteriophages. Dry powders containing viable encapsulated *Staphylococcus* phages (phiIPLA-RODI and phiIPLA88) diluted in skimmed milk and trehalose can be recovered at 12 months at 4°C. At 20°C, phiIPLA88 was stable for at least 12 months in trehalose but not in skimmed milk, while phiIPLA-RODI retained stability for only 6 months in either stabilizer (González-Menéndez et al. 2018). A powder matrix containing ≥40% trehalose provided a good survival rate for *Pseudomonas* phage PEV2 after 1 year storage at 4°C and at 0 and 22% relative humidity (Leung et al. 2017).

15.4 General Characteristics of Bacterial Viruses

Bacteriophages are viruses that infect bacterial cells. Like other representatives of this group, phages are obligate parasites able to metabolize and replicate only in host cells. New viral components are synthesized and assembled within the infected host cell. Phages are highly specific: they usually infect certain bacterial species; however, some bacterial viruses show specificity towards particular strains, while others infect several related species. Each phage particle (virion) contains one type of nucleic acid (DNA or RNA) usually enclosed in a protein coat (capsid). Some viruses also carry lipids in their capsids. The capsid protects the nucleic acid and takes part in the attachment of the virus to the surface of the host cell (Abedon 2016; Guttman et al. 2005).

The phage lifestyle can be divided into several types: lytic, chronic, lysogenic and pseudolysogenic. Virulent phages multiply through lytic cycles ending in the lysis of the host cell and the liberation of new phages (Figure 15.3). During chronic cycles, viruses are episodically or constantly released from cells by extrusion or budding without lethal effect.

Temperate phages additionally display a lysogenic cycle. In the prophage state, their genome integrates into the host genome or acts as a plasmid and is maintained for indefinite periods, being replicated as

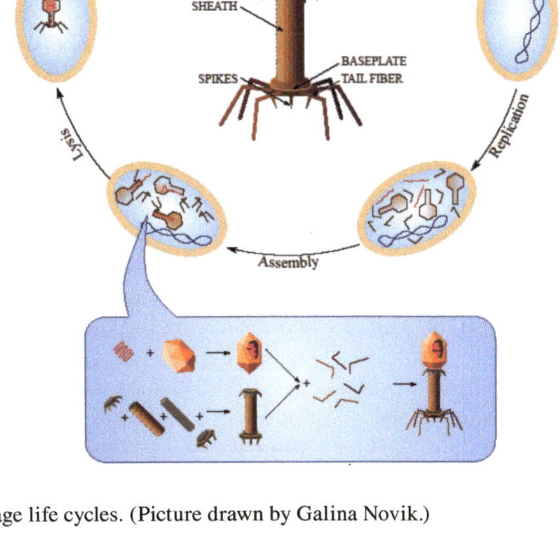

FIGURE 15.3 Bacteriophage life cycles. (Picture drawn by Galina Novik.)

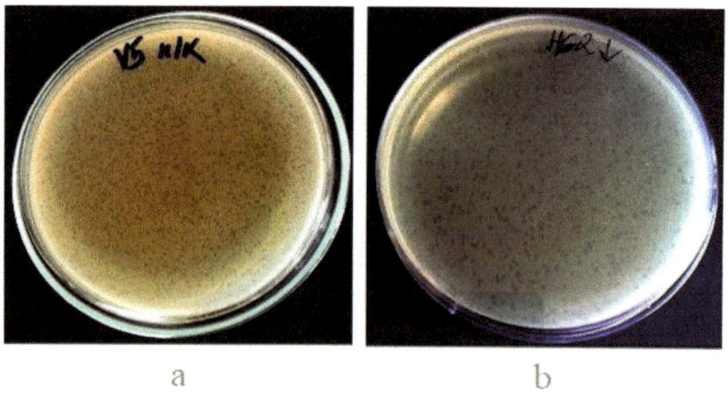

FIGURE 15.4 Appearance of temperate phage plaques on host bacterial cultures: (a) *Pseudomonas psychrotolerans* BIM B-1158 and (b) *Pantoea allii* BIM B-1401. (Photo Galina Novik and Victoria Savich.)

the host cell reproduces (Figure 15.4). In the pseudolysogenic cycle, the viral genome does not propagate (Sime-Ngando et al. 2018; Guttman et al. 2005).

15.5 Current Taxonomy and Classification of Bacteriophages: Virus Architecture

Historically, bacteriophage classification was based on phenotypic analysis. The nature of the nucleic acid and the morphological and physico-chemical properties of the virion were the main criteria for taxonomic affiliation (Ackermann 2011). The development of molecular-genetic methods of identification allowed part revision of phage classification.

Today, the bacteriophage taxonomy includes representatives of 1 realm, 1 order and 13 families (Table 15.2). The order Caudovirales (from Latin *cauda*, "tail") is the largest group of bacteriophages, comprising five families. For a long time, Caudovirales was divided into three families in accordance with its tail structure: Myoviridae (long contractile tail), Podoviridae (short non-contractile tail) and Siphoviridae (long non-contractile tail). Over 50% of the described tailed phages belong to Siphoviridae (Ackermann and Prangishvili 2012). Recently, the order was replenished with two novel families: Ackermannviridae and Herelleviridae. Previously, many representatives of these families were included in Myoviridae (ICTV 2019). Using a combination of genomic, proteomic and phylogenetic approaches for evaluating SPO1-like viruses associated with the subfamily Spounavirinae subordinate to Myoviridae, researchers transferred these phages from their current taxon to a new family of Caudovirales (Barylski et al. 2017). Based on detailed BLAST analyses, phylogeny and prokaryotic virus orthologous groups substantiated the relocation of 14 species from the genus *Vilvirus* into the novel family Ackermannviridae (Kropinski et al. 2017). Another potential family of tailed bacteriophages embraces 26 viruses with the characteristic features of phage Mu in the mode of transpositional replication and packaging from integrated genome copies. This group includes 8 representatives from Myoviridae and 18 from Siphoviridae. A new family Saltoviridae is proposed with subfamilies Myosaltovirinae and Siphosaltovirinae based on two morphological phage types (Hulo et al. 2015). However, this taxonomic group is currently under consideration.

Summing up, the order Caudovirales is distinguished by a virion consisting of two parts, the head and the tail, devoid of envelope. The head is a protein shell containing a single linear double-stranded (ds) DNA molecule, and the tail is a protein tube with a distal end that binds the surface receptors of susceptible bacterial cells. DNA migrates through the tail tube during injection into the cell. The heads usually have icosahedral symmetry or are shaped as elongated derivatives (Lavigne et al. 2011). The latter are aligned along a five-fold axis coincident with the axis of the phage tail. DNA packed with a high density of ~500 g/L exerts internal pressure of tens of atmospheres on the capsid walls (Fokine and Rossmann 2014).

The tails of Caudovirales have a very different size and morphology. Long tails consist of the tail tip complex, responsible for host recognition and the initiation of the infection process; the tail tube, a

TABLE 15.2

The Main Families of Bacteriophages

Family	Morphology	Nucleic Acid	Example
Realm Riboviria			
Cystoviridae	Enveloped icosahedral virion with two concentric protein layers	Three segments of linear dsRNA	*Pseudomonas* virus phi6
Leviviridae	Icosahedral virion	Linear ssRNA (+)	*Escherichia* virus MS2
Order Caudovirales			
Ackermannviridae	Icosahedral head with long contractile tail	Linear dsDNA	*Shigella* virus AG3
Herelleviridae	Icosahedral head with long contractile tail	Linear dsDNA	*Bacillus* virus SPO1
Myoviridae	Icosahedral head with long contractile tail	Linear dsDNA	*Escherichia* virus T4
Podoviridae	Icosahedral head with short non-contractile tail	Linear dsDNA	*Pseudomonas* virus phiKMV
Siphoviridae	Icosahedral head with long non-contractile tail	Linear dsDNA	*Escherichia* virus λ
Unassigned Families			
Corticoviridae	Icosahedral virion with protein-rich internal membrane	Supercoiled circular dsDNA	*Pseudoalteromonas* virus PM2
Inoviridae	Filaments or short rods	Circular ssDNA (+)	*Escherichia* virus M13
Microviridae	Icosahedral virion	Circular ssDNA (+)	*Escherichia* virus phiX174
Plasmaviridae	Quasi-spherical, pleomorphic virion	Supercoiled circular dsDNA	*Acholeplasma* virus L2
Sphaerolipoviridae	Icosahedral virion with internal lipid membrane	Circular dsDNA	*Thermus* virus P23-77
Tectiviridae	Icosahedral virion with protein-rich internal membrane	Linear dsDNA	*Pseudomonas* virus PRD1

conduit for the genomic DNA; and the terminator proteins finish the tail assembly and create a binding interface for the head attachment. Contractile tails also have a sheath surrounding the tail tube. Phages recognizing protein receptors for cell binding are equipped with conical tail tips, while phages recognizing polysaccharide receptors bear elaborate base plates at the distal end of the tail. In addition, phages usually carry side tail fibres or spikes, stemming from the periphery of the tail tip complex, as well as a central tail spike. The short tails of Siphoviridae contain a central tubular structure to which 6 or 12 appendages or fibres are attached. The tubular part is typically composed of two different proteins (Fokine and Rossmann 2014).

The order Caudovirales is represented by DNA phages. Other DNA viruses belong to families not assigned to superior ranks. The family Corticoviridae includes only two phage species infecting the bacterial genus *Pseudoalteromonas* (ICTV 2019). In 1968, *Pseudoalteromonas* virus PM2 was isolated from seawater and for a long time has been known as the sole member of the family. Its virion incorporates an icosahedral protein capsid and an internal protein-rich membrane enclosing the genome made up of highly supercoiled, circular dsDNA. The capsid consists of 200 major protein P2 trimers structurally arranged on a pseudo T=21 lattice. Spikes protruding about 8 nm from the capsid surface at the five-fold vertices are composed of the protein P1. The inner membrane contains host plasma membrane–derived phospholipids and the virus-encoded proteins P3 to P10 (Oksanen and ICTV Report Consortium 2017). *Pseudoalteromonas* phage Cr39582 shows 85% pairwise nucleotide identity with phage PM2 but lacks sequence homology in the spike protein. Pairwise nucleotide identities in the 5′ and 3′ regions flanking the spike protein gene are ~91 and ~96%, respectively (Leigh et al. 2018). The studies also demonstrated that integrated corticoviral genetic elements commonly reside within aquatic bacterial chromosomes (Krupovic and Bamford 2007).

Phages of the family Inoviridae are shaped as flexible filaments or short rods. Virions contain a circular, positive sense, single-stranded (ss) DNA within a helical array of thousands of copies of a major capsid protein. Members of the family do not lyse their hosts, but extrude virions (Day 2011). Currently, the family Inoviridae includes 33 species (ICTV 2019). However, 10,295 inovirus-like genomes from microbial genomes and metagenomes have been identified. These bacteriophages infect nearly all bacterial phyla across every ecosystem and may acts as potential viruses of archea. It is proposed to reclassify the Inoviridae family as an order divided into 6 families (Protoinoviridae, Vespertilinoviridae, Amplinoviridae, Paulinoviridae, Densinoviridae and Photinoviridae) and 212 subfamilies (Roux et al. 2019).

The family Microviridae is represented by non-enveloped viruses with T = 1 icosahedral symmetry and circular positive sense ssDNA. The family consists of two subfamilies: Bullavirinae and Gokushovirinae. Bullavirinae (former genus *Microvirus*) infects enterobacteria and is characterized by pentamers of a major spike protein decorating the five-fold axes of symmetry of the T = 1 lattice. Gokushovirinae unites viruses of obligate intracellular parasitic bacteria (*Bdellovibrio* and *Chlamydia*) and mollicutes (*Spiroplasma*) with mushroom-shaped protrusions at the three-fold axes of symmetry (Cherwa and Fane 2011).

Currently, the family Plasmaviridae includes only *Acholeplasma* virus L2; however, the related prophages reside in the genomes of different *Acholeplasma* species. Virions are quasi-spherical, slightly pleomorphic envelopes containing circular, negatively supercoiled dsDNA. Three distinct virion forms are produced during infection. These forms have the same protein composition, but differ in the relative amounts of each protein and vary in the number of genome copies (from 1 to 3). Virions are released by budding through the cell membrane without cell lysis (Krupovic and ICTV Report Consortium 2018).

The family Sphaerolipoviridae includes viruses infecting halophilic archaea and thermophilic bacteria. Sphaerolipoviridae comprises three genera: *Alphasphaerolipovirus*, *Betasphaerolipovirus* and *Gammasphaerolipovirus*. Only the latter is distinguished by bacterial viruses infecting extreme *Thermus thermophilus*. This group comprises two members, lytic phage P23-77 and temperate phage IN93, as well as several putative entries. Some related proviruses were also discovered as elements integrated in bacterial genomes of the families *Thermus* and *Meiothermus*. The capsid of *Gammasphaerolipovirus*, similar to other representatives of the family, is icosahedral crowned with spikes on its vertices. The inner lipid membrane is located between the capsid and the circular dsDNA genome. The lipids are selectively acquired from the host cell during virus assembly (Pawlowski et al. 2014).

Tectiviridae family members are morphologically close to Corticoviridae. Icosahedral virions lack an external envelope. The capsid encloses an inner membrane composed of approximately equal amounts of virus-encoded proteins and lipids derived from host cell plasma membrane. The linear DNA is coiled within this membrane. Virions are normally tailless, but produce tail-like tubes upon DNA release (Oksanen and Bamford 2011).

RNA bacteriophages embrace two families: Cystoviridae and Leviviridae, collected with RNA viruses of other groups in the realm Riboviria. Cystoviridae are enveloped icosahedral viruses. Their envelope surrounds a nucleocapsid, while turret-like extrusions of the underlying polymerase complex span the nucleocapsid surface shell layer at the five-fold symmetry positions. The genome of Cystoviridae contains three segments of linear dsRNA (Poranen et al. 2017). Leviviridae are icosahedral viruses without an envelope. The virion contains one molecule of positive sense ssRNA.

Interestingly, Picobirnaviridae constituents are considered dsRNA viruses of mammals, but analyses of viral genomes revealed that picobirnaviruses possess a classical bacterial sequence motif, the ribosomal binding site, in the 5′ untranslated regions. The enrichment of this motif is only observed in viral families infecting prokaryotes. Thus, Picobirnaviridae may be classified as a potential RNA family of bacteriophages (Krishnamurthy and Wang 2018; Adriaenssens et al. 2018).

Summing up, bacteriophages demonstrate low morphological diversity, especially compared to archaeal viruses (Pietilä et al. 2014).

15.6 Molecular-Genetic Identification of Bacteriophages

Bacteriophages are one of the most abundant organisms on Earth. More than 6,000 viruses have been described morphologically, but phage genomes require further studies (Ackermann and Prangishvili 2012). The size of the phage genome varies from 2,435 nucleotides in *Leuconostoc* phage L5 to almost 500 kbp in *Bacillus* virus G (GenBank 2019; NCBI 2019). One of significant features of phage genomes is their genetic diversity. The nucleotide sequences of phage genomes with non-overlapping host ranges are rarely distinguished by sequence similarity. However, phages infecting a common host also demonstrate enormous variety. Bacterial viruses can often be divided into groups sharing significant nucleotide sequence similarity, and no sequence similarity may be detected between the two different groups (Hatfull 2008). Bacteriophage genomes are characterized by a mosaic structure. Each genome can be considered a unique combination of modules capable of transposition within the microbial population. Usually, such modules correspond to single genes, while the junctions between segments match gene boundaries. Several events contribute to mosaicism. Homologous recombination of short conserved sequences at gene boundaries can lead to mosaicism, but it is more likely a minor mechanism, because most phage gene boundaries are not associated with sufficiently long conserved sequence segments for this process. The major contributor is illegitimate recombination coupled with functional gene selection. The genes are not engaged in mosaicism to the same degree. Non-core genes are transferred in and out of the phage genomes much faster than the core genes (Hatfull and Hendrix 2011).

For a long time, the classification of bacteriophages was based on their external characteristics, especially their morphology and the nature of their nucleic acid (Ackermann 2011). The development of molecular-genetic methods revealed that even phages infecting the same hosts may display low similarity. Thus, the families Ackermannviridae and Herelleviridae were distinguished from Myoviridae. While gene 16s rRNA is used for the classification of bacteria, phages do not contain such universal genes. Structural proteins could hypothetically serve as a basis for phage taxonomy, but they are extremely diverse and do not harbour conserved regions, making them of limited value as markers for biodiversity studies (Rohwer and Edwards 2002).

The first generation of DNA sequencing techniques was based on radio-/fluorescent-labelling ddNTP nucleotides to prevent further extension or chemical treatments applied to selectively remove the base from a small proportion of DNA sites. Bacteriophage phiX was one of the first DNA genomes sequenced in this way. However, reads slightly less than 1 kb in length are produced with these methods. Shotgun sequencing with overlapping DNA fragments cloned and sequenced separately allows the assembling of separate fragments into one long contiguous sequence (Heather and Chain 2016).

Phage genome sequencing meets several challenges: (1) phages rely on the host metabolism, hence additional disposal of bacterial DNA and cell debris is required; (2) some phages may carry highly methylated genomes, extreme guanine-cytosine (GC) content (48–51 mol %), a complex genomic structure and repeats interfering with polymerase chain reaction (PCR) analysis and sequencing. The full sequencing allows the establishment of taxonomic relations between various groups of phages as well as providing regulatory approval for further application of bacterial viruses in different areas. The second and third generation of sequencing technologies, based on cloning-free, massively parallel sequencing, aroused a deluge of genomic sequences of both prokaryotic and eukaryotic origin, including bacteriophages (Klumpp et al. 2012). Several technologies are used in the full-scale sequencing of phage genomes: 454 pyrosequencing (Illumina method); single molecule real-time sequencing (Pacific Biosciences); single molecule fluorescent sequencing (Helicos); Ion Torrent sequencing; and nanopore sequencing (Klumpp et al. 2012; Redondo et al. 2013; Rihtman et al. 2016; Harris et al. 2008; Castro-Wallace et al. 2017). The next generation of sequencing technologies will provide an opportunity to quickly and reliably recover bacteriophage genome sequences as well as decrease the costs of the process. However, most commercial companies focus on the genomes of bacteria or eukaryotes and generally possess little expertise regarding the specific requirements of phage genomes. The structure of the viral nucleic acid may also cause problems for sequencing. The introduction of high-tech innovations may solve the deadlock (Klumpp et al. 2012).

15.7 Physiology of Bacterial Viruses, Replication Cycle, Virus Interactions with the Host Cells

Bacteriophages are infectious agents that depend on the metabolism of their host cell for their life cycle. Viruses are notable for replicating themselves outside the cell. In a simple case, a bacteriophage infection consists of five steps: (1) adsorption; (2) penetration; (3) synthesis of viral components; (4) assembly of phage particles; and (5) release of new virions.

All viral infections start with the adsorption stage. A phage binds to specific molecules on the surface of the target bacteria. This step is reversible and the viral particle can be disengaged from the cell surface. The attachment of some phages involves two separate stages and two different receptors. External factors and the host physiological state may influence the adsorption rate and efficiency. At the next step, viral nucleic acid is injected into the bacterial cell, wherein viral genes are consistently transcribed. Initially, early gene products that take part in the protection of the phage genome and the reorientation of the host metabolism are synthesized. The following set of genes specializes in the production of novel nucleic acid, succeeded by the expression of late genes encoding the components of the phage particle. Subsequently, individual components are assembled into viral particles and crowned by lysis of the host cell. Lysin, the enzyme cleaving key bonds in the peptidoglycan matrix, and holin, a protein forming pores in the inner membrane, play their roles in the final step (Guttman et al. 2005). Lysis of Gram-negative bacteria is a three-step process due to the triplex composition of the cell wall. The inner membrane is permeabilized by either holin or pinholin. It allows a muralytic enzyme, either a canonical endolysin to escape from the cytoplasm, or a signal-arrest-release (SAR) endolysin, activated in the periplasm, to degrade the peptidoglycan. Spanin, another lytic protein, is required for disruption of the outer membrane (Young 2013). The optimal lysis time is shorter for phage strains with a high adsorption rate and vice versa (Shao and Wang 2008).

The above-mentioned cycle is characteristic for virulent phages. Temperate viruses have an additional lysogenic cycle with a prophage stage when a viral genome integrates with the host DNA or is maintained sometimes as a plasmid and replicates along with it. They can encode repressor proteins blocking the transcription of other phage genes, especially leading to the lytic cycle. The prophage stage may provide host bacterium with protection from other phages as well as impart new features such as antibiotic resistance or virulence. Under certain conditions, the lytic cycle of temperate phages can be restored (Guttman et al. 2005). Environmental factors may effect spontaneous prophage induction (Lunde et al. 2005). A multiplicity of infection, cell starvation and cell size may also affect the lysogeny (Shao et al. 2019). Phages of the spBeta group use a small-molecule communication system to coordinate lysis–lysogeny decisions. During infection, the virus produces a communication peptide that is released to the medium. In the course of subsequent infections, progeny phages measure the concentration of this peptide and lysogenize if the concentration is sufficiently high. Different phages encode various versions of this compound (Erez et al. 2017).

Environmental conditions and prokaryote physiology can strongly influence the dynamics of both the lysogenic and lytic bacteriophage replication cycles. Studies on Mediterranean lagoons revealed that the lytic cycle was always detected and showed seasonal patterns, while the lysogenic cycle was often not determined and highly variable. The lytic cycle was influenced by environmental and prokaryotic physiological cues, increasing with concentrations of dissolved organic carbon, chlorophyll a and the proportion of respiring cells, and decreasing with the proportion of damaged cells. In turn, lysogeny was dependent on the variability of the prokaryote physiology (Maurice et al. 2013).

Pseudolysogeny can be defined as the stage of stalled bacteriophage development in a host cell without either multiplication of the phage genome (in lytic process) or its replication synchronized with the cell cycle and stable maintenance in the cell line (in lysogenization), without viral genome degradation. It allows the subsequent restart of virus development. This phenomenon is usually caused by unfavourable growth conditions for the host cell. When growth parameters improve, true lysogenization or lytic growth is continued (Łoś and Węgrzyn 2012).

Some phages display a chronic cycle. In this case, virions are released over long intervals via extrusion or budding without substantial disruption to the host cells. The cycle can be found in both virulent and terminate phages (Hobbs and Abedon 2016).

15.8 Receptors on Bacterial Surface, Chemical Structure and Location

Every viral infection begins with the attachment of a virion to the surface of the host cell. This step is crucial for further development of infection. Specific attachment sites, or receptors, are essential for phage adsorption. Bacterial viruses target various receptors of different chemical specificities on the cell surface. Sometimes, more than one receptor is involved in the adsorption process. Phages are able to bind to components of the cell wall, capsule, pili and flagella.

The cell wall composition of Gram-positive and Gram-negative bacteria differs considerably, so that phages attach to completely diverse sites of the envelope. Peptidoglycan, or murein, is the main component of the bacterial cell wall, composed of multiple units of amino acids and sugar derivatives. The latter is connected through glycosidic bonds and forms glycan tetrapeptide sheets joined together by the cross-linking of amino acids. Peptidoglycan of Gram-positive bacteria, often involved in bacteriophage adsorption, generates a thick cell wall in contrast to Gram-negative species. Teichoic acids, polysaccharides composed of glycerol phosphate or ribitol phosphate and amino acids, are other vital components of the cell wall of Gram-positive bacteria also acting as attachment sites for viruses. However, the phage receptors of Gram-positive bacteria are less studied than their counterparts (Bertozzi Silva et al. 2016). Cell wall polysaccharides are surface receptors of the P335 and 936 phage groups infecting *Lactococcus lactis* strains (Ainsworth et al. 2014). Phage A511 with a wide host range uses the listerial peptidoglycan as the primary receptor, while *Listeria* phages A118 and A500 are dependent on cell wall teichoic acids (Wendlinger et al. 1996). Teichoic acids seem to be attachment sites for the viruses of *Staphylococcus* and *Bacillus* (Young 1967; Xia et al. 2011). Lipoteichoic acids are able to inactivate *Lactobacillus delbrueckii* phages (Räisänen et al. 2004).

Gram-negative bacteria are additionally equipped with an outer membrane surrounding a thin peptidoglycan layer. The outer membrane is an asymmetric phospholipid bilayer consisting of an internal sheet of glycerophospholipids and an external leaflet rich in lipopolysaccharides (LPS), which covers up to 75% of the cell surface. The bilayer also includes integral membrane proteins, such as porins. LPS usually demonstrate a three-domain structure: a lipophilic moiety termed lipid A; a hydrophilic glycan called the O-specific polysaccharide (O-chain or O-antigen); and a joining core oligosaccharide. The latter can be further separated into two regions, one proximal to lipid A (inner core) and another adjacent to the O-antigen (outer core). LPS comprising all three regions are termed "smooth" LPS, while those lacking the O-chain and/or portions of the core oligosaccharide are denoted as "rough" LPS. The O-antigen is the most variable portion of the LPS and it may be involved in interactions of bacteria with bacteriophages (Silipo and Molinaro 2011). Phages recognizing the O-antigen demonstrate a narrow host range, while viruses binding to a more conservative LPS core show a broader host range. Proteins embedded in the outer membrane can be divided into five groups: (1) structural proteins interacting with the peptidoglycan layer; (2) porins forming membrane channels; (3) enzymes; (4) substrate receptors with high affinity; and (5) transport proteins responsible for secretion (Rakhuba et al. 2010). The outer core region of the LPS is a receptor of phage SSU5, which can only infect rough strains of *Salmonella* (Kim et al. 2014). Bacteriophage λ binds to LamB protein, or maltoporin, required for growth on limiting concentrations of maltose (Chatterjee and Rothenberg 2012). Phage TLS of *E. coli* makes use of the inner core of the LPS as well as the outer membrane protein TolC, part of the bacterial efflux system, as receptors (German and Misra 2001). Bacteriophage T4 needs the presence of both the OmpC protein and the core-lipid A region of the LPS, including the heptose site. However, when the distal end of the lipopolysaccharide was removed to expose glucose residue, the OmpC protein was not required at all (Yu and Mizushima 1982).

Pili are other components of bacterial cells that can be recognized as viral receptors. Pili of various groups of bacteria contrast sharply in their morphogenesis. Pili in Gram-negative bacteria are typically formed by non-covalent homopolymerization of major pilus subunit proteins (pilins), generating the pilus shaft. In turn, the major pilus subunits of Gram-positive microorganisms are connected covalently and the pilus rod is then anchored to the cell wall by asortase enzyme. Pili are often described as adhesive organelles. Moreover, they can take part in phage binding, DNA transfer, biofilm formation, cell aggregation, host cell invasion and twitching motility (Proft and Baker 2009). Strains of *Pseudomonas aeruginosa* with functional type IV pilus consisting of a group IIa pilin were susceptible to phage PP7 (Kim et al. 2018). Bacteriophages DLP1 and DLP2 infecting both *Stenotrophomonas maltophilia* and *P. aeruginosa* strains utilize type IV pilus as the primary receptor for attachment. It promotes interaction with diverse bacterial species (McCutcheon et al. 2018). Filamentous bacteriophages (M13, fd, f1) infecting *E. coli* carrying the F episome coding for F-pilus, use the latter as the primary receptor. The F-pilus retracts upon phage binding, bringing the virus close to its coreceptor protein TolA (Karlsson et al. 2003).

The flagellum is an appendage that takes part in bacterial motility. The flagellum can be divided into at least three parts: the basal body as a rotary motor, the hook as a universal joint and the filament as a helical propeller. The appendage consists of about 30 different proteins highly conserved among bacterial species (Terashima et al. 2017). Both the *Salmonella* phages iEPS5 and Chi require the flagellum to rotate counterclockwise to infect bacteria. However, iEPS5 cannot infect the mutant that makes a polyhook without a flagellar filament, suggesting that these two flagellatropic phages utilize different infection mechanisms (Choi et al. 2013). Phages that attach to flagella also demonstrate a broader host range. Bacteriophage PV22 effectively infected *Proteus vulgaris* and was able to bind to *Campylobacter jejuni*, while phage phiOT8 isolated from *Serratia* infected the *Pantoea agglomerans* strain (Zhilenkov et al. 2006; Evans et al. 2010).

The capsule is a mucous layer outside the bacterial cell wall. It can be produced by many species of bacteria and is usually composed of polysaccharides (Wen and Zhang 2015). The capsule serves to protect cells from adverse environmental conditions and the impact of other organisms. The capsule may hide attachment sites on the cell surface and act as a receptor itself. K phages of *E. coli* were able to adsorb and lyse the capsular forms of the host bacteria (Stirm 1968). Phage F336 recognized O-methyl phosphoramidate moiety on the capsular polysaccharide of *C. jejuni* (Sørensen et al. 2011).

The plasma membrane of bacterial cells is usually covered by the cell wall, hence it rarely functions as a receptor. Phages c2, ml3, kh, l, h, 5 and 13 of *L. lactis* are adsorbed to the same sites on the host cell wall that are partly composed of rhamnose. The consecutive step envisages the attachment to the membrane protein PIP. Phages are unable to infect bacteria with defective protein, but can be inactivated in vitro by membranes from wild-type cells (Monteville et al. 1994). The membrane protein YueB of *B. subtilis* also acts as a receptor and appears as an orthologue of PIP (São-José et al. 2004). The bacteria of *Mollicutes* class are distinguished by the absence of a cell wall. However, they can still be targets of viruses. Virus P1 infecting some strains of *Mycoplasma pulmonis* requires surface antigen V-1 for cell adsorption (Dybvig et al. 1988). Phages MVL1, MVL2 and MVL3 of *Acholeplasma laidlawii* use various receptors on the cell surface. The essential receptor of MVL2 is the fragment of the oligosaccharide chain of lipoglycan plus membrane protein. MVL1 also recognizes protein and lipoglycan as attachment sites, while the total destruction of the MVL3 receptor after membrane treatment with proteases proves that protein is the main receptor for this virus (Al-Shammari and Smith 1982).

15.9 Phage Host Range and Efficiency of Restriction-Modification System: Restriction Enzymes

The host range usually defines the spectrum of organisms that viruses are able to infect. The term "broad host range" can be used to describe viruses acting against numerous conspecific strains (Anand et al. 2015), while in other cases it encompasses phages infecting various bacterial species. Two coliphages, AR1 and LG1, could infect many serotypes of *E. coli* and species of *Shigella dysenteriae*. Additionally, phage LG1 formed plaques on *P. mirabilis*, and AR1 lysed *Salmonella enterica* serovar Choleraesuis and *S. enterica* serovar Enteritidis (Goodridge et al. 2003). Recently, seven host range categories have been proposed to determine the host range of bacteriophages: adsorptive, penetrative, bactericidal, productive, plaquing, spotting and lysogenic. These categories are distinguished by the efficiency of the different steps of the infection process or the methods of evaluating the host range (Hyman and Abedon 2010). For a better description of the host range, several methods should be used to avoid false positive and false negative results (Hyman 2019).

A phage infection proceeds through several stages; however, it can be countered by bacterial defences at any stage. Bacteria resort to different mechanisms to achieve this purpose. They may prevent phage adsorption to the cell surface by masking, losing or altering the receptors. Uptake blocks hinder phage genomes from reaching the bacterial cytoplasm. If foreign DNA still gets inside the cell, other defence mechanisms, such as restriction–modification and the clustered regularly interspaced short palindromic repeats (CRISPR)–Cas systems, are switched on. If the above-mentioned protection fails to cope with a phage attack, abortive infection systems cause the death of the infected cell (Hyman and Abedon 2010; Labrie et al. 2010).

Restriction-modification systems are important components of bacterial defence often considered as primitive immune systems. The efficiency of these systems is defined by the activities of two contrasting enzymes: restriction endonuclease and methyltransferase. The former recognizes and cleaves foreign DNA sequences at specific sites, while the latter transfers methyl groups to the same specific DNA sequence within the host genome. There are four types of restriction-modification systems based on the enzymes involved (Vasu and Nagaraja 2013):

1. Type I enzymes make up a hetero-oligomeric protein complex with both restriction and modification activities, cleaving a DNA sequence from ~100 bp to tens of thousands of base pairs away from the target (EcoKI).
2. Type II systems include separate restriction endonuclease and methyltransferase enzymes. The constituent endonucleases are homodimeric or homotetrameric enzymes splitting DNA within or near their target site (EcoRI). These are the most thoroughly studied and extensively utilized nucleases in genetic engineering.
3. Type III enzymes are heterotrimers or heterotetramers demonstrating restriction, methylation and DNA-dependent NTPase activities (EcoP1I). The enzymes recognizing short asymmetric sequences of 5–6 bp are translocated along the DNA, and cleave the 3′ side of the target site at a distance of ~25 bp. A unique antirestriction mechanism is employed by T7 phage, by exploiting the cleavage mechanism of type III enzymes. EcoP1I, a well-characterized type III enzyme, requires two copies of its asymmetric recognition sites to be oriented in a head-to-head manner for successful cleavage. T7 phage exhibits strand bias; i.e., all the EcoP1I sites in T7 DNA are in the same orientation rather than in the head-to-head formation, which is required for cleavage. As a result, EcoP1I does not efficiently cleave the phage DNA. Thus, this example serves to illustrate the evolution of a phage genome to avoid the recognition sites of the host enzyme.
4. Type IV systems cleave only modified DNA substrates at specific sites (EcoKMcrBC).

A new defence system, defence island system associated with restriction-modification (DISARM), is widespread in defence islands across the microbial world and contains three core genes accompanied by methyltransferase and an additional gene. The modification module of DISARM containing methyltransferase, as in classic restriction-modification systems, modifies host CCWGG motifs, while the restriction module seems to be unique and requires multiple components and further characterization of the mechanism of action. DISARM most probably represents a new kind of restriction-modification system. The novel system may also restrict phages without CCWGG motif, suggesting that the restriction module of DISARM is not exclusively dependent on the presence of the motif (Ofir et al. 2018). Another phage resistance system, bacteriophage exclusion or BREX, demonstrates methylation of the host DNA at the fifth position of a non-palindromic 5′-TAGGAG-3′ hexamer motif. However, in contrast to the restriction-modification system, there is no sign of phage DNA cleavage or degradation, indicating that BREX might be responsible for the inhibition of phage replication and propagation (Barrangou and van der Oost 2015).

Besides anti-phage defence, restriction-modification systems could be engaged in other functions. They play a role in recombination, nutrition and upholding genetic diversity. They may regulate genomic flux, stabilize genomic islands or maintain methylation patterns (Vasu and Nagaraja 2013).

The CRISPR–Cas systems are alternative mechanisms aimed at the degradation of foreign nucleic acid. The CRISPR is an array of short repeated sequences separated by spacers derived from viruses and plasmids. The introduction of spacers enables the recognition of new viruses. CRISPR activity requires the presence of a set of CRISPR-associated (*cas*) genes encoding proteins essential for the immune response. A combination of RNA transcribed from CRISPR and Cas proteins allows the recognition and destruction of the target nucleic acid. There are three types of CRISPR–Cas systems. The Cas1 and Cas2 proteins involved in adaptation can be found in all types of systems. The Type I system additionally encodes the signature protein Cas3, with both the helicase and DNase domains responsible for degrading the target and the Cascade-like complex for binding CRISPR RNA and locating the target as well as for processing CRISPR RNA in most variants. Six subtypes of the Type I system were identified. The Type II system encodes the Cas9 signature protein and sometimes the fourth protein. Cas9 assists in adaptation, participates in RNA processing and cleaves the target DNA. The Type II system was partitioned into two subtypes, and a third was suggested. Type III CRISPR–Cas systems contain the signature protein Cas10. Most Cas proteins are destined for the Csm or Cmr complexes, similar to Cascade. Type III is divided into two groups capable of targeting RNA. Recently, a Type IV system was proposed, comprising several Cascade genes but no CRISPR, *cas1* or *cas2* (Rath et al. 2015).

15.10 Historical Background of Phage Cocktail Implementation in Medicine and Agriculture

The first mention of bacteriophages may be traced back to the study of Hankin who reported that the waters from Indian rivers showed antiseptic activity against bacteria. This activity was filterable and degradable by boiling. Hankin concluded that it was caused by some volatile chemical substance. D'Herelle suggested that antagonistic microorganism caused bacterial lysis in liquid culture and death in discrete patches (plaques) on the surface of agar and termed it "bacteriophage". He also advocated the use of phages as therapeutic agents in the pre-antibiotic era. The first report of phage therapy focused on the local treatment of cutaneous boils with a preparation of staphylococcal virus. D'Herelle also described phage administration to control epidemics of avian typhosis in chicken. Bacterial viruses were inoculated to cure bovine haemorrhagic septicemia in water buffalo populations (Summers 2005).

The discovery of bacteriophages triggered their application in the treatment of patients with dysentery, bubonic plague, cholera, chronic furunculosis, septicemia, staphylococcal meningitis, typhoid fever, etc., in the period between 1920 and the 1940s. The initial wave of success brought to the market commercial products based on phage preparations for the treatment of various diseases. However, further studies concerning phage therapy tended to wane. Some experiments with phages were controversial, phage products weren't always viable or safe and the true nature of bacterial viruses was unknown at that time. Moreover, the onset of antibiotics displaying a broad spectrum of action against bacteria exerted a strong negative impact on phage therapy research in the West. Western Europe and the United States carried out few experiments with phages, but the Soviet Union and countries of Eastern Europe were the main centres of work so that therapeutic phage production reached its apogee in the 1960s and on into the 1980s (Sulakvelidze and Kutter 2005).

Initially, phages were mostly used in human trials, but some experiments with animals were carried out. The inoculation of 10-day-old chicken eggs with *Sh. flexneri* (earlier *Sh. paradysenteriae*) led to almost 100% mortality of the embryos within 24–96 h (1 out of 39 survived). The addition of phages resulted in 32 live embryos from 44 (Rakieten and Rakieten 1943). Mice with experimental peritonitis due to the injection of *E. coli* were protected by bacteriophages. The injection of the phage protected an appreciable proportion of mice (56% compared to 11% survivors in the control group) (Walker 1929). Phage supplied immediately before the intraperitoneal inoculation of *Salmonella* protected all mice when injected intravenously or subcutaneously. The untreated control and animals treated with heated phage developed a fatal infection leading to 92.6% mortality. Mice remained resistant to infection for up to 24 h following intravenous phage injection (Fisk 1938). Rats immunized three times with phage 44 were able to withstand 1,000 times a fatal dose of the plague pathogen (Flu 1933).

As with human trials, some studies with animals demonstrated little or no effects on pathogenic bacteria. Experiments with rabbits infected with *Salmonella* revealed few survivors upon phage administration. However, animals were tested at different periods and under varying experimental conditions, so that technical factors, such as dosage, method, time of immunization procedure and the dose or method of infecting, could have provoked negative results (Kendrick 1933). Certain bacteriophage preparations exerted an unfavourable effect on the course and extent of experimental staphylococcal septicemia in rabbits. In general, the attempted therapy had no significant influence on the extent or persistence of the septicaemia (Sulkin et al. 1942). In contrast to d'Herelle's report, some experiments showed that phages were ineffective in protecting birds exposed to the *Salmonella* infection (Hall et al. 1949; Pyle 1926). The administration of bacteriophage failed to yield a protective effect against anthrax in mice, even allowing for preliminary contact between microorganism and virus before injection (Cowles and Hale 1931).

Bacteriophages of plant pathogens were also isolated, but data on phage application in plants were scarce at that time (Weindling et al. 1950). Aseptically prepared carrot disks demonstrated rotting upon inoculation with *Pectobacterium carotovorum* only in contrast to samples pretreated with bacteriophages. A similar protective action was found with potato slices inoculated with *Pe. atroseptica* and the phage, although the results in this experiment were not quite consistent (Coons and Kotilia 1925). Cultures of *Agrobacterium tumefaciens* subjected to bacteriophage action for 9 h failed to produce infection in tomato plants (Muncie and Patel 1929). Corn seeds inoculated with the phage of *P. stewartii* demonstrated a decrease in infection rate from 18 to 0.4% (Thomas 1935).

The "golden age" of antibiotics began in the 1940s and lasted for over four decades. Thereafter, the number of novel antibiotics steadily declined. Most were either modified or combined versions of previously known compounds. These antimicrobials have been used in medicine as well as in agriculture. Antibiotics are applied to promote the growth of livestock and treat crop and fish diseases. Large-scale use of agricultural antibiotics has released millions of tons of these chemicals into water effluents and aquatic bodies. The lack of appropriate wastewater treatment methods, the contribution of the pharmaceutical industry and the reduced distance between farmlands and municipal areas has augmented the discharge of antibiotics, their amassment and their persistence in the environmental media. The abuse of antibiotics has led to the appearance of resistant strains, some of which are not susceptible to several types of antibiotics. A group of bacterial species (*Enterococcus faecium*, *Staphylococcus aureus*, *Klebsiella pneumoniae*, *Acinetobacter baumannii*, *P. aeruginosa* and *Enterobacter* spp.), known as ESKAPE pathogens, are causal agents of the most frequent and severe healthcare-associated multidrug-resistant infections. The application of bacteriophages is now one of the ways to cope with the problem (Gordillo Altamirano and Barr 2019).

15.11 Current Trends in Allocation of Viral Genome Resources for Biotechnology, Medicine and Agriculture Management and Research: Market Situation and Strategies

The grave challenge of antibiotic-resistant strains has revived interest in bacteriophages. Phages have some preferential features over antibiotics (Sulakvelidze and Kutter 2005; Chhibber and Kumari 2012):

1. Viruses are host specific. Phages will normally lyse strains of one species or a subset of strains within a single species; however, some viruses infect diverse species. Phages target certain pathogens, bypassing desirable bacterial flora. On the other hand, the specificity of action demands the accurate identification of the pathogen and ensures that the phage preparations are highly lytic towards bacteria.
2. Phages don't affect human health in contrast to antibiotics. The latter may cause disorders and allergies. In some cases, phage action leads to the liberation of endotoxins, but such effects may also be observed when antibiotics are used.
3. One dose of phage preparation can be enough for viruses to maintain their number at the site of infection until the elimination of bacteria.
4. The natural coevolution of bacteriophages and their hosts, and the vast number of phages in nature, make it easy to identify new phages against any phage-resistant bacteria.

A better understanding of the viral nature of bacteriophages and a rekindled interest in them have spurred on developments of the corresponding market segment. In 2017, the global bacteriophage market was worth US$ 567.9 million and it is expected to grow at a compound annual growth rate of 3.9% during the forecast period from 2018 to 2026. In 2026, the market size is likely to reach US$ 797.2 million. Bacteriophages demonstrate potential as phage display systems, antibacterials, vehicles for vaccines and biocontrol agents. Despite considerable interest in phages in medicine as antimicrobials, currently food and beverages account for the largest revenue share in the global bacteriophage market. Bacterial viruses find their widest use in food biocontrol, offering a natural method of targeting pathogenic bacteria by lytic bacteriophages isolated from the environment. Studies engage bacteriophages to eliminate pathogens in ready-made meals, fresh fruits, vegetables and dairy products. Clinical applications will witness the fastest market growth during the forecast period. The increased use of bacteriophages in diagnostics and phage display technology and their potential use in drug discovery and development are the main contributors to the rapid growth of this segment (Credence Research 2018).

In terms of geography, in 2017, North America held the largest regional bacteriophage market with over 37% share of the global market in 2016. This correlates with attempts to overcome the burden of antibiotic-resistant infections, a strongly established research infrastructure and the existence of a potential pipeline for phage therapy. Europe is the second largest market with a wide application of bacteriophages in the food industry, the environmental field and phage therapy. The Asia Pacific region will witness the fastest progress related to its rapidly evolving biotechnology infrastructure and agriculture practices as well as its equally supportive government initiatives. The key players currently dominating the bacteriophage market include Federal State Scientific Industrial Company MICROGEN (Russia), AmpliPhi Biosciences Corporation (USA), EnBiotix (USA), Phage Biotech Ltd. (Israel), Fixed-Phage Limited (UK), InnoPhage (Portugal), Pherecydes Pharma (France) and TechnoPhage SA (Portugal) (Credence Research 2018).

Historically, wide clinical trials of phage therapy were carried out in the USSR. Phages were used in the treatment of dysentery, gas gangrene, skin ulcers, abscesses, lung and wound infections, etc. However, due to the language barrier these phage therapy–related studies from the former Soviet Union went unnoticed by the West (Sulakvelidze and Kutter 2005). Today, companies in Russia and Georgia, such as Microgen and Eliava BioPreparations Ltd., respectively, continue to produce phage and phage cocktail (mix of several bacterial viruses) preparations, such as "Intestibacteriophage" and "Pyobacteriophage", to treat infections caused by pathogenic bacteria (Microgen 2019; Eliava BioPreparations Ltd. 2019). Bohemia Pharmaceuticals produces Stafal®, a solution containing polyvalent phage particles that operates on a broad range of *St. aureus* strains; it is registered in Slovakia and is also available in the Czech Republic within the framework of the special treatment programme (Bohemia Pharmaceuticals 2019).

The application of phages is more research oriented in the West. The European Medicines Agency emphasizes that phages cannot be recommended for approval before their efficacy and safety have been proven on the basis of appropriately designed clinical trials. Currently, this is not the case with bacteriophage therapy for which very few randomized controlled clinical tests have been conducted (European Medicines Agency 2015). In 2019, the Food and Drug Administration (FDA) approved the first clinical trial of intravenously administered phage therapy in the United States (Voelker 2019). In June 2018, the Centre for Innovative Phage Applications and Therapeutics, the first phage therapy centre in North America, was founded in the UC San Diego School of Medicine (Centre for Innovative Phage Applications and Therapeutics 2019).

Some projects are carried out with phage preparations. The European collaborative project PhagoBurn funded by the European Commission has evaluated the tolerance and effectiveness of bacteriophages to treat skin infections in burn victims. A cocktail of 12 natural lytic anti-*P aeruginosa* bacteriophages (PP1131) or standard of health care (1% sulfadiazine silver emulsion cream), both were given as a treatment. The PP1131 cocktail was effective in 50% of the treated patients, despite the low dose administered on the infected burns. The action of phages was slower than that of silver sulfadiazine, but had less adverse effects (Jault et al. 2019). In addition to the formulation of common phage preparations, another approach envisages personalized treatment. For instance, a 68-year-old patient with necrotizing pancreatitis complicated by multidrug-resistant *A. baumannii* infection was administrated specially selected bacteriophages. A marked improvement in the patient's condition was observed within 48 h of starting intravenous therapy. Following 11 weeks of therapy, the patient made a full recovery (Schooley et al. 2017). Summing up, most studies testing bacteriophages in clinical practice against multidrug-resistant organisms, ESKAPE pathogens in particular, demonstrated high efficacy (87%) and safety (67%). However, few studies examined phage resistance (35%) (El Haddad et al. 2019). As an option, components of phages can be applied in the treatment of bacterial diseases. Staphefekt™, an endolysine product, specifically targets *St. aureus* that causes the majority of bacterial skin infections and aggravating skin conditions (Micreos 2019).

Today, bacteriophages are considered a solid alternative to antibiotics in agriculture. Phages are studied for the prevention and treatment of poultry diseases induced by *Campylobacter*, *Escherichia* and especially *Salmonella* bacteria. *Campylobacter* species cause food-borne illness worldwide. *C. jejuni* and *C. coli* cultures, responsible for human disease, readily colonize the poultry gut and transmit the impact on the whole flock, resulting in further contamination of poultry products. A phage cocktail containing two virulent phages, CP20 and CP30A, showed a cell titer reduction of 2.4 log CFU/g in *C. jejuni* relative to controls (Richards et al. 2019). Treatment using phages CP8 and CP34 orally administered in an antacid suspension resulted in *Campylobacter* counts falling from 0.5 to 5 CFU/g depending on the phage combination, the dose and the time elapsed after administration (Loc Carrillo et al. 2005). Intralytix has developed and licensed bacteriophage-based animal healthcare products that are effective against *Salmonella* (PLSV-1™) and *Clostridium perfringens* (INT-401™) in poultry (Intralytix 2019a). The oral administration of the latter preparation reduced the mortality of infected fowl by 92%. INT-401 was more effective than the toxoid vaccine against bacteria. Moreover, the preparation favourably influenced weight gain and feed conversion ratios (Miller et al. 2010).

Some strains of *E. coli* cause infections in poultry (avian colibacillosis). Phage ΦKAZ14 encapsulated in chitosan nanoparticles (C-ΦKAZ14 NP) and pure phage administered orally improved the body weight of infected chicks (600 ± 80 and 550 ± 80 g, respectively, compared to the control 240 ± 30 g). However, a C-ΦKAZ14 NP–treated group exhibited a lower mortality rate compared with the group that received pure ΦKAZ14 (mortality 16.7 and 50.0%, respectively). The therapy failure may be caused by the possible inactivation of pure ΦKAZ14 phages by gastric acid and enzymes. The mortality of the control animals reached 58.33% (Kaikabo et al. 2017). The application of single phage phi F78E in experimental rooms was dosage dependent. A concentration of 10^7 PFU/mL was insufficient to treat the infected chickens, while a concentration of 10^9 PFU/mL allowed a decrease of 25 and 43% in chicken mortality and morbidity, respectively. In turn, large-scale experiments showed the remarkable efficacy of the low titer phage cocktail (10^7 PFU/mL) composed of three phages (F78E, phi F258E and phi F61E) in diminishing the flock mortality rate to under 0.5% by 3 weeks, with no recidivism (Oliveira et al. 2010).

Salmonella species are zoonotic bacteria and are considered one of the four principal causes of diarrheal disease worldwide (WHO 2018). A source of *Salmonella* infection is the consumption of poultry products. A mix of three lytic bacteriophages was shown to reduce the incidence of *S. enteritidis* infection in the chicken group to 72.7% and decrease intestinal colonization by this bacterium (Borie et al. 2008). One-day-old infected broilers orally treated on the seventh day with a mixture of bacteriophages showed a decline by 3.5 orders of magnitude of the *S. enteritidis* titer (Fiorentin et al. 2005). The phage cocktail SalmoFREE® containing six lytic bacteriophages was mixed with drinking water and inhibited *Salmonella* incidence on a commercial poultry farm, not affecting the animals or the production parameters (Clavijo et al. 2019). BAFASAL® is another phage preparation for the control of *Salmonella* in poultry farming (Proteon Pharmaceuticals 2019).

Studies have been carried out on phage therapy of other livestock. Brucellosis caused by *Brucella abortus* is associated with abortion in the last trimester of gestation, delayed conception, temporary or permanent infertility in affected cattle, and weak newborn calves. Vaccine strains of *Brucella* sometimes display negative effects in animals. A single 2 mL subcutaneous dose of phage cocktail lysate of the vaccine strains *Brucella abortus* RB51 and S19 eliminated live virulent bacteria from brucellosis-infected cattle and blood tests were negative for the live pathogen during 3 months after immunization. Additionally, the S19 lysate induced a strong antibody response, while the lysate of RB51 stimulated cell-mediated immunity (Saxena and Raj 2018). The *E. coli* strain O157:H7 is a human pathogen that causes a range of different symptoms. In contrast to humans, cattle are a primary habitat for this pathogen, not-affecting bovine health. Studies are ongoing to reduce the carriage and prevalence of *E. coli* O157:H7 in live ruminants and, consequently, decrease the risk of human exposure to the pathogen. Phages SH1 and KH1 applied directly to the rectoanal junction mucosa and maintained in the drinking water reduced the average number of bacteria among treated steers compared to control animals, but did not eliminate the bacteria from the majority of steers (Sheng et al. 2006). Compared to rectal administration, the oral administration of bacteriophages resulted in fewer *E. coli* O157:H7-positive samples and a lower mean shedding level of bacteria, but phage therapy did not remove the pathogen completely (Rozema et al. 2009). Bovine mastitis is the inflammation of the udder and is usually caused by bacterial infection. *St. aureus* is one of the most frequent causative agents of bovine mastitis. The application of bacteriophages reduced bacterial contamination and inflammation of the mammary gland in a murine mastitis model (Iwano et al. 2018; Breyne et al. 2017). However, when phage K was used in the treatment of dairy cattle, the cure rate was 3 of 18 quarters (16.7%) in the phage-treated group, while none of the 20 control quarters was cured. This difference was not statistically significant (Gill et al. 2006).

Treatment with a bacteriophage cocktail reduced *Salmonella* contamination in pigs. The animals that received phages showed the highest level of *Salmonella* shedding at day 2 after infection, and thereafter, a continuous decrease in bacterial shedding was observed until the seventh day (Seo et al. 2018). Other experiments have also demonstrated the possibility of using phage cocktails to significantly reduce *Salmonella* concentrations (down to 99%) in small and market-weight pigs (Wall et al. 2010). The single phages GJ1-GJ6 as well as two viral combinations (GJ1, GJ2, GJ7 and GJ1, GJ6) were able to moderate the course of experimental *E. coli* O149:H10:F4 diarrhoea in weaned pigs when supplied prophylactically or therapeutically (Jamalludeen et al. 2009).

Rabbits orally infected with pathogenic *E. coli* can be cured by a sole phage ZRP1. The phage removed nearly all host cells in vivo and had few effects on other bacteria. However, the therapy was less effective than antibiotic treatment with ciprofloxacin lactate (Zhao et al. 2017). *St. aureus* causes a variety of diseases in humans and animals. *St. aureus* can induce necrotizing pneumonia in rabbits, resulting in a high mortality rate. Phage VB-SavM-JYL01 decreased bacterial counts, suppressed the bacterial production of Panton-Valentine leukocidin and alpha-toxin, reduced the production of inflammatory factors in the lungs, alleviated pathological manifestations of rabbit necrotizing pneumonia and improved the survival outcome (Ji et al. 2019).

The American Foulbrood is a bacterial disease caused by *Paenibacillus larvae* and affects honeybee larvae. Bacteriophages were able to control the development of the disease when administered directly, but indirect transmission to larvae through adult workers was not effective (Yost et al. 2016; Ribeiro et al. 2019).

Aquaculture is another field providing food to the increasing human population. Global aquaculture production in 2016 equalled 110.2 million tonnes. The total output included 80.0 million tonnes of food fish and 30.1 million tonnes of aquatic plants as well as 37,900 tonnes of non-food products. The contribution of aquaculture to the combined yield of capture fisheries and aquaculture has increased considerably, reaching 46.8% in 2016, up from 25.7% in 2000 (FAO 2018). One of the major reasons for financial loss in the fish farming industry is the outbreak of infections provoked by pathogenic bacteria, especially multidrug-resistant variants. Phage therapy is also considered a way to mitigate the effects. After 6 h of treatment, the phage AS-A inhibited the growth of *Aeromonas salmonicida*, the causative agent of furunculosis, both in batch culture and seawater in the presence of *Solea senegalensis* juveniles. After 72 h, fish juveniles treated with phages after exposure to the pathogen showed no lethal cases, contrary to the infected control fish youngsters, distinguished by a mortality rate of 36% (Silva et al. 2016). The addition of the bacteriophage HER 110 culture delayed by 7 days the onset of furunculosis caused by *A. salmonicida* HER 1107 in brook trout yearlings and greatly reduced fish mortality from 100 to 10% in aquariums. The pathogen also acquired phage resistance at the cost of reduced viability and pathogenicity, but these phage-resistant mutants can be lysed by many other phages (Imbeault et al. 2006). Phages may be used to suppress *P. plecoglossicida* infections. The mortality rates of ayu (*Plecoglossus altivelis*) growing in tanks and receiving phage PPpW-3, phage PPpW-4 and a mix of PPpW-3/W4, and that of control fish were 53.3, 40.0, 20.0 and 93.3%, respectively. The daily mortality of fish taking a phage mix feed in a pond decreased at a constant rate to one-third after a 2 week period (Park and Nakai 2003). Phage therapy effectively cured the ulcerative lesions caused by *P. aeruginosa* in infected *Clarias gariepinus* following 8–10 days of treatment, resulting in a seven-fold reduction in the lesion size when compared with the untreated control (Khairnar et al. 2013). *Flavobacterium columnare* induces columnaris disease in fish. Two host fish species, rainbow trout (*Oncorhynchus mykiss*) and zebrafish (*Daniorerio*) demonstrated a higher survival rate in the presence of the phage; 100% of the zebrafish and 50% of the rainbow trout survived the phage treatment (without phage exposure, 0 and 8.3%, respectively). The rainbow trout population can be rescued from infection by a single addition of the phage into the water of a flow-through fish tank system (Laanto et al. 2015). Studies demonstrated efficiency in the treatment of *Vibrio*-associated diseases (vibriosis) in marine aquaculture (Kalatzis et al. 2018). Intraperitoneally or orally administered phage PLgY-16 prevented experimental *L. garvieae* infection in yellowtail (*Seriola quinqueradiata*) (Nakai et al. 1999). Some preparations are based on phages. Lexia is such an antimicrobial preparation to eliminate *Vibrio parahemolyticus*, which causes acute hepatopancreatic necrosis syndrome in shrimp and leads to grave economic losses in Asian countries (Phagelux 2019). $CUSTUS^{®}_{YRS}$ contains bacteriophages that specifically infect and kill *Yersinia ruckeri* causing yersiniosis, or enteric red mouth disease, in fish (ACD Pharma 2019). The commercial bacteriophage cocktail BAFADOR®, containing three phages against *A. hydrophila* and four varieties against *P. fluorescens*, curtailed the mortality of rainbow trout. The application of the preparation by immersion 24 h after experimental infection with *A. hydrophila* and *P. fluorescens* reduced life loss by 25%. The therapy applied 72 h after the infection decreased the survival rate to 55%, which indicates only a 10% advantage over the non-treated group. Prophylactic tests showed a 36% higher survival rate vs. control animals (Schulz et al. 2019a). BAFADOR® therapy of the European eel resulted in a 40% higher survival rate compared with the infected control, while every 24 h of delay lowered the therapeutic efficiency by 10% (Schulz et al. 2019b).

Phages have also been tested for the biocontrol of plant pathogens. The treatment with phage PE204 at 10^8 PFU/mL concentration of *Ralstonia solanacearum* colonizing tomato rhizosphere completely inhibited the occurrence of bacterial wilt (Bae et al. 2012). The bacteriophage cocktail P1, containing six phage isolates, could protect 80% of potato plants from *R. solanacearum* bacterial wilt and kill 98% of live bacteria spiked in the sterilized soil 1 week after spraying (Wei et al. 2017). Phage treatment of bacterial spot disease of tomato incited by *Xanthomonas campestris* pv. *vesicatoria* decreased the severity of the infection. Wherein, the use of formulations that increase bacteriophage longevity could improve the efficacy of phage treatment. Skimmed milk gave the best results in greenhouse trials, while casecrete NH-400 (a water-soluble casein polymer) performed best of all in the field (Balogh et al. 2003). Lytic bacteriophage ΦD5 inhibited infection by *Dickeya solani*, reducing blackleg and soft rot by more than 50% in both tissue culture and compost-grown potato plants (Czajkowski et al. 2017). Another phage cocktail decreased both the disease incidence and the severity of potato soft rot caused by *Pe. atrosepticum* in tubers by 61 and 64% (Carstens et al. 2019). The application of bacteriophages reduced the severity of leaf blight in onion, caused by *X. axonopodis* pv. *allii*, by 26–50% under field conditions. However, weekly or biweekly bacteriophage treatment would cost two to three times more than conventional bactericide doses per season (Lang et al. 2007). A phage cocktail composed of four virulent phages could be used for the control of Pierce's disease in grapevines, caused by *Xylellafastidiosa* subsp. *fastidiosa*. The disease symptoms ceased to progress after 1 week of post-therapeutic treatment and were not observed in prophylactically treated grapevines (Das et al. 2015). The application of two specific bacteriophages significantly reduced the number of *Erwinia amylovora*, the fire blight pathogen, by at least 45% (Schwarczinger et al. 2011). Despite a number of studies concerning the use of phages for plant disease therapy, to date few commercial preparations have been developed. Various AgriPhage preparations are applied to prevent and control harmful bacteria on tomato and pepper plants and apple, pear and citrus trees. AgriPhage products are available to counter bacterial spot, bacterial speck, bacterial canker, fire blight and citrus canker (AgriPhage 2019). The Biolyse®-PB preparation provides for the specific treatment of the most common soft-rot bacteria of potatoes (APS Biocontrol Ltd. 2019).

Phage products are widely used in the food industry for the control of food-borne pathogens. Various preparations target *Listeria monocytogenes*, *E. coli*, *Salmonella* and *Shigella* species (Intralytix 2019b; PhageGuard 2019). Several reports are focused on the application of phages for the elimination of other food-borne pathogens. Phages have been shown to be effective in reducing *S. aureus* counts in both fresh and hard-type cheeses without compromising the performance of the starter culture or their physicochemical properties. In fresh cheeses, both a low pH and phages contributed to the clearance of the pathogen. In hard cheeses, the pathogen was not totally eliminated, but its level was kept far below the limit appropriate for toxin production (Bueno et al. 2012). Phages BCP1-1 and BCP8-2 were able to effectively eradicate *Bacillus cereus* from a soybean paste product in the presence of divalent cations (Bandara et al. 2012). A mix of phages had a strong bactericidal effect on *Cronobacter sakazakii* in different brands of infant formulas, depressing the microbial titer to below the limit of detection (Endersen et al. 2017).

Bacteriophages can be used in typing for intraspecies differentiation of bacteria as well as locating their specific host in a variety of surroundings. Few phage diagnostic clinical tools are commercially available, have gained FDA approval or are used to reveal the cause of infection (Schofield et al. 2012). Phage display technology allows the expression of exogenous (poly)peptides on the surface of phage particles. It has several applications such as analysis of protein–protein interactions, determination of the substrate specificity in enzymology and the development of modulators of both the active and allosteric sites of the enzyme, generation of target-specific antibodies, epitope mapping and the identification of receptor agonists and antagonists (Pande et al. 2010). Thus, bacteriophages demonstrate a broad spectrum of use, and some applications are laden with vivid commercial potential.

15.12 Perspectives of Phage Therapeutic and Agricultural Applications: Problems of Bacteriophage Cultivation in Bioreactor

The invasion of antibiotic-resistant bacteria resumed the interest in bacteriophages as antimicrobial agents. So far, many studies on phage application are still in progress; however, several commercial preparations are available for medicine, agriculture and the food industry. Additionally, bacterial viruses are widely used in biotechnology, especially in phage display. The phage market is constantly growing. Although page introduction into everyday practice is confronted by obstacles, such as clinical clearance, it is expected that ongoing trials will facilitate the access of viral products to various application fields.

Other issues concern the full-scale output of bacteriophage preparations. The conditions for optimal bacteriophage production may differ from the conditions for optimal virulence or optimal bacterial growth. The host concentration is presumed to reach a value near its theoretical maximum before population-wide lysis takes place. Many bacteriophages lyse faster-growing cells, so lysis should ideally occur before the population enters the transient stage between exponential growth and stationary phases. Multiple parameters affect bacteriophage production: host growth rate, density of bacteriophage receptors on cell surface, bacterial metabolic activity, stage in cell life cycle, phage adsorption rate, burst size, lysis time, temperature, pH, medium composition, aeration rate, agitation and the presence of ions or cofactors. Changing parameters may change final bacteriophages titers. Non-virulent bacterial strains but not drug-resistant ones are recommended as hosts in the production process on the industrial scale. However, the results obtained in the laboratory are not always applicable to the full-scale production process (Agboluaje and Sauvageau 2018; García et al. 2019).

Several modes of cultivation can be used for bacteriophage production. Most production schemes rely on batch operation. For this approach, the infection is started at a given initial multiplicity of infection, and the host and bacteriophage populations grow until the bacteriophage population takes over and population-wide lysis occurs. It is the simplest and most common method with relatively high titers obtained, but typically for batch bioreactor processes it is labour intensive. A single-stage continuous bioreactor operation compared to batch mode is characterized by the regular addition of nutrients and the collection of products. However, this process is not recommended for bacteriophage preparation. It may lead to the appearance of phage-resistant strains and demand compliance of two factors: correct time distribution and threshold population density. Two-stage processes demarcate bacterial production and the propagation of phages. The host is grown at the first stage, while infection takes place at the second stage. There are either two bioreactors connected in series with a constant flow through the system, or a semi-continuous system, where each stage functions similarly to a batch culture. Two-stage processes are characterized by improved robustness and titers as well as the decreased probability of coevolution and the appearance of phage-resistant strains (Agboluaje and Sauvageau 2018; García et al. 2019).

15.13 Conclusions and Prospects

Phage application such as phage therapy proceeded historically along alternative approval pathways. Initially regarded as potential antimicrobials after their discovery, bacteriophages were supplanted by antibiotics, except in the former USSR. For some time, phage studies were merely exploratory in nature. The wide distribution of antibiotic-resistant strains urged the consideration of other ways to deal with pathogenic microorganisms, and phages again became objects of keen interest. Coupled with a better understanding of bacteriophage biology, studies now focus on practical aspects, leading to the appearance of commercial phage preparations and the development of a phage market. To date, preparations of bacterial viruses have found limited use in medicine and agriculture, but are widespread in the food industry and biotechnology. Phage applications still face some challenges regarding their distribution; however, continuous research in different areas and the development of phage industrial cultivation methods plus a steadily growing market for phage products forecast the increasing role of bacterial viruses.

REFERENCES

Abedon ST (2016) Introduction to bacteriophages. http://phage-therapy.org/writings/bacteriophages.html Accessed 30 August 2019

ACD Pharm (2019) https://acdpharma.com/en/bacteriophage-products/custus-en/ Accessed 30 August 2019

Ackermann HW (2011) Bacteriophage taxonomy. *Microbiol Aust* 32:90–94.

Ackermann HW, Prangishvili D (2012) Prokaryote viruses studied by electron microscopy. *Arch Virol* 157(10):1843–1849. https://doi.org/10.1007/s00705-012-1383-y

Ackermann HW, Tremblay D, Moineau S (2004) Long-term bacteriophage preservation. *WFCC Newsletter* 38(1):35–40.

Adriaenssens EM, Farkas K, Harrison C, Jones DL, Allison HE, McCarthy AJ (2018) Viromic analysis of wastewater input to a river catchment reveals a diverse assemblage of RNA viruses. *mSystems* 3(3):e00025-18. https://doi.org/10.1128/mSystems.00025-18

Agboluaje M, Sauvageau D (2018) Bacteriophage production in bioreactors. *Methods Mol Biol* 1693:173–193. https://doi.org/10.1007/978-1-4939-7395-8_15.

AgriPhage (2019) www.agriphage.com/product-info/ Accessed 30 August 2019

Ainsworth S, Sadovskaya I, Vinogradov E, Courtin P, Guerardel Y, Mahony J, Grard T, Cambillau C, Chapot-Chartier MP, van Sinderen D (2014) Differences in lactococcal cell wall polysaccharide structure are major determining factors in bacteriophage sensitivity. *mBio* 5(3):e00880-14. https://doi.org/10.1128/mBio.00880-14.

Al-Shammari AJ, Smith PF (1982) Receptor sites for mycoplasmal viruses on *Acholeplasma laidlawii*. *Rev Infect Dis* 4(Suppl):S109–14. https://doi.org/10.1093/clinids/4.Supplement_1.S109

Anal AK, Singh H (2007) Recent advances in microencapsulation of probiotics for industrial applications and targeted delivery. *Trends Food Sci Technol* 18(5):240–251. https://doi.org/10.1016/j.tifs.2007.01.004

Anand T, Vaid RK, Bera BCh, Barua S, Riyesh T, Virmani N, Yadav N, Malik P (2015) Isolation and characterization of a bacteriophage with broad host range, displaying potential in preventing bovine diarrhoea. *Virus Genes* 51(2):315–321. https://doi.org/10.1007/s11262-015-1222-9

APS Biocontrol Ltd. (2019) www.apsbiocontrol.com/products Accessed 30 August 2019

Bae JY, Wu J, Lee HJ, Jo EJ, Murugaiyan S, Chung E, Lee SW (2012) Biocontrol potential of a lytic bacteriophage PE204 against bacterial wilt of tomato. *J Microbiol Biotechnol* 22(12):1613–1620. https://doi.org/10.4014/jmb.1208.08072

Balogh B, Jones JB, Momol MT, Olson SM, Obradovic A, King P, Jackson LE (2003) Improved efficacy of newly formulated bacteriophages for management of bacterial spot on tomato. *Plant Dis* 87(8):949–954. https://doi.org/10.1094/PDIS.2003.87.8.949

Bandara N, Jo J, Ryu S, Kim KP (2012) Bacteriophages BCP1-1 and BCP8-2 require divalent cations for efficient control of *Bacillus cereus* in fermented foods. *Food Microbiol* 31(1):9–16. https://doi.org/10.1016/j.fm.2012.02.003

Barrangou R, van der Oost J (2015) Bacteriophage exclusion, a new defense system. *EMBO J* 34(2):134–135. https://doi.org/10.15252/embj.201490620

Barylski J, Enault F, Dutilh BE, Schuller MB, Edwards RA, Gillis A, Lavigne R, Klumpp J, Knezevic P, Krupovic M, Kuhn JH, Oksanen HM, Sullivan MB, Wittmann J, Tolstoy I, Brister JR, Kropinski AM Adriaenssens EM (2017) Genomic, proteomic, and phylogenetic analysis of spounaviruses indicates paraphyly of the order Caudovirales. *bioRxiv* 220434. https://doi.org/10.1101/220434

Bertozzi Silva J, Storms Z, Sauvageau D (2016) Host receptors for bacteriophage adsorption. *FEMS Microbiol Lett* 363(4):fnw002. https://doi.org/10.1093/femsle/fnw002

Bohemia Pharmaceuticals (2019) www.bohemiapharm.eu/products/stafal-professional-community/?lang=en_us#tab-id-3 Accessed 10 August 2019

Borie C, Albala I, Sánchez P, Sánchez ML, Ramírez S, Navarro C, Morales MA, Retamales AJ, Robeson J (2008) Bacteriophage treatment reduces *Salmonella* colonization of infected chickens. *Avian Dis* 52(1):64–67. https://doi.org/10.1637/8091-082007-Reg

Breyne K, Honaker RW, Hobbs Z, Richter M, Żaczek M, Spangler T, Steenbrugge J, Lu R, Kinkhabwala A, Marchon B, Meyer E, Mokres L (2017) Efficacy and safety of a bovine-associated *Staphylococcus aureus* phage cocktail in a murine model of mastitis. *Front Microbiol* 8:2348. https://doi.org/10.3389/fmicb.2017.02348

Brum JR, Schenck RO, Sullivan MB (2013) Global morphological analysis of marine viruses shows minimal regional variation and dominance of non-tailed viruses. *ISME J* 7(9):1738–1751. https://doi.org/10.1038/ismej.2013.67

Bueno E, García P, Martínez B, Rodríguez A (2012) Phage inactivation of *Staphylococcus aureus* in fresh and hard-type cheeses. *Int J Food Microbiol* 158(1):23–27. https://doi.org/10.1016/j.ijfoodmicro.2012.06.012

Carstens AB, Djurhuus AM, Kot W, Hansen LH (2019) A novel six-phage cocktail reduces *Pectobacterium atrosepticum* soft rot infection in potato tubers under simulated storage conditions. *FEMS Microbiol Lett* 366(9):fnz101. https://doi.org/10.1093/femsle/fnz101

Castro-Wallace SL, Chiu CY, John KK, Stahl SE, Rubins KH, McIntyre ABR, Dworkin JP, Lupisella ML, Smith DJ, Botkin DJ, Stephenson TA, Juul S, Turner DJ, Izquierdo F, Federman S, Stryke D, Somasekar S, Alexander N, Yu G, Mason CE, Burton AS (2017) Nanopore DNA sequencing and genome assembly on the international space station. *Sci Rep* 7(1):18022. https://doi.org/110.1038/s41598-017-18364-0

Center for Innovative Phage Applications and Therapeutics (2019) https://medschool.ucsd.edu/som/medicine/divisions/idgph/research/center-innovative-phage-applications-and-therapeutics/pages/default.aspx Accessed 30 August 2019

Chatterjee S, Rothenberg E (2012) Interaction of bacteriophage l with its *E. coli* receptor, LamB. *Viruses* 4(11):3162–3178. https://doi.org/10.3390/v4113162

Cherwa JE Jr, Fane BA (2011) Family: Microviridae. Chapter Version: ICTV. *Ninth Report; 2009 Taxonomy Release.* https://talk.ictvonline.org/ictv-reports/ictv_9th_report/ssdna-viruses-2011/w/ssdna_viruses/147/microviridae Accessed 30 August 2019

Chhibber S, Kumari S (2012) Application of therapeutic phages in medicine. In: Kurtboke I (ed) *Bacteriophages.* InTechOpen, pp. 139–158. https://doi.org/10.5772/1065

Choi Y, Shin H, Lee JH, Ryu S (2013) Identification and characterization of a novel flagellum-dependent *Salmonella*-infecting bacteriophage, iEPS5. *Appl Environ Microbiol* 79(16):4829–4837. https://doi.org/10.1128/AEM.00706-13

Clavijo V, Baquero D, Hernandez S, Farfan JC, Arias J, Arévalo A, Donado-Godoy P, Vives-Flores M (2019) Phage cocktail SalmoFREE® reduces *Salmonella* on a commercial broiler farm. *Poultry Sci Pez* 251. https://doi.org/10.3382/ps/pez251v

Coons GH, Kotilia JE (1925) Transmissible lytic principle (bacteriophage) in relation to plant pathogens. *Phytopathology* 15:357–370.

Cowles PB, Hale WM (1931) Effect of bacteriophage on experimental anthrax in white mice. *J Infect Dis* 49(3):264–269. https://doi.org/10.1093/infdis/49.3.264

Credence Research (2018) *Bacteriophage Market by Product Type (dsDNA Bacteriophage, ssDNA Bacteriophage, ssRNA Bacteriophage), by Application (Clinical Application, Food & Beverages, Phage Display, Phage Therapy, Environmental Application, Veterinary) – Growth, Future Prospects & Competitive Analysis, 2018–2026.* www.credenceresearch.com/report/bacteriophage-market Accessed 30 August 2019

Czajkowski R, Smolarska A, Ozymko Z (2017) The viability of lytic bacteriophage ΦD5 in potato-associated environments and its effect on *Dickeya solani* in potato (*Solanum tuberosum* L.) plants. *PLoS One* 12(8):e0183200. https://doi.org/10.1371/journal.pone.0183200

Danovaro R, Corinaldesi C, Filippini M, Fischer UR, Gessner MO, Jacquet S, Magagnini M, Velimirov B (2008) Viriobenthos in freshwater and marine sediments: A review. *Freshwater Biol* 53(6):1186–1213. https://doi.org/10.1111/j.1365-2427.2008.01961.x

Das M, Bhowmick TS, Ahern SJ, Young R, Gonzalez CF (2015) Control of Pierce's disease by phage. *PLoS One* 10(6):e0128902. https://doi.org/10.1371/journal.pone.0128902

Day LA (2011) Family Inoviridae. Chapter Version: ICTV. *Ninth Report; 2009 Taxonomy Release.* https://talk.ictvonline.org/ictv-reports/ictv_9th_report/ssdna-viruses-2011/w/ssdna_viruses/145/inoviridae Accessed 30 August 2019

Dybvig K, Alderete J, Watson HL, Cassell GH (1988) Adsorption of mycoplasma virus P1 to host cells. *J Bacteriol* 170(9):4373–4375. https://doi.org/10.1128/jb.170.9.4373-4375.1988

El Haddad L, Harb CP, Gebara MA, Stibich MA, Chemaly RF (2019) A systematic and critical review of bacteriophage therapy against multidrug-resistant ESKAPE organisms in humans. *Clin Infect Dis* 69(1):167–178. https://doi.org/10.1093/cid/ciy947

Eliava BioPreparations (2019) http://phage.ge/products/Accessed 30 August 2019

Endersen L, Buttimer C, Nevin E, Coffey A, Neve H, Oliveira H, Lavigne R, O'Mahony J (2017) Investigating the biocontrol and anti-biofilm potential of a three phage cocktail against *Cronobacter sakazakii* in different brands of infant formula. *Int J Food Microbiol* 253:1–11. https://doi.org/10.1016/j.ijfoodmicro.2017.04.009

Erez Z, Steinberger-Levy I, Shamir M, Doron S, Stokar-Avihail A, Peleg Y, Melamed S, Leavitt A, Savidor A, Albeck S, Amitai G, Sorek R (2017) Communication between viruses guides lysis-lysogeny decisions. *Nature* 541(7638):488–493. https://doi.org/10.1038/nature21049

European Medicines Agency (2015) *Workshop on the Therapeutic Use of Bacteriophages*. www.ema.europa.eu/documents/other/workshop-therapeutic-use-bacteriophages-summary_en.pdf Accessed 30 August 2019

Evans TJ, Crow MA, Williamson NR, Orme W, Thomson NR, Komitopoulou E, Salmond GP (2010) Characterization of a broad-host-range flagellum-dependent phage that mediates high-efficiency generalized transduction in, and between, *Serratia* and *Pantoea*. *Microbiology* 156(Pt 1):240–247. https://doi.org/10.1099/mic.0.032797-0

FAO (2018) The state of world fisheries and aquaculture 2018. www.fao.org/3/i9540en/i9540en.pdf Accessed 30 August 2019

Fiorentin L, Vieira ND, Barioni W Jr (2005) Oral treatment with bacteriophages reduces the concentration of *Salmonella enteritidis* PT4 in caecal contents of broilers. *Avian Pathol* 34(3):258–63. https://doi.org/10.1080/01445340500112157

Fisk RT (1938) Protective action of typhoid phage on experimental typhoid infection in mice. *Exp Biol Med* 38(5):659–660. https://doi.org/10.3181/00379727-38-9973

Flu P (1933) Immunisation des rats contre la peste au moyen de suspension concentrfes des bacilles pesteux virulents lyses par le bactfriophage antipesteux. *Bull Soc Pathol Exot* 26:796–806.

Fokine A, Rossmann MG (2014) Molecular architecture of tailed double-stranded DNA phages. *Bacteriophage* 4(1):e28281. https://doi.org/10.4161/bact.28281

Fortier LC, Moineau S (2009) Phage production and maintenance of stocks, including expected stock lifetimes. *Methods Mol Biol* 501:203–219. https://doi.org/10.1007/978-1-60327-164-6_19

García R, Latz S, Romero J, Higuera G, García K, Bastías R (2019) Bacteriophage production models: An overview. *Front Microbiol* 10:1187. https://doi.org/10.3389/fmicb.2019.01187

GenBank (2019) www.genome.jp/dbget-bin/www:bget?genbank-phg:L06183 Accessed 30 August 2019

German GJ, Misra R (2001) The TolC protein of *Escherichia coli* serves as a cell-surface receptor for the newly characterized TLS bacteriophage. *J Mol Biol* 308(4):579–585. https://doi.org/10.1006/jmbi.2001.4578

Gill JJ, Pacan JC, Carson ME, Leslie KE, Griffiths MW, Sabour PM (2006) Efficacy and pharmacokinetics of bacteriophage therapy in treatment of subclinical *Staphylococcus aureus* mastitis in lactating dairy cattle. *Antimicrob Agents Chemother* 50(9):2912–2918. https://doi.org/10.1128/AAC.01630-05

González-Menéndez E, Fernández L, Gutiérrez D, Rodríguez A, Martínez B, García P (2018) Comparative analysis of different preservation techniques for the storage of *Staphylococcus* phages aimed for the industrial development of phage-based antimicrobial products. *PLoS One* 13(10):e0205728. https://doi.org/10.1371/journal.pone.0205728

Goodridge L, Gallaccio A, Griffiths MW (2003) Morphological, host range, and genetic characterization of two coliphages. *Appl Environ Microbiol* 69(9):5364–5371. https://doi.org/10.1128/AEM.69.9.5364-5371.2003

Gordillo Altamirano FL, Barr JJ (2019) Phage therapy in the postantibiotic era. *Clin Microbiol Rev* 32(2):e00066-18. https://doi.org/10.1128/CMR.00066-18

Guttman B, Raya R, Kutter E (2005) Basic phage biology. In: Kutter E, Sulakvelidze A (eds) *Bacteriophages: Biology and Applications*. CRC Press, Boca Raton, pp. 29–68.

Hall WJ, MacDonald AD, Legenhausen DH (1949) Studies on fowl typhoid: II. Control of the disease. *Poultr Sci* 28(6):789–801.https://doi.org/10.3382/ps.0280789

Harris TD, Buzby PR, Babcock H, Beer E, Bowers J, Braslavsky I, Causey M, Colonell J, Dimeo J, Efcavitch JW, Giladi E, Gill J, Healy J, Jarosz M, Lapen D, Moulton K, Quake SR, Steinmann K, Thayer E, Tyurina A, Ward R, Weiss H, Xie Z (2008) Single-molecule DNA sequencing of a viral genome. *Science* 320(5872):106–109. https://doi.org/10.1126/science.1150427

Hatfull GF (2008) Bacteriophage genomics. *Curr Opin Microbiol* 11(5):447–453. https://doi.org/10.1016/j.mib.2008.09.004

Hatfull GF, Hendrix RW (2011) Bacteriophages and their genomes. *Curr Opin Virol* 1(4):298–303. https://doi.org/10.1016/j.coviro.2011.06.009

Heather JM, Chain B (2016) The sequence of sequencers: The history of sequencing DNA. *Genomics* 107(1):1–8. https://doi.org/10.1016/j.ygeno.2015.11.003

Hobbs Z, Abedon ST (2016) Diversity of phage infection types and associated terminology: The problem with "Lytic or lysogenic". *FEMS Microbiol Lett* 363(7):fnw047. https://doi.org/10.1093/femsle/fnw047

Hulo C, Masson P, Le Mercier P, Toussaint A (2015) A structured annotation frame for the transposable phages: A new proposed family "*Saltoviridae*" within the Caudovirales. *Virology* 477:155–163. https://doi.org/10.1016/j.virol.2014.10.009

Hyman P (2019) Phages for phage therapy: Isolation, characterization, and host range breadth. *Pharmaceuticals (Basel)* 12(1):E35. https://doi.org/10.3390/ph12010035

Hyman P, Abedon ST (2010) Bacteriophage host range and bacterial resistance. *Adv Appl Microbiol* 70:217–248. https://doi.org/10.1016/S0065-2164(10)70007-1

ICTV (2019) Virus taxonomy: 2018b release. https://talk.ictvonline.org/taxonomy/ Accessed 30 August 2019

Imbeault S, Parent S, Lagacé M, Uhland CF, Blais JF (2006) Using bacteriophages to prevent furunculosis caused by *Aeromonas salmonicida* in farmed brook trout. *J Aquat Anim Health* 18(3):203–214. https://doi.org/10.1577/H06-019.1

Intralytix (2019a) Veterinary applications. www.intralytix.com/index.php?page=vet Accessed 30 August 2019

Intralytix (2019b) Bacteriophage products. www.intralytix.com/index.php?page=prod Accessed 30 August 2019

Iwano H, Inoue Y, Takasago T, Kobayashi H, Furusawa T, Taniguchi K, Fujiki J, Yokota H, Usui M, Tanji Y, Hagiwara K, Higuchi H, Tamura Y (2018) Bacteriophage ΦSA012 has a broad host range against *Staphylococcus aureus* and effective lytic capacity in a mouse mastitis model. *Biology (Basel)* 7(1):E8. https://doi.org/10.3390/biology7010008

Jamalludeen N, Johnson RP, Shewen PE, Gyles CL (2009) Evaluation of bacteriophages for prevention and treatment of diarrhea due to experimental enterotoxigenic *Escherichia coli* O149 infection of pigs. *Vet Microbiol* 136(1–2):135–141. https://doi.org/10.1016/j.vetmic.2008.10.021

Jault P, Leclerc T, Jennes S, Pirnay JP, Que YA, Resch G, Rousseau AF, Ravat F, Carsin H, Le Floch R, Schaal JV, Soler C, Fevre C, Arnaud I, Bretaudeau L, Gabard J (2019) Efficacy and tolerability of a cocktail of bacteriophages to treat burn wounds infected by *Pseudomonas aeruginosa* (PhagoBurn): A randomised, controlled, double-blind phase 1/2 trial. *Lancet Infect Dis* 19(1):35–45. https://doi.org/10.1016/S1473-3099(18)30482-1

Ji Y, Cheng M, Zhai S, Xi H, Cai R, Wang Z, Zhang H, Wang X, Xue Y, Li X, Sun C, Feng X, Lei L, Ur Rahman S, Han W, Gu J (2019) Preventive effect of the phage VB-SavM-JYL01 on rabbit necrotizing pneumonia caused by *Staphylococcus aureus*. *Vet Microbiol* 229:72–80. https://doi.org/10.1016/j.vetmic.2018.12.021

Kaikabo AA, AbdulKarim SM, Abas F (2017) Evaluation of the efficacy of chitosan nanoparticles loaded ΦKAZ14 bacteriophage in the biological control of colibacillosis in chickens. *Poult Sci* 96(2):295–302. https://doi.org/10.3382/ps/pew255

Kalatzis PG, Castillo D, Katharios P, Middelboe M (2018) Bacteriophage interactions with marine pathogenic vibrios: Implications for phage therapy. *Antibiotics (Basel)* 7(1):E15. https://doi.org/10.3390/antibiotics7010015

Karlsson F, Borrebaeck CA, Nilsson N, Malmborg-Hager AC (2003) The mechanism of bacterial infection by filamentous phages involves molecular interactions between TolA and phage protein 3 domains. *J Bacteriol* 185(8):2628–2634. https://doi.org/10.1128/jb.185.8.2628-2634.2003

Kendrick P (1933) The antigenic properties of bacteriophage lysates of *Salmonella suipestifer*. V. Protection tests with rabbits. General résumé. *Am J Epidemiol* 18(2):442–461. https://doi.org/10.1093/oxfordjournals.aje.a117961

Khairnar K, Raut MP, Chandekar RH, Sanmukh SG, Paunikar WN (2013) Novel bacteriophage therapy for controlling metallo-beta-lactamase producing *Pseudomonas aeruginosa* infection in catfish. *BMC Vet Res* 9:264. https://doi.org/10.1186/1746-6148-9-264

Kim ES, Bae HW, Cho YH (2018) A pilin region affecting host range of the *Pseudomonas aeruginosa* RNA phage, PP7. *Front Microbiol* 9:247. https://doi.org/10.3389/fmicb.2018.00247

Kim M, Kim S, Park B, Ryu S (2014) Core lipopolysaccharide-specific phage SSU5 as an auxiliary component of a phage cocktail for *Salmonella* biocontrol. *Appl Environ Microbiol* 80(3):1026–1034. https://doi.org/10.1128/AEM.03494-13

Klumpp J, Fouts DE, Sozhamannan S (2012) Next generation sequencing technologies and the changing landscape of phage genomics. *Bacteriophage*. 2(3):190–199. https://doi.org/10.4161/bact.22111

Kotsonis SE, Powell IB, Pillidge CJ, Limsowtin GK, Hillier AJ, Davidson BE (2008) Characterization and genomic analysis of phage asccphi28, a phage of the family Podoviridae infecting *Lactococcus lactis*. *Appl Environ Microbiol* 74(11):3453–3460. https://doi.org/10.1128/AEM.02379-07

Krishnamurthy SR, Wang D (2018) Extensive conservation of prokaryotic ribosomal binding sites in known and novel picobirnaviruses. *Virology* 516:108–114. https://doi.org/10.1016/j.virol.2018.01.006

Kropinski AM, Anany H, Kuhn JH, Tolstoy I, Kutter E, Adriaenssens EM (2017) To create a new bacteriophage family, Ackermannviridae, containing two (2) new subfamilies including four (4) genera. Report number: ICTV [International Committee for Taxonomy of Viruses] Proposal (Taxoprop) 2017.001B. https://doi.org/10.13140/RG.2.2.29173.88800

Krupovic M, ICTV Report Consortium (2018) ICTV virus taxonomy profile: Plasmaviridae. *J Gen Virol* 99:617–618. https://doi.org/10.1099/jgv.0.001060

Krupovic M, Bamford DH (2007) Putative prophages related to lytic tailless marine dsDNA phage PM2 are widespread in the genomes of aquatic bacteria. *BMC Genomics* 8:236. https://doi.org/10.1186/1471-2164-8-236

Kwan T, Liu J, Dubow M, Gros P, Pelletier J (2006) Comparative genomic analysis of 18 *Pseudomonas aeruginosa* bacteriophages. *J Bacteriol* 188(3):1184–1187. https://doi.org/10.1128/JB.188.3.1184-1187.2006

Laanto E, Bamford JK, Ravantti JJ, Sundberg LR (2015) The use of phage FCL-2 as an alternative to chemotherapy against columnaris disease in aquaculture. *Front Microbiol* 6:829. https://doi.org/10.3389/fmicb.2015.00829

Labrie SJ, Samson JE, Moineau S (2010) Bacteriophage resistance mechanisms. *Nat Rev Microbiol* 8(5):317–327. https://doi.org/10.1038/nrmicro2315

Lang JM, Gent DH, Schwartz HF (2007) Management of *Xanthomonas* leaf blight of onion with bacteriophages and a plant activator. *Plant Dis* 91(7):871–878. https://doi.org/10.1094/PDIS-91-7-0871

Lavigne R, Molineux IJ, Kropinski AM (2011) Order Caudovirales. Chapter Version: ICTV. *Ninth Report; 2009 Taxonomy Release*. https://talk.ictvonline.org/ictv-reports/ictv_9th_report/dsdna-viruses-2011/w/dsdna_viruses/67/caudovirales Accessed 30 August 2019

Leigh BA, Breitbart M, Oksanen HM, Bamford DH, Dishaw LJ (2018) Genome sequence of PM2-like phage Cr39582, induced from a *Pseudoalteromonas* sp. isolated from the gut of *Ciona robusta*. *Genome Announc* 6(21):e00368-18. https://doi.org/10.1128/genomeA.00368-18

Leung SSY, Parumasivam T, Gao FG, Carter EA, Carrigy NB, Vehring R, Finlay WH, Morales S, Britton WJ, Kutter E, Chan HK (2017) Effects of storage conditions on the stability of spray dried, inhalable bacteriophage powders. *Int J Pharm* 521(1–2):141–149. https://doi.org/10.1016/j.ijpharm.2017.01.060

Leung V, Szewczyk A, Chau J, Hosseinidoust Z, Groves L, Hawsawi H, Anany H, Griffiths MW, Ali MM, Filipe CD (2017) Long-term preservation of bacteriophage antimicrobials using sugar glasses. *Acs Biomater Sci Eng* 4(11):3802–3808. https://doi.org/10.1021/acsbiomaterials.7b00468

Łobocka MB, Głowacka A, Golec P (2018) Methods for bacteriophage preservation. *Methods Mol Biol* 1693:219–230. https://doi.org/10.1007/978-1-4939-7395-8_17

Loc Carrillo C, Atterbury RJ, el-Shibiny A, Connerton PL, Dillon E, Scott A, Connerton IF (2005) Bacteriophage therapy to reduce *Campylobacter jejuni* colonization of broiler chickens. *Appl Environ Microbiol* 71(11):6554–6563. https://doi.org/10.1128/AEM.71.11.6554-6563.2005

Łoś M, Węgrzyn G (2012) Pseudolysogeny. *Adv Virus Res* 82:339–349. https://doi.org/10.1016/B978-0-12-394621-8.00019-4

Lunde M, Aastveit AH, Blatny JM, Nes IF (2005) Effects of diverse environmental conditions on {phi}LC3 prophage stability in *Lactococcus lactis*. *Appl Environ Microbiol* 71(2):721–727. https://doi.org/10.1128/AEM.71.2.721-727.2005

Maurice CF, Bouvier C, de Wit R, Bouvier T (2013) Linking the lytic and lysogenic bacteriophage cycles to environmental conditions, host physiology and their variability in coastal lagoons. *Environ Microbiol* 15(9):2463–2475. https://doi.org/10.1111/1462-2920.12120

McCutcheon JG, Peters DL, Dennis JJ (2018) Identification and characterization of type IV pili as the cellular receptor of broad host range *Stenotrophomonas maltophilia* bacteriophages DLP1 and DLP2. *Viruses* 10(6):E338. https://doi.org/10.3390/v10060338

Micreos (2019) Micreos human health. www.micreos.com/content/human-health.aspx Accessed 30 August 2019

Microgen (2019) www.microgen.ru/en/products/bakteriofagi/ Accessed 30 August 2019

Miller RW, Skinner EJ, Sulakvelidze A, Mathis GF, Hofacre CL (2010) Bacteriophage therapy for control of necrotic enteritis of broiler chickens experimentally infected with *Clostridium perfringens*. *Avian Dis* 54(1):33–40. https://doi.org/10.1637/8953-060509-Reg.1

Monteville MR, Ardestani B, Geller BL (1994) Lactococcal bacteriophages require a host cell wall carbohydrate and a plasma membrane protein for adsorption and ejection of DNA. *Appl Environ Microbiol* 60(9):3204–3211.

Muncie JH, Patel MK (1929) Potency and specificity of a lytic principle (bacteriophage) obtained from *Pseudomonas tumefaciens*. *Phytopath* 19:98.

Nakai T, Sugimoto R. Park K-H, Matsuoka S, Mori K, Nishioka T, Maruyama K (1999) Protective effects of bacteriophage on experimental *Lactococcus garvieae* infection in yellowtail. *Dis Aquat Organ* 37:33–41. http://doi.org/10.3354/dao037033

NCBI (2019) www.ncbi.nlm.nih.gov/genomes/GenomesGroup.cgi?taxid=10239&host=bacteria# Accessed 30 August 2019

Nguyen S, Baker K, Padman BS, Patwa R, Dunstan RA, Weston TA, Schlosser K, Bailey B, Lithgow T, Lazarou M, Luque, A., Rohwer F, Blumberg RS, Barr JJ (2017) Bacteriophage transcytosis provides a mechanism to cross epithelial cell layers. *mBio* 8(6):e01874-17. http://doi.org/10.1128/mBio.01874-17

Ofir G, Melamed S, Sberro H, Mukamel Z, Silverman S, Yaakov G, Doron S, Sorek R (2018) DISARM is a widespread bacterial defence system with broad anti-phage activities. *Nat Microbiol* 3(1):90–98. http://doi.org/10.1038/s41564-017-0051-0

Oksanen HM, Bamford DH (2011). Family: Tectiviridae. Chapter Version: ICTV. *Ninth Report; 2009 Taxonomy Release*. https://talk.ictvonline.org/ictv-reports/ictv_9th_report/dsdna-viruses-2011/w/dsdna_viruses/133/tectiviridae Accessed 30 August 2019

Oksanen HM, ICTV Report Consortium (2017) ICTV virus taxonomy profile: Corticoviridae. *J Gen Virol* 98:888–889. http://doi.org/10.1099/jgv.0.000795

Oliveira A, Sereno R, Azeredo J (2010) In vivo efficiency evaluation of a phage cocktail in controlling severe colibacillosis in confined conditions and experimental poultry houses. *Vet Microbiol* 146(3–4):303–308. http://doi.org/10.1016/j.vetmic.2010.05.015

Pande J, Szewczyk MM, Grover AK (2010) Phage display: Concept, innovations, applications and future. *Biotechnol Adv* 28(6):849–858. http://doi.org/10.1016/j.biotechadv.2010.07.004

Park SC, Nakai T (2003) Bacteriophage control of *Pseudomonas plecoglossicida* infection in ayu *Plecoglossus altivelis*. *Dis Aquat Organ* 53(1):33–39. http://doi.org/10.3354/dao053033

Pawlowski A, Rissanen I, Bamford JK, Krupovic M, Jalasvuori M (2014) *Gammasphaerolipovirus*, a newly proposed bacteriophage genus, unifies viruses of halophilic archaea and thermophilic bacteria within the novel family Sphaerolipoviridae. *Arch Virol* 159(6):1541–1554. http://doi.org/10.1007/s00705-013-1970-6

PhageGuard (2019) https://phageguard.com Accessed 30 August 2019

Phagelux (2019) www.phagelux.com/en/single/menu_84.htm?menuid=84 Accessed 30 August 2019

Pietilä MK, Demina TA, Atanasova NS, Oksanen HM, Bamford DH (2014) Archaeal viruses and bacteriophages: Comparisons and contrasts. *Trends Microbiol* 22(6):334–344. http://doi.org/10.1016/j.tim.2014.02.007

Poranen MM, Mäntynen S, ICTV Report Consortium (2017) ICTV virus taxonomy profile: *Cystoviridae*. *J Gen Virol* 98(10):2423–2424. http://doi.org/10.1099/jgv.0.000928

Proft T, Baker EN (2009) Pili in gram-negative and gram-positive bacteria: Structure, assembly and their role in disease. *Cell Mol Life Sci* 66(4):613–635. http://doi.org/10.1007/s00018-008-8477-4

Proteon Pharmaceuticals (2019) www.proteonpharma.com/products/phage-products/ Accessed 30 August 2019

Pyle NJ (1926) The bacteriophage in relation to *Salmonella pullora* infection in the domestic fowl. *J Bacteriol* 12(4):245–261.

Räisänen L, Schubert K, Jaakonsaari T, Alatossava T (2004) Characterization of lipoteichoic acids as *Lactobacillus delbrueckii* phage receptor components. *J Bacteriol* 186(16):5529–5532. https://doi.org/10.1128/JB.186.16.5529-5532.2004

Rakhuba DV, Kolomiets EI, Dey ES, Novik GI (2010) Bacteriophage receptors, mechanisms of phage adsorption and penetration into host cell. *Pol J Microbiol* 59(3):145–155. http://doi.org/10.1016/j.micres.2015.01.008.1.94

Rakieten TL, Rakieten ML (1943) Bacteriophagy in the developing chick embryo. *J Bacteriol* 45(5):477–484.

Rath D, Amlinger L, Rath A, Lundgren M (2015) The CRISPR-Cas immune system: Biology, mechanisms and applications. *Biochimie* 117:119–128. http://doi.org/10.1016/j.biochi.2015.03.025

Redondo RA, Kupczok A, Stift G, Bollback JP (2013) Complete genome sequence of the novel phage MG-B1 infecting *Bacillus weihenstephanensis*. *Genome Announc* 1(3):e00216-13. http://doi.org/10.1128/genomeA.00216-13

Ribeiro HG, Correia R, Moreira T, Vilas Boas D, Azeredo J, Oliveira A (2019) Bacteriophage biodistribution and infectivity from honeybee to bee larvae using a T7 phage model. *Sci Rep* 9(1):620. http://doi.org/10.1038/s41598-018-36432-x

Richards PJ, Connerton PL, Connerton IF (2019) Phage biocontrol of *Campylobacter jejuni* in chickens does not produce collateral effects on the gut microbiota. *Front Microbiol* 10:476. http://doi.org/10.3389/fmicb.2019.00476

Rihtman B, Meaden S, Clokie MR, Koskella B, Millard AD (2016) Assessing Illumina technology for the high-throughput sequencing of bacteriophage genomes. *PeerJ* 4:e2055. http://doi.org/10.7717/peerj.2055

Rohwer F, Edwards R (2002) The phage proteomic tree: A genome-based taxonomy for phage. *J Bacteriol* 184(16):4529–4535. http://doi.org/10.1128/jb.184.16.4529-4535.2002

Roux S, Krupovic M, Daly RA, Borges AL, Nayfach S, Schulz F, Cheng JF, Ivanova NN, Bondy-Denomy J, Wrighton KC, Woyke T, Visel A, Kyrpides N, Eloe-Fadrosh EA (2019) Cryptic inoviruses are pervasive in bacteria and archaea across Earth's biomes. *bioRxiv* 548222. https://doi.org/10.1101/548222

Rozema EA, Stephens TP, Bach SJ, Okine EK, Johnson RP, Stanford K, McAllister TA (2009) Oral and rectal administration of bacteriophages for control of *Escherichia coli* O157:H7 in feedlot cattle. *J Food Prot* 72(2):241–250. https://doi.org/10.4315/0362-028x-72.2.241

São-José C, Baptista C, Santos MA (2004) *Bacillus subtilis* operon encoding a membrane receptor for bacteriophage SPP1. *J Bacteriol* 186(24):8337–8346. https://doi.org/10.1128/JB.186.24.8337-8346.2004

Saxena HM, Raj S (2018) A novel immunotherapy of brucellosis in cows monitored non-invasively through a specific biomarker. *PLoS Negl Trop Dis* 12(4):e0006393. https://doi.org/10.1371/journal.pntd.0006393

Schofield DA, Sharp NJ, Westwater C (2012) Phage-based platforms for the clinical detection of human bacterial pathogens. *Bacteriophage* 2(2):105–283. https://doi.org/10.4161/bact.19274

Schooley RT, Biswas B, Gill JJ, Hernandez-Morales A, Lancaster J, Lessor L, Barr JJ, Reed SL, Rohwer F, Benler S, Segall AM, Taplitz R, Smith DM, Kerr K, Kumaraswamy M, Nizet V, Lin L, McCauley MD, Strathdee SA, Benson CA, Pope RK, Leroux BM, Picel AC, Mateczun AJ, Cilwa KE, Regeimbal JM, Estrella LA, Wolfe DM, Henry MS, Quinones J, Salka S, Bishop-Lilly KA, Young R, Hamilton T (2017) Development and use of personalized bacteriophage-based therapeutic cocktails to treat a patient with a disseminated resistant *Acinetobacter baumannii* infection. *Antimicrob Agents Chemother* 61(10):e00954-17. https://doi.org/10.1128/AAC.00954-17

Schulz P, Pajdak-Czaus J, Robak S, Dastych J, Siwicki AK (2019a) Bacteriophage-based cocktail modulates selected immunological parameters and post-challenge survival of rainbow trout (*Oncorhynchus mykiss*). *J Fish Dis* 42:1151–1160. https://doi.org/10.1111/jfd.13026

Schulz P, Robak S, Dastych J, Siwicki AK (2019b) Influence of bacteriophages cocktail on European eel (*Anguilla anguilla*) immunity and survival after experimental challenge. *Fish Shellfish Immunol* 84:28–37. https://doi.org/10.1016/j.fsi.2018.09.056

Schwarczinger I, Kiss E, Süle S, Tóth M, Hevesi M (2011) Control of fire blight by bacteriophages on apple flowers. *Acta Hortic* 896:457–462 https://doi.org/10.17660/ActaHortic.2011.896.66

Seo BJ, Song ET, Lee K, Kim JW, Jeong CG, Moon SH, Son JS, Kang SH, Cho HS, Jung BY, Kim WI (2018) Evaluation of the broad-spectrum lytic capability of bacteriophage cocktails against various *Salmonella* serovars and their effects on weaned pigs infected with *Salmonella typhimurium*. *J Vet Med Sci* 80(6):851–860. https://doi.org/10.1292/jvms.17-0501

Shao Q, Trinh JT, Zeng L (2019) High-resolution studies of lysis–lysogeny decision-making in bacteriophage lambda. *J Biol Chem* 294(10):3343–3349. https://doi.org/10.1074/jbc.TM118.003209

Shao Y, Wang IN (2008) Bacteriophage adsorption rate and optimal lysis time. *Genetics* 180(1):471–482. https://doi.org/10.1534/genetics.108.090100

Sheng H, Knecht HJ, Kudva IT, Hovde CJ (2006) Application of bacteriophages to control intestinal *Escherichia coli* O157:H7 levels in ruminants. *Appl Environ Microbiol* 72(8):5359–5366. https://doi.org/10.1128/AEM.00099-06

Shkoporov AN, Hill C (2019) Bacteriophages of the human gut: The "known unknown" of the microbiome. *Cell Host Microbe* 25(2):195–209. https://doi.org/10.1016/j.chom.2019.01.017

Silipo A., Molinaro A (2011) Lipid a structure. In: Knirel Y, Valvano M (eds) *Bacterial Lipopolysaccharides*. Springer, Vienna, pp. 1–20. https://doi.org/10.1007/978-3-7091-0733-1_1

Silva YJ, Moreirinha C, Pereira C, Costa L, Rocha RJ, Cunha Â, Gomes NCM, Calado R, Almeida A (2016) Biological control of *Aeromonas salmonicida* infection in juvenile Senegalese sole (*Solea senegalensis*) with phage AS-A. *Aquaculture* 450:225–233. https://doi.org/10.1016/j.aquaculture.2015.07.025

Sime-Ngando T, Bertrand JC, Bogusz D, Brugère JF, Franche C, Fardeau ML, Froussart E, Geiger A, Goñi-Urriza MS, Ollivier B, O'Toole PW (2018) The evolution of living beings started with prokaryotes and in interaction with prokaryotes. In: Bertrand JC, Normand P, Ollivier B, Sime-Ngando T (eds) *Prokaryotes and Evolution*. Springer, Cham, pp. 241–338. https://doi.org/10.1007/978-3-319-99784-1_5

Sørensen MC, van Alphen LB, Harboe A, Li J, Christensen BB, Szymanski CM, Brøndsted L (2011) Bacteriophage F336 recognizes the capsular phosphoramidate modification of *Campylobacter jejuni* NCTC11168. *J Bacteriol* 193(23):6742–6749. https://doi.org/10.1128/JB.05276-11

Stirm S (1968). *Escherichia coli* K bacteriophages. I. Isolation and introductory characterization of five *Escherichia coli* K bacteriophages. *J Virol* 2(10):1107–1114.

Sulakvelidze A, Kutter E (2005) Bacteriophage therapy in humans. In: Kutter E, Sulakvelidze A (eds) *Bacteriophages: Biology and Applications*. CRC Press, Boca Raton, pp. 382–436.

Sulkin SE, Douglass DD, Bronfenbrenner J (1942) Bacteriophage therapy. IV. Effect of bacteriophage in experimental staphylococcal septicemia in rabbits. *J Infect Dis* 70(1):92–95. https://doi.org/10.1093/infdis/70.1.92

Summers WC (2005) Bacteriophage research: Early history. In: Kutter E, Sulakvelidze A (eds) *Bacteriophages: Biology and Applications*. CRC Press, Boca Raton, pp 5–27.

Suttle CA (2007) Marine viruses-major players in the global ecosystem. *Nat Rev Microbiol* 5(10):801–812. https://doi.org/10.1038/nrmicro1750

Terashima H, Kawamoto A, Morimoto YV, Imada K, Minamino T (2017) Structural differences in the bacterial flagellar motor among bacterial species. *Biophys Physicobiol* 14:191–198. https://doi.org/10.2142/biophysico.14.0_191

Thomas RC (1935) A bacteriophage in relation to Stewart's disease of corn. *Phytopathology* 25:371–372.

Vasu K, Nagaraja V (2013) Diverse functions of restriction-modification systems in addition to cellular defense. *Microbiol Mol Biol Rev* 77(1):53–72. https://doi.org/10.1128/MMBR.00044-12

Voelker R (2019) FDA approves bacteriophage trial. *JAMA*. 321:638. https://doi.org/10.1001/jama.2019.0510

Walker JE (1929) The protective effect of bacteriophage against the simultaneous injection of colon bacilli. *J Infect Dis* 45(1):73–78. https://doi.org/10.1093/infdis/45.1.73

Wall SK, Zhang J, Rostagno MH, Ebner PD (2010) Phage therapy to reduce preprocessing *Salmonella* infections in market-weight swine. *Appl Environ Microbiol* 76(1):48–53. https://doi.org/10.1128/AEM.00785-09

WDCM (2019) www.wdcm.org Accessed 30 August 2019

Wei C, Liu J, Maina AN, Mwaura FB, Yu J, Yan C, Zhang R, Wei H (2017) Developing a bacteriophage cocktail for biocontrol of potato bacterial wilt. *Virol Sin* 32(6):476–484. https://doi.org/10.1007/s12250-017-3987-6

Weindling R, Katznelson H, Beale HP (1950) Antibiosis in relation to plant diseases. *Annu Rev Microbiol* 4:247–260. https://doi.org/10.1146/annurev.mi.04.100150.001335

Wen Z, Zhang JR (2015) Bacterial capsules. In: Tang YW, Sussman M, Liu D, Poxton I, Schwartzman J (eds.) *Molecular Medical Microbiology*. Academic Press, Elsevier, Cambridge, MA, pp. 33–53.

Wendlinger G, Loessner MJ, Scherer S (1996) Bacteriophage receptors on *Listeria monocytogenes* cells are the N-acetylglucosamine and rhamnose substituents of teichoic acids or the peptidoglycan itself. *Microbiology* 142(Pt 4):985–992. https://doi.org/10.1099/00221287-142-4-985

WFCC (2019) www.wfcc.info Accessed 30 August 2019

WHO (2018) *Salmonella*. www.who.int/news-room/fact-sheets/detail/salmonella-(non-typhoidal) Accessed 30 August 2019

Williamson KE, Fuhrmann JJ, Wommack KE, Radosevich M (2017) Viruses in soil ecosystems: An unknown quantity within an unexplored territory. *Annu Rev Virol* 4(1):201–219. https://doi.org/10.1146/annurev-virology-101416-041639

Xia G, Corrigan RM, Winstel V, Goerke C, Gründling A, Peschel A (2011) Wall teichoic acid-dependent adsorption of staphylococcal siphovirus and myovirus. *J Bacteriol* 193(15):4006–4009. https://doi.org/10.1128/JB.01412-10

Yost DG, Tsourkas P, Amy PS (2016) Experimental bacteriophage treatment of honeybees (*Apis mellifera*) infected with *Paenibacillus larvae*, the causative agent of American Foulbrood disease. *Bacteriophage* 6(1):e1122698. https://doi.org/10.1080/21597081.2015.1122698

Young FE (1967) Requirement of glucosylated teichoic acid for adsorption of phage in *Bacillus subtilis* 168. *Proc Natl Acad Sci U S A* 58(6):2377–2384. https://doi.org/10.1073/pnas.58.6.2377

Young R (2013) Phage lysis: Do we have the hole story yet? *Curr Opin Microbiol* 16(6):790–797. https://doi.org/10.1016/j.mib.2013.08.008

Yu F, Mizushima S (1982) Roles of lipopolysaccharide and outer membrane protein OmpC of *Escherichia coli* K-12 in the receptor function for bacteriophage T4. *J Bacteriol* 151(2):718–722.

Zhao J, Liu Y, Xiao C, He S, Yao H, Bao G (2017) Efficacy of phage therapy in controlling rabbit colibacillosis and changes in cecal microbiota. *Front Microbiol* 8:957. https://doi.org/10.3389/fmicb.2017.00957

Zhilenkov EL, Popova VM, Popov DV, Zavalsky LY, Svetoch EA, Stern NJ, Seal BS (2006) The ability of flagellum-specific *Proteus vulgaris* bacteriophage PV22 to interact with *Campylobacter jejuni* flagella in culture. *Virol J* 3:50. https://doi.org/10.1186/1743-422X-3-50

16

The Interaction between Microorganisms and Soil Processes and Their Effects on Nutrient Availability and Plant Production

Arash Hemati, Nasser Delangiz, Behnam Asgari Lajayer and Mansour Ghorbanpour

CONTENTS

16.1 Introduction

The inoculation of plants with beneficial bacteria can be traced back centuries. Farmers knew from experience that plant growth improved when a legume plant was mixed with a soil. At the end of the 19th century, using the technique of soil mixing with grain, natural inoculation was defined as a method for legume inoculation in America. Ten years later, the first inoculum for soybean inoculation with the bacterium *Rhizobium* sp. was patented under the name "Nitragin".

An important issue addressed in this chapter is the functional importance of organisms and their interactions with each other for ecosystem processes, especially the nutrient cycle. Emphasis is placed on the availability of nitrogen and phosphorus, which are considered as two limiting elements in crop production in nature and under managed ecosystems (Chapin et al., 2002). These organisms are considered biofertilizers with the ability to stabilize nitrogen and increase phosphorus uptake. This chapter examines the relationships of these organisms to plants as well as the factors that influence them.

16.2 Microbial Control of Soil Nutrient Availability

Of the total raw material that enters the soil, 80–90% is the dead root components of plants. Bacteria and fungi are the primary degraders of these substances. The initial separation and digestion of fresh organic matter by dentivores (feeders from dead plant tissues) increase the effectiveness level of these compounds for microbial colonization, but the microbes produce more chemical changes by enzyme production. Microbes have the capacity to produce a wide range of enzymes that can break down all compounds of plant origin. In many ecosystems, fungi are the major and most abundant primary degraders. They work more efficiently than bacteria to decompose insoluble plant materials and compounds, so their hyphae may readily penetrate into the new plant materials. They can develop and spread among the dead plant cells. They can also produce a wide range of enzymes capable of decomposing complex cell wall components such as lignin, ultimately producing more unstable compounds. The other advantage of fungi over bacteria is their ability to transport nutrients through their hyphae to areas where they are used. This means that the fungi have the ability to extract new substrates, even under conditions of nutrient restriction such as nitrogen and phosphorus (Boddy, 1999). Conversely, bacteria within the soil system are very slow or inactive. Therefore, they tend to evacuate and extract substrates that are more accessible in small habitats and also passively present in resources, for example, tending to use soluble carbon compounds from the newly grown roots of plants.

16.2.1 Nitrogen Mineralization

The mineralization of nutrients is a process in which soil microbes break down soluble and insoluble organic matter into forms that are critical to ecosystem functioning, so that in many ecosystems these mineral constituents can be directly accessed by plants (Nadelhoffer et al., 1985). Much of the soil nitrogen in dead organic matter is in complex insoluble polymers such as proteins, nucleic acids and chitin. These polymers are too large to pass through microbial membranes, so the microbes produce extracellular enzymes (e.g. proteinases, ribonucleases and chitinases) that convert them into smaller components and water-soluble subunits (amino acids) that can be absorbed by microbes.

These materials are called soluble organic nitrogen compounds and can form a large part of the total soluble nitrogen reserves, especially in infertile and arid ecosystems with poor soils such as coniferous forests and the polar tundra (Jones and Kielland, 2002; Northup et al., 1995). Even in agricultural soils that regularly receive nitrogen fertilizers from the top of the plants, soil soluble organic nitrogen concentrations can be equal to or even higher than inorganic nitrogen (Bardgett et al., 2003).

Most of the soluble organic nitrogen compounds in the soil solution are absorbed by the soil-releasing microbes, utilizing carbon and nitrogen in the soil for their growth. How microbes use soluble organic nitrogen compounds is largely dependent on the microbes being exposed to carbon or nitrogen conditions. Microbes use carbon molecules exist in soluble organic nitrogen compounds under conditions that in situations where easy-to-reach carbon resources are limited. under these situations microbes excrete ammonium as a byproduct. When soluble organic nitrogen compounds are insufficient to meet the microbial nitrogen requirements, microbes absorb additional inorganic nitrogen compounds from the soil solution, including nitrogen-depleted nitrogen. In addition to its uptake in plants, ammonium produced by microbial mineralization has many other uses: it can be absorbed into the negatively charged surface on clay minerals and organic matter and it can be converted to nitric acid by bacteria when oxygen levels are high. Both the immobilization and mineralization of nitrogen occur simultaneously within the soil, but pure mineralization occurs when microbes are in a carbon-deficient condition, while pure inactivation occurs in a nitrogen-deficient condition. A range of factors determine the equilibrium between the two phenomena of nitrogen immobilization, such as the availability of soluble organic nitrogen and predation by soil organisms; however, a factor of high importance is the ratio of carbon to nitrogen (C/N) of the organic substrates that are used by microbes. This ratio determines the need of microbes for carbon and nitrogen. Heterotrophic bacteria and fungi contain C/N ratios ranging from 4:1 to 12:1, but when they break down and decompose organic matter, they consume approximately 50% of the carbon contained in these compounds for respiration and 50% for biomass production (Kaye and Hart, 1997). Therefore, a wide range of critical substrates have a C/N ratio of about 30:1, which provides the microbes with nitrogen. When the C/N ratio of the substrate is higher, which usually occurs in dead plant materials, the microbes are restricted by the nitrogen element and so they use external sources of mineral nitrogen that are decomposed in the environment (Kaye and Hart, 1997). In such conditions, plants and microbes must compete theoretically for the nitrogen element in the soil solution.

16.2.2 Nitrogen Fixation

The process of nitrogen fixation is another way in which plants obtain their nitrogen requirement. This is especially the case in nature, in non-polluted arid ecosystems, where nitrogen fixation is often the primary route for nitrogen entry into the system (Cleveland et al., 1997). In undergoing the nitrogen fixation

TABLE 16.1

Rhizobia and Their Host Plants

Host Plant	Species	Genus
Alfalfa	*meliloti*	*Rhizobium*
Vetch (bean or pea)	*leguminosarum*	
Clover (*Trifolium*)		
Beans (*Faseolus*)	*loti*	
Three leaf clover	*fredii*	
Soybean (*Glycine*)	*japonicum*	*Bradyrhizobium*
Tropical legumes		
Stem nodes (*Sesbania*)	*caulinodans* spp.	*Azorhizobium*
Non-legume (*Parasponia*)		

process, microbes must be able to restore molecular nitrogen (N2) to ammonia (NH3) and combine it with amino acids for protein synthesis. Bacteria that are symbiotic with plants form nodules in their roots and use part of the carbohydrate that the plant absorbs to provide energy for themselves. In turn, these microbes provide nitrogen to plants by reducing nitrogen to amino acids. The most well-known nitrogen fixers are legume plants which are symbiotic with rhizobia bacteria (Table 16.1). Non-legume plants, such as alder, are symbiotic with actinomycetes-type Frankia (Table 16.2). These symbiotic relationships import large amounts of nitrogen into the soils of natural and agricultural ecosystems, the benefits of which are well known in soil fertility from ancient times. For example, in agricultural meadows, leguminous plants can provide up to 150 kg/ha of nitrogen per year (Newbould, 1982), and in natural situations, especially in early sequences, they are often one of the major contributors to nitrogen accumulation (Walker, 1993). The primary pathway by which nitrogen enters the soil from legume plants is through the breakdown of nitrogen-rich plant materials. Other points of entry are nitrogen-rich extractions from the roots.

The formation of a symbiotic relationship, the biochemistry of the nitrogen fixation process and the factors that influence the rate of nitrogen fixation have been studied by Paul and Clark (1996).

Another potential and important pathway is nitrogen fixation through the release of bacteria that can stabilize nitrogen by decomposing plant materials and soil organic matter (Vitousek and Hobbie, 2000).

These non-symbiotic nitrogen stabilizers are ubiquitous, dispersed in arid ecosystems and stabilize relatively small but significant amounts of nitrogen (3 kg/ha per year) (Cleveland et al., 1999). In harsh environments with scattered vascular plants, cyanobacteria are often the main sources of nitrogen. These nitrogen-fixing organisms are particularly abundant in arid or cold regions, where the growth of vascular plants is limited. For example, in arid ecosystems, they are found in soil shells composed of cyanobacteria, lichens and mosses (Belnap, 2003).

Similarly, cyanobacteria are abundant in the layers of surface soil in Antarctica (Wynn-Williams, 1993) as well as in lands where their freezing surfaces are distributed, as are those found in glaciers (Chapin et al., 1994). In all these situations, cyanobacteria are vital for the maintenance of fertility and

TABLE 16.2

Distribution of Actinorizal Plants

Continent	Native Genus
North America	*Alnus, Ceanothus, Cerocarpus, Chamaebatia, Comptonia, Coraria, Cowania, Datisca, Dryas,Elaeagnus, Myrica, Purshia, Shepherdia*
South America	*Alnus, Colletia, Coriaria, Discaria, Kentrothamnus, Myrica, Retanilla, Talguenea, Trevoa*
Africa	*Myrica*
Asia	*Alnus, Coriaria, Datisca, Dryas, Eleagnus,Hippophae, Myrica*
Oceania, Australia	*Allocasuarina, Casuarina, Ceuthostoma,Coriaria, Discaria, Gymnostoma, Myrica*

often stabilize high amounts of nitrogen (1–10 kg/ha per year) (Vitousek et al., 2002); cyanobacteria also play an important role in fixation and stability. Recent findings also point to cyanobacteria as the primary source of nitrogen in coniferous forests of the northern leaf, where nitrogen fixation generally places major constraints on growth in the biosphere.

16.2.3 Microbial Mineralization Phosphorus

Despite the nitrogen element, most of the phosphorus in soil is in insoluble inorganic form, such as blocked phosphorus, which is inaccessible to plants. As a result, many of the reactions that control the availability of phosphorus to plants are of a more geochemical nature than a biological one. The availability of phosphorus in the soil is also strongly influenced by previous land management practices. For example, phosphorus fertilizer has major long-term effects on the availability of soil phosphorus, as it greatly increases the phosphate concentrations available to plants, especially in surface horizons. However, soil microbes also actively participate in the phosphorus cycle. They take part in the dissolution of inorganic phosphorus and in the mineralization of organic phosphorus, and due to low phosphorus availability, in the production of plant and microbial phosphates. These enzymes release phosphate by breaking ester bonds in organic matter, which can then be adsorbed into the phosphate by plant or microbial biomass.

Similar to the nitrogen element, the mineralization of organic phosphorus is partly regulated by the ratio of carbon to phosphorus (C/P) substrates. Generally, when the C/P ratio is above 100, phosphorus is produced by microbes that have relatively high phosphorus requirements (1.5–2.5% dry weight compared to 0.05–0.5% for plants). The phosphorus becomes immobilized (absorbed), resulting in microbes competing with plants for the available phosphorus in the soil. The importance of microbial phosphorus immobilization (microbial uptake of phosphorus) is illustrated by the fact that microbes account for more than 20–30% of the total organic phosphorus reserves in the soil (Jonasson et al., 1999). It is much higher than the carbon (1–2%) or nitrogen (2–10%) held by microbes. Also, the capacity of the soil microbial biomass as a phosphorus-rich reservoir is severely affected by environmental stresses such as drought and moisture cycles, and this has been shown to result in the death of part of the soil microbial biomass and thus the potential for the formation of accessible phosphorus and probably other elements in the soil (Turner and Haygarth, 2001). Therefore, microbial phosphorus is an important and potential source of phosphorus in soil. Mycorrhizal fungi also play an important role in the transfer of phosphorus to soil systems, which will be described in the next section.

16.2.4 The Role of Mycorrhizal Fungi in Plant Nutrition

Mycorrhizal fungi are scattered over a wide range of ecosystems, but there is a clear pattern to the distribution of different mycorrhizal species with respect to biomass, soil type and limiting resources (Read, 1996). Ericoid mycorrhizal fungi are generally dominant in ecosystems at high latitudes and elevations that are spanned by shrubby and short-shrub vegetation on nitrogen-restricted peat soils. Ectomycorrhizal fungi are also distributed in northern forest ecosystems and in broadleaf forests that are commonly located in brown soils and podzol soils, while the forest vegetation in tropical regions, grasslands and deserts often coexist with arbuscular mycorrhizal fungi.

The primary advantage of a plant's symbiotic relationship with mycorrhizal fungi is the increased availability of nutrients, especially nitrogen and phosphorus. In fact, mycorrhizal fungi are one of the largest carbon sources of fertility and primary production (Harley, 1971), so the contribution of these fungi increases, especially when nutrients are in limited supply (Johnson et al., 2003).

It has long been known that mycorrhizal fungi increase ammonium uptake by soil, and it has recently been shown that both erythroid and ectomycorrhizal fungi can directly adsorb soluble organic nitrogen from the soil in the form of amino acids, pure proteins and even hard-to-decompose proteins, which are fused with tannins. Mycorrhizal fungi can utilize these highly complex proteins because they produce carboxyl proteinase enzymes capable of separating proteins into their structural components, namely amino acids. Producing and activation of these enzymes at pH 2–4, are at maximum level, where mycorrhizal fungi are commonly present. Another way in which mycorrhizal fungi can affect soil nitrogen availability is to increase nitrogen fixation by assisting nitrogen-fixing microbes with plants (Egerton-Warburton and Allen, 2000).

Arbuscular mycorrhizal fungi can also transmit stabilized nitrogen from legume plants through the soil and transport it to adjacent plants using their hyphae networks (Gardner et al., 1984).

Mycorrhizal fungi play an important role in soil phosphorus uptake by host plants. These fungi grow deep within the soil matrix, accessing the soil phosphorus, which is farther from the roots of the plant. They do this by using three important mechanisms. First, they increase the contact area and, more specifically, they increase the level of phosphorus uptake from the soil. Second, they produce phosphatase enzymes that break down the ester bonds that bind phosphorus to carbon in organic matter. These enzymes release phosphate, which can also be absorbed by the fungi and transferred inside the plant. Third, mycorrhizal fungi produce low-molecular-weight organic acids such as oxalates, which increase the availability of the soil phosphorus by increasing the rate of aeration of phosphorus-containing clay minerals.

16.3 Effect of Mycorrhizal Symbiosis on Plant Growth and Yield

Researchers have stated that one of the most important reasons for using mycorrhizal fungi is to increase crop yields, especially in low fertility soil. This increase in yield is due to increased root uptake through the infiltration of mycelium into the soil followed by crop access to greater soil volume. Research has shown that the major role of mycorrhizal fungi is to provide phosphorus for plant growth, as phosphorus is an extremely sedimentary element in soil. Even if phosphorus is added to the soil in soluble form, it rapidly stabilizes in the form of calcium phosphate or other forms and is immobilized (Raiesi and Ghollarata, 2006). The researchers reported that mycorrhizal symbiosis physically increased the uptake of phosphorus in the vegetative body of fennel by improving the growth of fungal hyphae in the soil pores, which subsequently improved the phosphorus content of fennel seed by increasing the plant dry weight (Smith and Read, 1997). Research has also shown that the use of mycorrhizal fungi increases plant growth and biomass. These fungi increase the biomass by increasing the uptake of water and minerals and producing growth hormones. It has been proven that these fungi can produce growth hormones such as auxins, cytokinins and gibberellins (Swift, 2004). Swift (2004) reported that increasing the percentage of mycorrhizal root colonization increases plant growth, root growth and shoot dry weight through improving the absorption of water and minerals. Azcon and Atrach (1996) stated that the inoculation of onion seedlings with mycorrhizal fungi not only increased phosphorus concentration, but also increased the concentration of sodium, chlorine, zinc, potassium, copper and total nitrogen concentrations. Mycorrhiza symbiosis affects several physiological aspects of the plant such as root, nutrient acquisition, plant protection and food cycle. It is also important for key ecological processes to improve plant organs and soil quality (Smith, 1997).

16.4 Conclusion

Microbes play important roles in the ecosystem processes of nutrient degradation, which can determine the availability of nutrients to plants. However, we need to consider the interactions between microbes and other organisms in the soil food web to fully understand the role of soil organisms in controlling and directing these processes. The amount of bioavailable trace elements in the soil is determined by the relationships between the plant and the microbes. Rhizobacterial and mycorrhizal relationships are the most effective and can provide most of a plant's needs. Strengthening a plant's defence system increases both the growth factors and the applicability of these microorganisms. Thus, this group of microorganisms is highly efficient in biofertilizer production and this ability increases their use. However, to increase the consumption rate of biofertilzers, producing whit this kind of microorganisms it is necessary to identity more efficient microbes, which are better and more adaptable than native ones.

REFERENCES

Azcon, R and Atrach, EF (1996). Influence of arbuscular mycorrhizae and phosphorus fertilization on growth, nodulation and N2 fixation in *Medicago sativa* at four salinity levels. *Biology and Fertility of Soils* 24, 81–86.

Bardgett, RD, Streeter, T and Bol, R (2003). Soil microbes compete effectively with plants for organic nitrogen inputs to temperate grasslands. *Ecology* 84, 1277–1287.

Belnap, J (2003). The world at your feet: Desert biological soil crusts. *Frontiers in Ecology and the Environment* 1, 181–189.

Boddy, L (1999). Saprotrophic cord-forming fungi: Meeting the challenge of heterogeneous environments. *Mycologia* 91, 13–32.

Chapin, FS, Matson, PA and Mooney, HA (2002). *Principles of Terrestrial Ecosystem Ecology*. Springer, New York.

Chapin, FS, Walker, LR, Fastie, CL and Sharman, LC (1994). Mechanisms of primary succession following deglaciation at Glacier Bay, Alaska. *Ecological Monographs* 64, 149–175.

Cleveland, CC, Townsend, AR, Schimel, DS, Fisher, H, Howarth, RW, Hedin, LO, Perakis, SS, Latty, EF, von Fischer JC, Eleseroad A & Wasson, MF (1999). Global patterns of terrestrial nitrogen (N2) fixation in natural ecosystems. *Global Biogeochemical Cycles* 13, 623–645.

Egerton-Warburton, LM and Allen, EB (2000). Shifts in arbuscular mycorrhizal communities along an anthropogenic nitrogen deposition gradient. *Ecological Applications* 10, 484–496.

Gardner, IC, Clelland, DM and Scott, A (1984). Mycorrhizal improvement in non-leguminous nitrogen fixing associations with particular reference to *Hippophae rhannoides* L. *Plant and Soil* 78, 189–199.

Harley, JL (1971). Fungi in ecosystems. *Journal of Applied Ecology* 8, 627–642.

Johnson, NC, Rowland, DL, Corkidi, L, Egerton-Warburton, LM and Allen, EB (2003). Nitrogen enrichment alters mycorrhizal allocation at five mesic to semi-natural grasslands. *Ecology* 84, 1895–1908.

Jonasson, S, Michelsen, A and Schmidt, IK (1999). Coupling of nutrient cycling and carbon dynamics in the arctic, integration of soil microbial and plant processes. *Applied Soil Ecology* 11, 135–146.

Jones, DL and Kielland, K (2002). Soil amino acid turnover dominates the nitrogen flux in permafrost-dominated taiga forest soils. *Soil Biology and Biochemistry* 34, 209–219.

Kaye, JP and Hart, SC (1997). Competition for nitrogen between plants and soil microorganisms. *Trends in Ecology and Evolution* 12, 139–143.

Nadelhoffer, KL, Aber, JD and Melillo, JM (1985). Fine roots, net primary production and soil nitrogen availability. *Ecology* 66, 1377–1390.

Newbould, P (1982). Biological nitrogen fixation in upland and marginal areas. *Philosophical Transactions of the Royal Society of London Series B* 296, 405–417.

Northup, RR, Yu, Z, Dahlgren, RA and Vogt, KA (1995). Polyphenol control on nitrogen release from pine litter. *Nature* 377, 227–229.

Paul, EA and Clark, FE (1996). Soil Microbiology and Biochemistry, 2nd edn. Academic Press, Inc., New York.

Raiesi, F and Ghollarata, M (2006). Interactions between phosphorous availability and an AM fungus (*Glomus intraradices*) and their effects on soil microbial respiration, biomass and enzyme activities in a calcareous soil. *Pedobiologia* 50, 413–425.

Read, DJ (1996). Plant-microbe mutualisms and community structure. In: *Biodiversity and Ecosystem Function* (eds. ED Shulze and HA Mooney), Springer-Verlag, Berlin, pp. 181–209.

Smith, SE and Read, DJ (1997). *Mycorrhizal Symbiosis*, 2nd Edition. Academic Press, London.

Swift, CE (2004). *Mycorrhiza and Soil Phosphorus Levels*. Area Extension Agent. www.colostate. edu/ Depts/Coop Ext/TRA/Plants/mycorrhiza.

Turner, BL and Haygarth, PM (2001). Phosphorus solubilisation in rewetted soils. *Nature* 411, 258.

Vitousek, PM and Hobbie, A (2000). Heterotrophic nitrogen-fixation in decomposing litter: Patterns and regulation. *Ecology* 81, 2366–2376.

Vitousek, PM, Cassman, K, Cleveland, C, Crews, T, Field, CB, Grimm, NB, et al. (2002). Towards an ecological understanding of biological nitrogen fixation. *Biogeochemistry* 57, 1–45.

Walker, LR (1993). Nitrogen fixers and species replacements in primary succession. In: *Primary Succession on Land* (eds. J Miles and DWH Walton), Blackwell Scientific, Oxford, pp. 249–272.

Wynn-Williams, DD (1993). Microbial processes and initial stabilization of fell field soil. In: *Primary Succession on Land* (eds. J Miles and DWH Walton), Blackwell Scientific, Oxford, pp. 17–32.

17

Agricultural Bioeconomy and Sustainable Proposals for Turkey

İrfan Kalaycı

CONTENTS

17.1 Introduction

Agriculture is a series of processes carried out by blending four elements, namely soil, water, air and fire, or by harmonizing the four factors of production, namely nature, labour, capital and entrepreneurship. Agriculture comes first in the hierarchy of sectors because it is identical with life itself. Moreover, aside from its contribution to the economy and employment, agriculture has many dimensions ranging from food security to rural development and environmental policies.

With its territory spanning three continents, Turkey is the world's only Afro-Eurasian country. Turkey thanks to its large agricultural land and production is one of the few countries worldwide. If Turkey had not have a high share of agricultural gross domestic product (GDP), she might not be among the world's 20 largest economies (G-20).

As Sindir (2008) emphasizes, "agriculture is the food sovereignty of a country". However, despite this, the fact that a country (society) is an agricultural country (society) could be considered a criticism and even an adjective of humiliation in the modern world. In reality, old agricultural countries (society) such as China, Korea, Brazil and Turkey have become today's star industrialized countries (communities) and they owe this development to the accumulation of income from agriculture. It has come to a point that agriculture, which transferred its resources to industry and thus served it, has now turned into a sector from which industry has been transferred and now serves. The modern world has been faced with the phenomenon of industrializing agricultural country (society) for some time. Being an industrialized agricultural country (society) means creating visibility and/or awareness in the competitive world. Proof of this is the field of biotechnology, which is developing and expanding as a sub-branch or a variation of bioeconomy.

Turkey has a Mediterranean climate, a wide geography, various products, rich ecological and strong geo-economic structures and large agricultural basins. Also Turkey, carries some features of the agricultural society, can be considered comparatively advantageous in the agricultural sector, because of using low levels of chemical fertilizers and pesticides. To maintain this advantage, the governments of the period subjected the agricultural sector to a number of transformation policies. These policies, the foundations of which were laid at the 1923 Izmir Economic Congress, are generally implemented through interventionist and protective paradigms. The variability of nature and market fluctuations have made agriculture an open door to interventions by the state.

17.2 Agriculture in Turkey: Strengths, Weaknesses, Opportunities and Threats (SWOT) Analysis

The various dynamics that have led to the structural transformation of agriculture in Turkey are

a. *Internal dynamics* such as the transition from empire to nation state, the urbanization of the peasant society, the 1980 structural stabilization measures and movements of localization and democratization, privatization and deregulation.
b. *External dynamics* such as the Great Collapse of 1929, Keynesian policies, World War II, Marshall Aid, the liberalization of world agricultural trade, globalization, Keynesianism in agriculture, protectionism, foreign capital inflows through MAI,* structural policies of the International Monetary Fund (IMF) and the World Bank, the General Agreement on Tariffs and Trade (GATT) laws, the European Customs Union and the EU negotiation process.

A paradigm shift in agricultural practices is essential for Turkey's future. The name of the new paradigm is the "sustainable agricultural bioeconomy" and intensive and determined efforts are being made to foster this paradigm. Thus, Turkey, thanks to this new paradigm, will be able to produce and market globally original agricultural products which have highly the potential revenue and brand equity. In this sense, an agricultural sector that can hold onto the world market can only become a locomotive sector.

Although the sociological and political base of agriculture in Turkey is of great importance, its neglect and sometimes marginalization in terms of economic development has become a serious topic of discussion. In this regard, agriculture is called "Turkey's forgotten titan". When an unbiased and versatile evaluation of Turkey's agricultural sector is made, not only are there strong (G) and weak (Z) aspects, but also a number of opportunities (F) and threats (T) exist (for SWOT, see Table 17.1).

17.3 Key to Agricultural Bioeconomy: Sustainability

The 21st century is the age of the bioeconomy as well as the age of finance and integration. The bioeconomy is associated with sustainability (see Figure 17.1) and the agriculture sector. It seems that the economic destiny of humanity will be determined by sustainable agricultural policies. These policies of the contemporary world have been influenced by the bioeconomy and/or biotechnology, emerging as a synthesis of "biology + agroculture + technology", with sustainability indexed in almost every economic activity, including agriculture, since 1990.

The bioeconomy promotes renewable resources from land and sea and the sustainable production of food, bio-based products, biofuel and energy. Bioeconomics encompasses a range of sciences (e.g. life sciences, agriculture, ecology, forestry, fisheries and social sciences) and uses industrial technologies (e.g. biotechnology, nanotechnology and information and communication technologies) (Pilzecker, 2011). Therefore, if the bioeconomy/bioeconomics is compared to a clock, the hour hand is the axis of food, non-food and biomass and the minute hand points to climate change, food security, sustainable production, energy security and public health, especially natural resources.

According to Stoor et al. (2011), where the bioeconomy is concerned, sustainability is a key concept. If the bioeconomy ensures that resources are consumed rationally and strategically and that these resources turn into high value-added products, then it can be considered as sustainable. The bioeconomy creates an opportunity for new knowledge-intensive industry, as well as improving a country's security of its key sources such as food and fuel. Any bioeconomic solution for food and fuel safety is actually sustainable. At this point, the definition in the Brundtland Commission under the Organization for Economic Co-operation and Development (OECD) in 1987 stands out: "Sustainability is to meet the needs of the present generation without compromising their power to meet the needs of future generations".

* Multilateral Agreement on Investment (in OECD).

TABLE 17.1
Short SWOT Analysis of Turkey's Agriculture

S-Strong Sides (++)	W-Weak Sides (-)
++ Ecology (climate and soil conditions) suitable for agricultural production. ++ There are high numbers of producer organizations (agricultural cooperative, chamber of agriculture). ++ There are trade exchanges. ++ High irrigable land potential. ++ Ready supply of workers. ++ Agriculture-based industry exists. ++ The importance of agricultural input use and technological innovations is known. ++ There are agricultural markets. ++ If the agricultural cooperatives are supported by partners and the state, it will be easier to organize the markets and to plan production. ++ There are many agricultural faculties and agricultural engineers.	- Our country is not self-sufficient in many products such as rice and sunflower. - Producer organizations are not active, there are financial problems and no state support. - The number of exchanges is very high and ineffective. - Irrigation networks and irrigated agriculture are insufficient, irrigation efficiency is low. - The labour force is unqualified, unemployment is excessive. - Input use is not optimal, technological developments are not utilized sufficiently. - Since agricultural public enterprises have been privatized and agricultural cooperatives are not supported by the state, agricultural markets are independent and cannot plan for production. - The employment problem of agricultural engineers is serious.
O-Opportunity Sides (+)	**T-Threat Sides (--)**
+ There may be accession to the EU. The policies and political development implemented in this direction. + There are producers, cooperatives and business people with an entrepreneurial spirit who can take risks. + Lands and ecology are very suitable for organic agriculture, foreign markets are ready. + Existing exchanges can work more effectively if appropriate legal conditions and producer participation are provided. + Product marketing, price stability, input and credit problems can be solved if producer organizations and their characteristics are provided with a suitable environment and support for agricultural cooperatives. + Manufacturers' cooperatives, unions, etc., willing to establish autonomous producer organizations.	-- EU legislation and common agricultural policy (CAP) cannot be applied exactly as in the EU, compliance with EU norms is not easy. -- Surplus rural population to benefit from FEOGA[a] is required by CAP; unemployment in the country is a serious obstacle to entering the EU. -- Government's intervention in productive cooperative management, insufficient financial support. -- The existing laws do not help manufacturers establish autonomous and democratic cooperatives and producer organizations. -- The state's current approach to agricultural cooperatives is essential with changes that will help producers solve their product marketing, input and credit problems in the near future. -- There is no specific standardization or understanding of quality in agricultural and food products. -- Producers uninformed about organic farming; there are no marketing organizations.

Source: Gaytancıoglu (2009: 184–185).
[a] FEOGA: Fonds europèen d'orientation et de garantie agricole (The European Agricultural Guidance and Guarantee Fund—EAGGF).

According to McDonald (2010), the agricultural economy of the future focuses on bioproducts that continue food and feed production in the axis of the bioeconomy while the traditional agricultural economy addresses food and feed producers and the middle class. The bioeconomy encompasses bioproducts made from renewable sources through existing and new technologies and processes. These products are biofuels (ethanol, butanol, biodiesel), bioenergy (biogas, heat, power), biomaterials (bioplastics, biochemicals, biofibres) and biodrugs and biocosmetics (functional foods, nutraceuticals, bioactive compounds).

According to Omafra (2012) and Sundar (2012), who discussed ecologically sustainable economic development, the concept of sustainability can be multifunctional. For example, its first function is biological, mainly in terms of supporting life and human activities. Its second function is techno-economics as a tool for productivity, and its third function is socio-cultural. All are interdependent and interact. Bioproducts have significant potential (environmental) sustainability benefits.

On the other hand, as Simon (2003) observes, sustainability indicators are now being designed; an example in the context of sustainable economic prosperity is the "Green gross national product" (gGNP)

FIGURE 17.1 The main approaches to sustainability. (Samsung, 2019.)

which includes natural and manufactured capital (stock of capital). These concepts and other expansions show us that the paradigm called sustainability is an inspiration for all agricultural policies and practices. In its report, Global GAP (2011), a private sector organization that certifies agricultural products, states that "Sustainability is a key; good agricultural practices are a way to achieve it".

Herndl et al. (2012), in linking the relationship between sustainability and the bioeconomy, state that the sustainability of human, natural communities and raw material production directs research and policy that design and regulate the new bioeconomy. According to them, a sustainable bioeconomy asks us to move from output production programmes to a complex system of scientific, social and economic components.

According to the IFPRI (2007), bioeconomics is a developing platform for a sustainable future. It can be said that even if bioeconomic agricultural systems do not completely "substitute" for conventional agricultural systems, they will be "assertively" reinforced. The world population will reach 9.5 billion people by 2050. In accordance with this demographic projection, it is expected that the 21st century will have all the features of the bioeconomic age, which has started to be observed in the provision of natural resource efficiency and security (BMBF, 2011).

17.4 Sustainable Action Proposals for Turkey

Various proposals on the potential to move Turkey's agriculture sector forward can be improved. Here, several sustainable proposals requiring simple but feasible action are presented. It should be noted that some of these proposals are new and some are based on tried and/or similar examples in the literature.

Action 1: *Turkey must see the bioeconomy strategy as a geostrategy. One cannot replace the other.* Thus, to be an advanced economy, Turkey can increase the agricultural potential and superiority of its biological diversity through the bioeconomy and/or biotechnology.

Action 2: *Turkey is rapidly industrializing, but it should not sacrifice the agricultural sector to industry, otherwise it may catch the "Japanese syndrome" (JS)* as described by Brown (2004). If Turkey does not want to enter the JS swirl, she must plan her grain production and stocks economically and she should protect the farmer with a base price policy.

Action 3: *New, specialized and efficient "gene banks" should be established in the biotechnology field to make seed diversity and richness sustainable.* GenBank (in Ankara), which contains hundreds of thousands of seeds, was established by the Ministry of Agriculture. Previously, Turkey imported almost all seeds, but today Turkey has reached the point of exporting dozens of seeds to the world.

Action 4: *The financial sector should extend credit to the agriculture sector and it should be made stable.* Banks should activate agricultural credit schemes that would offer more credit to farmers, and should give priority to agricultural demands, especially micro-credit applications. However, banks should be wary of farmers producing fruits and vegetables within the scope of cancer-causing "genetically modified organisms" (GMOs) and act according to ethics committee reports.

Action 5: *The agricultural sector is very suitable for informal activities which should be limited.* There is much informality in the agricultural sector, from hiring uninsured workers to illegal slaughtering. The state should play the role of a "hard investigator" when it comes to public health and tax, and a "welfare state" when it comes to broadening its social security base.

Action 6: *Agriculture in Turkey is perceived as a female sector. While developing agricultural policies, the phenomenon of the feminization of agriculture should also be evaluated in terms of "gender equality".* Although they are the highest contributors of labour, women in the agricultural sector in Turkey are one of the most disadvantaged groups in terms of welfare. Therefore, it is imperative to urgently solve labour-related issues such as working hours, wages and insurance, with a positive discrimination approach towards rural women. Socio-economic policies focused on the goal of women's development against the feminization of agriculture should be implemented.

Action 7: *An economic development strategy based on the bioeconomy and/or biotechnology should be designed and implemented.* The concept of a sustainable agricultural bioeconomy requires the realization of new ideas on many subjects and the innovation of these ideas, from agricultural e-commerce to agricultural museums; from Turkey farmers with exchange programmes between the EU to agri-food technopolis; from the mechanisms preventing agricultural work accidents to biotechnology universities; from humanitarian agriculture consultancy to the balance of agricultural payments.

Action 8: *Ecological literacy and/or its complementary, bioeconomic literacy should be popularized in Turkey.* Both for sustainable living can significantly help us to become conscious of natural resource consumption, the risk of hunger and thirst, pollution of the land and even global warming.

17.5 Conclusion

If people were as humble as the land, then the agricultural world would not be facing so many problems such as a food crisis. We know that agriculture is a chain of cross-territorial activities. It consists of unique value rings that link to chains of socio-economic value. Here is a new ring, bioeconomy/biotechnology; it is the new necklace that hangs on the neck of humanity. It is priceless. The accumulation, experience and transformation of agricultural production factors play a great role in making this necklace.

While putting the bioeconomy under the command of agriculture, let us not harm the emotional structure of nature; because nature cries rain and laughs when in bloom. If we do not want nature to leave us hungry and thirsty, we have to make agriculture sustainable with bioeconomic tools. In order to make it sustainable, we should transform our mindset into goodness, beauty and justice as much as agriculture. And this should be done not only in Turkey, but all over the world.

REFERENCES

BMBF—Federal Ministry of Education and Research (2011), *Public Relations Division, National Research Strategy BioEconomy 2030 Our Route towards a Biobased Economy*, Berlin, www.bmbf.de/pub/bio-economy_2030.pdf

Brown, Lester R (2004), "Pushing Beyond the Earth's Limits: Growth: The Japan Syndrome", in *Outgrowing the Earth: The Food Security Challenge in an Age of Falling Water Tables and Rising Temperatures*, Earth Policy Institute, www.earth-policy.org/books/out/ote1_5

Gaytancıoglu, Okan (2009), *Türkiye'de ve Dünyada Tarımsal Destekleme Politikası*, İTO Yayın No: 2009–14, İstanbul.

Global GAP (2011), *Annual Report 2011: A Year of Initiatives and Advences*, www.globalgap.org/cms/upload /Resources/Publications/Newsletter/120321_AR11_web-FINAL. pdf

Herndl, Carl, Rob Anex, Robert Brown, Rick Cruse, Steve Fales, and Gene Takle (2012), *Defining a Future Bioeconomy for the Nation: Bio-Renewable Resources and Complex Systems*, Iowa State University, Dep.of Agronomy, www.agron.iastate.edu/.../White%20pap%2

IFPRI—The International Food Policy Research Institute (2007), "Guiding the Economic Development and Poverty Reduction Strategy and Supporting the Strategic Plan for Agricultural Transformation", in *Long-Term Framework for the Implementation of the Comprehensive Africa Agriculture Development Programme (CAADP) in Rwanda*, Kigali, March 29–31, www.ifpri.org/ publication/comprehensive -africa-agriculture-development-programme-caadp

McDonald, Ian (2010), *Agricultural/Forestry Opportunities in the Bioeconomy with Biomass Applied Research Coordinator*, Ontario Ministry of Agriculture, Food and Rural Affairs, www.indianag.on.ca /5.Ian%20McDonald,%20OMAFRA%20AgriculturalForestry%20 Opportunities%20.pdf

OMAFRA—Ontario Ministry of Agriculture, Food and Rural Affairs (2012), "Bioeconomy – Industrial Uses", in *Strategic Research Themes—Priorities for 2008–2012*, www.omafra. http://gov.on.ca/english/ research/priorities/bioeconomy.htm

Pilzecker, Andreas (2011), "Biomass for the Bioeconomy, High-Level Policy-Maker", in *Conference on Biorefineries, European Commission (Directorate-General for Agriculture)*, October 5, Brussels.

Samsung (2019), *Sustainability Report*, www.samsung.com/us/about samsung/sustainability / report-and-policy/

Simon, Sandrine (2003), "Sustainability Indicators", in E Neumayer (ed.) *Online Encyclopaedia of Ecological Economics (OEEE)*, International Society for Ecological Economics (ISEE).

Sındır, K Okyay (2008), *Tarım solma, üretici yok olma noktasında*, Tarım Merkezi, www.tarimmerkezi.com /yazar_kose.php?hid=13814

Stoor, Robert, Anastasia Tsvetkova, and Magnus Gustafsson (2011), *Sustainable Bio-economy: Potential, Challenges and Opportunities in Finland*, PBI Research Institute, March, www.sitra.fi/maamerkit

Sundar, I. (2012), "Bioeconomics and Sustainable Development", in *International Conference on Management, Applied and Social Sciences (ICMASS'2012)*, March 24–25, Dubai.

18

Utilization of Agricultural Waste in Agri-Based Biogas Production and Sustainable Development

Manoj Kumar Chitara, Sadhna Chauhan and Rajesh Pratap Singh

CONTENTS

18.1 Introduction

Many countries around the world are paying great attention to the production of bioenergy and other forms of renewable energy due to the increase in the cost of producing fossil fuels as well as climate protection. Today, sustainable bioenergy production is considered an important alternative to achieve low emissions of greenhouse gases, a global issue of concern (Diesendorf and Diesendorf, 2007). The agriculture sector is considered an artificial, vast and diversified ecosystem of nature, in which various processes are continuously transferring energy from one state to another, for example crops are converting photosynthetic accumulates into grain formation (Black et al., 1995). The global community follows the principles of energy conservation which is based on energy that is neither created nor destroyed but can be converted from one form to another.

Globally, the primary energy share of bioenergy is 35% from oil, 25% from coal, 21% from gases, 13% from renewable energy and 6% from nuclear energy production (Bioenergy, IEA Report, 2010); however,

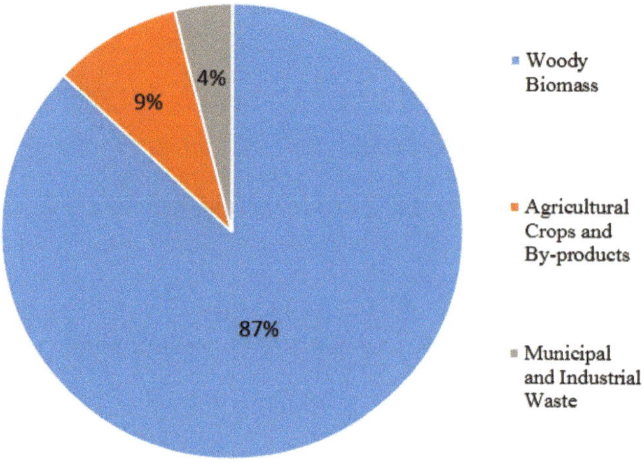

FIGURE 18.1 Share of bioenergy in the global primary energy supply.

only renewable energy is sustainably produced from agricultural waste products, while oil, coal, gases and nuclear bioenergy are not produced sustainably. All primary bioenergy supplies except renewable energy are considered harmful to nature because they emit greenhouse gases after burning (Van Ruijven and Van Vuuren, 2009; Burnham et al., 2012), so their current utilization is not considered beneficial for living entities or the environment. The percentage share of renewable energy sources (13%) includes 77% from bioenergy, 15% from hydro energy and 8% from other sources. The major type of renewable energy source is bioenergy, which has 87% woody biomass, 9% crops and byproducts and 4% municipal and industrial waste (Figure 18.1) (Bioenergy, IEA Report, 2010).

Recently, microbial-based sustainable bioenergy production has been popularized and has become an important tool in sustainable bioenergy production such as biogas production, in which the microbial conversion of energy crops and organic wastes to biogas has become one of the most attractive technologies for energy production, resource recovery and waste treatment (Rikkonen and Tapio, 2009). This technique's sustainably produced product is environmentally friendly and does not harm the surrounding environment, unlike emissions of greenhouse gases from burning petroleum fossil fuels, etc. The products and byproducts from microbial-based bioenergy production are safe to use and eco-friendly, and are utilized in many ways such as for electricity generation (Wang et al., 2014; Pant et al., 2010), heat generation (Akbulut, 2012) and fuel for vehicles (Subramanian et al., 2013). Agricultural waste is utilized as an input for biogas production, so it plays a very crucial role in bioenergy production. Generally, agricultural waste, for example crops, stubble, grasses, raw vegetable and fruits parts and livestock product, are readily available in huge amounts (Cuéllar and Webber, 2008;Deublein and Steinhauser, 2011; Parris, 2004).

18.2 Role of Agriculture in Bioenergy Production

Agriculture is the backbone of many countries because of its major contribution to economic growth and global exchange (Addison, 2005). It is a multicomponent system in which each system interacts with each other in a complex way to influence sustainable production (Walters et al., 2016). In potential bioenergy production, the choice of agriculture crop species must also be taken into account according to the local climatic conditions (Heinsoo et al., 2015). Agricultural farms are considered the basic producing unit of agriculture, delivering local energy in terms of producing raw materials such as crop straw, stubble, raw vegetables and fruits and wood biomass for bioenergy production (Rikkonen and Tapio, 2009).

The agricultural bioenergy production system is a self-sufficient system that includes woody/agro biomass which is utilized in the production of heat in the form of electricity, fuel, biodiesel and biogas at the farm level. It provides biomass/energy to the refinery industry and other customers, from cultivated varieties of turnip, rape, sugar beet, maize, barley, etc., and for heating, the cultivated agro-biomass includes reed canary grass, crop residues (straw) and willow (Rikkonen and Tapio, 2009; Aziz et al., 2019). Both agricultural and animal waste is utilized for biogas production on a small and medium scale, and some of this biogas also goes to industrial uses (Hernandez et al., 2018).

18.3 Biogas Production

Biogas is a renewable energy source. It is a popular substitute for fossil fuels in the generation of power and heat, and it can be used as fuel for vehicles. It is a methane-rich gas, also known as biomethane, and can replace natural gas as a basic raw material in the production of chemicals and materials (Weiland, 2010). Biogas production is an anaerobic process that is based on the microbial degradation of a complex molecule into a simple molecule. The biogas production process consists of the breakdown of organic matter in the absence of oxygen into methane and carbon dioxide (Karellas et al., 2010). Microbes such as methanogens (*Methanosarcina*, *Thermogymnomonas*, *Methanoculleus* genera and *Archaeon*) and sulfate-reducing bacteria (*Desulfomicrobium*, *Desulfobulbus* and *Desulfovibrio*) are involved in this process (Kushkevych et al., 2017). The criteria for the assessment of biogas are given in Table 18.1.

18.4 Process of Biogas Production

The process of biogas production includes the anaerobic transformation of agricultural waste into biogas. It is totally microbial dependent and is divided into four stages: hydrolysis, acidogenesis, acetogenesis and methanogenesis (Bochmann and Montgomery, 2013; Kim et al., 2013) (Figure 18.2). Each stage plays a significant role in biogas production.

18.4.1 Hydrolysis

Hydrolysis is the first step in biogas production, in which the complex long chain of the carbohydrate, protein and fat molecule is broken down into a simple molecule in the presence of an enzyme (cellulase, amylase, protease and lipase). In this process, the complex carbohydrate, polysaccharide, converts to a simple monosaccharide, the long-chain fatty acids covert to simple fatty acids and protein converts to peptide and amino acids (Mulat et al., 2018).

18.4.2 Acidogenesis

The next step in biogas production is acidogenesis in which the simple monomer molecule is further broken down following hydrolysis. Acidogenic bacteria produce an acidic condition in the fermenter

TABLE 18.1

Significant Criteria for the Assessment of Biogas

–	Energy density (per mass or volume): 1.15–1.25 kg/m^3; heat value: 30–60 MJ/m^3; calorific value: 20–26 MJ/m^3
–	Major constituents: Methane (CH_4): 50–65%, CO_2 (30–50%); minor constituents: H_2S, H_2 and NH_3: 1–2% O_2: in trace amount
–	Application potential: Direct combustion system (broilers, turbine or fuels cells). Flexibility: Highly flexible and long-term process stability at high OLRs of up to 6 kg VS m^3 per day. Handling: Fixed and economic. Combustion characteristics: Lower heat release rate than conventional fuel
–	Transportation characteristics: Concentrated and compressed biogas can be used in vehicle transportation. Storage: 2–200 psi
–	Efficiency in production/upgrading process: Methane (60%) and CO_2 (29%)
–	Availability of raw material/resources: Agricultural waste, manure, municipal waste, plant material, sewage, green waste and food waste
–	Production cost: USD 50–75 per m^3. Price: USD 250–275. Market potential: USD 70 billion (2018 base year), USD 110 billion (2019–2025)

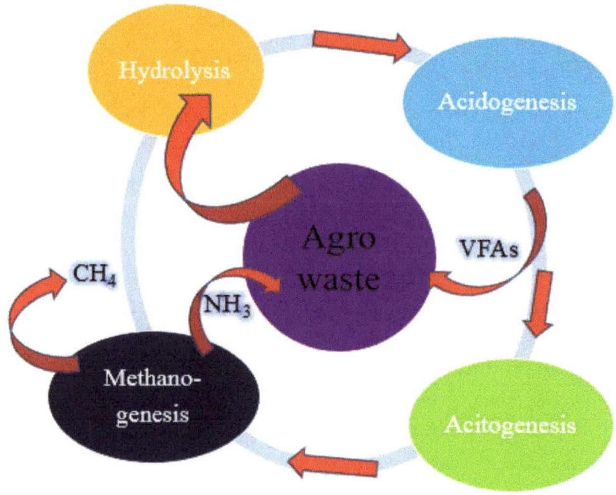

FIGURE 18.2 Schematic representation of the process of biogas production.

while creating NH_3, H_2, CO_2, H_2S, short-chain volatile fatty acids (VFAs) and organic acids (acetic acid, butyric acid and propionic acid) (Elsayed et al., 2019).

18.4.3 Acetogenesis

The third step is acetogenesis which includes the production of acetate from acetic acid, CO_2 and H_2 in the presence of acetogenic bacteria (Elsayed et al., 2019).

18.4.4 Methanogenesis

The final step in biogas production is methanogenesis, in which the acetogenesis product (acetic acid, CO_2 and H_2) is converted into methane (CH_4) and carbon dioxide (CO_2) (Bo et al., 2014).

$$CO_2 + 4H_2 \rightarrow CH_4 + 2H_2O$$

$$CH_3COOH \rightarrow CH_4 + CO_2$$

18.5 Anaerobic Digestion of Carbohydrate, Protein and Fat

Anaerobic digestion of the carbohydrate, protein and fat molecule is as follows:

18.5.1 Carbohydrate Digestion

The anaerobic digestion process in biogas production requires agricultural waste as a substrate for degradation, which is crop residues, fruit and vegetables as a carbohydrate food waste including sugars and the disaccharide compound which is easily degraded by the methanogenesis process into simple volatile fatty acid substances (Figure 18.3) (Daniels et al., 1984). Increasing the levels of simple sugars produces rapid and more volatile fatty acids in the production unit due to the accumulation of VFAs, the level of the pH decreases and the process of methanogenesis is slowed down; thus, it is recommended that a large amount of simple carbohydrates waste is mixed with a lower content of easily degradable organic

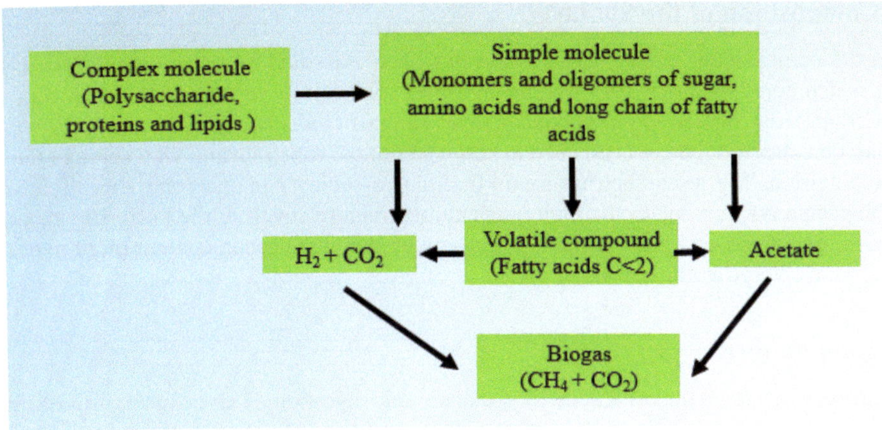

FIGURE 18.3 Flow diagram of biogas production.

components as a feedstock/input to balance the pH (Hagos et al., 2017). All plant-derived materials are carbohydrate-rich substrates.

18.5.2 Protein Digestion

Likewise, protein-rich substrate has been found in agricultural waste, such as butchery waste products, and low amounts of protein have been found in domestic waste. These protein waste products are rich in energy and their microbial degradation produces a high amount of methane. The microbial degradation of protein results in the formation of ammonia (NH_3) and ammonium (NH_4^+), which are in equilibrium to maintain the stability of the process of microbial-based protein degradation that depends on the pH and temperature (Wagner et al., 2013). Released ammonia (NH_3) gas is a strong inhibitor of methanogenic bacteria, which, in high concentrations, can inhibit methanogens (Chen et al., 2008). To maintain this situation, it is recommended that a high amount of N substrate is added to a high amount of a carbon-rich substrate to minimize the risk (Kovács et al., 2015).

18.5.3 Fat Digestion

Fat-rich substrates have been found in agricultural waste from activities such as the food processing industry and wastewater from slaughterhouses, and from edible oil, dairy products and olive oil (Li et al., 2011) with a high fat content. The process of anaerobic digestion of fat-rich substrates produces a high amount of biogas due to their ease of degradation. During microbial degradation of the fat-rich lipid substrate in anaerobic digestion, the high amount of fat causes problems related to blocking, the adsorption to biomass and microbial inhibition, because of the formation of high amounts of long-chain fatty acids (stearic and oleic acids) which inhibit microbial growth (Angelidaki and Ahring, 1992). For this, it is recommended to mix the carbohydrate-rich materials with the fat-rich materials to balance the whole process.

All of the above-mentioned practices of microbial degradation of carbohydrates, proteins and fats are recommended for the following advantages: balancing nutrition, enriching microorganisms, reducing the accumulation of inhibitors and increasing the stability and efficiency of biogas production and methane yield.

18.6 Factors Affecting Biogas Production

Six major factors, viz. the composition of the sublayer, temperature, retention time, pressure, pH and volatile fatty acids, play an important role in biogas production.

18.6.1 Composition of the Sublayer

Based on the composition, anaerobic digestion has been classified into two categories: dry anaerobic digestion which contains 2–15% dry matter and total solid >15%; and wet anaerobic digestion which contains 25–45% dry matter and total solid <15% (Kothari et al., 2014; Dobre et al., 2014). In the case of dry anaerobic digestion, there is no need to add water to the substrate and vice versa in the case of wet anaerobic digestion. The researcher has found that the co-sublayer in anaerobic digestion improves the yield of biogas production due to the high nutrient and manure covering the buffering capacity and the risk of microbial inhibition by ammonia is overcome by the high carbon content plant material balance in the C/N ratio (Dobre et al., 2014).

18.6.2 Temperature

Temperature is considered the critical factor for anaerobic digestion of agro-waste, influencing the quality as well as the quantity of biogas production. Based on the temperature, the participation of microbial communities in biogas production can be classified into three categories: psychrophilic (12–24°C), mesophilic (22–40°C) and thermophilic (50–60°C) (Nozhevnikova et al., 2001). In the anaerobic digestion process, the reduced retention time of the thermophilic methanogen is longer compared to the mesophilic and psychrophilic methanogens, a result of a faster and more efficient degradation process as well increase the yield of the methane, increases the solubility of the soild sublayers in anaerobic digestion process and help in separation of liquid and solid fraction of the substrate. Thermophilic methanogens are sensitive to lower temperature (±10°C) while mesophilic methanogens tolerate temperature fluctuations of ±30°C, without significantly influencing biogas production (Dobre et al., 2014). Thermophilic methanogens have high energy consumption due to their high-temperature regimes and this is associated with the risk of ammonia inhibition because high temperatures directly influence ammonia production. Thus, to avoid inhibition, a temperature below 500°C is required (Angelidaki and Ellegaard, 2002).

18.6.3 Retention Time

The retention time for sludge, also known as the sludge retention time (SRT), plays an important role in biogas production. If the substrate/feedstock remains in the fermenter for less than 8 days (SRT ≤8) then the biogas production is less due to acidogenic conditions, and if the substrate/feedstock remains in the fermenter for more than 8 days (SRT >8) then the biogas production is more due to methanogenic conditions (Miron et al., 2000). This means that degradation of the carbohydrates, proteins and lipids increases with an increased sludge retention time (Si et al., 2015)

18.6.4 Pressure of the Fermenter

In anaerobic digestion, if the pressure of the fermenter increases above 4–5 kPa then biogas production ceases due to high pressure (Dobre et al., 2014). This means that lower pressure in the fermenter provides a greater yield of biogas compared to a higher pressure (Sarkar et al., 2019).

18.6.5 pH of the Medium

In biogas production, the pH is an important factor in the methanogenesis process, ranging from 6.5–8.5. Fluctuations of the pH may hamper the methanogenesis process (Azman et al., 2015), and if the pH falls below 6, the process ceases.

18.6.6 Volatile Fatty Acids

The long-chain volatile fatty acid has an inhibitory effect on methanogens compared to short-chain VFAs (Azman et al., 2015). VFAs that accumulate in biogas plants include acetic acid, butyric acid, propionic acid, valeric acid and caproic acid (Siegert and Banks, 2005)

18.7 Advantages

Biogas production from renewable energy sources is one of the best substitutes for petroleum-derived products, and is one of the safest, cleanest and free energy sources (Börjesson and Mattiasson, 2008). It reduces the risks of non-renewable energy sources and decreases the incidence of respiratory-related problems in humans by reducing the level of pollution in the aerial environment (Haines and Kammen, 2002). This type of renewable energy source also helps to reduce forest tree cutting for firewood and cooking food and helps to reduce air pollution due to coal burning, greenhouse gases and deforestation (Nathaniel and Bekun, 2020). Thus, biogas production helps to support and maintain the ecological balance of the environment (Manolis et al., 2019).

18.8 Disadvantages

Biogas production has some disadvantage, including the lower energy efficiency of biogas compared to a simple kerosene lamp. It has a low energy density per volume so it is not easily stored. Thus, biogas does not fulfill the total energy requirements nor the requirement of a continuous supply of energy. It cannot replace all types of energy.

18.9 Future Challenge

Biogas production from the anaerobic digestion of agro-waste is an environmentally and economically feasible technology for the conservation of renewable energy resources in a sustainable manner (Abbas et al., 2017). The use of biogas is increasing daily because it is a clean, safe and environmentally friendly energy source and the raw materials for biogas production are easily available (Alexander et al., 2019). This indicates that there are many possibilities to increase biogas production units at the industrial level by using co-digestion of different substrate/feedstock wastes. The development of a biogas production unit at an industrial level is challenging due to the need to maintain the fragile microbial system as well as increasing its buffering capacity and enhancing the biogas/methane production rate, microbiological stability and nutritional balance.

Improving anaerobic digestion for biogas production requires input assessment, the development of microbial processes and accessible and easy to degrade substrates. Factors that control the gas production process include the organic materials, heterogeneous microbial activities, biodegradability, accessibility and determining accurate limitations and factors. Therefore, advanced techniques need to be developed to characterize biodegradability, chemical compositions, the vigorous nature of microorganisms and the inherent properties of the entire anaerobic co-digestion system. Ultimately, the main objective is to improve the process of the biogas production by the implementation of the novel strategy and new challenges for target achievement as well as research on anaerobic co-digestion helps to recognize the significant improvement of biogas production as a renewable source and economically feasible energy.

REFERENCES

Abbas, T, Ali, G, Adil, SA, Bashir, MK and Kamran, MA (2017). Economic analysis of biogas adoption technology by rural farmers: The case of Faisalabad district in Pakistan. *Renewable Energy*, 107: 431–439.

Addison, T (2005). *Agricultural Development for Peace*, WIDER Research Paper, No. 2005/07, United Nations University World Institute for Development Economics Research (UNU-WIDER), Helsinki, ISBN 929190676X

Akbulut, A (2012). Techno-economic analysis of electricity and heat generation from farm-scale biogas plant: Çiçekdağı case study. *Energy*, 44(1): 381–390.

Alexander, S, Harris, P and McCabe, BK (2019). Biogas in the suburbs: An untapped source of clean energy?. *Journal of Cleaner Production*, 215: 1025–1035.

Angelidaki, I and Ahring, BK (1992). Effects of free long-chain fatty acids on thermophilic anaerobic digestion. *Applied Microbiology and Biotechnology*, 37(6): 808–812.

Angelidaki, I and Ellegaard, L (2002). Anaerobic digestion in Denmark: Past, present and future. In *7th FAO/SREN-Workshop* (pp. 129–138). Department of Chemical Enzymology, Chemistry Faculty, Moscow State University.

Aziz, NIHA, Hanafiah, MM and Ali, MYM (2019). Sustainable biogas production from agrowaste and effluents–A promising step for small-scale industry income. *Renewable Energy*, 132: 363–369.

Azman, S, Khadem, AF, Van Lier, JB, Zeeman, G and Plugge, CM (2015). Presence and role of anaerobic hydrolytic microbes in conversion of lignocellulosic biomass for biogas production. *Critical Reviews in Environmental Science and Technology*, 45(23): 2523–2564.

Bioenergy, IEA (2010). "Bioenergy, land use change and climate change mitigation". Report for policy advisors and policymakers. *IEA Bioenergy: ExCo* 3, 1–61.

Black, CC, Tu, ZP, Counce, PA, Yao, PF and Angelov, MN (1995). An integration of photosynthetic traits and mechanisms that can increase crop photosynthesis and grain production. *Photosynthesis Research*, 46(1–2): 169–175.

Bo, T, Zhu, X, Zhang, L, Tao, Y, He, X, Li, D and Yan, Z (2014). A new upgraded biogas production process: Coupling microbial electrolysis cell and anaerobic digestion in single-chamber, barrel-shape stainless steel reactor. *Electrochemistry Communications*, 45: 67–70.

Bochmann, G and Montgomery, LF (2013). Storage and pre-treatment of substrates for biogas production. In *The Biogas Handbook* Weillinger, A, Murphy, J and Baxter, T (Eds) (pp. 85–103). Woodhead Publishing.

Börjesson, P and Mattiasson, B (2008). Biogas as a resource-efficient vehicle fuel. *Trends in Biotechnology*, 26(1): 7–13.

Burnham, A, Han, J, Clark, CE, Wang, M, Dunn, JB and Palou-Rivera, I (2012). Life-cycle greenhouse gas emissions of shale gas, natural gas, coal, and petroleum. *Environmental Science & Technology*, 46(2): 619–627.

Chen, Y, Cheng, JJ and Creamer, KS (2008). Inhibition of anaerobic digestion process: A review. *Bioresource Technology*, 99(10): 4044–4064.

Cuéllar, AD and Webber, ME (2008). Cow power: The energy and emissions benefits of converting manure to biogas. *Environmental Research Letters*, 3(3): 034002.

Daniels, L, Sparling, R and Sprott, GD (1984). The bioenergetics of methanogenesis. *Biochimica et Biophysica Acta (BBA)-Reviews on Bioenergetics*, 768(2): 113–163.

Deublein, D and Steinhauser, A (2011). *Biogas From Waste and Renewable Resources: An Introduction*. John Wiley & Sons.

Diesendorf, M and Diesendorf, M (2007). *Greenhouse Solutions with Sustainable Energy* (p. 252). Sydney: University of New South Wales Press.

Dobre, P, Nicolae, F and Matei, F (2014). Main factors affecting biogas production: An overview. *Romanian Biotechnological Letters*, 19(3): 9283–9296.

Elsayed, M, Abomohra, AEF, Ai, P, Jin, K, Fan, Q and Zhang, Y (2019). Acetogenesis and methanogenesis liquid digestates for pretreatment of rice straw: A holistic approach for efficient biomethane production and nutrient recycling. *Energy Conversion and Management*, 195, 447–456.

Hagos, K, Zong, J, Li, D, Liu, C and Lu, X (2017). Anaerobic co-digestion process for biogas production: Progress, challenges and perspectives. *Renewable and Sustainable Energy Reviews*, 76: 1485–1496.

Haines, A and Kammen, DM (2002). Sustainable energy and health. *Blueprint for the Clean, Sustainable Energy Age*, 79, 79–89.

Heinsoo K, Melts I and Kull T (2015). Agricultural bioenergy production. In: Lichtfouse E (eds) *Sustainable Agriculture Reviews. Sustainable Agriculture Reviews* (pp. 77–106) vol 18. Springer, Cham.

Hernandez, SC, Jimenez, LD and Biernat, K (2018). The potential for biogas production from agriculture wastes in Mexico. In *Biofuels—State of Development* Biernat, K (Ed) (pp. 15–36). IntechOpen.

Karellas, S, Boukis, I, and Kontopoulos, G (2010). Development of an investment decision tool for biogas production from agricultural waste. *Renewable and Sustainable Energy Reviews*, 14(4): 1273–1282.

Kim, MS, Cha, J and Kim, DH (2013). *Biohydrogen: Chapter 11. Fermentative Biohydrogen Production from Solid Wastes* (pp. 259–283). Elsevier Inc. Chapters.

Kothari, R, Pandey, AK, Kumar, S, Tyagi, VV and Tyagi, SK (2014). Different aspects of dry anaerobic digestion for bio-energy: An overview. *Renewable and Sustainable Energy Reviews*, 39: 174–195.

Kovács, E, Wirth, R, Maróti, G, Bagi, Z, Nagy, K, Minárovits, J Rákhely, G. and Kovács, KL. (2015). Augmented biogas production from protein-rich substrates and associated metagenomic changes. *Bioresource Technology*, 178: 254–261.

Kushkevych, I, Vítězová, M, Vítěz, T and Bartoš, M (2017). Production of biogas: Relationship between methanogenic and sulfate-reducing microorganisms. *Open Life Sciences*, 12(1): 82–91.

Li, C, Champagne, P and Anderson, BC (2011). Evaluating and modeling biogas production from municipal fat, oil, and grease and synthetic kitchen waste in anaerobic co-digestions. *Bioresource Technology*, 102(20): 9471–9480.

Manolis, EN, Zagas, TD, Karetsos, GK and Poravou, CA (2019). Ecological restrictions in forest biomass extraction for a sustainable renewable energy production. *Renewable and Sustainable Energy Reviews*, 110: 290–297.

Miron, Y, Zeeman, G, Van Lier, JB and Lettinga, G (2000). The role of sludge retention time in the hydrolysis and acidification of lipids, carbohydrates and proteins during digestion of primary sludge in CSTR systems. *Water Research*, 34(5): 1705–1713.

Mulat, DG, Dibdiakova, J and Horn, SJ (2018). Microbial biogas production from hydrolysis lignin: Insight into lignin structural changes. *Biotechnology for Biofuels*, 11(1): 61.

Nathaniel, SP and Bekun, FV (2020). Environmental management amidst energy use, urbanization, trade openness, and deforestation: The Nigerian experience. *Journal of Public Affairs*, 20(2): e2037.

Nozhevnikova, AN, Simankova, MV, Parshina, SN and Kotsyurbenko, OR (2001). Temperature characteristics of methanogenicarchaea and acetogenic bacteria isolated from cold environments. *Water Science and Technology*, 44(8): 41–48.

Pant, D, Van Bogaert, G, Diels, L and Vanbroekhoven, K (2010). A review of the substrates used in microbial fuel cells (MFCs) for sustainable energy production. *Bioresource Technology*, 101(6): 1533–1543.

Parris, K (2004). *Agriculture, Biomass, Sustainability and Policy: An Overview in OECD, Biomass and Agriculture, Sustainability, Markets and Policies* (pp. 27–36).

Rikkonen, P and Tapio, P (2009). Future prospects of alternative agri-based bioenergy use in Finland—constructing scenarios with quantitative and qualitative Delphi data. *Technological Forecasting and Social Change*, 76(7): 978–990.

Sarker, S, Lamb, JJ, Hjelme, DR and Lien, KM (2019). A review of the role of critical parameters in the design and operation of biogas production plants. *Applied Sciences*, 9(9): 1915.

Si, B, Li, J, Li, B, Zhu, Z, Shen, R, Zhang, Y and Liu, Z (2015). The role of hydraulic retention time on controlling methanogenesis and homoacetogenesis in biohydrogen production using upflow anaerobic sludge blanket (UASB) reactor and packed bed reactor (PBR). *International Journal of Hydrogen Energy*, 40(35): 11414–11421.

Siegert, I and Banks, C (2005). The effect of volatile fatty acid additions on the anaerobic digestion of cellulose and glucose in batch reactors. *Process Biochemistry*, 40(11): 3412–3418.

Subramanian, KA, Mathad, VC, Vijay, VK and Subbarao, PMV (2013). Comparative evaluation of emission and fuel economy of an automotive spark ignition vehicle fuelled with methane enriched biogas and CNG using chassis dynamometer. *Applied Energy*, 105: 17–29.

Van Ruijven, B and Van Vuuren, DP (2009). Oil and natural gas prices and greenhouse gas emission mitigation. *Energy Policy*, 37(11): 4797–4808.

Wagner, AO, Lins, P, Malin, C, Reitschuler, C and Illmer, P (2013). Impact of protein-, lipid-and cellulose-containing complex substrates on biogas production and microbial communities in batch experiments. *Science of the Total Environment*, 458: 256–266.

Walters, JP, Archer, DW, Sassenrath, GF, Hendrickson, JR, Hanson, JD, Halloran, JM, Vadas, P and Alarcon, VJ (2016). Exploring agricultural production systems and their fundamental components with system dynamics modelling. *Ecological Modelling*, 333: 51–65.

Wang, X, Chen, Y, Sui, P, Gao, W, Qin, F, Wu, X, and Xiong, J (2014). Efficiency and sustainability analysis of biogas and electricity production from a large-scale biogas project in China: An energy evaluation based on LCA. *Journal of Cleaner Production*, 65: 234–245.

Weiland, P (2010). Biogas production: Current state and perspectives. *Applied Microbiology and Biotechnology*, 85(4): 849–860.

19

Secreted Antigens of Bacilli: The Methodology for Immunochemical Investigation of Dynamics of Concentration per Bacterial Cell

E.P. Kiseleva, K.I. Mikhailopulo, D. Kołożyn-Krajewska and G.I. Novik

CONTENTS

19.1 Introduction

Bacilli are Gram-positive, aerobic, spore-forming bacteria that are found in soil and water and detected in transit in human and animal gut (Hong et al., 2005). There are 377 species of the genus *Bacillus* (last update in January 2019) (http://www.bacterio.net/bacillus.html) belonging to the *Bacillus cereus* and *B. subtilis* groups. Human relationships with members of the *B. cereus* group are ambiguous (Fritze, 2004). On the one hand, *B. anthracis* is the causative agent of anthrax, and pathogenic strains of *B. cereus* cause food-borne disease (Bennett et al., 2013; Ceuppens et al., 2012; Senesi and Ghelardi, 2010). On the other hand, non-pathogenic strains of *B. cereus* are effective in protecting health: they are used as probiotic feed additives in agriculture and aquaculture (Latorre et al., 2015; Li et al., 2015; Lodemann et al., 2008) and are the active principle in veterinary drugs (Williams et al., 2009) and pharmaceuticals for disbacteriosis treatment (Ceuppens et al., 2012).

Members of *B. subtilis* are not pathogenic and are conventionally used along with non-pathogenic strains of *B. cereus* in the production of enzymes (Westers et al., 2004) and surfactants (Gudiña et al., 2015). Waste from processing fruits and vegetables can be used as a medium for the growth of bacteria, which makes it possible to treat the process as biodegradation (Stabnikova et al., 2010).

Bacilli belong to the group of plant growth-promoting rhizobacteria. A number of *Bacillus* strains promote improvements in plant health and productivity using three mechanisms (Ahemad and Kibret, 2014; Choudhary and Johri, 2009; Niu et al., 2016; Perchat et al., 2005; Romeiro et al., 2010; Shafi et al., 2017): (i) antagonism of pest and pathogens (direct antagonism and action of toxins [peptide/lipopeptide antibiotics, insecticides] and catabolic enzymes [proteases, chitinases, glucanases, etc.]); (ii) promotion of host nutrition and growth (so-called biofertilization) by converting nutrients to plant-available form; and (iii) stimulation of plant host defences by increasing the production of growth-promoting hormones and defence-related enzymes in plants and by triggering induced systemic resistance to plant pathogens.

The known health protection mechanisms of bacilli used in animal/poultry husbandry and aquaculture are (i) the production of enzymes increasing the absorption of food (Latorre et al., 2015); (ii) the competitive exclusion of pathogens (La Ragione et al., 2001; La Ragione and Woodward, 2003) and the increase in the diversity of the intestinal microbiota (Hu et al., 2014); and (iii) the modulation of the host immune system (Hosoi et al., 2003; Huttunen et al., 2003; Sebastian and Keerthi, 2014; Urdaci et al., 2004; Xu et al., 2012). Due to the synthesis of biosurfactants and certain enzymes, selected bacilli strains are applicable for the bioremediation of hydrocarbon- and nitrogen-polluted soils and water (Borah and Yadav, 2014; Yao et al., 2013), as well as the biodegradation of pesticides (Singh et al., 2013) and polyethylene (Vimala and Mathew, 2016).

The above-listed areas of practical use of non-pathogenic bacilli strains, on the one hand, and the need to detect toxins of pathogenic bacilli strains in foodstuff, on the other hand, explain the relevance of the methods of detection and quantification of the secreted substances of these bacteria. Enzyme-linked immunosorbent assay (ELISA) is a suitable method for the detection/quantification of secreted biopolymers with antigenic properties because it is specific, sensitive, fast, simple, gives reproducible results and employs stable reagents and inexpensive equipment. The ability to monitor the biotechnological process via the detection/quantification of secreted biopolymers is advantageous in the case of (i) bacteria that are unavailable without interrupting the process; (ii) the procedure for detecting a secreted end-product is labour-intensive and time-consuming and requires expensive equipment; and (iii) bacteria are in a medium that have a complex composition and contain other microorganisms.

The aims of this chapter are (i) the elaboration of a methodological approach to investigating the dynamics of a specific concentration (C_{sp}) of secreted antigens per cell of bacilli and (ii) the presentation of the approach using six strains of bacilli and pre-obtained rabbit polyclonal antibodies (PAb) against genera-specific antigens of bacilli as an example.

19.2 Materials and Methods

Physiologically active cultures of *B. cereus* BIM B-491 (*Bc* 491), BIM B-169 (*Bc* 169), BIM B-205 (*Bc* 205), BIM B-206 (*Bc* 206), *B. mycoides* BIM B-179 (*Bm* 179) and *B. subtilis* BIM B-25 (*Bs* 25) after the second 24 h incubation were used as inoculums. All strains were incubated 6, 12 and 24 h in 50 mL of meat-peptone broth (MPB) (HiMedia, India) (medium 1) and 50 mL of MPB 2-fold diluted with distilled water (medium 2) for studying the dynamics of C_{sp}.

The immunization of rabbits with a sample containing lyophilized vegetative cells and spores (cells/spores) of *Bc* 491 to obtain $PAb_{anti\text{-}Bc\ 491}$ and the determination of $PAb_{anti\text{-}Bc\ 491}$ cross-reactivity were performed as shown in our study (Kiseleva et al., 2017).

The purification of $PAb_{anti\text{-}Bc\ 491}$: using the method of Perosa et al. (1990), total rabbit immunoglobulins (Ig) enriched with $PAb_{anti\text{-}Bc\ 491}$ ($Ig_{anti\text{-}Bc\ 491}$) were isolated from the antiserum of rabbits immunized as described above (Kiseleva et al., 2017). An optical density at 280 nm of $Ig_{anti\text{-}Bc\ 491}$ solution in 0.01 M sodium phosphate, pH 7.5 (Buffer 1) was measured using a spectrophotometer (Specord, Carl Zeiss, Germany). The concentration of $Ig_{anti\text{-}Bc\ 491}$ was calculated using $A_{280\ nm,\ 1\ cm,\ 1\ mg/mL} = 1.35$.

Cell fractions used for studying the dynamics of C_{sp} and the interpretation of ELISA data: cells/spores of bacilli (6 strains × 2 media × 3 times = 36 samples) were separated from a cultural liquid (CL, a term used for the designation of a cultural medium after bacteria growth) by centrifugation at 2000 g for 20 min. A portion (3 mL) of each CL specimen was collected in a separate vial. Cells/spores were washed (3 × 50 mL) with Buffer 1, then suspended in Buffer 1 (1 mL) to obtain a suspension of approximately 30–50% and a protease inhibitor cocktail (50 μL) was added (Kiseleva et al., 2017). Cells/spores were manually homogenized (Wheaton, USA) and ultrasonicated on ice using the Branson Digital Sonifier 450 (Branson, USA), 50% duty cycle, 4 × 1 min with 1.5 min pauses. The fraction containing cell walls and spores (CW/sp) was sedimented by centrifugation at 13,000 g for 30 min using the Eppendorf Mini

Spin plus (Eppendorf AG, Germany). The supernatant (cell-free fraction, CFF) was collected in a separate tube.

Pre-collected CL specimens (CL_6, CL_{12}, CL_{24}; the number in the lower register corresponds to the duration of culture growth [h]) were lyophilized using a freeze dryer (FreeZone Labconco; США). Each CW/sp specimen (CW/sp_6, CW/sp_{12}, CW/sp_{24}) was weighted using an analytical balance (CPA225D, Sartorius, Germany). The value of the optical density at 260 nm in a cuvette with a 1 cm optical path length ($A_{260\,nm,\,1\,cm}$) of each CFF specimen (CFF_6, CFF_{12}, CFF_{24}) was measured using a spectrophotometer (Specord, Carl Zeiss, Germany). Thirty-six pieces of each CFF specimens and lyophilized CL specimens were stored at –20°C for ELISA.

ELISA was performed using 96-well high-binding microplates (Microlon, Greiner Bio-One, Germany). *Buffer solutions:* Buffer 1, Buffer 2 (Buffer 1 supplemented with 0.2 M NaCl) and either Buffer 3 (Buffer 2 supplemented with 1 g/L of bovine serum albumin [BSA]) or Buffer 4 (Buffer 2 supplemented with 2 g/L of BSA) were used for immobilizing the antigens and washing wells (3×200 μL) and as an incubation buffer, respectively.

A study of the dynamics of the C_{sp} of secreted antigens: (a) *Immobilization:* CFF specimens ($n=36$; 6 specimens for each tested strain) pre-diluted to obtain $A_{260\,nm,\,1\,cm}=0.025$ (each) were immobilized by passive adsorption (0.1 mL per well, each well in duplicate) at 4°C overnight in an appropriate series of wells (36 series, 12 wells in each series); pairs of wells were designated B_0–B_5. (b) *Preparation for the first stage of ELISA:* lyophilized CL specimens ($n=36$) were diluted in 300 μL of Buffer 3 (each) using a vortex mixer (Bio Vortex V-1 plus, BioSan, Latvia) to obtain 10-fold concentrates of CL specimens (designated 10×CL). Subsequently, 36 series of samples were prepared by serial dilution of 10×CL specimens with Buffer 1; samples B_1–B_5 with a concentration (5–80%) of 10×CL in each series. (c) *First stage of ELISA:* samples B_1–B_5 were added into appropriate wells; 50 μL per well. Buffer 1 was added into two wells (designated as B_0) of each series; 50 μL per well. $Ig_{anti\text{-}Bc\,491}$ solution (15 mg/L) in Buffer 4 was added into each well; 50 μL per well. ELISA plates were incubated for 1 h at 37°C. (d) *Second stage of ELISA:* performed with a conjugate of sheep PAb against rabbit Ig with horseradish peroxidase (ShAR–HRP) (Xema Co. Ltd., Russia) (Kiseleva et al., 2017). The enzymatic peroxidation reaction was initiated and stopped, and A_{450} values were measured in each well, as shown in our recent study (Kiseleva et al., 2017).

A study of the dynamics of the C_{sp} of intracellular antigens: (a) *Immobilization:* CFF specimens ($n=36$; 6 specimens for each tested strain) pre-diluted to obtain $A_{260\,nm,\,1\,cm}=0.1$ (each) were immobilized in an appropriate series of wells (36 series, 12 wells in each series) as described above. (b) *First stage of ELISA:* $Ig_{anti\text{-}Bc\,491}$ solutions in Buffer 3 ($n=6$, [0.4–104.0] mg/L) were added into appropriate pairs of wells of each series and incubated for 1 h at 37°C. ELISA was completed as described above.

19.3 Results

The methodological approach to investigating the dynamics of the C_{sp} of secreted antigens per cell of bacilli detailed in the following sections includes (i) the method of selecting bacteria suitable for ELISA with the available Ab; (ii) the principles of preparing CL, CFF and CW/sp specimens; the criteria for the quantitative assessment of CFF and CW/sp; (iii) the design of ELISA; the methods of data processing (graphical representations followed by calculations); and (iv) the principles of interpreting the results. In our recent study, we used a similar approach to studying the dynamics of the C_{sp} of cell wall antigens of bacilli (Kiseleva et al., 2017). The vital difference between our recent and current study relates to points (iii) and (iv). We show our experimental data in accordance with points (i)–(iv) of the methodology for simplicity of presentation and perception; in connection with the foregoing, points (i) and (ii) are described briefly and points (iii) and (iv) are detailed.

19.3.1 (i) The Method of Selecting Bacteria Suitable for ELISA with the Available Ab

In our recent study (Kiseleva et al., 2017), we used lyophilized vegetative cells and spores of *Bc* 491 as an immunogen to produce $PAb_{anti\text{-}Bc\,491}$. The cross-reactivity of $PAb_{anti\text{-}Bc\,491}$ with antigens of related bacteria of the *B. cereus* and *B. subtilis* groups was determined by competitive ELISA. Of 15 tested strains, 9 strains had cross-reactivity values in the range 9–183%; 2 strains: 3%; 4 strains: less than 1% (Kiseleva et al., 2017). On this principle, we concluded that $PAb_{anti\text{-}Bc\ 491}$ interact with genus-specific antigens of bacilli. $PAb_{anti\text{-}Bc\ 491}$ were considered suitable as a key reagent in the ELISA of antigens of preselected strains of bacilli.

A cross-reaction of not less than 10% was proposed as a criterion for selecting strains suitable for ELISA with the available Ab ($PAb_{anti\text{-}Bc\ 491}$ and any other Ab, depending on the purpose of the study) (Kiseleva et al., 2017).

In the current study, we used strain-immunogen and five preselected strains of bacilli having the highest $PAb_{anti\text{-}Bc\ 491}$ cross-reactivity values, viz., 97, 183, 55, 118 and 40% for *Bc* 169, *Bc* 205, *Bc* 206, *Bm* 179 and *Bs* 25, respectively (Kiseleva et al., 2017). High cross-reactivity values suggest that five strains have antigens that are identical/similar to antigens of *Bc* 491. The possible explanations for cross-reactivity values above 100% are as follows: (i) tested strain has antigens that are identical to antigens of *Bc* 491, but their concentration per cell is higher; (ii) tested strain has antigens that are similar but not identical to antigens of *Bc* 491 and interact with $PAb_{anti\text{-}Bc}$ with higher affinity than antigens of *Bc* 491.

19.3.2 (ii) The Principles of Preparing CL, CFF and CW/sp Specimens; the Criteria for the Quantitative Assessment of CFF and CW/sp

Selected strains were grown in two media: the first was MPB used for the cultivation of bacilli by tradition (medium 1) and the second was MPB 2-fold diluted with distilled water (medium 2) to reduce the concentration of nutrients. The duration of growth was 6, 12 and 24 h. CFF specimens and CL specimens, 36 pieces of each (6 strains × 2 media × 3 time points), were used in ELISA tests.

The time dependence of the total nucleic acids (tNA) amount in CFF and the weight of CW/sp were determined for each strain of bacilli (Table 19.1). The parameter proportional to the amount of total nucleic acids in CFF (referred to as the amount of tNA) was calculated as $A_{260\ nm,\ 1\ cm}$ (AU), multiplied by the corresponding CFF volume (mL); the parameter is proportional to the number of vegetative cells of bacilli

TABLE 19.1

Dynamics of tNA Amount in CFF and Weight of CW/sp of Six Strains of Bacilli (6–24 h)—Duration of Culture Growth

		tNA Amount in CFF, CU			CW/sp weight, mg		
#	**Strain**	**6 h**	**12 h**	**24 h**	**6 h**	**12 h**	**24 h**
Medium 1							
1	*Bc* 491	13.7 ± 0.3	31.9 ± 1.2	30.2 ± 0.9	17.5 ± 0.5	96.1 ± 3.9	162.5 ± 2.2
2	*Bc* 205	21.6 ± 0.5	20.1 ± 1.7	17.9 ± 0.7	24.8 ± 0.6	118.4 ± 5.0	**178.9 ± 2.4**
3	*Bc* 206	23.1 ± 0.3	36.4 ± 1.9	35.7 ± 1.3	12.6 ± 0.3	71.7 ± 2.4	159.4 ± 2.5
4	*Bc* 169	25.9 ± 0.4	40.6 ± 2.1	**53.1 ± 1.8**	49.9 ± 0.7	95.8 ± 3.4	162.6 ± 3.5
5	*Bm* 179	11.8 ± 0.2	33.3 ± 1.7	33.0 ± 1.5	19.7 ± 0.5	93.8 ± 2.5	124.7 ± 2.7
6	*Bs* 25	4.5 ± 0.1	15.3 ± 0.4	25.8 ± 1.6	7.1 ± 0.2	38.1 ± 0.7	39.2 ± 0.7
Medium 2							
1	*Bc* 491	6.9 ± 0.1	6.7 ± 0.1	10.1 ± 0.4	8.4 ± 0.2	35.2 ± 0.6	**99.1 ± 3.1**
2	*Bc* 205	9.9 ± 0.3	7.6 ± 0.2	**11.6 ± 0.4**	23.5 ± 0.5	51.4 ± 1.2	70.1 ± 2.5
3	*Bc* 206	8.4 ± 0.2	5.8 ± 0.1	6.9 ± 0.2	21.7 ± 0.4	61.1 ± 0.9	93.8 ± 2.7
4	*Bc* 169	10.6 ± 0.5	7.9 ± 0.3	8.3 ± 0.3	20.1 ± 0.5	61.3 ± 1.1	91.5 ± 3.7
5	*Bm* 179	8.2 ± 0.2	9.5 ± 0.3	9.0 ± 0.4	13.2 ± 0.4	39.9 ± 0.5	78.8 ± 2.9
6	*Bs* 25	6.2 ± 0.1	6.1 ± 0.1	7.0 ± 0.2	13.4 ± 0.2	18.3 ± 0.3	22.2 ± 2.1

Note: Values are means ± SD. The amount of total nucleic acids in CFF (tNA) (conventional units, CU) was calculated as the result of multiplying $A_{260\ nm,\ 1\ cm}$ (AU) by CFF volume (mL) and is proportional to the number of bacteria cells that were the source of CFF. The weight of CW/sp is proportional to the total number of vegetative cells and spores. Highest values of tNA amount in CFF and CW/sp weight are indicated in italics for each medium.

(Kiseleva et al., 2017). The weight of CW/sp is proportional to the total number of vegetative cells and spores. A comparison of the dynamics of the two parameters allows (i) estimating the ratio of cells to spores to determine the phase of culture growth (e.g. in medium 1 and medium 2, the exponential phase of strain *Bc* 491 lasts up to 12 h and 6 h, respectively) and (ii) interpreting the ELISA data as will be shown below.

19.3.3 (iii) The Design of ELISA; the Methods of Data Processing (Graphical Representations Followed by Calculations)

19.3.3.1 *The Dynamics of C_{sp} of Secreted Antigens*

To compare six strains of bacilli in terms of the dynamics of the C_{sp} of secreted antigens located in CL, we used the following approach. Considering that after centrifuging each culture, the volume of the CL was greater than the volume of the sediment containing cells and spores, and also in view of the known data that no more than 4% of the bacillus proteins appear in the external environment (Tjalsma et al., 2000), we assumed that the concentration of antigens in CL would be much lower than the concentration of antigens in CFF. For this reason, we previously lyophilized CL specimens, and then dissolved them with water to obtain 10-fold concentrates (designated above 10×CL) and thus increase the concentration of secreted antigens.

To prepare a series of B_1–B_5 samples, 10×CL specimens ($n=36$) were used. These samples (36×5=180) as well as probe B_0 (Buffer 1) were added into wells with the appropriate CFF specimens (36 items) pre-immobilized in wells from solutions with values $A_{260\ nm,\ 1\ cm}$ equal to 0.025 (Figure 19.1). Equal values of $A_{260\ nm,\ 1\ cm}$ were used to equalize the number of cells of each strain from which CFF antigens were obtained. Indeed, $Ig_{anti\text{-}Bc\ 491}$ solution (15 mg/L) was added into all wells. At the first stage of ELISA, CL antigens (located in the liquid phase) competed with CFF antigens (immobilized) for binding with $Ig_{anti\text{-}Bc\ 491}$. The amount of immune complexes formed on the solid phase was inversely proportional to the quantity of antigens in the 10×CL specimens. Typical data of competitive ELISA are shown in Figure 19.1 for strain *Bm* 179 as an example. Similar graphs of competitive ELISA were obtained for five other bacilli strains (data not shown).

The value $A_{260\ nm,\ 1\ cm}$ in the solutions for the immobilization of CFF specimens (equal for all CFF specimens), the $Ig_{anti\text{-}Bc\ 491}$ concentration and volumes of 10×CL specimens used for the preparation of B_1–B_5 samples were initially selected; the selection criteria were (i) sigmoid shape of curves in coordinates B_n/B_0, % (OY)–volume of 10×CL specimens, μL (OX) and (ii) the presence along each curve of at least one experimental point with values of B_n/B_0 less than 50%.

The data presented in Figure 19.1 for *Bm* 179, and similar data obtained for five other bacilli strains (not shown), were processed as follows. In each graph for each of the three curves, we determined the volume of a 10×CL specimen (μL), corresponding to the experimental point with a value of B_n/B_0, equal to 50%; the values obtained were designated y_1–y_3, where numerals 1–3 in the lower register correspond to 6–24 h of culture growth. Then, we calculated $1/y_1$–$1/y_3$ values. Each of the calculated values was divided by the corresponding value of the tNA amount in CFF, viz. (n_1–n_3) (data of Table 19.1); thus, we "equalized" the number of cells that were used as the source of antigens located in CL specimens and converted the data of ELISA to C_{sp} (per cell). A maximum of three values $1/y_1n_1$–$1/y_3n_3$ was taken as 100% and other values $1/y_nn_n$ were converted into percentages. The calculated values proportional to the C_{sp} of secreted antigens are shown in Figure 19.2.

Thus, the number of cells that were the source of the immobilized antigens (CFF antigens) was equalized (by diluting all CFF specimens to equal $A_{260\ nm,\ 1\ cm}$ values), and the number of cells that were the source of the antigens located in the liquid phase in stage 1 of the ELISA (CL antigens) was equalized by recalculation.

An alternative way to equalize the number of cells that are the source of CL antigens is the preliminary dilution of 10×CL concentrates in the calculated number of times to prepare sample B_1. The calculation of the dilution factor for each 10×CL specimen should be done as follows. The CL specimen respective to the CFF specimen with a minimum value of tNA amount should be designated a_1, the dilution factor of the corresponding 10×CL specimen should be assumed to be 1 (i.e. this 10×CL specimen is used without dilution). Other CFF specimens ($n=35$) should be designated a_2–a_{35}. The dilution factors should be calculated as the ratio (tNA amount in CFF specimen a_n/tNA amount in CFF specimen a_1) for 10×CL specimens respective to (a_2–a_{35}) CFF specimens.

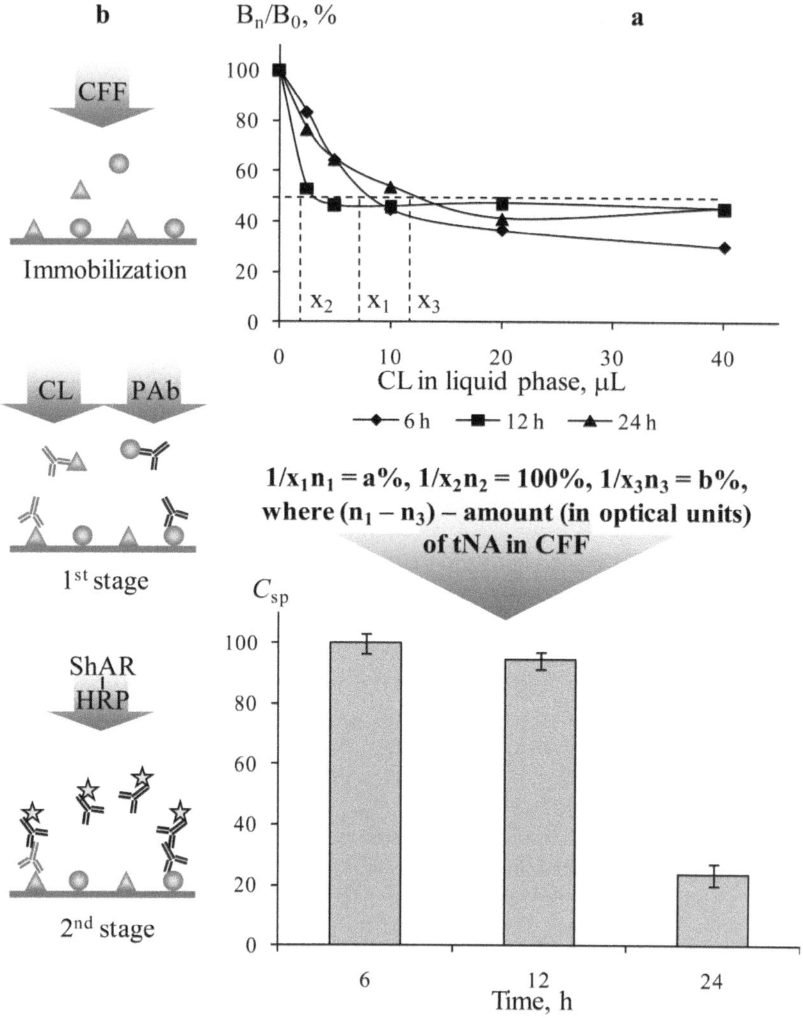

FIGURE 19.1 The dynamics of the C_{sp} of secreted antigens exemplified by strain *Bm* 179 grown in medium 1. (A) A typical graph of competitive ELISA and the calculation of C_{sp} values (per cell of bacteria). The average data of three independent experiments are shown. SD values are in the range ±5%. (B) Scheme of competitive ELISA.

Each pre-diluted 10×CL specimen ($n = 35$) and 10×CL specimen designated a_1 (used as is) should be designated as sample B_1, and B_2–B_5 samples should be prepared by serial dilution for use in the liquid phase in competitive ELISA. The data of ELISA should be done in coordinates B_n/B_0, % (OY)–volumes of 10×CL specimens, μL (OX). In each graph for each of the three curves the volume of pre-diluted 10×CL specimen (μL), corresponding to the experimental point with a value of B_n/B_0, equal to 50%, should be detected. The values obtained should be designated y_1–y_3, where numerals 1–3 in the lower register correspond to 6–24 h of culture growth, and $1/y_1$–$1/y_3$ values should be calculated. The maximum of three values $1/y_1$–$1/y_3$ should be taken as 100% and other values $1/y_n$ should be converted into percentages. The calculated $1/y_1$–$1/y_3$ values are proportional to the C_{sp} of the secreted antigens.

19.3.3.2 *The Dynamics of C_{sp} of Intracellular Antigens*

All bacilli antigens secreted into the environment were originally located inside the cell. In this regard, we studied the dynamics of the C_{sp} of intracellular antigens (antigens of CFF) for a comparison with the

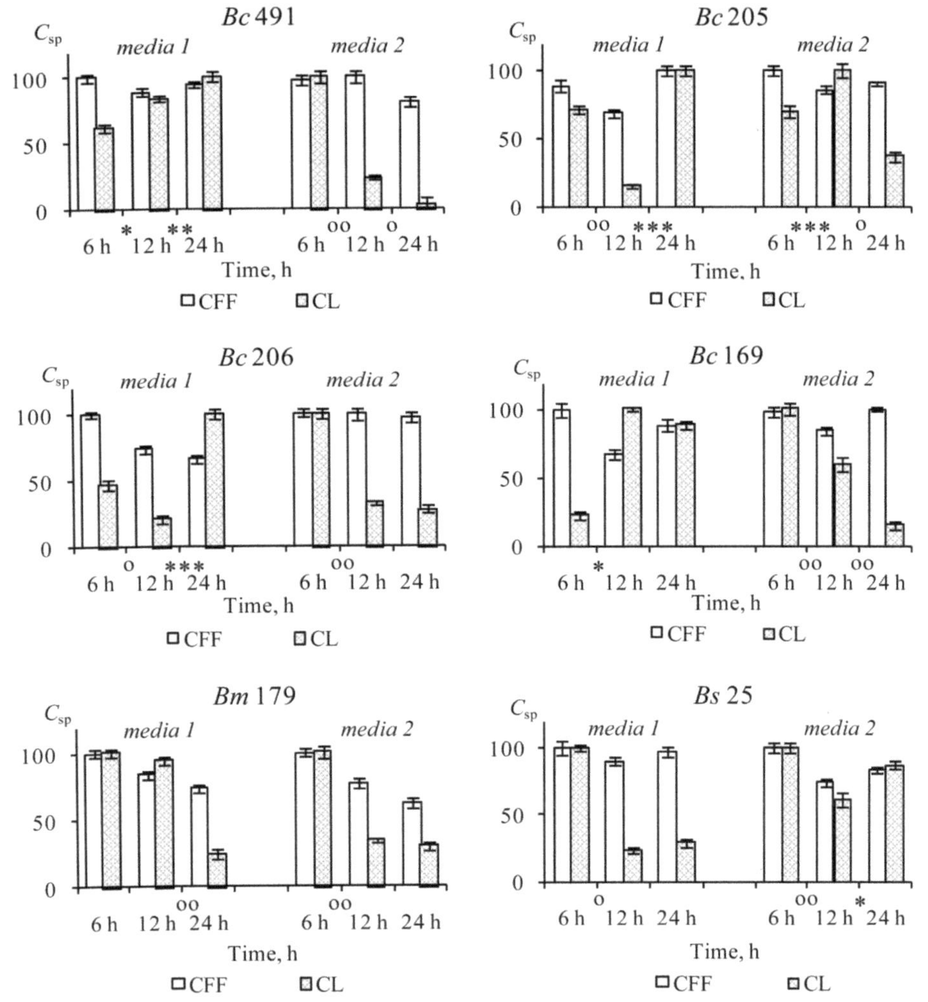

FIGURE 19.2 Dependence of the C_{sp} (per cell) of intracellular antigens (CFF) and secreted antigens (CL) on the duration of the culture growth in medium 1 and medium 2. Possible reasons for an increase in the C_{sp} of secreted antigens: * release of cytoplasmic proteins and cell wall proteins due to cell wall turnover; ** secretion of true exoproteins, exopolysaccharides and antimicrobial proteins; and *** transition to CL of mother cell components during spore formation. Possible reasons for a decrease in the C_{sp} of secreted antigens: ° predominance of the growth rate of the culture over the intensity of secretion; and °° loss of native structure and immunoreactivity of antigens in CL due to proteolysis, enzymatic hydrolysis of carbohydrate antigens or aggregation.

dynamics of the C_{sp} of secreted antigens. In other words, we compared the dynamics of the synthesis and the secretion of substances with antigenic properties.

To compare the dynamics of the C_{sp} of intracellular antigens of six strains of bacteria, the following approach was used. CFF specimens ($n=36$) were immobilized from solutions with $A_{260\ \text{nm},\ 1\ \text{cm}}$ values equal to 0.1, thereby equalizing the number of cells of each strain from which CFF antigens were obtained. $\text{Ig}_{\text{anti-}Bc\ 491}$ solutions prepared by serial dilution were added into all wells. The amount of immune complexes formed on the solid phase at the first stage of ELISA was directly proportional to the quantity of antigens in the immobilized CFF specimens.

The value $A_{260\ \text{nm},\ 1\ \text{cm}}$ in the solutions for the immobilization of CFF specimens (equal for all CFF specimens) and the range of $\text{Ig}_{\text{anti-}Bc\ 491}$ concentrations were initially selected; the selection criterion was to obtain the classical saturation curves with a pronounced plateau characteristic of a specific interaction

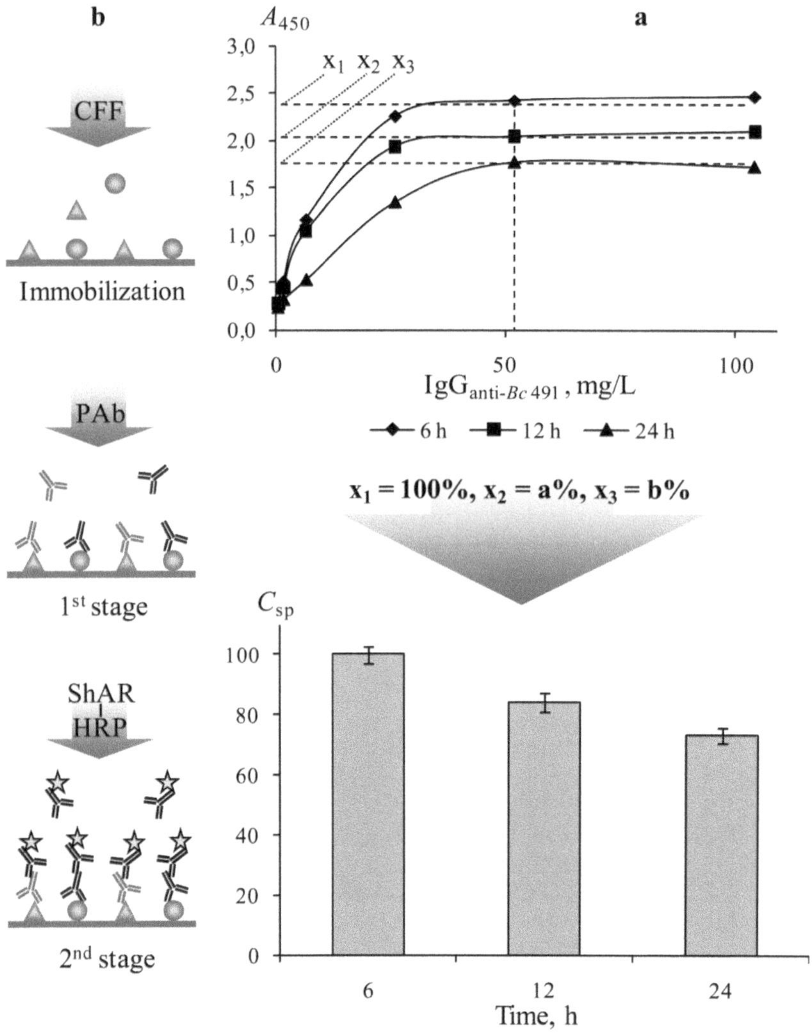

FIGURE 19.3 The dynamics of the C_{sp} of intracellular antigens exemplified by strain *Bm* 179 grown in medium 1. (A) A typical graph of ELISA and the calculation of C_{sp} values (per cell of bacteria). The average data of three independent experiments are shown. SD values are in the range ±5%. (B) Scheme of ELISA.

in coordinates A_{450}, AU (OY)–$Ig_{anti\text{-}Bc\,491}$ concentrations, mg/L (OX). Typical data obtained for strain *Bm* 179 are shown in Figure 19.3 as an example.

These data and similar data obtained for five other bacilli strains (not shown) were processed as follows. In each figure, similar to Figure 19.3, for each of the three plots, an A_{450} value corresponding to 52 mg/L of $Ig_{anti\text{-}Bc\,491}$ (the "plateau" of each of the three curves) was determined. For each strain, the values obtained were designated x_1–x_3, where numerals 1–3 in the lower register correspond to 6–24 h of culture growth. The maximum of x_1–x_3 values was taken as 100% and the other two x_n values were converted to percentages. The calculated values proportional to the C_{sp} of intracellular antigens (antigens of CFF) are shown in Figure 19.2.

Based on the results of the study of the dynamics of the C_{sp} of intracellular antigens (located in CFF) and secreted antigens (located in CL) per bacterial cell, six bacilli strains can be grouped in the following way (Table 19.2). A type of C_{sp} change is similar in bacteria belonging to groups 1 and 1* (the coincident strain is only *Bc* 491), 2 and 2* (the coincident strain is only *Bm* 179), 4 and 4* and 5 and 5* (there are no coinciding strains). Bacteria belonging to groups 3 and 6 (medium 1) and groups 7* and 8* (medium 2) do not have analogues among the groups of bacteria regimented for another medium.

TABLE 19.2

Systematization of the Results of the Study of the Dynamics of Specific Concentration (C_{sp}) of Intracellular Antigens (CFF) and Secreted Antigens (CL) per Cell of Six Bacilli Strains Grown in Two Media

	Medium 1			Medium 2								
	CFF		CL	CFF	CL							
	Group of Bacteria											
	1	2	3	4	5	6	1*	2*	4*	5*	7*	8*
Bc 491	+					+	+			+		
Bc 169			+			+	+			+		
Bc 205			+	+			+					+
Bc 206		+		+			+				+	
Bm 179		+			+			+			+	
Bs 25	+				+			+	+			
Type of C_{sp} changes	6 h/24 h – const (change within 10%)	6 h/12 h - ↓, 12 h/24 h - ↓; change by ~30% from the initial value	6 h/12 h - ↓, 12 h/24 h - ↑, difference min and max values is ~30%	6 h/12 h - ↓, 2/4-fold decrease; 12 h/24 h - ↑, increase is no less than 80%	6 h/12 h - ↓, 12 h/24 h - ↓; ~75% decrease from the initial value	6 h/12 h - ↑; 12 h/24 h ~ const (↑/↓ not more than 20%)	C_{sp} ~ const (change not more than 20%)	Tendency to C_{sp} decrease; change by more than 25%	6 h/12 h - ↓, 12 h/24 h - ↑; the difference between min and max values is 40%	6 h/12 h - ↓, 12 h/24 h - ↓, decrease by (70–90)%	6 h/12 h - ↓, 45% from the initial value; 12 h/24 h - const	6 h/2 h - ↑, 12 h/24 h - ↓; the difference between min and max values is ~60%

Note: Symbol * was used to designate groups of bacteria growing in media 2. Groups of bacteria with a similar type of C_{sp} changes (bold type) were denoted by the same numbers: 1 and 1*; 2 and 2*; 4 and 4*; 5 and 5*. Groups 3 and 6 (medium 1) and groups 7* and 8* (medium 2) have no analogues among the groups of bacteria regimented for another medium.

TABLE 19.3

Possible Causes of Increase/Decrease in C_{sp} of CL Antigens

	Combinations of changes in the two parameters—tNA in CFF and the weight of CW/sp[1]	The reasons for the entry[2]/disappearance[3] of antigens in CL
1	tNA ↑, CW/sp ↑ (either exponential growth phase or spore germination)	[2]Release of cytoplasmic proteins and cell wall proteins due to cell wall turnover
2	tNA const, CW/sp ↑ (stationary growth phase)	[2]Secretion of true exoproteins, exopolysaccharides and antimicrobial proteins; [3]secretion of endoproteases leading to protein hydrolysis
3	tNA ↓, CW/sp ↑ (late stationary growth phase/sporulation)	[2]Transition to CL of mother cell components during spore formation

Note: [1]Data from Table 19.1; [2] and [3]—reasons for the entry/disappearance of antigens in CL, respectively.

19.3.4 (iv) The Principles of Interpreting the Results

To find out the possible causes of the increase/decrease in the C_{sp} of CL antigens (designated by symbols below the histograms in Figure 19.2), we compared the dynamics of two parameters, viz., tNA amount in CFF and the weight of CW/sp (data of Table 19.1). Given the fact that the values of each of the parameters can increase, decrease or remain unchanged, nine different combinations are theoretically possible. In fact, we observed only three combinations of the two parameters and used the known scientific data presented in Section 19.4 to explain the possible reasons for the entry/disappearance of antigens in CL (Table 19.3).

19.4 Discussion

ELISA is a suitable method for the detection/quantification of bacterial biopolymers with antigenic properties because it is specific, sensitive, fast, simple, gives reproducible results and employs stable reagents and inexpensive equipment. Key reagents in ELISA are specific antibodies that recognize antigens for their high affinity and non-covalent binding. In addition to their high specificity, another advantage of Abs that allows their use as probes for the detection of molecules of interest is their relatively uniform and well-characterized protein structure that enables them to be purified, labelled and detected predictably and reproducibly by generalized methods. Procedures for generating, purifying and modifying antibodies for use as antigen-specific probes were developed during the 1970s and 1980s (Harlow and Lane, 1988) and have remained relatively unchanged to date. The updated *Thermo Scientific Pierce Antibody Production Technical Handbook* (Anonymous, 2010) is a user's guide for working with Abs. The handbook provides technical information on the procedures, reagents and tools used to produce, purify, fragment and label antibodies.

Unfortunately, only some antigens are immunogens, i.e. molecules capable of stimulating an immune response to induce specific antibody synthesis. To be immunogenic, a substance must have three characteristics: (i) foreignness, (ii) high molecular weight and (iii) chemical complexity. (i) Foreignness means that the immunized animal does not recognize and ignores the substance as "self". (ii) Most compounds with a molecular weight greater than 6,000 are immunogenic. (iii) Even high molecular weight homopolymers of amino acids and simple polysaccharides are seldom good immunogens because they lack chemical complexity.

Among the four major classes of macromolecules (viz., proteins, carbohydrates, nucleic acids and lipids), proteins are the strongest immunogens. Carbohydrates are immunogenic only if they have a relatively complex polysaccharide structure or are part of glycoproteins. Indeed, the immune response to carbohydrates is T-cell independent and results in the production of low affinity IgM antibodies. Nucleic acids are poor immunogens, and lipids are not usually immunogenic. Peptides and other small molecules are not immunogens in accordance with points (ii) and (iii), although they are able to act as recognition sites for specific antibodies and are referred to as haptens. Haptens can be made immunogenic by coupling them to a suitable protein to obtain a conjugate "hapten–carrier". Similarly to small antigens, the non-immunogenic and low-immunogenic macromolecular antigens referred to above become good immunogens as a result of coupling them to a suitable protein (Haji-Ghassemi et al., 2015; Heimburg-Molinaro and Rittenhouse-Olson, 2009; Hu et al., 2014; Plescia and Braun, 1967). A robust arsenal of new anti-carbohydrate and anti-nucleic acid commercial antibodies is currently available.

Available databases of commercial antibodies to bacilli antigens include PAbs (rabbit, goat, chicken) and MAbs (mouse) of varying degrees of specificity, the preparation of which was based on using as immunogens of (i) spores; (ii) spore extracts; (iii) chromatographic fractions with a selected Mr, resulting from the fractionation of spore extracts; (iv) purified bacillus proteins; and (v) peptides obtained by genetic engineering methods, homologous to proteins or fragments of bacilli proteins. The most represented are antibodies to *B. anthracis* virulence factors. Vegetative cells of *B. cereus* are rarely used as an immunogen (Charni et al., 2000; Zhu et al., 2016), and commercial kits for whole cells of *B. cereus* strains were not available before 2016 (Zhu et al., 2016).

Methods (iii) to (v) make it possible to obtain highly specific PAbs having a low cross-reactivity with other antigens or bacilli of other species. Other highly specific antibodies are MAbs obtained as a result of immunization according to points (i) or (ii), but with the subsequent selection of a clone with the desired specificity, as well as PAbs obtained as a result of fractionating the antiserum by affinity chromatography with the target bacillus antigen. In addition, a theoretically possible way of obtaining highly specific antibodies is the use of selective extraction methods (ii), which make it possible to obtain a mono-substance for use as an immunogen. On the contrary, PAbs obtained by method (i) or (ii) (without selective extraction) have a wide specificity, bind numerous antigens of the strain-immunogen and cross-react with other species of bacilli. Such PAbs are universal reagents useful for a specialist working with related species and strains of bacilli, on condition that they are separately cultivated. The PAbs used in our study (viz., $PAb_{anti\text{-}Bc\ 491}$) are just such; they were obtained using vegetative cells and spores of selected bacilli strain, viz., *Bc* 491, as an immunogen (Kiseleva et al., 2017). $PAb_{anti\text{-}Bc\ 491}$ are genus specific; they allow the detection of any bacilli antigen identical or homologous to that of *Bc* 491 and are suitable for the detection/quantification of intracellular antigens (note, cells should be disrupted before ELISA and CFF is obtained note, cells should be desrupted and CFF should be obtained for ELISA) and antigens secreted by bacilli into the external environment (Kiseleva et al., 2017). In our study, we used six strains of bacilli including *Bc* 491 (strain-immunogen) and other preselected bacilli strains ($n=5$) for which the cross-reactivity values of $PAb_{anti\text{-}Bc\ 491}$ were 40% and more. We did not try to find out exactly which antigens interact with $PAb_{anti\text{-}Bc\ 491}$. Theoretically, these can be proteins, glycoproteins, lipoproteins, carbohydrates (including teichoic acids; all within the limitations mentioned above) and nucleic acids (in the form of natural complexes with protein) of bacilli.

The answers to a number of questions, which are listed below,allow us to define the scope of our methodology and to carry out the last point of the methodology, namely, the interpretation of the results. Which bacilli compounds, particularly extracellular substances, produced by bacilli are of interest in biotechnologies (including biotechnologies for agriculture) and research studies? What substances are secreted by bacilli into their environment? Which are antigens detectable by means of ELISA? Is there strong dependence between the maximum yield of the bacilli antigen and the medium composition/growth stage?

Numerous high and low molecular substances of bacilli are simultaneously of interest in biotechnologies and have antigenic properties, making them suitable for detection by means of ELISA. From our point of view, the most interesting for biotechnologists are proteins (especially enzymes), polysaccharides and substances with antimicrobial properties. Next, we briefly indicate the possible areas of their application and discuss the probability of the presence of these substances in the environment, taking into account the stage of culture growth.

In the first place, there are numerous exported (i.e. transported from the cytoplasm) proteins of bacilli. Exported proteins are either secreted directly into the growth medium/environment (true exoproteins) or retained at the membrane/cell wall interface (Tjalsma et al., 2000). Both have a signal peptide and the proteins of the second group have a retention signal in addition to an export signal (Tjalsma et al., 2000). As will be discussed below in more detail, bacilli are able release in medium (i) true exoproteins; (ii) proteins that do not have a signal peptide but do have a retention signal, and are likely to be localized at the membrane/cell wall interface; (iii) proteins that do not have a signal peptide or a retention signal, and are likely to be localized in the cytoplasm (Tjalsma et al., 2004). For this reason, in our study we focus on proteins belonging to the three above-mentioned groups.

Biotechnologically, the most interesting bacilli proteins are enzymes. They are key reagents in a large number of biotechnological processes based on the use of live bacilli. For example, live bacilli that have well defined enzymes are used for bioremediation, in particular, to eliminate oil spills and fatty contamination (Borah and Yadav, 2014; Liu, 2015), water purification from nitrites and nitrates (Yao et al., 2013; Zhu et al., 2008), as well as the biodegradation of pesticides (Singh et al., 2013) and polyethylene (Sowmya et al., 2014).

One of the known health protection mechanisms of bacilli used in animal husbandry/poultry is the production of enzymes that increase the absorption of food (Latorre et al., 2015), along with the competitive exclusion of pathogens (La Ragione et al., 2001; La Ragione and Woodward, 2003), an increase in the diversity of the intestinal microbiota (Hu et al., 2014) and the modulation of the immune system (Hosoi et al., 2003; Huttunen et al., 2003; Sebastian and Keerthi, 2014; Urdaci et al., 2004; Xu et al., 2012). In particular, the dietary supplement of the bacillus *B. subtilis* strains with the high activity of the hydrolytic enzymes cellulase (EC 3.2.1.4.) and xylanase (EC 3.2.1.8) reduces the likelihood of necrotic enteritis of broilers. This disease, which is a serious problem in poultry, is caused by *Clostridium perfringens* and *Salmonella enterica* serotype *enteritidis* type A strains, and feed-based grain with a high content of polysaccharides other than starch is a predisposing factor (Xu et al., 2012). The same mechanisms of probiotic action of bacilli including an increase in the degree of absorption of food due to bacterial enzymes are used in the aquaculture of white-legged shrimp (Ninawe and Selvin, 2009).

A number of catabolic enzymes (proteases, chitinases, glucanases, etc.) of *Bacillus* strains promote improvements in plant health and productivity as anti-pest/pathogen weapons. For example, the 42-kDa enzyme beta-1,3-glucan-hydrolase BgIM of *B. circulans* IAM1165 degrades the cell wall of *Aspergillus oryzae*, a known plant pathogen (Asano et al., 2002), and *B. pumilus* ribonuclease possesses antiviral activity against plant RNa-viruses and produces virus-free plants (Sharipova et al., 2015). The serine protease Sepl of *B. firmus* DS-1 has nematicidal activity and degrades multiple intestinal and cuticle-associated proteins of root-knot nematodes, e.g. *Meloidogyne incognita* and *Caenorhabditis elegans* (Geng et al., 2016).

Bacilli are implicated in biofertilization by releasing enzymes for converting nutrients to plant-available form. For example, under conditions of phosphate limitation and in the presence of phytate, extracellular bacterial phytase (an enzyme catalyzing the dephosphorylation of *myo*-inositol hexakisphosphate [phytate], EC 3.1.3.8 for 3-phytase and EC 3.1.8.26 for 6-phytase) can contribute to the plant growth-promoting activity of *B. amyloliquefaciens* FZB (Makarewicz et al., 2006); indeed, enzymatic reaction leads to the loss of the chelate-forming properties of phytate and makes available the nutritionally important minerals Zn^{2+}, Fe^{2+} and Ca^{2+} (Choudhary and Johri, 2009).

Selected *B. subtilis* strains are industrial producers of 60% of the commercial enzymes used in research studies, green chemistry, etc. (Verma and Satyanarayana, 2013; Westers et al., 2004). It is noteworthy that industrial enzymes produced by bacilli are usually prepared directly from fermentation supernatant.

The *B. subtilis* genome sequencing project and the availability of programmes for the identification of signal peptides and transmembrane segments in large collections of protein sequences through worldwide web servers have now made it possible to predict the most likely location of all 4,106 annotated proteins (i.e. the proteome) of this organism (BSORF; http://bacillus.genome.jp/and SubtiList; http://genolist.pasteur.fr/SubtiList/) (Kunst et al., 1997). Informatic analysis of the databases predicted 1,609 cytoplasmic soluble proteins, 1,490 certain integral membrane proteins, 635 putative membrane proteins, 260 secretory proteins and lipoproteins as well as about 100 short peptides such as pheromones and antibiotics (Yamane et al., 2004).

Using a genomic approach (Tjalsma et al., 2000), it was shown that approximately 75% of the proteins of *B. subtilis* lack an amino-terminal signal peptide or membrane anchor, and are likely to be localized in the cytoplasm. A large portion (~21%) of *B. subtilis* proteins remain linked to the membrane as transmembrane proteins (~18%), as lipid-modified proteins (~2.5%) that remain linked to the extracytoplasmic surface of the membrane by their lipid moieties, or as pilin-like structures (<0.1%). A small portion of the putative exported proteins most likely remain specifically attached to the cell wall (~0.5%), whereas most of the remaining exported proteins (~4%) have the potential to pass through the cell wall and are secreted into the environment. Thus, the current survey of the *B. subtilis* secretome (Tjalsma et al., 2000) predicts approximately 300 proteins with the potential to be exported and describes four distinct pathways for protein export from the cytoplasm.

The availability of the genome sequence significantly stimulated the proteome research performed by a combination of high-resolution two-dimensional gel electrophoresis (2-DE), computer-based analysis of the complex protein patterns and high-sensitivity, high-throughput mass spectrometric (MS) analysis. In particular, 693 proteins of *B. subtilis* wild-type strain 168 expressed in exponentially growing cells were detected (Eymann et al., 2004). In exponentially growing cells, most of the cellular components (i.e. amino acids) need to be synthesized because they are not available for uptake in the culture medium (Eymann et al., 2004), and there is no need to eliminate competitors; for this reason, the bacterium "pays little attention" to secretion and the extracellular proteome in the exponential growth phase is poorer compared to the stationary phase in terms of a set of proteins and their concentration. At the late-exponential growth phase, *B. subtilis* 168 produces from 150 to 180 proteins in a culture medium (Hirose et al., 2000; Tjalsma et al., 1999.).

One more approach to studying bacilli proteome is non-gel-based online 2D LC-MS/MS technology. For example, the proteome of *B. thuringiensis* 4.0718 strain at three phases (middle vegetative, early sporulation and late sporulation, designated T_1–T_3, respectively) was analyzed and totals of 918, 703 and 778 proteins were identified, respectively (Huang et al., 2012). The commonly expressed proteins of the T_1–T_2, T_2–T_3 and T_1–T_3 phases were 70, 122, and 111, respectively. The numbers of commonly expressed proteins across the three phases was 308, which comprised 33.55, 43.81 and 39.59% of the proteins in the T_1, T_2 and T_3 phases, respectively. On the other hand, the phase-specific proteins from the T_1–T_3 phases were 429, 203 and 237, which comprised 46.73, 28.88 and 30.46% of the total proteins in the same order.

Below are four examples showing that genome-based predictions reflect the actual composition for approximately 50% of the extracellular proteome. Major problems are currently encountered with the prediction of extracellular proteins lacking signal peptides and extracellular proteins having transmembrane segments/lipoproteins. In a study, extracellular proteins were extracted from a *B. pumilus* SCU11 late-exponential phase culture supernatant and identified by liquid chromatography-tandem mass spectrometry (LC MS/MS)-based shotgun proteomic analysis (Wang et al., 2016). Of the 513 identified extracellular proteins, 413 were classified as intracellular because of their lack of signal peptides. It was assumed that intracellular proteins detected in medium were secreted through an atypical signal peptide or an alternative, yet unrecognized pathway (Wang et al., 2016).

Secondly, Tjalsma et al. (2004) identified 90 extracellular proteins from *B. subtilis* 168 at the early stationary phase in rich medium (LB), and an additional 23 extracellular proteins from the same strain in a very rich (2×SNB) medium or under phosphate starvation conditions. Further signal peptide analysis of 113 extracellular proteins indicated only 54 proteins that were predicted to be secreted (Antelmann et al., 2006).

Thirdly, using a hyper-secreting mutant, approximately 200 extracellular proteins of *B. subtilis* were visualized by two-dimensional gel electrophoresis, of which 82 were identified by mass spectrometry (Antelmann et al., 2001). As expected, these include 41 proteins that have a potential signal peptide with a type I signal peptidase cleavage site, and lack a retention signal. Strikingly, the remaining 41 proteins were previously predicted to be cell associated because of the apparent absence of a signal peptide (22 items) or the presence of a specific cell retention signal in addition to an export signal (19 items) (Antelmann et al., 2001). Models for the release of proteins with transmembrane segments and the shedding of lipoproteins were proposed (Antelmann et al., 2001). Indeed, it was shown that the relative amounts of 60 extracellular proteins are significantly increased during the stationary growth phase as compared with the exponential growth phase (Antelmann et al., 2001), which conforms to the data of Eymann et al. (2004).

Fourthly, it was shown that even though the cell wall proteins of *B. subtilis* are characterized by specific cell wall retention signals, some of these are also components of the extracellular proteome. For example, in a study, five predicted cell wall–binding proteins (viz., LytD[CwlG], YvcE, YwtD, WapA and WprA) were identified among the proteins secreted by *B. subtilis* wild-type strain 168 grown in complete medium (Antelmann et al., 2002). In contrast to the majority of extracellular proteins, cell wall–binding proteins are present in the extracellular proteome during the exponential phase and disappear in the stationary phase. A model for the degradation and release of cell wall–binding proteins in *B. subtilis* was proposed (Antelmann et al., 2002). Briefly, during exponential growth, cell wall–binding proteins accumulate in the medium because of the high cell wall turnover (Doyle et al., 1988). Turnover is defined as any enzymatic process that leads to the excision of fragments from pre-existing insoluble cell wall material. The most popular explanation of turnover is that the cell wall material "spread" as it aged (Pooley, 1976). The rate of turnover of wild-type cells is proportional to the growth rate. During exponential turnover, approximately 50% of the side walls are shed each generation. The onset of the stationary phase is characterized by a substantial decrease in the synthesis of the cell wall material up to sporogenesis (Doyle et al., 1988). The first increase in the rate of peptidoglycan synthesis (and rate of turnover) occurs during the formation of the future germ cell wall (stage III of sporogenesis); the second indicates the synthesis of the cortical endogenous cell wall (stage IV) (Doyle et al., 1988). The final stage of sporulation (the release of spore upon lysis [death] of the mother cell by autolysins) is also characterized by high turnover (Doyle et al., 1988).

Because extracellular proteases are expressed and secreted only at a very low level during the exponential growth phase (see below for more details), these cell wall–binding proteins localized in medium are not degraded. In concert with the reduced cell wall turnover during the stationary phase, the release of the cell wall–binding proteins is also reduced. In addition, the expression and secretion of extracellular proteases are increased in the stationary phase, which leads to the degradation of the cell wall–binding proteins in the medium. In sum, in contrast to the majority of true extracellular proteins, the cell wall–binding proteins disappear from the extracellular proteome during the stationary phase since they are subjected to proteolysis. Epr functions as a "cleaning protease" involved in the degradation of secreted cell wall–binding proteins and cytoplasmic proteins which leave the cell via lysis (Antelmann et al., 2002).

The four examples given above show that predicted intracellular proteins and cell wall–binding proteins exist extensively in the extracellular proteome. These proteins can be isolated from fermentation supernatant to reduce the cost price. One more conclusion is that the composition of the bacilli proteome (especially the extracellular proteome) is growth stage–dependent, and significantly more extracellular proteins are characteristic of the stationary growth phase than the exponential phase. There are exceptions to this "rule". For example, in contrast to the majority of true exoproteins, cell wall–binding proteins disappear from the extracellular proteome during the stationary phase since they are subjected to proteolysis. Indeed, the proteomes of growing cells are different from the proteomes of non-growing cells (late stationary phase) because the stress/starvation that triggers the non-growing state normally induces a huge number of stress/starvation-specific genes (Eymann et al., 2004).

One of the greatest problems for detecting/quantifying original bacilli proteins in medium and for producing many heterologous proteins at commercially significant concentrations is proteolytic degradation, as mentioned above. Bacilli have proteases secreted in the growth medium; their major function is supplying amino acids for growth via degradation of the extracellular proteins. Indeed, bacilli have membrane-/cell wall-associated proteases that are part of the quality control system for exported proteins. Incidentally, proteases represent one of the three largest groups of industrial enzymes and account for about 60% of total worldwide sales of enzymes (Badhe et al., 2016). *B. subtilis* secretes at least eight extracellular proteases at the end of the exponential growth phase; two major extracellular proteases are also produced during the onset of sporulation: the alkaline serine protease subtilisin (AprE) and a neutral protease (NprE) (Park et al., 2004). Another study showed that among the exported proteins of *B. subtilis*, at least 10 proteases degrade many native membrane proteins, lipoproteins and secreted proteins (Krishnappa et al., 2014). One more problem in the detection/quantification of original bacilli proteins in medium is that not all extracellular proteins can be visualized. A considerable part of the extracellular proteome remains silent because particular environmental conditions are necessary for the expression of certain proteins, as shown above.

Interestingly, *B. subtilis* has a much smaller genome (Helgason et al., 2000) than that of *B. cereus* and contains no plasmids. The data are consistent with a study showing that the extracellular proteome of *B. cereus* contains about 500 proteins, which is more than twice the amount found for *B. subtilis* (Antelmann et al., 2001). Moreover, 12 proteins, most of which seem to be completely absent in the proteome of *B. subtilis*, contributed to more than 80% of the total amount of *B. cereus* extracellular proteins. Many of these highly abundant proteins were identified as collagenases, phospholipases, hemolysins, proteases, enterotoxins and flagellins, all of which are potential virulence factors (Gohar et al., 2002). Virulence factors are abundant components of the *B. cereus* exoproteome (Omer et al., 2015). Among them are the three diarrheal enterotoxins (hemolysin BL, non-hemolytic enterotoxin and cytotoxin K) and numerous PlcR-independent virulence factors, such as InhA1 and HlyII (Omer et al., 2015). This is consistent with the study data of Madeira et al. (2015) showing that the *B. cereus* exoproteome is dominated by toxin-related proteins (~35% of the exoproteome, as estimated by spectral count) and degradative enzymes plus adhesins (~35% of the exoproteome), which are recognized as major virulence factors. The other components of the *B. cereus* exoproteome comprise components of the flagellar apparatus (~15% of the exoproteome), as well as proteins that lack export signal sequences (~15% of the exoproteome). These moonlighting proteins, found more abundantly in the cytoplasm, include metabolic enzymes (mainly glycolytic enzymes), translation-related proteins, molecular chaperones and antioxidant enzymes such as catalase, hydroperoxide reductase and superoxide dismutase.

Tjalsma et al. (2004) made a first inventory of the mechanisms applied by *B. subtilis* to (i) secrete proteins directly into its environment and (ii) release membrane/cell wall proteins, lypoproteins and extracellular proteins without typical export signals. Briefly, the main mechanisms for the extracellular accumulation of proteins are (i) protein secretion via the Sec and Tat pathways; (ii) release of membrane proteins by proteolysis (e.g. proteins Yfnl and YflE, YfkN and YhcR, PbpA and HtrA); (iii) release of lipoproteins (at least nine predicted lipoproteins of *B. subtilis* 168) by proteolytic shaving and/or shedding; (iv) release of cell wall proteins by proteolytic shaving and cell wall turnover; (v) release of extracellular proteins without typical export signals (at least 23 items) via (a) flagellar export machinery (thereupon proteins are released from the [damaged] flagella); (b) the holin systems (PBSX prophage-encoded holin XhlB, the SPβ prophage-encoded holins BhlA and BhlB, the holin homologue YqxH); (c) other unidentified export systems; and (d) as a result of cell lysis (Tjalsma et al., 2004; Tjalsma, 2007).

This study provides strong evidence for the non-classical secretion of large amounts of cytoplasmic proteins in the stationary growth phase and excludes cell autolysis (see point [iv] above [Tjalsma et al., 2004]) as the cause of this phenomenon in *B. subtilis* (Yang et al., 2011). Indeed, it was shown that signal-less protein secretion is not significantly mediated by membrane vesicles (membrane shedding) (Yang et al., 2011).

Regardless of the mechanism of secretion, the presence of several cytoplasmic proteins such as GroEL, SodA and enolase with defined physiological cytoplasmic functions in extracellular proteome (Yang et al., 2011) raises a question: what is the significance of their presence in the medium? Firstly, many of these enzymes are involved in energy consumption, and their "dumping" provides a way to reorientate the metabolism for sporulation. Alternatively, these enzymes are able to have "moonlighting" functions.

Polysaccharides of bacilli are the second group of high-molecular substances with antigenic properties (Aithal et al., 2012) that are of interest in biotechnologies. They can be generally classified as intracellular storage polysaccharides, capsular polysaccharides and extracellular bacterial polysaccharides (Schmid and Sieber, 2015). Polysaccharides that are secreted out of the cell are defined as exopolysaccharides; they have a role in cell protection, adhesion and cell-to-cell interactions (Nicolaus et al., 2010). Over recent decades, exopolysaccharides associated with high-value applications have received considerable research attention because of their biocompatibility, biodegradability and both environmental and human compatibility. Due to their rheological properties, exopolysaccharides of bacilli are valued in the dairy industry as viscosifying, stabilizing, gelling and emulsifying agents. Indeed, some exopolysaccharides are gaining increased attention as health-promoting agents due to their anti-inflammatory, anti-oxidative, antimicrobial, anti-tumoral, hepatoprotective, heavy metal toxicity–lowering, blood cholesterol–lowering, antifibrotic, antidiabetic and hypoglycemic activities (Liu et al., 2010; Malick et al., 2017), and their immunostimulatory activity in animals and plants (Jiang et al., 2016; Liu et al., 2010; Sathishkumar et al., 2018; Van Dyk et al., 2012). They are applicable as prebiotics in the formulation of health-promoting functional foods (Malick et al., 2017). For example, purified exopolysaccharides of plant growth-promoting rhizobacterium, including *B. amyloliquefaciens* and *B. polymyxa*, showed the ability to activate multiple plant defence mechanisms against wheat powdery mildew and leaf rust (Haggag et al., 2014). Using a plant-pathogen system (*Arabidopsis* and *Pseudomonas syringae* pathovar *tomato* strain DC3000), it was shown that exopolysaccharide of rhizobacterium *B. cereus* AR156 containing mannose, galactose and glucose in the proportion 70.97: 17.59: 11.45% can act as a micro-associated molecular pattern (MAMP) and facilitate the induction of systemic plant resistance to pathogens through the MAPK- and SA-signalling pathways (Jiang et al., 2016).

Antimicrobial compounds of bacilli are the third group of high-molecular substances of interest in biotechnologies. These substances are directed against competitive microorganisms, thereby generating a selective advantage for their producers (Abriouel et al., 2011). A review contains a full classification of bacilli antimicrobial metabolites based on their biosynthetic pathways and chemical nature, i.e. ribosomal peptides, volatile compounds, polyketides, non-ribosomal peptides and hybrids of the last two groups (Caulier et al., 2019). The most studied are ribosomally synthesized peptides, commonly referred to as bacteriocins and bacteriocin-like inhibitory substances; the last term is used when ribosomal synthesis is only presumed. Most known modern classifications of bacteriocins were proposed in review articles (Abriouel et al., 2011; Cotter et al., 2013; Rea et al., 2011), with a review specifically dedicated to bacilli bacteriocins (Abriouel et al., 2011). The second well-studied class of antimicrobial compounds comprises small antimicrobial peptides synthesized enzymatically by non-ribosomal pathways. They have the properties of biosurfactants (Baruzzi et al., 2011; Sarwar et al., 2018). Biosurfactants are amphiphilic molecules that tend to accumulate at the interfaces between fluid phases with different polarities (e.g. oil–water or air–water), reducing surface and interfacial tensions (Gudiña et al., 2013a). The surfactants of bacilli are lipopeptides, i.e. cyclic molecules consisting of a fatty acid of variable length (hydrophobic moiety) linked to a short peptide chain (hydrophilic moiety) of 7 or 10 amino acids.

Recently, the production of antimicrobial peptides by *Bacillus* strains has been increasingly characterized because they are suitable for health, food, livestock and environmental applications (Abriouel et al., 2011). Compared with conventional antibiotics, which are mostly active against bacteria or fungi, antimicrobial peptides are considered to be antibacterial, antifungal and antiviral drugs that may also be suitable in the treatment of parasitic infections and cancer (e.g. mersacidin) (Baindara et al., 2018; Kaur and Kaur, 2015; Sumi et al., 2015). For example, *B. amyloliquefaciens* subsp. *plantarum* FZB42 (Chowdhury et al., 2015) produces cyclic lipopeptides with antiviral activity (surfactin and fengitsin) and antifungal activity (bacillomycin D). Indeed, the strain produces the ribosomally synthesized peptides plantazolicin (active against *B. anthrax* and nematodes) and amylocyclicin (active against related bacteria). In food applications, bacteriocins of *Bacillus* are superior to bacteriocins of lactic acid bacteria since most of them are stable to heat and pH conditions occurring during food processing, can be degraded by human proteases and show more broad spectra of antimicrobial activities.

The potential livestock applications of antimicrobial peptides are exemplified by BioPlus2B which contains a mixture of *B. licheniformis* and *B. subtilis* strains (Mutus et al., 2006) and spores *B. amyloliquefaciens* CECT 5940 (Ecobiol®, Norel & Nature Nutrition) (Ahmed et al., 2014); both are used in poultry. Environmental applications of antimicrobial peptides are illustrated by the following examples. Lipopeptides iturin A, fenicillin and surfactin A from *B. subtilis* CMB32 can fight against plant blotch caused by the fungus *Colletotrichum gloeosporioides* (Kim et al., 2010;); fengycin of *B. subtilis* NCD-2 is promising for controlling cotton disease caused by the bacterium *Rhizoctonia solani* (Guo et al., 2014); numerous surfactants produced by *Bacillus* spp. are active against various *Fusarium* and *Trichoderma* species (Sarwar et al., 2018); and three homologs of iturin A of *B. subtilis* Z-14 exhibit a broad spectrum of antifungal activities (Zhang et al., 2017). *B. thuringiensis* produces a number of soluble unidentified substances with insecticidal activities (products of *cry4C/Da1*, *cry30Ea*, *cry39A*, *cry40* and *cry54* genes) that are active against *Ceratitis capitata*, a fruit plants pest (Elleuch et al., 2015). The antimicrobial peptide of *B. firmus* strain H 2 O-1 offers potential use as a biocide in the petroleum industry and pipeline-cleaning technologies to inhibit biofilm formation and reduce biocorrosion (Korenblum et al., 2005, 2008).

Biosurfactants are considered a high-value microbial product due to their low toxicity, specificity of action, simplicity of preparation and extensive applicability (Elazzazy et al., 2015; Geys et al., 2014; Rebello et al., 2014; Youssef et al., 2007; Yun et al., 2018). Biosurfactants have many advantages over synthetic ones, including bioavailability and biodegradability, structural diversity, specific activity at extreme salinity, temperatures and pH, and the possibility of production from renewable resources (Gudiña et al., 2015; Pereira et al., 2013). They can be used as moistening agents, dispersing agents, emulsifiers, foaming agents, beneficial food elements and detergents in organic chemicals, pharmaceuticals and cosmetics, beverages and foods, metallurgy, mining, petroleum, petrochemicals, biological control and management (Banat et al., 2000; Perfumo et al., 2010; Vedaraman and Venkatesh, 2011). The role of bacilli biosurfactants in the bioremediation of oil pollution is well known (Karlapudi et al., 2018; Parthipan et al., 2017). Indeed, biosurfactants are potentially useful for therapeutic purposes as cancer therapeutics or as constituents of drug delivery systems (Gudiña et al., 2013b). Some of them demonstrate immunomodulatory properties (Giri et al., 2016). Interestingly, Ab against surfactin were produced and used for antigen detection/quantification (Lin et al., 1994). Indeed, the detection of other biosurfactants by ELISA has already been described (Cao et al., 2009; Satpute et al., 2016).

Obtaining substances of practical interest (enzymes, exopolysaccharides, antimicrobial substances, etc.) in pure form to meet the requirements of researchers and biotechnologists requires the selection of the strain, the optimization of the conditions for its cultivation, the selection of the growth stage ensuring the maximum yield of the target product and the selection and control of the purification procedure. The optimization of biotechnological processes based on the use of live bacteria per se is also based on the detection/quantification of cell components/secreted products that are responsible for the expected result, at least for the choice of strain.

The need to select the growth stage and the medium to ensure maximum yield of the target product is illustrated by the following examples.

Some proteins can be detected in extracellular proteome during exponential growth and disappear from the exoproteome of starved cells and vice versa (Antelmann et al., 2001). Under most culture conditions, *Bacillus* species produce extracellular proteases during post-exponential and stationary phases (Bhunia et al., 2012). Specifically, a cryV-type insecticidal protein of *B. thuringiensis* is expressed early in the stationary phase (Kostichka et al., 1996), while the *wprA* gene of *B. subtilis* 168 encoding a cell wall–associated protease is expressed during exponential growth (Margot and Karamata, 1996), and two major extracellular proteases of *B. subtilis* are produced during the onset of sporulation (Park et al., 2004). Linear growth in the concentration of the protease from a wild strain of *B. subtilis*, which is potentially useful as an additive in the detergent industry, occurs over a period of 6–24 h with maximum yield of the target product at 24 h (Badhe et al., 2016). Hemolysin BL, non-hemolytic enterotoxin and cytotoxin K are secreted by *B. cereus* in growth medium in the early stationary phase (Laouami et al., 2014). It is worth noting that the proteins secreted by *B. subtilis* in the late stationary phase accumulated in the cytoplasm in the early stationary phase (Yang et al., 2011).

A Venn diagram–based comparison of the exoproteins ($n=229$) of *B. cereus* ATCC 14579 identified in the three growth conditions (aerobically and anaerobically grown cells; with low-oxidoreduction potential [ORP] and high-ORP for the latter) was performed (Madeira et al., 2015). The number of "unique" proteins missing at other conditions is 54, 16 and 12, respectively. In anaerobiosis, EntA and EntC, NheA and NheC, and HblL1 are detected as extracellular proteins of *B. cereus* ATCC 14579; in aerobiosis, the list is supplemented by CytK, EntFM, HblB, HblB', HblL1, HblL2, HlyII and NheB (Laouami et al., 2014). In the medium of *B. subtilis* 168 phosphate-starved cells, eight additional extracellular proteins were identified as compared with cells cultivated in Luria–Bertani broth (Tjalsma et al., 2004). Five extracellular proteins of *B. subtilis* strain K-1 were specifically induced by growth in xylan-containing medium (Chu et al., 2000). When *B. subtilis* cells were grown in minimal medium, much lower levels of extracellular proteins were detected than in traditional medium (Tjalsma et al., 2004). Extracellular protease production is strongly influenced by the presence of some easily metabolizable sugars, nitrogen sources and metal ions, and several physical factors, such as aeration, inoculum density, pH and temperature (Bhunia et al., 2012).

Maximum yields of exopolysaccharides of bacilli correspond to the stationary and late stationary phases. For example, the incubation period required for maximum yields of exopolysaccharides is 24 h (*B. thuringiensis* strain 27 ([Wang et al., 2011]); 48 h (selected *Bacillus* strains [Malick et al., 2017]); 48–80 h (*B. velezensis* strain KY471306 [Moghannem et al., 2018]) and 96 h (*B. amyloliquefaciens* BPRGS [Rao et al., 2013] and *B. licheniformis* strain QS5 [Berekaa, 2014]).

A variety of antibiotics and antimicrobials are pumped out of the cell at the stationary phase in an effort to outcompete other nearby microbial species as a last but one (before sporulation) probability to survive (Phillips and Strauch, 2002). For example, *B. subtilis* strain SK.DU.4 produced a bacteriocin-like antimicrobial peptide (5324 Da) with a maximum yield corresponding to 24 h (Baindara et al., 2013). In other studies, the maximum yield of bacilli biosurfactants corresponded to 50 h (Anvari et al., 2015), 72 h (Mehmood et al., 2018), 80 h (Elazzazy et al., 2015), 96 h (Aparna et al., 2012, Varadavenkatesan and Murty, 2013) and 7 days (Nayarisseri et al., 2018).

The above is the basis for developing a methodological approach to investigating the dynamics of the C_{sp} of secreted antigens (per cell of bacilli) to achieve the maximum yield of the target substance in biotechnologies and research studies. The approach detailed in Section 19.3 includes four points and the most important comments on each point are detailed next.

Point (i)—The method of selecting bacteria suitable for the study. It is important to note that the elaborated methodological approach to investigating the dynamics of the C_{sp} (per cell) of bacilli antigens is universal and suitable for any bacilli-specific, commercially available/self-produced Abs, including PAbs and monoclonal Abs to the secreted antigen of interest. Criterion for strain selection is cross-reactivity values calculated for Abs used. They should be not less than 10% to indicate that the selected strains have antigens that interact with these Abs (Kiseleva et al., 2017). In our study, we used $PAb_{anti\text{-}Bc\ 491}$ interacting with genera-specific bacilli antigens (Kiseleva et al., 2017) and six strains of bacilli as an example to simplify the statement of the elaborated methodological approach.

Point (ii)—The principles of preparing CL, CFF and CW/sp specimens; the criteria for the quantitative assessment of CFF and CW/sp. Selected strains were grown in two media, "rich" (MPB, medium 1) and "poor" (MPB 2-fold diluted with distilled water, medium 2) for 6, 12 and 24 h. CL specimens were collected. Cells were disrupted by ultrasound and fractionated by ultracentrifugation to obtain CFF specimens and CW/sp specimens. A quantitative assessment of the CFF specimens was performed by calculating the parameter proportional to the tNA amount (in conventional units). It was calculated as $A_{260\ nm,\ 1\ cm}$ (AU) multiplied by the corresponding CFF volume (mL); the parameter is proportional to the number of vegetative cells (Kiseleva et al., 2017). This method is suitable for comparing the number of vegetative cells of spore-forming bacteria (as an alternative to A_{600} detection [AU], traditionally used for measuring cell concentration in suspension by optical density; http://www.linkedin.com/in/scottvwsutton), since cells are disrupted by ultrasound while spores are not. A quantitative assessment of CW/sp specimens was performed by their weighting. The weight of CW/sp is proportional to the total number of vegetative cells and spores. A comparison of the dynamics of two parameters (tNA amount in CFF and weight of CW/sp) performed for each strain of bacilli and each medium (Table 19.1) allows (i) estimating the ratio of cells to spores to determine the phase of culture growth and (ii) interpreting ELISA data as will be shown below.

It was obvious that the ratio of cells to spores depends on the medium composition and strain (Table 19.1). In particular, CW/sp weight gain in medium 1 occurs mainly due to the increase in the cell number of strain *Bc* 169 (6–24 h) and strains *Bc* 206, *Bm* 179 and *Bs* 25 (6–12 h), which is evidenced by a significant increase in the tNA amount in CFF. On the contrary, strains *Bc* 491, *Bc* 206 and *Bm* 179 (12–24 h) and strain *Bc* 205 (6–24 h) give a CW/sp weight gain in medium 1 due to the increase in spores number, which is evidenced by the absence of an increase in the tNA amount in CFF (Table 19.1). It was also obvious that the decrease in the concentration of nutrients in the medium predictably limits the growth of all tested strains and reduces the duration of the logarithmic growth phase (corresponding to a period of pronounced increase in tNA amount) (Table 19.1). In particular, the log phase in medium 2 is completed after 6 h for each strain; an increase in the tNA amount in the CFF of *Bc* 491, *Bc* 205, *Bc* 206 and *Bs* 25 in the period 12–24 h after a decrease in the values of this parameter in the period 6–12 h can be attributed to spore germination, outgrowth and subsequent vegetative growth (Ceuppens et al., 2012).

Point (iii)—The design of ELISA; the methods of data processing (graphical representations followed by calculations). CFF and CL specimens were used in ELISA tests. The principles of sample preparation, the graphical representation of the ELISA data and subsequent calculations are described in detail in Section 19.3 and shown in Figures 19.1 and 19.3. Here, we draw attention to three important points. Firstly, we previously lyophilized CL specimens and then dissolved them with water to obtain 10-fold concentrates designated above 10×CL, thereby raising the concentration of secreted antigens. Secondly, to obtain the exact C_{sp} (per cell), the number of cells that were the source of the immobilized antigens (CFF antigens) in direct and competitive ELISA was equalized (by diluting all CFF samples to equal $A_{260\ nm,\ 1\ cm}$ values), and the number of cells that were the source of the antigens located in the liquid phase (CL antigens) in competitive ELISA was equalized by recalculation. An alternative way to equalize

CL antigens in competitive ELISA (in fact versus by recalculation) was recognized as acceptable and discussed. Thirdly, we indicated which parameters should be initially selected for optimizing ELISA. The purpose of the selection is to obtain graphs of the "correct form", which is a necessary condition for performing subsequent calculations.

Obviously, the C_{sp} of CFF antigens changes by no more than 30%, except for strain *Bm* 179 growing in medium 2 (Figure 19.2). We assumed that C_{sp} changes in CFF antigens were directly proportional to the rate of their synthesis and inversely proportional to the rate of their secretion into the external environment and "integration" in the membrane/cell wall. Generally, the changes in the C_{sp} of CL antigens were more expressed than the changes in the C_{sp} of CFF antigens, especially in strains related to groups 4 and 5 (medium 1) and 5* and 8* (medium 2) (Table 19.2). However, strains related to group 6 (medium 1) showed changes in the C_{sp} of CL antigens that did not exceed 20% (Table 19.2).

Point (iv)—Principles of interpreting the results. Combinations of changes in the two parameters—tNA in CFF and the weight of CW/sp—were used to discuss the causes of an increase/decrease in the C_{sp} of CL antigens (Table 19.3), which are represented by symbols below the histograms in Figure 19.2.

In accordance with the data in Table 19.1 and Table 19.3, in medium 1 the increase in the C_{sp} of CL antigens of strains *Bc* 491 and *Bc* 169 in the period 6–12 h (Figure 19.2) occurs as a result of the release of cytoplasmic proteins and cell wall proteins due to cell wall turnover; and in the antigens of strain *Bc* 205 and strain *Bc* 206 in the period 12–24 h as a result of the secretion of true exoproteins, exopolysaccharides and antimicrobial proteins (Figure 19.2). The medium 2 increase in the C_{sp} of CL antigens of strain *Bc* 205 during the period 6–12 h occurs as a result of the transition of the cell components into the solution during spores formation; antigens of the same strain in the period 12–24 h appear in the CL as a result of spore germination and the beginning of the second exponential growth phase which can be very short due to depleted environmental resources (Figure 19.2).

The decrease in the values of the C_{sp} of CL antigens (Figure 19.2) can be explained by two reasons: (i) the increase in the number of cells occurs faster than the increase in the concentration of the antigens in the CL, hence the value of C_{sp} calculated per cell decreases; (ii) the antigens present in the CL lose their native structure (e.g. due to proteolysis, enzymatic hydrolysis of carbohydrate antigens or aggregation), and therefore they do not interact with $PAb_{anti\text{-}Bc\ 491}$ in ELISA.

Systematization of the results of the study of the dynamics of the C_{sp} of intracellular antigens and secreted antigens per cell of six bacilli strains grown in two media (Table 19.3) suggests that the type of change in the C_{sp} depends on the strain and culture medium, is unpredictable and must be determined individually for each object. In particular, when isolating extracellular antigens, some of the bacilli should be cultivated in the MPB (medium 1) for no more than 6 h (e.g. *Bs* 25 [Figure 19.2]), and others for at least 24 h (e.g. *Bc* 206 [Figure 19.2]) to achieve the maximum yield of the target substance.

19.5 Conclusion

In our study, the methodological approach to investigating the dynamics of the specific concentration (C_{sp}) (per bacterial cell) of antigens secreted by bacilli has been elaborated. ELISA was selected as a method for detecting/quantifying secreted biopolymers with antigenic properties because it is specific, sensitive, fast, simple, gives reproducible results, employs stable reagents and uses inexpensive equipment. The approach includes (i) the method of selecting bacteria suitable for ELISA with the available Ab; (ii) the principles of preparing CL, CFF and CW/sp specimens; the criteria for the quantitative assessment of CFF and CW/sp; (iii) the design of ELISA; the methods of data processing (graphical representations followed by calculations); and (iv) the principles of interpreting the results.

We exemplified our approach using rabbit PAb against genera-specific antigens of bacilli (Kiseleva et al., 2017) and six strains of bacilli including *Bc* 491 (strain-immunogen) and other preselected bacilli strains ($n = 5$) satisfying the criterion for the production of antigens that are identical/similar to antigens of *Bc* 491. It was shown that the tested bacilli strains contain intracellular and secreted antigens. The C_{sp} (per cell) of both types of antigens varies depending on the bacillus strain, the stage of culture growth and the composition of the medium. The dynamics of the synthesis and the secretion of antigens as well as their distribution in the cell-environment system are unpredictable and must be determined on a case-by-case basis for each object. In general, during the observation period, the C_{sp} of secreted antigens changed to a much greater extent than the C_{sp} of antigens inside the bacterial cell.

It is important to note that the elaborated methodological approach to investigating the dynamics of the C_{sp} (per cell) of antigens secreted by bacilli is universal and suitable with other bacilli-specific Abs, including monoclonal Abs to the secreted antigen of interest.

The advantage of the approach to monitoring the biotechnological process via the detection/quantification of secreted biopolymers by ELISA is obvious in the case of (i) a bacterium that is unavailable without interrupting the process (e.g. the assessment of the need to replace the carrier with bacteria in wastewater treatment); (ii) a procedure for detecting a secreted end-product is labour-intensive and time-consuming and requires expensive equipment (for example, the production of a secreted enzyme, to assess the yield and activity of which a multistage purification procedure is required); and (iii) a medium that has a complex composition and contains other microorganisms (e.g. biogas production, biodegradation of food and agricultural waste). The important condition relating to point (ii) is that either the product has antigenic properties or the relationship between the yield of the product and the concentration of the specific antigen is pre-established.

The data will find application in biotechnology and research studies.

REFERENCES

Abriouel H, Franz CM, Omar NB, Gálvez A (2011) Diversity and applications of *Bacillus* bacteriocins. *FEMS Microbiol Rev* 35(1):201–232. doi:10.1111/j.1574-6976.2010.00244.x

Ahemad M, Kibret M (2014) Mechanisms and applications of plant growth promoting rhizobacteria: current perspective. *J King Saud Univ – Sci* 26:1–20. doi:10.1016/j.jksus.2013.05.001

Ahmed ST, Islam MdM, Mun H-S, Sim H-J, Kim Y-J, Yang C-J (2014) Effects of *Bacillus amyloliquefaciens* as a probiotic strain on growth performance, cecal microflora, and fecal noxious gas emissions of broiler chickens. *Poult Sci* 93:1963–1971. doi:10.3382/ps.2013-03718

Aithal A, Sharma A, Joshi S, Raghava GPS, Varshney GC (2012) PolysacDB: a database of microbial polysaccharide antigens and their antibodies. *PLoS ONE* 7(4):e34613. doi:10.1371/journal.pone.0034613

Anonymous (2010) *Thermo Scientific Pierce Antibody Production and Purification Technical Handbook*, Thermo Fisher Scientific Inc., 77p, https://tools.thermofisher.com/content/sfs/brochures/1601975-Antibody-Production-Purification-Guide.pdf

Antelmann H, Tjalsma H, Voigt B, Ohlmeier S, Bron S, van Dijl JM, Hecker M (2001) A proteomic view on genome-based signal peptide predictions. *Genome Res* 11(9):1484–1502. doi:10.1101/gr.182801

Antelmann H, Van Dijl JM, Bron S, Hecker M (2006) Proteomic survey through secretome of *Bacillus subtilis*. *Methods Biochem Anal* 49:179–208. doi:10.1002/0471973165.ch12

Antelmann H, Yamamoto H, Sekiguchi J, Hecker M (2002) Stabilization of cell wall proteins in *Bacillus subtilis*: a proteomic approach. *Proteomics* 2(5):591–602. doi:10.1002/1615-9861(200205)2:5<591::AID-PROT591>3.0.CO;2-8

Anvari S, Hajfarajollah H, Mokhtarani B, Noghabi KA (2015) Physiochemical and thermodynamic characterization of lipopeptide biosurfactant secreted by *Bacillus tequilensis* HK01. *RSC Adv* 5:91836–91845. doi:10.1039/C5RA17275F

Aparna A, Srinikethan G, Hegde S (2012) Isolation, screening and production of biosurfactant by *Bacillus clausii* 5B. *Res Biotechnol* 3(2):49–56.

Asano T, Taki J, Yamamoto M, Aono R (2002) Cloning and structural analysis of bglM gene coding for the fungal cell wall-lytic beta-1,3-glucan-hydrolase BglM of *Bacillus circulans* IAM1165. *Biosci Biotechnol Biochem* 66(6):1246–1255. doi:10.1271/bbb.66.1246

Badhe P, Joshi M, Adivarekar R (2016) Optimized production of extracellular proteases by *Bacillus subtilis* from degraded abattoir waste. *J BioSci Biotechnol* 5(1):29–36. http://www.jbb.uni-plovdiv.bg

Baindara P, Korpole S, Grover V (2018) Bacteriocins: perspective for the development of novel anticancer drugs. *Appl Microbiol Biotechnol* 102(24):10393–10408. doi:10.1007/s00253-018-9420-8

Baindara P, Mandal SM, Chawla N, Kumar Singh P, Pinnaka AK, Korpole S (2013) Characterization of two antimicrobial peptides produced by a halotolerant *Bacillus subtilis* strain SK.DU.4 isolated from a rhizosphere soil sample. *AMB Express* 3(1):2. doi:10.1186/2191-0855-3-2

Banat IM, Makkar RS, Cameotra SS (2000) Potential commercial applications of microbial surfactants. *Appl Microbiol Biotechnol* 53(5):495–508. doi:10.1007/s002530051648

Baruzzi F, Quintieri L, Morea M, Caputo L (2011) Antimicrobial compounds produced by *Bacillus* spp. and applications in food. In: Méndez-Vilas A (ed) *Science against Microbial Pathogens: Communicating Current Research and Technological Advances*, Formatex, Badajoz. pp 1102–1111. http://www.formatex.info/microbiology3/book/1102-1111.pdf

Bennett SD, Walsh KA, Gould LH (2013) Foodborne disease outbreaks caused by *Bacillus cereus, Clostridium perfringens*, and *Staphylococcus aureus* – United States, 1998–2008. *Clin Infect Dis* 57(3):425–433. doi:10.1093/cid/cit244

Berekaa MM (2014) Improved exopolysaccharide production by *Bacillus licheniformis* strain-QS5 and application of statistical experimental design. *Int J Curr Microbiol Appl Sci* 3(4):876–886.

Bhunia B, Basak B, Dey A (2012) A review on production of serine alkaline protease by *Bacillus* spp. *J Biochem Technol* 3(4):448–457.

Borah D, Yadav RNS (2014) Optimization of BH medium for efficient biodegradation of diesel, crude oil and used engine oil by a newly isolated *Bacillus cereus* strain DRDU1 from an automobile engine. *Biotechnology* 13(4):181–185. doi:10.3923/biotech2014.181.185

Cao XH, Liao ZY, Wang CL, Yang WY, Lu MF (2009) Evaluation of a lipopeptide biosurfactant from *Bacillus natto* TK-1 as a potential source of anti-adhesive, antimicrobial and antitumor activities. *Braz J Microbiol* 40(2):373–379. doi:10.1590/S1517-838220090002000030

Caulier S, Nannan C, Gillis A, Licciardi F, Bragard C, Mahillon J (2019) Overview of the antimicrobial compounds produced by members of the *Bacillus subtilis* group. *Front Microbiol* 10:302. doi:10.3389/fmicb.2019.00302

Ceuppens S, Uyttendaele M, Drieskens K, Heyndrickx M, Rajkovic A, Boon N, Van de Wiele T (2012) Survival and germination of *Bacillus cereus* spores without outgrowth or enterotoxin production during in vitro simulation of gastrointestinal transit. *Appl Environ Microbiol* 78(21):7698–7705. doi:10.1128/AEM.02142-12

Charni N, Perissol C, le Petit J, Rugani N (2000) Production and characterization of monoclonal antibodies against vegetative cells of *Bacillus cereus*. *Appl Environ Microbiol* 66(5):2278–2281. doi:10.1128/AEM.66.5.2278-2281.2000

Choudhary DK, Johri BN (2009) Interactions of *Bacillus* spp. and plants – with special reference to induced systemic resistance (ISR). *Microbiological Research* 164:493–513. doi:10.1016/j.micres.2008.08.007

Chowdhury SP, Hartmann A, Gao X, Borriss R (2015) Biocontrol mechanism by root-associated *Bacillus amyloliquefaciens* FZB42 – a review. *Front Microbiol* 6:780. doi:10.3389/fmicb.2015.00780

Chu PW, Yap MN, Wu CY, Huang CM, Pan FM, Tseng MJ, Chen ST (2000) A proteomic analysis of secreted proteins from xylan-induced *Bacillus* sp. strain K-1. *Electrophoresis* 21(9):1740–1745. doi:10.1002/(SICI)1522-2683(20000501)21:9<1740::AID-ELPS1740>3.0.CO;2-N

Cotter PD, Ross RP, Hill C (2013) Bacteriocins — a viable alternative to antibiotics? *Nat Rev Microbiol* 11(2):95–105. doi:10.1038/nrmicro2937

Doyle RJ, Chaloupka J, Vinter V (1988) Turnover of cell walls in microorganisms. *Microbiol Rev* 52(4):554–567

Elazzazy AM, Abdelmoneim TS, Almaghrabi OA (2015) Isolation and characterization of biosurfactant production under extreme environmental conditions by alkali-halo-thermophilic bacteria from Saudi Arabia. *Saudi J Biol Sci* 22(4):466–475. doi:10.1016/j.sjbs.2014.11.018

Elleuch J, Tounsi S, Ben Hassen NB, Lacoix MN, Chandre F, Jaoua S, Zghal RZ (2015) Characterisation of novel *Bacillus thuringiensis* isolates against *Aedes aegypti* (Diptera: Culicidae) and *Ceratitis capitata* (Diptera: Tephridae). *J Invertebr Pathol* 124:90–97. doi:10.1016/j.jip.2014.11.005

Eymann C, Dreisbach A, Albrecht D, Bernhardt J, Becher D, Gentner S, Tam LT, Büttner K, Buurman G, Scharf C, Venz S, Völker U, Hecker M (2004) A comprehensive proteome map of growing *Bacillus subtilis* cells. *Proteomics* 4:2849–2876. doi:10.1002/pmic.200400907

Fritze D (2004) Taxonomy of the genus *Bacillus* and related genera: the aerobic endospore-forming bacteria. *Phytopathology* 94(11):1245–1248. doi:10.1094/PHYTO.2004.94.11.1245

Geng C, Nie X, Tang Z, Zhang Y, Lin J, Sun M, Peng D (2016) A novel serine protease, Sep1, from *Bacillus firmus* DS-1 has nematicidal activity and degrades multiple intestinal-associated nematode proteins. *Sci Rep* 6: art. 25012. doi:10.1038/srep25012

Geys R, Soetaert W, Van Bogaert I (2014) Biotechnological opportunities in biosurfactant production. *Curr Opin Biotechnol* 30:66–72. doi:10.1016/j.copbio.2014.06.002

Giri SS, Sen SS, Jun JW, Sukumaran V, Park SC (2016) Role of *Bacillus subtilis* VSG4-derived biosurfactant in mediating immune responses in *Labeo rohita. Fish Shellfish Immunol* 54:220–229. doi:10.1016/j.fsi.2016.04.004

Gohar M, Okstad OA, Gilois N, Sanchis V, Kolsto AB, Lereclus D (2002) Two-dimensional electrophoresis analysis of the extracellular proteome of *Bacillus cereus* reveals the importance of the PlcR regulon. *Proteomics* 2:784–791. doi:10.1002/1615-9861(200206)2:6<784::AID-PROT784>3.0.CO;2-R

Gudiña EJ, Fernandes EC, Rodrigues AI, Teixeira JA, Rodrigues LR (2015) Biosurfactant production by *Bacillus subtilis* using corn steep liquor as culture medium. *Front Microbiol* 6:59. doi:10.3389/fmicb.2015.00059

Gudiña EJ, Pereira JF, Costa R, Coutinho JA, Teixeira JA, Rodrigues LR (2013a) Biosurfactant-producing and oil-degrading *Bacillus subtilis* strains enhance oil recovery in laboratory sand-pack columns. *J Hazard Mater* 261:106–113. doi:10.1016/j.jhazmat.2013.06.071

Gudiña EJ, Rangarajan V, Sen R, Rodrigues LR (2013b) Potential therapeutic applications of biosurfactants. *Trends Pharmacol Sci* 34(12):667–675. doi:10.1016/j.tips.2013.10.002

Guo Q, Dong W, Li S, Lu X, Wang P, Zhang X, Wang Y, Ma P (2014) Fengycin produced by *Bacillus subtilis* NCD-2 plays a major role in biocontrol of cotton seedling damping-off disease. *Microbiol Res* 169(7–8):533–540. doi:10.1016/j.micres.2013.12.001

Haggag WWM, Hussein MM, Mehanna HM, El-Moneim D (2014) Bacteria polysaccharides elicit resistance of wheat against some biotic and abiotic stress. *Int J Pharm Sci Rev Res* 29(2):292–298, Article number 50.

Haji-Ghassemi O, Blackler RJ, Martin Young N, Evans SV (2015) Antibody recognition of carbohydrate epitopes. *Glycobiology* 25(9):920–952. doi:10.1093/glycob/cwv037

Harlow E, Lane D (1988) *Antibodies: A Laboratory Manual.* Cold Spring Harbor Laboratory, Cold Spring Harbor, p. 726.

Heimburg-Molinaro J, Rittenhouse-Olson K (2009) Development and characterization of antibodies to carbohydrate antigens. *Methods Mol Biol* 534:341–357. doi:10.1007/978-1-59745-022-5_24

Helgason E, Okstad OA, Caugant DA, Johansen HA, Fouet A, Mock M, Hegna I, Kolsto A-B (2000) *Bacillus anthracis*, *Bacillus cereus*, and *Bacillus thuringiensis*—one species on the basis of genetic evidence. *Appl Environ Microbiol* 66(6):2627–2630. doi:10.1128/AEM.66.6.2627-2630.2000

Hirose I, Sano K, Shioda I, Kumano M, Nakamura K, Yamane K (2000) Proteome analysis of *Bacillus subtilis* extracellular proteins: a two-dimensional protein electrophoretic study. *Microbiology* 146(Pt 1):65–75. doi:10.1099/00221287-146-1-65

Hong HA, Ducle H, Cutting SM (2005) The use of bacterial spore formers as probiotics. *FEMS Microbiol Rev* 29(4):813–835. doi:10.1016/j.femsre.2004.12.001

Hosoi T, Hirose R, Saegusa S, Ametani A, Kiuchi K, Kaminogawa S (2003) Cytokine responses of human intestinal epithelial-like Caco-2 cells to the nonpathogenic bacterium *Bacillus subtilis* (natto). *Int J Food Microbiol* 82(3):255–264. doi:10.1016/S0168-1605(02)00311-2

Huang S, Ding X, Sun Y, Yang Q, Xiao X, Cao Z, Xia L (2012) Proteomic analysis of *Bacillus thuringiensis* at different growth phases by using an automated online two-dimensional liquid chromatography-tandem mass spectrometry strategy. *Appl Environ Microbiol* 78(15):5270–5279. doi:10.1128/AEM.00424-12

Huttunen K, Hyvärinen A, Nevalainen A, Komulainen H, Hirvonen MR (2003) Production of proinflammatory mediators by indoor air bacteria and fungal spores in mouse and human cell lines. *Environ Health Perspect* 111(1):85–92. doi:10.1289/ehp.5478

Hu Y, Dun Y, Li S, Zhao S, Peng N, Liang Y (2014) Effects of *Bacillus subtilis* KN-42 on growth performance, diarrhea and faecal bacterial flora of weaned piglets. *Asian-Australas J Anim Sci* 27(8):1131–1140. doi:10.5713/ajas.2013.13737

Hu Z, Leppla SH, Li B, Elkins CA (2014) Antibodies specific for nucleic acids and applications in genomic detection and clinical diagnostics. *Expert Rev Mol Diagn* 14(7):895–916. doi:10.1586/14737159.2014.931810

Jiang C-H, Fan Z-H, Xie P, Guo J-H (2016) *Bacillus cereus* AR156 extracellular polysaccharides served as a novel micro-associated molecular pattern to induced systemic immunity to Pst DC3000 in arabidopsis. *Front Microbiol* 7:664. doi:10.3389/fmicb.2016.00664

Karlapudi AP, Venkateswarulu TC, Tammineedi J, Kanumuri L, Ravuru BK, Dirisala VR, Kodali VP (2018) Role of biosurfactants in bioremediation of oil pollution—a review. *Petroleum* 4(3):241–249. doi:10.1016/j.petlm.2018.03.007

Kaur S, Kaur S (2015) Bacteriocins as potential anticancer agents. *Front Pharmacol* 6:272. doi:10.3389/fphar.2015.00272

Kim PI, Ryu J, Kim YH, Chi YT (2010) Production of biosurfactant lipopeptides Iturin A, fengycin and surfactin A from *Bacillus subtilis* CMB32 for control of *Colletotrichum gloeosporioides. J Microbiol Biotechnol* 20(1):138–145. doi:10.4014/jmb.0905.05007

Kiseleva E, Mikhailopulo K, Ladutska A, Novik G (2017) Methodological approach to the study of dynamics of specific concentration of cell wall antigens per cell of *Bacillus* species and examples of its application. *EuroBiotech Journal* 1(1):71–77. doi:10.24190/ISSN2564-615X/2017/01.10

Korenblum E, Sebastián GV, Paiva MM, Coutinho CM, Magalhães FC, Peyton BM, Seldin L (2008) Action of antimicrobial substances produced by different oil reservoir *Bacillus* strains against biofilm formation. *Appl Microbiol Biotechnol* 79(1):97–103. doi:10.1007/s00253-008-1401-x

Korenblum E, Der Weid I, Santos AL, Rosado AS, Sebastián GV, Coutinho CM, Magalhães FC, Paiva MM, Seldin L (2005) Production of antimicrobial substances by *Bacillus subtilis* LFE-1, *B. firmus* HO-1 and *B. licheniformis* T6-5 isolated from an oil reservoir in Brazil. *J Appl Microbiol* 98(3):667–675. doi:10.1111/j.1365-2672.2004.02518.x

Kostichka K, Warren GW, Mullins M, Mullins AD, Craig JA, Koziel MG, Estruch JJ (1996) Cloning of a cryV-type insecticidal protein gene from *Bacillus thuringiensis*: the cryV-encoded protein is expressed early in stationary phase. *J Bacteriol* 178(7):2141–2144

Krishnappa L, Monteferrante CG, Neef J, Dreisbach A, van Dijl JM (2014) Degradation of extracytoplasmic catalysts for protein folding in *Bacillus subtilis. Appl Environ Microbiol* 80(4):1463–1468. doi:10.1128/AEM.02799-13

Kunst F, Ogasawara N, Moszer I and 148 other authors (1997) The complete genome sequence of the Gram-positive bacterium *Bacillus subtilis. Nature* 390:249–256. doi:10.1038/36786

Lin SC, Minton MA, Sharma MM, Georgiou G (1994) Structural and immunological characterization of a biosurfactant produced by *Bacillus licheniformis* JF-2. *Appl Environ Microbiol* 60(1):31–38.

Laouami S, Clair G, Armengaud J, Duport C (2014) Proteomic evidence for rex regulation of metabolism in toxin-producing *Bacillus cereus* ATCC 14579. *PLoS ONE* 9(9):e107354. doi:10.1371/journal.pone.0107354

La Ragione RM, Woodward MJ (2003) Competitive exclusion by *Bacillus subtilis* spore of *Salmonella enterica* serotype enteritidis and *Clostridium perfringens* in young chickens. *Vet Microbiol* 94(3):245–256.

La Ragione RM, Casula G, Cutting SM, Woodward MJ (2001) *Bacillus subtilis* spores competitively exclude *Escherichia coli* 070:K80 in poultry. *Vet Microbiol* 79(2):133–142.

Latorre JD, Hernandez-Velasco X, Kuttappan VA, Wolfenden RE, Vicente JL, Wolfenden AD, Bielke LR, Prado-Rebolledo OF, Morales E, Hargis BM, Tellez G (2015) Selection of *Bacillus* spp. for cellulase and xylanase production as direct-fed microbials to reduce digesta viscosity and *Clostridium perfringens* proliferation using an *in vitro* digestive model in different poultry diets. *Front Vet Sci* 2:art. 25. doi:10.3389/fvets.2015.00025

Li J, Xu Y, Jin L, Li X (2015) Effects of a probiotic mixture (*Bacillus subtilis* YB-1 and *Bacillus cereus* YB-2) on disease resistance and non-specific immunity of sea cucumber, *Apostichopus japonicus* (Selenka). *Aqua Res* 46(12):3008–3019. doi:10.1111/are.12453

Liu C, Lu J, Lu L, Liu Y, Wang F, Xiao M (2010) Isolation, structural characterization and immunological activity of an exopolysaccharide produced by *Bacillus licheniformis* 8-37-0-1. *Bioresour Technol* 101(14):5528–5533. doi:10.1016/j.biortech.2010.01.151

Liu JF (2015) Chemical structure, property and potential applications of biosurfactants produced by *Bacillus subtilis* in petroleum recovery and spill mitigation. *Int J Mol Sci* 16(3):4814–4837. doi:10.3390/ijms16034814

Lodemann U, Lorenz BM, Weyrauch KD, Martens H (2008) Effects of *Bacillus cereus* var. *toyoi* as probiotic feed supplement on intestinal transport and barrier function in piglets. *Arch Anim Nutr* 62(2):87–106. doi:10.1080/17450390801912068

Madeira J-P, Alpha-Bazin B, Armengaud J, Duport C (2015) Time dynamics of the *Bacillus cereus* exoproteome are shaped by cellular oxidation. *Front Microbiol* 6:342. doi:10.3389/fmicb.2015.00342

Makarewicz O, Dubrac S, Msadek T, Borriss R (2006) Dual role of the PhoP-P response regulator: *Bacillus amyloliquefaciens* FZB45 phytase gene transcription is directed by positive and negative interaction with the *phy C* promoter. *J Bacteriol* 188(19):6953–6965. doi:10.1128/JB.00681-06

Malick A, Khodaei N, Benkerroum N, Karboune S (2017) Production of exopolysaccharides by selected *Bacillus* strains: optimization of media composition to maximize the yield and structural characterization. *Int J Biol Macromol* 102:539–549. doi:10.1016/j.ijbiomac.2017.03.151

Margot P, Karamata D (1996) The *wprA* gene of *Bacillus subtilis* 168, expressed during exponential growth, encodes a cell-wall-associated protease. *Microbiology* 142:3437–3444.

Mehmood N, Iqbal M, Asgher M (2018) Production and optimization of biosurfactants from *Bacillus subtilis*. *EC Microbiol* 14(5):211–224.

Moghannem SAM, Farag MMS, Shehab AM, Azab MS (2018) Exopolysaccharide production from *Bacillus velezensis* KY471306 using statistical experimental design. *Braz J Microbiol* 49(3):452–462. doi:10.1016/j.bjm.2017.05.012

Mutus R, Kocabagli N, Alp M, Acar N, Eren M, Gezen SS (2006) The effect of dietary probiotic supplementation on tibial bone characteristics and strength in broilers. *Poultr Sci* 85(9):1621–1625. doi:10.1093/ps/85.9.1621

Nayarisseri A, Singh P, Singh SK (2018) Screening, isolation and characterization of biosurfactant producing *Bacillus subtilis* strain ANSKLAB03. *Bioinformation* 14(6):304–314. doi:10.6026/97320630014304

Nicolaus B, Kambourova M, Oner ET (2010) Exopolysaccharides from extremophiles: from fundamentals to biotechnology. *Environ Technol* 31(10):1145–1158. doi:10.1080/09593330903552094.

Ninawe AS, Selvin J (2009) Probiotics in shrimp aquaculture: avenues and challenges. *Crit Rev Microbiol* 35(1):43–66. doi:10.1080/10408410802667202

Niu D, Wang X, Wang Y, Song X, Wang J, Guo J, Zhao H (2016) *Bacillus cereus* AR156 activates PAMP-triggered immunity and induces a systemic acquired resistance through a NPR1-and SA-dependent signaling pathway. *Biochem Biophys Res Commun* 469(1):120–125. doi:10.1016/j.bbrc.2015.11.081

Omer H, Alpha-Bazin B, Brunet J-L, Armengaud J, Duport C (2015) Proteomics identifies *Bacillus cereus* EntD as a pivotal protein for the production of numerous virulence factors. *Front Microbiol* 6:1004. doi:10.3389/fmicb.2015.01004

Park CH, Lee SJ, Lee SG, Lee WS, Byun SM (2004) Hetero- and autoprocessing of the extracellular metalloprotease (Mpr) in *Bacillus subtilis*. *J Bacteriol* 186(19):6457–6464. doi:10.1128/JB.186.19.6457-6464.2004

Parthipan P, Preetham E, Machuca LL, Rahman PKSM, Murugan K, Rajasekar A (2017) Biosurfactant and degradative enzymes mediated crude oil degradation by bacterium *Bacillus subtilis* A1. *Front Microbiol* 8:193. doi:10.3389/fmicb.2017.00193

Perchat S, Buisson C, Chaufaux J, Sanchis V, Lereclus D, Gohar M (2005) *Bacillus cereus* produces several nonproteinaceous insecticidal exotoxins. *J Invertebr Pathol* 90(2):131–133. doi:10.1016/j.jip.2005.08.002

Pereira JFB, Gudiña EJ, Costa R, Vitorino R, Teixeira JA, Coutinho JAP, Rodrigues LR (2013) Optimization and characterization of biosurfactant production by *Bacillus subtilis* isolates towards microbial enhanced oil recovery applications. *Fuel* 111:259–268. doi:10.1016/j.fuel.2013.04.040

Perfumo A, Smyth TJP, Marchant R, Banat IM (2010) Production and roles of biosurfactants and bioemulsifiers in accessing hydrophobic substrates. In: Timmis KN (ed.) *Handbook of Hydrocarbon and Lipid Microbiology*. Springer, Berlin, pp. 1501–1512. doi:10.1007/978-3-540-77587-4_103

Perosa F, Carbone R, Ferrone S, Dammacco F (1990) Purification of human-immunoglobulins by sequential precipitation with caprylic-acid and ammonium-sulfate. *J Immunol Methods* 128(1):9–16.

Phillips ZEV, Strauch MA (2002) *Bacillus subtilis* sporulation and stationary phase gene expression. *Cell Mol Life Sci* 59(3):392–402. doi:10.1007/s00018-002-8431-9

Plescia OJ, Braun W (1967) Nucleic acids as antigens. *Adv Immunol* 6:231–252. doi:10.1016/S0065-2776(08)60523-4

Pooley HM (1976) Turnover and spreading of old wall during surface growth of *Bacillus subtilis*. *J Bacteriol* 125(3):1127–1138

Rao BP, Kasirajan S, Sekaran RCHG, Mandal AB (2013) Characterization of exopolysaccharide from *Bacillus amyloliquefaciens* BPRGS for its bioflocculant activity. *Int J Sci Eng Res* 4(10):1696–1704.

Rea MC, Ross RP, Cotter PD, Hill C (2011) Classification of bacteriocins from Gram-positive bacteria. In: Drider D, Rebuffat S (eds) *Prokaryotic Antimicrobial Peptides: From Genes to Application*. Springer, New York, pp. 29–47. doi:10.1007/978-1-4419-7692-5_3

Rebello S, Asok AK, Mundayoor S, Jisha MS (2014) Surfactants: toxicity, remediation and green surfactants. *Environ Chem Lett* 12:275–287. doi:10.1007/s10311-014-0466-2

Romeiro RS, Filho RL, Macagnan D, Garcia FAO, Silva HSA (2010) Evidence that the biocontrol agent *Bacillus cereus* synthesizes protein that can elicit increased resistance of tomato leaves to *Corynespora cassiicola*. *Trop Plant Pathol* 35(1):11–15. doi:10.1590/S1982-56762010000100002

Sarwar A, Brader G, Corretto E, Aleti G, Abaidullah M, Sessitsch A, Hafeez FY (2018) Qualitative analysis of biosurfactants from *Bacillus* species exhibiting antifungal activity. *PLoS ONE* 13(6):e0198107. doi:10.1371/journal.pone.0198107

Sathishkumar R, Ananthan G, Senthil L, Moovendhan M, Arun J (2018) Structural characterization and anticancer activity of extracellular polysaccharides from ascidian symbiotic bacterium *Bacillus thuringiensis*. *Carbohyd Polymers* 190:113–120. doi:10.1016/j.carbpol.2018.02.047

Satpute SK, Kulkarni GR, Banpurkar AG, Banat IM, Mone NS, Patil RH, Cameotra SS (2016) Biosurfactant/s from *Lactobacilli* species: properties, challenges and potential biomedical applications. *J Basic Microbiol* 56(11):1140–1158. doi:10.1002/jobm.201600143

Schmid J, Sieber V (2015) Enzymatic transformations involved in the biosynthesis of microbial exopolysaccharides based on the assembly of repeat units. *Chembiochem* 16(8):1141–1147. doi:10.1002/cbic.201500035

Sebastian AP, Keerthi TR (2014) Immunomodulatory effect of probiotic strain *Bacillus subtilis* MBTU PBBMI spores in Balb/C mice. *Int J Eng Tech Res* 2(11):258–260.

Senesi S, Ghelardi E (2010) Production, secretion and biological activity of *Bacillus cereus* enterotoxins. *Toxins (Basel)* 2(7):1690–1703. doi:10.3390/toxins2071690

Shafi J, Tian H, Ji M (2017) *Bacillus* species as versatile weapons for plant pathogens: a review. *Biotechnol Biotechnol Equip* 31(3):446–459. doi:10.1080/13102818.2017.1286950

Sharipova M, Rockstroh A, Balaban N, Mardanova A, Toymentseva A, Tikhonova A, Vologin S, Stashevsky Z (2015) Antiviral effect of ribonuclease from *Bacillus pumilus* against phytopathogenic RNA-viruses. *Agri Sci* 6:1357–1366. doi:10.4236/as.2015.611130

Singh B, Kaur J, Singh K (2013) Bioremediation of malathion in soil by mixed *Bacillus* culture. *Adv Biosci Biotechnol* 4(5):674–678. doi:10.4236/abb.2013.45088

Sowmya HVR, Krishnappa M, Thippeswamy B (2014) Biodegradation of polyethylene by *Bacillus cereus*. *Advances in Polymer Science and Technology: An Int J* 4(2):28–32.http://www.urpjournals.com

Stabnikova O, Wang J-Y, Ivanov V (2010) Value-added biotechnological products from organic wastes. In: Vang LK, Ivanov V, Tay J-H, Hung Y-T (eds) *Environmental Biotechnology*, Springer, New York, pp. 344–394. doi:10.1007/978-1-4419-7692-5_3

Staudt C, Horn H, Hempel DC, Neu TR (2004) Volumetric measurements of bacterial cells and extracellular polymeric substance glycoconjugates in biofilms. *Biotechnol Bioeng* 88(5):585–592. doi:10.1002/bit.20241

Sumi CD, Yang BW, Yeo I-C, Hahm YT (2015) Antimicrobial peptides of the genus *Bacillus*: a new era for antibiotics. *Can J Microbiol* 61(2):93–103. doi:10.1139/cjm-2014-0613

Tjalsma H (2007) Feature-based reappraisal of the *Bacillus subtilis* exoproteome. *Proteomics* 7:73–81. doi:10.1002/pmic.200600520

Tjalsma H, Antelmann H, Jongbloed JD, Braun PG, Darmon E, Dorenbos R, Dubois J-YF, Westers HA, Zanen G, Quax WJ, Kuipers OP, Bron S, Hecker M, van Dijl JM (2004) Proteomics of protein secretion by *Bacillus subtilis*: separating the "secrets" of the secretome. *Microbiol Mol Biol Rev* 68(2):207–233. doi:10.1128/MMBR.68.2.207-233.2004

Tjalsma H, Bolhuis A, Jongbloed JDH, Bron S, van Dijl JM (2000) Signal peptide-dependent protein transport in *Bacillus subtilis*: a genome-based survey of the secretome. *Microbiol Mol Biol Rev* 64(3):515–547 .doi:10.1128/MMBR.64.3.515-547.2000

Tjalsma H, Kontinen VP, Pragai Z, Wu H, Meima R, Venema G, Bron S, Sarvas M, van Dijl JM (1999) The role of lipoprotein processing by signal peptidase II in the Gram-positive eubacterium *Bacillus subtilis*: signal peptidease II is required for the efficient secretion of a-amylase, a non-lipoprotein. *J Biol Chem* 274(3):1698–1707.

Urdaci MC, Bressollier P, Pinchuk I (2004) *Bacillus clausii* probiotic strains: antimicrobial and immunomodulatory activities. *J Clin Gastroenterol* 38(6):S86–S90. doi:10.1097/01.mcg.0000128925.06662.69

Van Dyk JS, Low Ah Kee N, Frost CL, Pletschke B (2012) Extracellular polysaccharide production in *Bacillus licheniformis* SVD1 and its immunomodulatory effect. *Bioresources* 7(4):4976–4993. doi:10.15376/biores.7.4.

Varadavenkatesan T, Murty VR (2013) Production of a lipopeptide biosurfactant by a novel *Bacillus* sp. and its applicability to enhanced oil recovery. *ISRN Microbiol*, 2013:Article ID 621519, 8p. doi:10.1155/2013/621519

Vedaraman N, Venkatesh N (2011) Production of surfactin by *Bacillus subtilis* MTCC 2423 from waste frying oils. *Brazil J Chem Eng* 28(2):175–180. doi:10.1590/S0104-66322011000200001

Verma D, Satyanarayana T (2013) Production of cellulase-free xylanase by the recombinant *Bacillus subtilis* and its applicability in paper pulp bleaching. *Biotechnol Prog* 29(6):1441–1447. doi:10.1002/btpr.1826

Vimala PP, Mathew L (2016) Biodegradation of polyethylene using *Bacillus subtilis*. *Procedia Technol* 24:232–239. doi:10.1016/j.protcy.2016.05.031

Wang C, Yu S, Song T, He T, Shao H, Wang H (2016) Extracellular proteome profiling of *Bacillus pumilus* SCU11 producing alkaline protease for dehairing. *J Microbiol Biotechnol* 26(11):1993–2005.doi:10.4014/jmb.1602.02042

Wang ZR, Sheng JP, Tian XL, Wu TT, Liu WZ, Shen L (2011) Optimization of the production of exopolysaccharides by *Bacillus thuringiensis* 27 in sand biological soil crusts and its bioflocculant activity. *Afr J Microbio Res* 5(16):2359–2366. doi:10.1111/j.1472-765X.2009.02777.x

Westers L, Westers H, Quax WJ (2004) *Bacillus subtilis* as cell factory for pharmaceutical proteins: a biotechnological approach to optimize the host organism. *Biochimica et Biophysica Acta* 1694(1–3):299–310. doi:10.1016/j.bbamcr.2004.02.011

Williams LD, Burdock GA, Jiménez G, Castillo M (2009) Literature review on the safety of Toyocerin, a non-toxigenic and non-pathogenic *Bacillus cereus* var. *toyoi* preparation. *Regul Toxicol Pharmacol* 55(2):236–246. doi:10.1016/j.yrtph.2009.07.009

Xu X, Huang Q, Mao Y, Cui Z, Li Y, Huang Y, Rajput IR, Yu D, Li W (2012) Immunomodulatory effects of *Bacillus subtilis* (natto) B4 spore on murine macrophages. *Microbiol Immunol* 56(12):817–824. doi:10.1111/j.1348-0421.2012.00508.x

Yamane K, Bunai K, Kakeshita H (2004) Protein traffic for secretion and related machinery of *Bacillus subtilis*. *Biosci Biotechnol Biochem* 68(10):2007–2023. doi:10.1271/bbb.68.2007

Yang C-K, Ewis HE, Zhang XZ, Lu C-D, Hu H-J, Pan Y, Abdelal AT, Tai PC (2011) Nonclassical protein secretion by *Bacillus subtilis* in the stationary phase is not due to cell lysis. *J Bacteriol* 193(20):5607–5615. doi:10.1128/JB.05897-11

Yao RL, Qiu LN, Zhang WW, Cai H (2013) Isolation and characteristics of heterotrophic nitrification-aerobic denitrification bacterium, *Bacillus cereus* X7 at high salinity. *Adv Mater Res* 864–867:111–114. doi:10.4028/www.scientific.net/AMR.864-867.111

Youssef N, Simpson DR, Duncan KE, McInerney MJ, Folmsbee M, Fincher T, Knapp RM (2007) In situ biosurfactant production by *Bacillus* strains injected into a limestone petroleum reservoir. *Appl Environ Microbiol* 73(4):1239–1247. doi:10.1128/AEM.02264-06

Yun J-H, Cho Dae-Hyun, Lee B, Kim Hee-Sik, Chang Yong Keun (2018) Application of biosurfactant from *Bacillus subtilis* C9 for controlling cladoceran grazers in algal cultivation systems. *Sci Rep* 8: article number 5365. doi:10.1038/s41598-018-23535-8

Zhang D, Gao T, Li H, Lei B, Zhu B (2017) Identification of antifungal substances secreted by *Bacillus subtilis* Z-14 that suppress *Gaeumannomyces graminis* var. *tritici*. *Biocontrol Sci Technol* 27(2):237–251. doi:10.1080/09583157.2016.1275522

Zhu G, Peng Y, Li B, Guo J, Yang Q, Wang S (2008) Biological removal of nitrogen from wastewater. *Rev Environ Contam T* 192:159–195

Zhu L, He J, Cao X, Huang K, Luo Y, Xu W (2016) Development of a double-antibody sandwich ELISA for rapid detection of *Bacillus cereus* in food. *Sci Rep* 6: article number 16092. doi:10.1038/srep16092

20

Increasing the Tolerance of Water-Deficit Stress in Plants with Integrated Application of Organic and Inorganic Fertilizers

Solmaz Kazemalilou, Behnam Asgari Lajayer and Mansour Ghorbanpour

CONTENTS

20.1 Introduction

Water-deficit stress is one of the important abiotic factors affecting the physiological and biochemical process in plants and limiting crop production (Ghassemi et al., 2018; Khoshmanzar et al., 2019). Water stress affects plants when the amount of water absorption is lower than crop evapotranspiration (Andrew et al., 2000; Islam et al., 2011). Drought stress increases the generation of reactive oxygen species (ROS) due to the accumulation of energy in stressed plants which use less light energy through photosynthetic carbon fixation (Asada, 2006). In many countries, water and nutrients are the two most important factors influencing crop production and both interact with each other (Ahmadian et al., 2011). There is a positive and strong relationship between moisture and the nutrients available to plants. Nutrient uptake by plants increases with increasing soil moisture content due to changes in the availability of the nutrients (Misra and Tyler, 1999). Plant nutrition has a critical role in increasing plant tolerance to drought stress.

Therefore, an important mechanism to moderate drought damage is integrated plant nutrient management (IPNM) (Haefele and Bouman, 2009). Under drought stress, a plant's nutrient uptake is reduced, thus its growth and leaf area are decreased. Additionally, as a result of a reduction in its light absorption and photosynthetic capacity, plant growth and yield are reduced (Hassani, 2005). Drought stress causes the stomata to close, reducing the plant height, its specific area and number of leaves. Also, the reduction in water potential due to drought stress induces stomatal closures, which in turn can reduce the photosynthesis rate, leaf growth and yield (Anjum et al., 2011). Bahar et al. (2009) reported that there is a positive correlation between stomatal conductance and crop yield. Soil water is also necessary for maintaining the activity and growth of soil microorganisms and the development of roots and shoots (Stockdale et al., 2002). Drought stress reduces the efficiency of physiological processes such as protein synthesis and photosynthesis, which inhibits the activity of many enzymes and accelerates leaf senescence (El-Refaee et al., 2008). Drought stress during flowering and grain-filling periods decreases the number of seeds and the total seed yield by decreasing vegetative growth. The occurrence of drought stress at the pollination stage decreases photosynthesis, reduces assimilate, increases the production of sterile pollen grain and, ultimately, decreases the number of seeds per plant (El Sabagh et al., 2018). By increasing external osmolality, drought stress causes water to exit from the interior to the exterior, which reduces turgor pressure in the cell and cytoplasmic volume (Zhang et al., 2009). Faced with drought stress, some plants use different biochemical, physiological and anatomical strategies. However, many plants are unable to cope with stress conditions through internal mechanisms. Therefore, humans can help plants to overcome stress by using a variety of water-absorbing materials and organic and chemical fertilizers (Ali et al., 2007). A water deficit disturbs the nutritional balance in plants (Marschner, 1995). It is well known that an adequate and balanced supply of nutrients such as phosphorus is more important for drought tolerance in plants because phosphorus increases root growth, provides access to subsoil moisture and maintains nutrient uptake during drought stress (Marschner, 1995; Huang et al., 1997).

Additionally, phosphorus availability helps in stomatal closure, photosynthesis, higher cell membrane turgidity, high leaf water potential and water-use efficiency and increases the assimilation of nitrate under drought stress conditions (Waraich et al., 2011). However, over-applying phosphorus and nitrogen fertilizers in agricultural soils increases production costs and is harmful to the environment because of potential water pollution and eutrophication (Lewis et al., 2011). Continuous use of mineral fertilizers can destroy the soil's chemical and physical characteristics and reduce biological activity and soil health (Mahajan et al., 2008). The application of organic fertilizers increases plant growth by improving the available macro and micronutrients and the soil's physical (such as soil structure), chemical (such as soil organic matter [OM] and buffering capacity) and biological properties (Dejene and Lemlem, 2012). It is clear that adding organic matter to soil improves the soil structure, porosity and texture, and increases the soil's water-holding capacity (Stockdale et al., 2002; Asgari Lajayer et al., 2019). Ibrahim and El-Samad (2009) concluded that water use by plants decreased as soil organic fertilizer increased. However, organic fertilizers alone cannot meet crop nutrient demands over large areas because they contain small amounts of available nutrients (Jobe, 2003). Thus, it is critical to find an effective approach for increasing nutrient uptake such as the integrated application of organic, chemical and biofertilizers (Bagayoko et al., 2000; Akdeniz et al., 2006; Zarabi et al., 2011), which can effectively alleviate the adverse effects of drought using a number of mechanisms. IPNM enhances soil productivity through improving nutrient cycling and decreasing fertilizer loss rate, enabling nutrients to be available in sufficient and balanced quantities for plants and sustainable productivity. IPNM is a more effective way as it is less costly and is an accepted practice for many crops (Conacher and Conacher, 1998). One of the beneficial effects of the integrated application of organic and chemical fertilizers is a reduction in the deficiency of some secondary and micronutrients that have continuously received NPK fertilizers over the long term without adding any organic fertilizer (Chand et al., 2006).

20.2 Soil Fertility

Adequate and balanced nutrient application can be efficient in obtaining optimal yield. The impact on nutrients added to the soil through fertilizers and other sources is as follows:

- They are absorbed by the plants and removed from the soil through plant harvesting
- They remain available in the soil and are used by the next crop (residual)
- They are fixed and thereby released in nutrient cycling over a long period
- They are removed from the soil system in the form of volatilization, denitrification, leaching and soil and water erosion (Simonne and Hochmuth, 2003)

The use of legumes improves nitrogen concentration in the soil, in others words the use of legumes alone or with crop rotation will increase soil fertility. Both organic and chemical fertilizers are suitable for replenishing soil nutrients. The nutrient replenishment strategy is only valid in cases of naturally good soil fertility for sustaining high yields and is not applicable on inherently poor soils, since fertilization on the basis of removals only can further deplete such soils.

20.2.1 Properties of Good Fertile Soil

- Optimum soil pH within a practical range
- Adequate organic matter, nutrient availability, high water storage capacity and the activities of soil organisms
- Optimal soil structure, good drainage and permeability without a compact layer
- Water availability, especially during drought stress
- Neutralization of toxic substances, especially in highly acidic soils

20.3 IPNM in Both Drought Stress and Non-Stress Conditions

Moradi et al. (2018) concluded that the integrated application of cow manure + NPK improved the soil's physical structure and increased its water retention capacity and chemical balance, thereby providing, to some extent, the water and nutrient requirements of black cumin (*Nigella sativa* L.) in both drought stress and non-stress conditions. Heidarian et al. (2018) concluded that the integrated application of farmyard manure (FYM) and urea enhanced sugar beet yield and reduced the contamination of the soil and water. Randhawa et al. (2017) reported that the impact of the combined use of NPK fertilizers with FYM is an effective strategy to increase drought tolerance in maize. An increase in plant growth with the integrated application of NPK and FYM might be due to the increase in the soil's biological properties and water-holding capacity (Sohu et al., 2015). Khalid (2016) concluded that the integrated use of phosphorus (160 kg ha^{-1}) + poultry manure + seed inoculation with biofertilizers could increase maize yield in semiarid regions. Khalid also mentioned that the increase in biomass yield due to the better growth and development of the plant is as a result of the adequate and balanced availability of nutrients, with plants absorbing more water and nutrients throughout the growing period. Nyalemegbe (2009) reported that the combined use of poultry manure and chemical fertilizers increased rice yield. Chemura (2014) concluded that the application of organic fertilizers increases plant growth under low water conditions but the use of chemical fertilizers improves plant growth under optimum water levels. Khodaei-Joghan et al. (2018) concluded that the application of organic fertilizers in combination with zeolite is an effective method for reducing chemical fertilizer consumption and achieving sustainable agriculture. Chemura (2014) indicated that the application of organic manure improves the growth and yield of coffee plants under low soil moisture conditions, but the use of chemical fertilizers increases plant growth under optimal moisture conditions. Ibrahim et al. (2008) and Faraji et al. (2015) found that the integrated application of manures and chemical fertilizers is more effective than pure fertilizers. Azab (2016) reported that the use of biofertilizers alone under drought stress conditions is efficient but under optimum irrigation conditions integrated application with half of the recommended NPK fertilizers is recommended. Amanolahi-Baharvand et al. (2014) claimed that the integrated use of biofertilizers and chemical fertilizers increased maize yield under water stress conditions. Akbari et al. (2009) reported that a combination of chemical and biological fertilizers significantly increased the seed yield and harvest index of sunflowers. According to Crecchio et al. (2001), the balanced nutrition management of organic and chemical fertilizers increases nutrient mobility and improves the contribution of nutrients to vegetative and reproductive organs.

20.4 Components of IPNM Systems

The combined use of organic and inorganic sources of nutrients is required for sustainable crop production. The application of organic fertilizers alone or in combination with chemical fertilizers increases crop nutrient uptake and maintains soil fertility (Talashiker and Rinal, 1986). Hussain et al. (1988) reported that the application of organic fertilizers increased the efficiency of chemical fertilizers. In fact, organic fertilizers had beneficial effects on crop production through the soil's improved physical properties and fertility (Singh and Sarivastore, 1971). The different components of IPNM are shown in Figure 20.1.

20.4.1 Inorganic or Chemical Fertilizers

The constant use of chemical fertilizers in agriculture causes nutritional imbalances in the soil solution which have adverse effects on plant yield and soil health. Moreover, the excessive use of chemical fertilizers causes negative environmental impacts and public health issues such as soil, water and air pollution (Ghost and Bhat, 1998). In addition, such use may also have a negative effect on the activity of soil microorganisms and mycorrhizal fungi (Plencette et al., 2005). The excessive use of chemical fertilizers can kill beneficial soil microbes, damage the soil's natural fertility and destroy the crop's biological resistance to pests and diseases. The consumption of organic fertilizers can reduce the use of

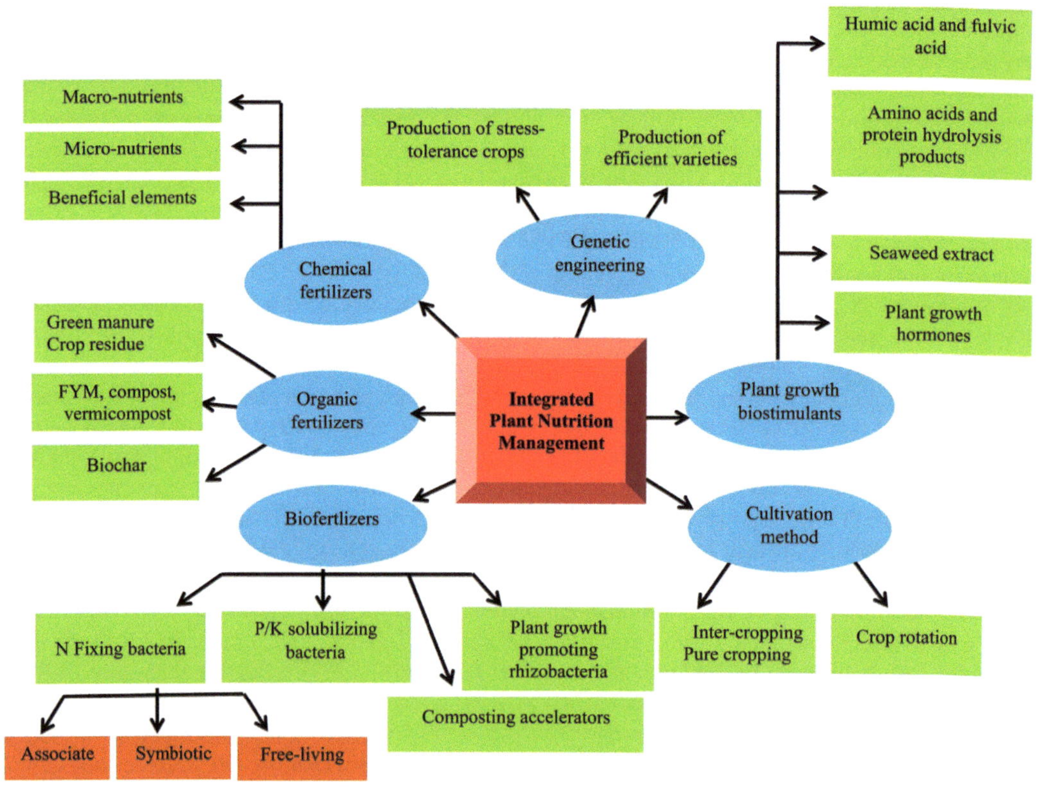

FIGURE 20.1 The components of integrated plant nutrition management (IPNM).

chemical fertilizers and reduce negative environmental effects (Shirkhani et al., 2016). Murillo-Amador et al. (2015) reported that the partial or complete replacement of chemical fertilizers may be achieved using organic fertilizers and biofertilizers, which are economical for farmers and produce healthy and quality crops.

20.4.2 Beneficial Elements

Beneficial elements such as sodium, nickel and silicon are needed in very small amounts and are often supplied by the soil. Silicon has positive effects on plant growth under drought stress. The application of silicon enhances the production of antioxidants, decreases the production of ROS, reduces the risk of photo-oxidative damage and maintains the integrity of the chloroplast membrane integrity, thus drought tolerance in crops is increased (Waraich et al., 2011).

20.4.3 Organic or Natural Fertilizers

Several types of organic manures originate from plants, animals and humans, such as agricultural residues, green manure, farmyard manure, composts and vermicomposts, sewage sludge (SS) and animal, urban and industrial wastes (slow-release fertilizers). The composting process often increases their value as sources of plant nutrients. Crop residues of the legume have a low C:N ratio and are rich in plant nutrients, which are rapidly mineralized and available to the plant. Organic fertilizers contain variable amounts of macro and micronutrients, vitamins, enzymes and plant growth regulators. Organic fertilizers can be used as soil amendments (soil aeration and root penetrability) and nutrients, which enhance nutrient uptake by the plant and stimulate plant growth. It has been found that the use of organic fertilizers such as vermicompost alone or in combination with other organic or chemical fertilizers increases the growth and yield of different plants (Simsek-Ersahin, 2011; Singh et al., 2011). Vermicomposts can increase soil porosity, drainage, water-holding capacity and soil microbial activity. Humic acids obtained from vermicompost decomposition can increase the tolerance of plants to drought stress because they can stimulate protein synthesis and enzyme activity in plants (Muscolo et al., 2007). The gradual release of nitrogen and other nutrients from organic fertilizers increases photosynthesis and assimilate production during the plant reproductive period. The continuous availability of nutrients by organic fertilizers decreases the leaching of mineral elements into groundwater. Additionally, through their gradual release of nutrients, organic fertilizers increase the humus content of the soil, which provides an excellent substrate for plant growth. Organic fertilizers also retain water in the soil for a long time (Elhabet, 2018). Auge (2001) reported an increase in root colonization with mycorrhizal fungi in a soil with low fertility under drought stress conditions. Therefore, the uptake of phosphorus and potassium by solubilization and the expansion of the rhizosphere are increased due to the presence of fungal hyphae. The use of organic fertilizers, especially in areas with low moisture and soil fertility, can increase the crop yield (Saidia and Mrema, 2017). An integrated application of mycorrhizal fungi and nitrogen-fixing bacteria increased drought tolerance in legume plants. Also, the consumption of *Pseudomonas* spp. under drought stress increased the content of antioxidants and photosynthetic pigments in plants (Liddycoat et al., 2009).

Organic fertilizers are important for retaining soil organic matter and providing nutrients for plants, but may not be sufficient for balanced plant nutrition and for achieving optimal yield (Stockdale et al., 2002). Efficient fertilizer management and balanced plant nutrition are important factors that increase plant growth and production under drought conditions (Amanullah et al., 2016). Soil organic matter also plays an important role in the soil's water retention capacity by increasing the number of micropores and macropores in the soil. Soil moisture is important in the decomposition of organic matter and the release of available nutrients from organic and mineral fertilizers to the plants (El-Kader et al., 2010). The decomposition of organic fertilizers improves the soil's fertility and water retention, which increases nutrient availability (El-Refaee, 2006). The content of phosphorus in manures and plant residues is normally insufficient to meet the needs of plants. However, the decomposition of organic matter may produce organic acids that can dissolve phosphate rock (Palm, 1995). Thus, in soils with deficiency of organic matter such as Iran, the combined use of organic and chemical fertilizers is recommended.

Iron chlorosis is one of the most deficient micronutrients to correct in the field. It should be consumed with the plant nutritional needs simultaneously. Iron fertilizer spraying is carried out either as inorganic salts or through iron chelates with EDTA (useful only in slightly acid soil) and EDDHA (a wide pH range of 4–9). The concentration of ferrous sulphate used as a foliar spray is 0.5–2% and should be repeated several times at 10–15 day intervals. Ferrous sulphate ($FeSO_4.7H_2O$) is the most common iron fertilizer for soil application. However, soil application is not recommended due to the rapid oxidation and immobilization of ferrous ions to ferric ions in the soil. The efficiency of ferrous sulphate increases when it is mixed with an organic fertilizer and applied to the soil (Shand, 2007).

20.4.4 Green Manure

Green manure as a source of organic matter and plant nutrient returns whatever it has absorbed from the soil and thus, it also aids in the recycling of nutrients from the lower depths to the topsoil. The net benefit is only in the case of biological nitrogen fixation (a net N input). Green manure can be either planted in situ and added to the field or planted outside (ex situ) and ploughed into the soil to serve as manure. The physical and chemical benefit of using organic manures in the soil is that they increase the crop yield (Stone and Elioff, 1998). Legume plants are suitable for green manure in practical farming. In general, legume plants require only a very low starter dose of nitrogen at the beginning of growth. This is because these plants are able to obtain large amounts of nitrogen through nitrogen fixation under optimal nodulation conditions. When incorporated into the soil before planting rice, a good green manure crop such as *Sesbania* (non-grain legumes) can add 50–60 kg N ha^{-1}.

When selecting a green manure crop, the most important characteristics are (i) rapid growth and production of large amounts of biomass per unit time/area; (ii) local plant adaptation; (iii) resistance to environmental and soil stresses such as drought; (iv) plant tolerance to pests; and (v) rapid decomposition.

In most cases, the growth of a legume plant in a rotation has a significant effect on the nitrogen nutrition of the plants that follow. Mulch application in areas with a shortage of organic matter and hot climates can reduce soil water loss and soil temperature, in addition to providing protection for the soil and as a nutrient source.

20.4.5 Application Time of Organic Fertilizer

Organic fertilizers can be applied to all soils and almost all crops. In addition to supplying nutrients, the application of organic fertilizers increases the soil structure, water storage, cation exchange capacity (CEC) and biological activity. Organic manures should be consumed when these are most effective. Organic fertilizers should be applied several weeks before planting to convert into soluble and plant available forms in the soil, distributed uniformly over the field and quickly ploughed into the soil to prevent ammonia emissions. However, legume green manures can be incorporated into puddled soil only a few days before transplanting the main plant because of their rapid rate of degradation. Ground phosphate rock and elemental sulphur must be applied before sowing (Shand, 2007).

20.5 Biofertilizers

Biofertilizer is a term widely used for products containing dead or living microorganisms such as bacteria (atmospheric nitrogen-fixing or phosphate-solubilizing bacteria [PSB]), fungi, actinomycetes, protozoa, viruses and algae alone or in combination. These organisms are present in all soils but not in equal amounts. The number of bacteria depends on the amount of soil oxygen, the presence or absence of calcium and phosphorus, the amount of trace elements, the amount of organic matter, the presence of toxic materials, the soil's physical and chemical properties and the plant species (Shirkhani et al., 2016). Soil microorganisms are essential for decomposing organic matter in the soil (improving the availability of macro and micronutrients to plants) and increasing plant tolerance to different environmental stresses such as drought (Amanullah, 2015). Among soil microorganisms, only nitrogen-fixing bacteriaprovide an additional source of nutrient Other biofertilizers do not contain nutrients in themselves and simply solubilize nutrients that are already present in the soil. The use of beneficial microorganisms (biofertilizers) can reduce environmental pollution by decreasing the application of chemical fertilizers. Biofertilizers are eco-friendly, non-hazardous and non-toxic (Adhikary, 2012).

Biofertilizers are classified into six categories:

- Nitrogen-fixing bacteria include *Rhizobium*, *Azotobacter*, *Azospirillum*, *Clostridium*, *Acetobacter* and *Azolla* which convert molecular N_2 in the atmosphere into ammonia through biological processes (Choudhury and Kennedy, 2004).
- Composting accelerators: cellulolytic and lignolytic microorganisms.
- Plant growth promoting rhizobacteria (PGPR): *Pseudomonas* species increases plant growth and performance by mobilizing nutrients in soils and providing phytohormones and anti-metabolites. PGPR decompose the soil organic matter which helps in the mineralization process, increasing nutrient availability and plant yield (Bhattacharyya and Jha, 2012).
- Potassium-mobilizing/solubilizing biofertilizer (KMB).
- Sulphur-oxidizing biofertilizer (SOB).
- Phosphate-solubilizing bacteria: include *Bacillus*, *Pseudomonas* and *Aspergillus*. However, mycorrhizae are nutrient-mobilizing fungi. These organisms secrete organic acids that decrease the pH of the surroundings rhizosphere, and dissolve insoluble phosphate compounds to their soluble forms (Chen et al., 2006). Mycorrhizae fungi can also enhance the mobility of nutrients such as P, Zn, Cu and S through greater root exploration in the soil by the mycorrhizal hyphae. Mycorrhiza transfers nutrients from the fungus to the root system and vesicles store phosphorus as phospholipids. Mycorrhiza has a symbiotic association with the majority of agricultural crops.

20.5.1 Mycorrhiza

Mycorrhiza is a mobilizer of soil nutrients and promotes root growth to take up plant nutrients. Mycorrhizal fungal spores are used to produce inoculum. The relationship between plant roots and mycorrhizae improves the ability of plants to explore the soil and take up nutrients by increasing the root surface area. Mycorrhizae have special structures called vesicles and arbuscules. The arbuscules assist in the transfer of nutrients from the fungus to the plant roots, and the vesicles, which have a sac-like structure, store phosphorus as phospholipids. The survival of mycorrhizae depends on the host plant, cultivation practices, soil fertility and biological and environmental factors; maximum root colonization and sporulation occur in poor soils (Jakobsen et al., 2005). Under drought stress conditions, mycorrhizae act as a lateral root system, increasing the surface area and providing water and nutrients from long distances to the roots of plants (Graf and Frei, 2013). The widespread use of biofertilizers such as mycorrhiza as an eco-friendly method reduces production costs and increases crop yield compared to the continuous application of chemical fertilizers alone (Sharma et al., 2013). Abdel-Razzak et al. (2013) reported that the combined use of mycorrhiza inoculation with a chemical fertilizer (superphosphate) increased the plant root growth and yield of sweet potato compared to a single fertilizer application.

Precautions before applying biofertilizers:

- Never apply biofertilizers to seed treated with fungicides, pesticides, insecticides and chemical fertilizers
- Biofertilizers should be kept away from direct sunlight
- Biofertilizers should be stored in a dry place at room temperature not less than 0°C and not more than 35°C
- Any used solution should be discarded (Hari and Perumal, 2010)

20.6 Abscisic Acid (ABA)

It is clear that under stress conditions, the use of plant growth hormones such as abscisic acid can increase plant resistance. For example, Abdelaal et al. (2017) reported that the use of abscisic acid significantly increased the activity of antioxidant enzymes and grain yield in maize under drought stress conditions.

20.7 Integrated Water and Nutrient Management

Plants need water and nutrients for their survival. Water is not only essential for the growth of plants but it is also the medium through which elements are transported and absorbed by the plant roots. An adequate water supply improves the nutritional status of plants (through uptake and the availability and efficiency of nutrients for plant growth) and sufficient nutrient storage saves water. Under drought stress conditions, the rate of root development is limited in soils with low fertility and the plant cannot use deeper moist horizons in the soil. In most soils, the topsoil has the highest nutrient content, but this horizon dries out. Although some plants have the ability to absorb some water from the subsoil, this may not be sufficient for optimal growth. Phosphorus plays an important role in the growth and development of the root system. If phosphorus is applied to the soil before planting, the plant develops a deep and vigorous root system before drought occurs. Even if the surface soil becomes dry, such roots can absorb water and nutrients from deeper layers. Thus, the application of phosphorus can be an insurance against drought stress. On the other hand, soil biological activities, the breakdown and degradation of organic matter and the mineralization of organic forms of nutrients such as nitrogen into mineral forms are reduced under drought stress conditions (Shand, 2007).

20.8 Balanced Plant Nutrition

The nitrogen uptake by crops is increased with balanced plant nutrition, and less nitrate is leached below the root development area for groundwater contamination or denitrification. IPNM identifies and utilizes all available nutrient sources in the field and helps reduce the loss of nutrients to the environment.

20.8.1 Water and Mobility of Nutrients

In dry soils, cations are strongly retained by soil colloids and are not easily exchangeable, thus they are less available to plants. In addition, when the volume of the soil solution decreases, the amount of less-soluble nutrients such as phosphorus decreases and plants are unable to absorb them at the required levels. For nutrients such as nitrogen that only move a small distance in the soil solution to reach the root surface by mass flow require an adequate soil moisture. The movement of nutrients, such as phosphorus and potassium, by diffusion in the soil solution is important and requires sufficient soil moisture. In addition, when plants have sufficient water, the uptake of nutrients by plants and the transport of absorbed nutrients within the plants are also increased. However, under insufficient water availability, metabolic activities and biomass production and yield in plants are restricted. It is also important that all production inputs and management factors are adequate during the period of plant growth. Excess water can cause nutrient losses, and water shortage at a critical stage can limit growth and yield (Shand, 2007).

20.8.2 Tolerance to Water Stress by Plant Nutrition

Optimizing plant nutrition makes plants resistant to drought stress. The role of some nutrients will be discussed in detail. Well-balanced nutrients enable plants to explore a larger volume of soil to obtain water and nutrients. When exposed to drought stress, plants may also suffer from nutrient stress due to the close association between nutrients and water availability.

20.8.2.1 Nitrogen

Hermanson et al. (2000) found a direct relationship between nitrogen availability and crop yield. Plants overfertilized with nitrogen have more vegetative growth relative to their roots. This results in a rapid loss of water from the plant canopy, which discharges soil water more rapidly than the plant can receive balanced nutrition. Such plants are very susceptible to drought stress. In legume plants, nitrate reductase activity protein synthesis and nitrogen fixation reduce under drought stress conditions (Shand, 2007).

20.8.2.2 Phosphorous

Phosphorus significantly increases plant root growth, promotes flowering and accelerates seed germination, and should be used at planting. In the absence of phosphorus, plants are unable to access water from deeper soil layers due to poor root systems. Thus, such plants are more susceptible to drought stress than plants with sufficient phosphorus that have strong root systems. The availability of phosphorus in the soil is often insufficient due to low availability of phosphorus in the soil or slow mobility towards the roots (especially in soils with high $CaCO_3$ and soil pH). Phosphorus is essential for root growth and, as deep rooting can be crucial for crop survival under drought conditions, an adequate phosphorus supply is important beyond its actual role as a nutrient (Khalid, 2016). Phosphorus increases drought tolerance in crops through an increase in stomatal conductance, photosynthesis, cell membrane stability and water relationships (Waraich et al., 2011).

20.8.2.3 Potassium

Potassium plays an osmotic role in plants and helps plant tissue to retain water. The movement of potassium in and outside of the guard cells around the stomata is responsible for the opening and closing of these cells on the plant leaves, which helps to reduce moisture loss when the plant is exposed to drought stress. When plants are deficient in potassium, the stomata cannot perform their tasks properly, and water losses from the plant may be high. The application of potassium increases plant resistance to drought stress and can help the plant maintain a higher growth rate (Marschner, 1995). Potassium deficiency may increase the production of ROS and decrease photosynthesis, because potassium is required to maintain the photosynthetic CO_2 fixation. An adequate potassium supply is essential to reduce transpiration losses from plants. However, most arid soils in dryland agriculture have sufficient available potassium, magnesium and sulphur for at least low to medium yield (Waraich et al., 2011).

20.8.3 Irrigation and Crop Production

Irrigation provides a vital input of water for crop production and the uptake of nutrients. It stimulates the mineralization of soil organic matter and the solubilization and transport of nutrients from poorly water-soluble forms to available inorganic forms. Irrigation by diluting the soil solution reduces osmotic pressure but the disadvantage is a decrease in the concentration of nutrients that cannot quickly adapt. Irrigation also increases the concentration of monovalent cations such as potassium in the soil solution caused by cation exchange. An increase in the potassium concentration may temporarily decrease the supply of magnesium and calcium concentrations. Under drought conditions, mobile and soluble nutrients are converted to the reserve fraction (immobilization). Under drought stress phosphorus precipitates and its availability in the soil is reduced. Also, under drought stress, iron and magnesium are oxidized and their availability in the soil is reduced. Potassium is strongly adsorbed by clay minerals in the soil, and is thus less available. Adding organic matter to soil reduces potassium leaching because organic matter has a large cation exchange capacity and can retain potassium effectively (Shand, 2007).

20.9 Conclusion

Drought stress and low concentrations of organic matter in soil are two major problems for crop production. The integrated application of fertilizers can supply the necessary nutrients to crops, retain moisture in the soil, increase water-use efficiency, reduce production costs and prevent contamination of the soil and the environment. Therefore, increasing the crop yield using IPNM may be related to a good balance of nutrients, an increase in the root surface area, photosynthesis and water-use efficiency and higher water in the root zone.

REFERENCES

Abdelaal, KA, YM Hafez, AEL Sabagh, and H Saneoka, 2017. Ameliorative effects of abscisic acid and yeast on morphophysiological and yield characters of maize (*Zea mays* L.) plants under water deficit conditions. *Fresenius Environmental Bulletin*, 26(12): 7372–7383.

Abdel-Razzak, HS, AG Moussa, MA Abd-El-Fattah, and GA El-Morabet, 2013. Response of sweet potato to integrated effect of chemical and natural phosphorus fertilizer and their levels in combination with mycorrhizal inoculation. *Journal of Biological Sciences*, 13(3): 112–122.

Adhikary, S, 2012. Vermicompost, the story of organic gold: a review. *Agricultural Sciences*, 3(7): 905–917.

Ahmadian, A, A Ghanbari, and BA Siahsar, 2011. Study of the yield and its components of chamomile (*Matricaria chamomilla* L.) under drought stress and organic and inorganic fertilizers using and their residue. *Journal of Agroecology*, 3(3): 383–395. [In Persian with English summary].

Akdeniz, H, I Yilmaz, MA Bozkurt, and B Keskin, 2006. The effects of sewage sludge and nitrogen applications on grain sorghum grown (*Sorghum vulgare* L.) in Van-Turkey. *Polish Journal of Environmental Studies*, 15: 19–26.

Ali, Q, M Ashraf, and HUR Athar, 2007. Exogenously applied proline at different growth stages enhances growth of two maize cultivars grown under water deficit conditions. *Pakistan Journal of Botany*, 39(4): 1133–1144.

Amanolahi-Baharvand, Z, H Zahedi, and M Rafiee, 2014. Effect of vermicompost and chemical fertilizers on growth parameters of three corn cultivars. *Journal of Applied Science and Agriculture*, 9(9): 22–26.

Amanullah, I, 2015. The role of beneficial microbes (bio-fertilizers) in increasing crop productivity and profitability. *EC Agriculture*, 2(6): 504.

Amanullah, I, A Irfanullah, and Z Hidayat, 2016. Potassium management for improving growth and grain yield of maize (*Zea mays* L.) under moisture stress condition. *Scientific Report*, 6: 34627.

Andrew, KB, GL Hammer, and RG Henzell, 2000. Does maintaining green leaf area in sorghum improve yield under drought? II. Dry matter production and yield. *Crop Science*, 40(4): 1037–1048.

Anjum, SA, LC Wang, M Farooq, M Hussain, LL Xue, and CM Zou, 2011. Brassinolide application improves the drought tolerance in maize through modulation of enzymatic antioxidants and leaf gas exchange. *Journal of Agronomy and Crop Science*, 197(3): 177–185.

Asada, K, 2006. Production and scavenging of reactive oxygen species in chloroplasts and their functions. *Plant Physiology*, 141(2): 391–396.

Asgari Lajayer, B, N Najafi, E Moghiseh, M Mosaferi, and J Hadian, 2019. Effects of gamma irradiated and non-irradiated sewage sludge on growth characteristics, leaf chlorophyll index, and macronutrients concentrations in basil. *Journal of Soil Science and Plant Nutrition*, 19:580–591.

Auge, RM, 2001. Water relations, drought and vesicular-arbuscular mycorrhizal symbiosis. *Mycorrhiza*, 11(1): 3–42.

Azab, E, 2016. Effect of water stress and biological fertilization on maize growth, chemical composition and productivity in calcareous soil. *American Journal of Plant Physiology*, 11: 1–11.

Bagayoko, M, E George, V Römheld, and A Buerkert, 2000. Effects of mycorrhizae and phosphorus on growth and nutrient uptake of millet, cowpea and sorghum on a West African soil. *Journal of Agricultural Science*, 135: 399–407.

Bahar, B, M Yildirim, and C Barutçular, 2009. Relationships between stomatal conductance and yield components in spring durum wheat under Mediterranean conditions. *Notulae Botanicae HortiAgro Botanici Cluj-Napoca*, 37(2): 45–48.

Bhattacharyya, PN and DK Jha, 2012. Plant growth-promoting rhizobacteria (PGPR): emergence in agriculture. *World Journal of Microbiology and Biotechnology*, 28(4): 1327–1350.

Chemura, A, 2014. The growth response of coffee (*Coffea arabica* L) plants to organic manure, inorganic fertilizers and integrated soil fertility management under different irrigation water supply levels. *International Journal of Recycling of Organic Waste in Agriculture*, 3(2): 59.

Chen, YP, PD Rekha, AB Arunshen, WA Lai, and CC Young, 2006. Phosphate solubilizing bacteria from subtropical soil and their tri-calcium phosphate solubilizing abilities. *Applied Soil Ecology*, 34(1): 33–41.

Choudhury, MA, and IR Kennedy, 2004. Prospect and potentials for system of biological nitrogen fixation in sustainable rice production. *Biological Fertilizer and Soil*, 39(4): 219–227.

Conacher, J, and A Conacher, 1998. Organic farming and the environment, with particular reference to Australia: a review. *Biological Agriculture and Horticulture*, 16: 145–171.

Crecchio, C, M Curci, R Mininni, P Ricciuti, and P Ruggiero, 2001. Short term effects of municipal solid waste compost amendments on soil carbon and nitrogen content, some enzyme activities and genetic diversity. *Biology and Fertility of Soils*, 34(5): 311–318.

Elhabet, H, 2018. Effect of organic and inorganic fertilizers on rice and some nutrients availability under different water regimes. *Journal of Agricultural Science and Food Research*, 9: 4.

El-Kader, A, S Shaaban, and M El-Fattah, 2010. Effect of irrigation levels and organic compost on okra plants (*Abelmoschus esculentus* L.) grown in sandy calcareous soil. *Agriculture and Biology Journal of North America*, 1(3): 225–231.

El-Refaee, IS, 2006. The interaction effect between water regimes and potassium levels on growth, grain yield and water productivity in rice. *Egyptian Journal of Agricultural Research*, 85(4): 1399–1421.

El-Refaee, IS, AM Al-Khtyar, and AA El-Gohary, 2008. Improving rice productivity under irrigation intervals and nitrogen fertilizer. *Second Field Crops Conference*, Egypt.

El Sabagh, A, A Hossain, C Barutçular, K Abdelaal, S Fahad, F Anjorin, MS Islam, D Ratnasekera, F Kizilgeçi, G Yadav, and M Yildirim, 2018. Sustainable maize (*Zea mays* L.) production under drought stress by understanding its adverse effect, survival mechanism and drought tolerance indices. *Journal of Experimental Biology and Agricultural Sciences*, 6(2): 282–295.

Faraji, S, M Rafieiolhossaini, and A Abbasi Soorki, 2015. The effect of solitary and combined application of organic and biological manure and chemical fertilizer on some of the qualitative and quantitative properties of sugar beet. *Agricultural Crop Management*, 17: 789–800.

Ghassemi S, Farhangi-Abriz S, Faegi-Analou R, Ghorbanpour M, and Asgari Lajayer B, (2018) Monitoring cell energy, physiological functions and grain yield in field-grown mung bean exposed to exogenously applied polyamines under drought stress. *Journal of Soil Science and Plant Nutrition*, 18(4): 1108–1125.

Ghosh, BC, and R Bhat, 1998. Environmental hazards of nitrogen loading in wetland rice fields. *Environmental Pollution*, 102(1): 123–126.

Graf, F, and M Frei, 2013. Soil aggregate stability related to soil density, root length and mycorrhiza using site-specific *Alnus incana* and *Melanogaster variegatus* sl. *Ecological Engineering*, 57: 314–332.

Hari, M, and K Perumal, 2010. *Booklet on Bio-Fertilizer (Phosphabacteria)*. Shri AMM Murugappa Chettiar Research Center, Taramani, p.113.

Hassani, A, 2005. Effect of water deficit on growth, yield and essential oil herb Badrshbov. *Iranian Journal of Medicinal Aromatic Plants*, 22(3): 256–261. [In Persian with English Summary]

Heidarian, F, A Rokhzadi, and F Mirahmadi, 2018. Response of sugar beet to irrigation interval, harvesting time and integrated use of farmyard manure and nitrogen fertilizer. *Environmental and Experimental Biology*, 16: 169–175.

Hermanson, RW, C Pan, R Perillo, R Stevans, and C Stockle, 2000. *Nitrogen Use by Crop and the Fate of Nitrogen in the Soil and Vadose Zone*. Washington State University and Washington Department of Ecology, Interagency Agreement. No. C9600177.

Huang, B, RR Duncan, and RN Carrow, 1997. Drought-resistance mechanisms of seven warm-season turfgrasses under surface soil drying: II. Root aspects. *Crop Science*, 37(6): 1863–1869.

Hussain, T, G Jilani, and MZ Iqbal, 1988. Integrated use of organic and inorganic N fertilizer in rice-wheat cropping system. *Pakistan Journal of Soil Science*, 3: 1923.

Ibrahim, A, and G El-Samad, 2009. Effect of different irrigation regimes and partial substitution of N-mineral by organic manures on water use, growth and productivity of Pomegranate trees. *European Journal of Scientific Research*, 38(2): 199–218.

Ibrahim, M, A Hassan, M Iqbal, and EE Valeem, 2008. Response of wheat growth and yield to various levels of compost and organic manure. *Pakistan Journal of Botany*, 40(5): 2135–2141.

Islam, MS, MM Akhter, A EL Sabagh, LY Liu, NT Nguyen, A Ueda, and H Saneoka, 2011. Comparative studies on growth and physiological responses to saline and alkaline stresses of Foxtail millet (*Setaria italica* L.) and Proso millet (*Panicum miliaceum* L.). *Australian Journal of Crop Science*, 5(10): 1269–1277.

Jakobsen, I, B Chen, L Munkvold, T Lundsgaard, and YG Zhu, 2005. Contrasting phosphate acquisition of mycorrhizal fungi with that of root hairs using the root hairless barley mutant. *Plant, Cell and Environment*, 28(7): 928–938.

Khalid, S, 2016. Integrated use of phosphorus, animal manures and biofertilizers improve maize productivity under semiarid condition. In *Organic Fertilizers-From Basic Concepts to Applied Outcomes*. IntechOpen.

Khodaei-Joghan, A, M Gholamhoseini, AA Majid, F Habibzadeh, A Sorooshzadeh, and A Ghalavand, 2018. Response of sunflower to organic and chemical fertilizers in different drought stress conditions. *Acta Agriculturae Slovenica*, 111(2): 271–284.

Khoshmanzar, E, Aliasgharzad, N, Neyshabouri, MR, Khoshru, B, Arzanlou, M, and Lajayer, BA, 2019. Effects of *Trichoderma* isolates on tomato growth and inducing its tolerance to water-deficit stress. *International Journal of Environmental Science and Technology*. http://dx.doi.org/10.1007/s13762-019-02405-4.

Lewis, WM, WA Wurtsbaugh, and HW Paerl, 2011. Rationale for control of anthropogenic nitrogen and phosphorus to reduce eutrophication of inland waters. *Environmental Science and Technology*, 45: 300–305
Liddycoat, SM, BM Greenberg, and DJ Wolyn, 2009. The effect of plant growth promoting rhizobacteria on asparagus seedlings and germinating seeds subjected to water stress under greenhouse conditions. *Canadian Journal of Microbiology*, 55(4): 388–394.
Marschner, H, 1995. *Mineral Nutrition of Higher Plants*. (2nd Ed.). Academic Press, USA.
Misra, A, and G Tyler, 1999. Influence of soil moisture on soil solution chemistry and concentrations of minerals in the calcicoles *Phleum phleoides* and *Veronica spicata* grown on a limestone soil. *Annals of Botany*, 84(3): 401–410.
Moradi, R, N Poorghasemian, and M Naghizadeh, 2018. Effect of different deficit irrigation levels and nutritional recourses on some quantitative and qualitative characteristics of black cumin (*Nigella sativa* L.) in Bardsir climate. *Environmental Stresses in Crop Sciences*, 11(1): 35–46. [In Persian with English Summary.]
Mukhongo, RW, JB Tumuhairwe, P Ebanyat, AHA Gadir, M Thuita, and C Masso, 2017. Combined application of biofertilizers and inorganic nutrients improves sweet potato yields. *Frontiers in Plant Science*, 8: 219.
Murillo-Amador, B, LE Morales-Prado, E Troyo-Diéguez, MV Córdoba-Matson, LG Hernández-Montiel, EO Rueda-Puente, and A Nieto-Garibay, 2015. Changing environmental conditions and applying organic fertilizers in *Origanum vulgare* L. *Frontiers in Plant Science*, 6: 549.
Muscolo, A, M Sidari, E Attina, O Francioso, V Tugnoli, and S Nardi. 2007. Biological activity of humic substances is related to their chemical structure. *Soil Science Society of America Journal*, 71(1): 75–85.
Nyalemegbe, K, J Oteng, and S Asuming-Brempong, 2010. Integrated organic–inorganic fertilizer management for rice production on the vertisols of the Accra Plains of Ghana. *West African Journal of Applied Ecology*, 16(1): 23–33.
Palm, CA, 1995. Contribution of agroforestry trees to nutrient requirements of intercropped plants. *Agroforestry Systems*, 30: 105–124.
Plencette, C, C Clermont-Dauphin, JM Meynard, and JA Fortin, 2005. Managing arbuscular mycorrhizal fungi in cropping systems. *Canadian Journal of Plant Science*, 85(1): 31–40.
Randhawa, MS, M Maqsood, MA Shehzad, MU Chattha, MB Chattha, F Nawaz, S Yasin, T Abbas, MM Nawaz, RD Khan, and U Zulfiqar, 2017. Light interception, radiation use efficiency and biomass accumulation response of maize to integrated nutrient management under drought stress conditions. *Turkish Journal of Field Crops*, 22(1): 134–142.
Saidia, PS, and JP Mrema, 2017. Effects of farmyard manure and activated effective microorganisms on rainfed upland rice in Mwanza, Tanzania. *Organic Agriculture*, 7(2): 83–93.
Shand, C, 2007. Plant nutrition for food security. A guide for integrated nutrient management. By RN Roy, A Finck, GJ Blair and HLS Tandon. Rome: Food and Agriculture Organization of the United Nations (2006), pp. 348, US $70.00. ISBN 92-5-105490-8. *Experimental Agriculture*, 43(1): 132.
Sharma, SB, RZ Sayyed, MH Trivedi, and TA Gobi, 2013. Phosphate solubilizing microbes: sustainable approach for managing phosphorus deficiency in agricultural soils. *Springer Plus*, 2(1): 587.
Shirkhani, A, and S Nasrolahzadeh, 2016. Vermicompost and *Azotobacter* as an ecological pathway to decrease chemical fertilizers in the maize, *Zea mays*. *Bioscience Biotechnology Research Communications*, 9(3): 382–390.
Simonne, E, and G Hochmuth, 2003. *Supplemental Fertilizer Application in the BMP Area for Vegetable. UF/IFAS, Fla.* Cooperative Extension Service, HS906.
Simsek-Ersahin, Y, 2011. The use of vermicompost products to control plant diseases and pests. In: Karaca A (ed.) *Biology of Earthworms, Soil Biology*. Springer, Berlin.
Singh, BK, KA Pathak, AK Verma, VK Verma, and BC Deka, 2011. Effects of vermicompost, fertilizer and mulch on plant growth, nodulation and pod yield of French bean (*Phaseolus vulgaris* L.). *Vegetable Crops Research Bulletin*, 74(1): 153–165.
Singh, BK, and OP Sarivastore, 1971. Effect of organic manure in soil fertility as shown by nutrient availability and crop yield response in potato. *International Symposium on Soil Fertility Evaluation*, pp. 815–820.
Sohu, I, AW Gandahi, GR Bhutto, MS Sarki, and R Gandahi, 2015. Growth and yield maximization of chickpea (*Cicer arietinum*) through integrated nutrient management applied to rice-chickpea cropping system. *Sarhad Journal of Agriculture*, 31(2): 131–138.

Stockdale, EA, MA Shepherd, S Fortune, and SP Cuttle, 2002. Soil fertility in organic farming systems–fundamentally different? *Soil Use and Management*, 18: 301–308.

Stone, DM, and JD Elioff, 1998. Soil properties and Aspen development five years after compaction and forest floor removal. *Canadian Journal of Soil Science*, 78(1): 51–58.

Talashiker, SC, and OP Rinal, 1986. Studies on increasing in combination with city solid waste. *Journal of the Indian Society of Soil Science*, 34: 780–784.

Waraich, EA, R Ahmad, AM Saifullah, and Y Ehsanullah, 2011. Role of mineral nutrition in alleviation of drought stress in plants. *Australian Journal of Crop Science*, 5(6): 764–777.

Zarabi, M, I Alahdadi, GA Akbari, and GA Akbari, 2011. A study on the effects of different biofertilizer combinations on yield, its components and growth indices of corn (*Zea mays* L.) under drought stress condition. *African Journal of Agricultural Research*, 6: 681–685.

Index